IDEAL AND INCOMPRESSIBLE FLUID DYNAMICS

Mathematics and its Applications

Series Editor: G. M. BELL, Professor of Mathematics, King's College (KQC), University of London

Mathematics and its applications are now awe-inspiring in their scope, variety and depth. Not only is there rapid growth in pure mathematics and its applications to the traditional fields of the physical sciences, engineering and statistics, but new fields of application are emerging in biology, ecology and social organisation. The user of mathematics must assimilate subtle new techniques and also learn to handle the great power of the computer efficiently and economically.

The need of clear, concise and authoritative texts is thus greater than ever and our series will endeavour to supply this need. It aims to be comprehensive and yet flexible. Works surveying recent research will introduce new areas and up-to-date mathematical methods. Undergraduate texts on established topics will stimulate student interest by including applications relevant at the present day. The series will also include selected volumes of lecture notes which will enable certain important topics to be presented earlier than would otherwise be possible.

In all these ways it is hoped to render a valuable service to those who learn, teach, develop and use mathematics.

For full series list see end of book.

IDEAL AND INCOMPRESSIBLE FLUID DYNAMICS

M. E. O'NEILL, B.Sc., M.Sc., Ph.D.
Reader in Mathematics
University College London

and

F. CHORLTON, B.Sc., M.Sc., D.I.C., F.I.M.A.
Honorary Senior Visiting Lecturer
Aston University, Birmingham

ELLIS HORWOOD LIMITED
Publishers · Chichester

Halsted Press: a division of
JOHN WILEY & SONS
New York · Chichester · Brisbane · Toronto

First published in 1986 by
ELLIS HORWOOD LIMITED
Market Cross House, Cooper Street, Chichester, West Sussex, PO19 1EB, England

The publisher's colophon is reproduced from James Gillison's drawing of the ancient Market Cross, Chichester.

Distributors:
Australia and New Zealand:
Jacaranda-Wiley Ltd., Jacaranda Press,
JOHN WILEY & SONS INC.
GPO Box 859, Brisbane, Queensland 4001, Australia
Canada:
JOHN WILEY & SONS CANADA LIMITED
22 Worcester Road, Rexdale, Ontario, Canada
Europe and Africa:
JOHN WILEY & SONS LIMITED
Baffins Lane, Chichester, West Sussex, England
North and South America and the rest of the world:
Halsted Press: a division of
JOHN WILEY & SONS
605 Third Avenue, New York, NY 10158, USA

British Library Cataloguing in Publication Data
O'Neill, M.E.
Ideal and incompressible fluid dynamics. —
(Ellis Horwood series in mathematics and its applications)
1. Fluid dynamics — Mathematics
I. Title II. Chorlton, Frank
532'.051'072 QA911

Library of Congress Card No. 85–30194

ISBN 0–85312–825–1 (Ellis Horwood Limited — Library Edn.)
ISBN 0–85312–977–0 (Ellis Horwood Limited — Student Edn.)
ISBN 0–470–20273–4 (Halsted Press)

Printed in Great Britain by The Camelot Press, Southampton

Contents

Preface

The aim of this book is to provide a readily accessible introductory text on the theory of fluid mechanics which is as far as possible 'self-contained' with regard to mathematical methods and techniques. The development of fluid mechanics has accelerated greatly over recent years, resulting in an expansion of those branches of science and engineering in which a knowledge of the subject is now almost essential. However, to provide a text which aims to give essential basic mathematical techniques and methods while at the same time encompassing the mechanics of a comprehensive range of fluids, both viscous and inviscid, compressible and incompressible, would have resulted in an extremely bulky book. We therefore decided to restrict the material within this volume to that relating to an ideal (frictionless) fluid. This has the advantage that the book is physically shorter and smaller in size while at the same time covering material found in an introductory honours degree course on fluid mechanics such as is given at many universities, including the University of Aston and University College London. In addition, the techniques learned in the study of ideal fluid motion do provide the student with an excellent background prior to studying the mechanics of viscous and compressible fluids. Furthermore, the subject is not without important successes in modelling fluid motions encountered in the real world, notably water waves and the flight of an aeroplane.

Although the book is primarily a text in the theoretical description of ideal incompressible flow, it has been written in a style which makes it accessible to engineers and scientists as well as mathematical specialists. The text is presented in a form which is independent of any specific system of units for measuring physical quantities, but the use of units in general, and the widely accepted Système International d'Unités (SI) in particular, is discussed at length in the Appendix. Because problem solving is an extremely useful aid in understanding the concepts and mastering the techniques of fluid mechanics, many worked examples are incorporated into the text. A complete set of unsolved problems, with answers provided, appears at the end of each chapter. Many of these problems are taken from past examination papers set by the Universities of Bristol, Cambridge, Leeds, Liverpool, London, Manchester and Surrey. We are most grateful to these institutions for granting permission to use their questions and also to the Master of Trinity College Cambridge for permission to use questions set at Trinity College Cambridge. We also acknowledge permission granted by Macmillan, London and Basingstoke, for inclusion of problems taken from *Theoretical Hydrodynamics* by L. M. Milne-Thomson, and also Bell & Hyman, London, for inclusion of problems taken from *A Treatise on Hydromechanics* by A. S. Ramsey.

The authors are indebted to Dr E. R. Johnson of the Department of Mathematics, University College London, for constructive discussion on various aspects of the text, and to Mrs M. Hasleton and Mrs B. Lankester for their expert typing of the manuscript. We should also like to thank the publishing staff of Ellis Horwood Limited for their help and encouragement throughout the various stages of publication.

Michael E. O'Neill
Frank Chorlton

1

Vector and tensor methods

1.1 SCALARS AND VECTORS

Physical quantities such as distance, mass, volume, temperature, electric potential, energy, etc., require for their description the specification of a single basic *unit of measurement* such as metre, kilogram, litre, degree Celsius, volt, etc., and a *pure number* which indicates precisely how many of the basic units make up the particular quantity being described. For instance, to say that the mass of a body is five kilograms (5 kg) means that the unit of mass is the kilogram and five such units make up the total mass of the body. A description of various systems of units is given in the Appendix, but if no specific unit is mentioned, it will be assumed that the now almost universal international system (SI) of units is operable. An important property of the quantities named above is that they are quite independent of the notion of direction in defining them completely. Such quantities are classified as *scalars*. The real numbers may also be regarded as scalars.

However, it is not difficult to realise that certain other physical quantities depend very profoundly on direction. An example is force and it can be readily appreciated that the effect produced by applying a force to a body is very much dependent on the direction in which the force is applied. Other examples of quantities whose character is dependent on direction are velocity, displacement electric field intensity etc. Such quantities are classified as *vectors*. A velocity of 60 metres per second in the north-easterly direction is an example of a vector quantity and it may be written as 60 m/s NE for brevity. Here the *magnitude* of the vector is 60, the unit of measurement being 1 m/s. and the *direction* of the vector is north-easterly. There are other physical quantities such as stress which require more than one direction for their full description and these are called *tensors*. An account of such quantities will be given in Section 1.21.

A vector therefore is a quantity possessing both magnitude and direction. It is denoted in written work by $\underset{\sim}{a}$ or $\vec{a}$ while in printed texts it is written as **a** using bold type face. The magnitude of **a** is denoted by a or $|\mathbf{a}|$ and the direction of **a** by $\hat{\mathbf{a}}$. Thus

$$\mathbf{a} = a\,\hat{\mathbf{a}} = |\mathbf{a}|\,\hat{\mathbf{a}}.$$

Provided $a \neq 0$, we may write

$$\hat{\mathbf{a}} = \mathbf{a}/a = \mathbf{a}/|\mathbf{a}|,$$

so clearly $\hat{\mathbf{a}}$ is a *unit vector*, i.e. a vector with unit magnitude. It also follows that the direction of a vector **a** is found by dividing **a** by its magnitude $|\mathbf{a}|$. If $a = 0$, then **a** can be thought of as a *zero vector*, i.e. a vector with zero magnitude and arbitrary direction and denoted by **0**.

Just as a scalar can be represented geometrically by a line segment – this is exactly what happens when graphs of physical quantities are plotted, a length of 1 cm or 1 inch representing 1 second or 1 newton or whatever the basic unit of measurement of the physical quantity is – so also can a vector be represented geometrically by a *directed* line segment. The length of the line segment represents the magnitude of the vector while the direction of the directed line segment is exactly the direction in space of the physical quantity. This representation means that the directed line segment $\overrightarrow{OA}$, drawn from O to A, represents a certain given vector **a** in both magnitude and direction, though not necessarily in position (Fig. 1.1). If the physical vector does actually pass through the point O, it is said to be *localised* at O, an obvious example being a force acting at a point. If the physical vector does not pass through any specific point it is said to be *non-localised* or *free*. In either case we write

$$\overrightarrow{OA} \equiv \mathbf{a}.$$

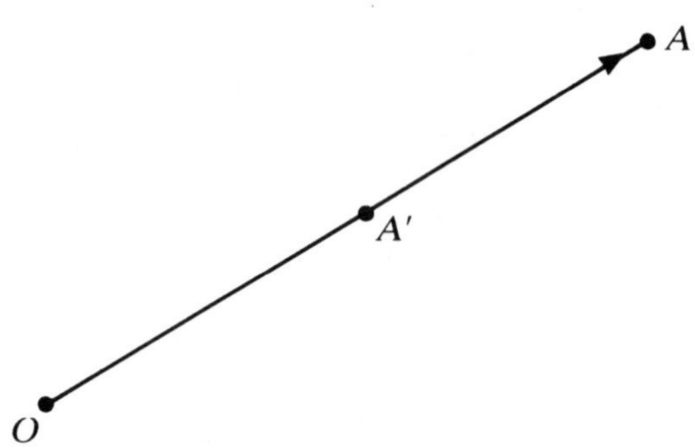

Fig. 1.1

In Fig. 1.1, the length $OA = |\mathbf{a}|$ and if A' is a point on OA (or OA produced) such that the length OA' is one unit in measuring $|\mathbf{a}|$, it is clear that

$$\overrightarrow{OA}' \equiv \hat{\mathbf{a}}.$$

It is also clear from Fig. 1.1 that $\overrightarrow{AO}$, is a vector of equal magnitude but *opposite direction* to $\overrightarrow{OA}$. $\overrightarrow{AO}$ may therefore be thought of as the *negative* of $\overrightarrow{OA}$ and thus of **a** and so we write

$$\overrightarrow{AO} \equiv -\mathbf{a}.$$

It follows from the above that *equality of two vectors* **a** and **b** requires:

(i) $|\mathbf{a}| = |\mathbf{b}|$;
(ii) $\hat{\mathbf{a}} = \hat{\mathbf{b}}$.

Thus two *different* vectors can be equal, since equality only requires that they have equal magnitudes and coparallel directions. This is in sharp contrast to the equality of scalars.

1.2 ADDITION AND SUBTRACTION OF VECTORS

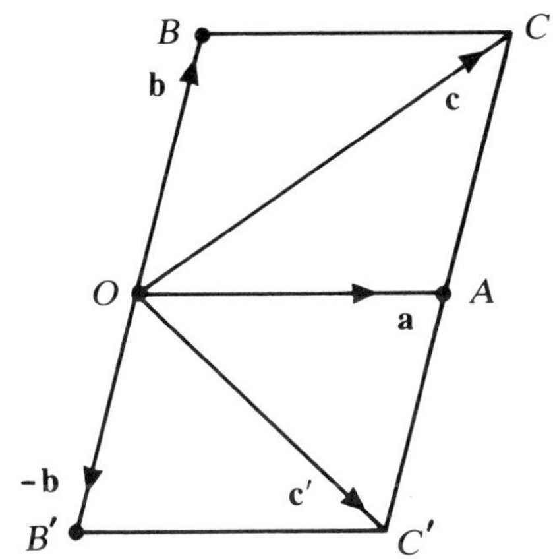

Fig. 1.2

In Fig. 1.2, $\overrightarrow{OA} \equiv \mathbf{a}$, $\overrightarrow{OB} \equiv \mathbf{b}$, are two given vectors. The parallelogram $OACB$ is completed and we write $\overrightarrow{OC} \equiv \mathbf{c}$. Then **c** is defined to be the *sum of the vectors* **a** and **b** and we write

$$\mathbf{a} + \mathbf{b} = \mathbf{c}.$$

In geometric notation this gives the *Parallelogram Law of Vector Addition* in the form

$$\vec{OA} + \vec{OB} = \vec{OC}.$$

Essentially this is a generalization of the method of finding the resultant of two forces encountered in elementary statics. As $\vec{AC} \equiv \mathbf{b}$, we can also write

$$\vec{OA} + \vec{AC} = \vec{OC},$$

which is the *Triangle Law of Vector Addition* and is also met in statics. In this last form, observe that in adding $\vec{OA}$ to $\vec{AC}$, the inner repeated A is dropped and the vector sum is contracted to $\vec{OC}$. This case illustrates an advantage achieved in using the notation of the geometrical representation. The method is extensible to summing any numbers of vectors. Clearly the sum of the three vectors, $\vec{AB}$, $\vec{BC}$, $\vec{CD}$ is

$$\vec{AB} + \vec{BC} + \vec{CD} = \vec{AD},$$

the inner repeated letters being dropped.

Reverting to Fig. 1.2, produce $\vec{BO}$ to B' so that O is the mid-point of BB'. Then $\overrightarrow{OB} \equiv -\mathbf{b}$. When the parallelogram $OB'\,C'\,A$ is completed and we write $\vec{OC}' \equiv \mathbf{c}'$, the rule of vector addition gives

$$\mathbf{c}' = \mathbf{a} + (-\mathbf{b})$$

or

$$\mathbf{c}' = \mathbf{a} - \mathbf{b}$$

following the convention of elementary scalar algebra. Thus we have a device for finding the *difference of two vectors*.

The reader should be able to extend the method of addition of two vectors by the parallelogram and/or triangle laws to several vectors using the *Polygon Law of Vector Addition*. Also to establish such results as

$$\underbrace{\mathbf{a} + \mathbf{a} + \ldots + \mathbf{a}}_{n \text{ terms}} = n\mathbf{a}$$

and the association and distributive laws

$$m(n\mathbf{a}) = n(m\mathbf{a}) = mn\mathbf{a};$$

$$(m + n)\mathbf{a} = m\mathbf{a} + n\mathbf{a}.$$

These results are all treated in detail in Ref. 3, pp. 15–18.

1.3 USE OF CARTESIAN COORDINATES

In Fig. 1.3 the three mutually perpendicular lines OX, OY, OZ form a *right-handed tri-rectangular frame of coordinate axes*. This means that the three axes are mutually perpendicular at O and that the rotation of a right-handed screw from OX to OY – a *positive rotation* – advances the screw along $\vec{OZ}$. Similarly a positive rotation from OY to OZ advances it along $\vec{OX}$ and one from OZ to OX

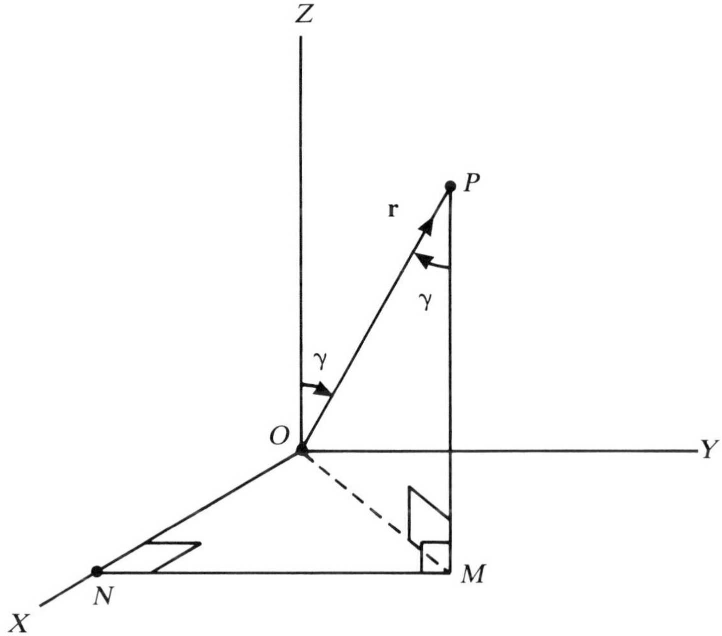

Fig. 1.3

advances it along $\overrightarrow{OY}$. The reader can easily verify that if in Fig. 1.3 the direction of any one the three coordinate axes is reversed leaving the other two *in situ*, then a left-handed frame would be obtained for which negative rotations from OX to OY would advance a right-handed screw along $\overrightarrow{OZ}$, etc.

Let $ON = x$, $NM = y$, $MP = z$. Then (x, y, z) are the *Cartesian coordinates of* P in the frame and these respective coordinates are the distances of P from the three coordinate planes YOZ, ZOX, XOY. Now let $\mathbf{i}, \mathbf{j}, \mathbf{k}$ denote the unit vectors in the axes $\overrightarrow{OX}$, $\overrightarrow{OY}$, $\overrightarrow{OZ}$, so that $\overrightarrow{ON} \equiv x\mathbf{i}$, etc. By the polygon law of vector addition,

$$\overrightarrow{ON} + \overrightarrow{NM} + \overrightarrow{MP} = \overrightarrow{OP}$$

or

$$x\mathbf{i} + y\mathbf{j} + z\mathbf{k} = \mathbf{r}. \tag{1.1}$$

From the right-angled triangles OMP, ONM,

$$OP^2 = OM^2 + MP^2 = ON^2 + NM^2 + MP^2$$

or

$$r^2 = x^2 + y^2 + z^2. \tag{1.2}$$

The statement (1.1) will sometimes be written $\mathbf{r} = [x, y, z]$, meaning that x, y, z

are respectively the components of $\mathbf{r}$. If $\mathbf{r}$ is non-zero, $r \neq 0$ and the unit vector in the direction of $\mathbf{r}$ is

$$\hat{\mathbf{r}} = \mathbf{r}/r = [x/r, y/r, z/r]. \tag{1.3}$$

If α, β, γ denote the angles which $\overrightarrow{OP}$ makes with the respective axes $\overrightarrow{OX}$, $\overrightarrow{OY}$, $\overrightarrow{OZ}$, then Fig. 1.3 shows that

$$\cos\gamma = z/r.$$

Similarly $\cos\alpha = x/r$, $\cos\beta = y/r$. The quantities $[l, m, n]$ defined by

$$[l, m, n] = [\cos\alpha, \cos\beta, \cos\gamma] \tag{1.4}$$

are called the *direction cosines* of $\overrightarrow{OP}$ in the coordinate frame. Clearly they satisfy the relation

$$l^2 + m^2 + n^2 = 1. \tag{1.5}$$

Any set of numbers $[kl, km, kn]$ are called the *direction ratios* of the line and they have the property of being proportional to the direction cosines.

Now suppose that $P_1(x_1, y_1, z_1)$; $P_2(x_2, y_2, z_2)$ are two distinct points in the space of Fig. 1.3. By moving the origin of coordinates from O to P_1 while keeping the directions of the coordinate axes $\overrightarrow{OX}$, $\overrightarrow{OY}$, $\overrightarrow{OZ}$ unchanged, it is easy to see that the new coordinates of the points become $(0, 0, 0)$ for P_1 and $(x_2 - x_1, y_2 - y_1, z_2 - z_1)$ for P_2. If now $\overrightarrow{P_1P_2} \equiv \mathbf{r}$, then

$$\mathbf{r} = [x_2 - x_1, y_2 - y_1, z_2 - z_1], \tag{1.6}$$

$$r = |\mathbf{r}| = \sqrt{\{(x_2 - x_1)^2 + (y_2 - y_1)^2 + (z_2 - z_1)^2\}}, \tag{1.7}$$

$$\hat{\mathbf{r}} = \frac{\mathbf{r}}{r} = \left[\frac{(x_2 - x_1),\ (y_2 - y_1),\ (z_2 - z_1)}{\sqrt{\{(x_2 - x_1)^2 + (y_2 - y_1)^2 + (z_2 - z_1)^2\}}}\right]. \tag{1.8}$$

Also the components of $\hat{\mathbf{r}}$, as given by (1.8), are the direction cosines of the line $\overrightarrow{P_1P_2}$: they are the cosines of the angles made by $\overrightarrow{P_1P_2}$ with the directions $\overrightarrow{OX}$, $\overrightarrow{OY}$, $\overrightarrow{OZ}$, respectively and the components of $\mathbf{r}$ in (1.6) may be taken as a set of direction ratios of $\overrightarrow{P_1P_2}$.

If now $P_1(x_1, y_1, z_1)$ is a given point in space and if $[l, m, n]$ denote the direction cosines of a line through P_1 then we can find the equations of the line. Since $l^2 + m^2 + n^2 = 1$ for direction cosines, $[l, m, n]$ are the Cartesian coordinates of a unit vector $\hat{\mathbf{a}}$ in the line through P_1. Let $\overrightarrow{OP_1} \equiv \mathbf{r}_1$ and take any other point $P(x, y, z)$ on the line through P_1 letting $\overrightarrow{OP} \equiv \mathbf{r} = [x, y, z]$. Then since

$$\overrightarrow{OP} = \overrightarrow{OP_1} + \overrightarrow{P_1P},$$

$$\mathbf{r} = \mathbf{r}_1 + d\hat{\mathbf{a}}, \tag{1.9}$$

where d is the distance P_1P so that $\overrightarrow{P_1P} \equiv d\hat{\mathbf{a}}$. Then (1.9) is the *vector equation of the line through P_1* along a *specified direction* $\hat{\mathbf{a}}$. Otherwise expressed, (1.9) gives

$$[x, y, z] = [x_1, y_1, z_1] + d[l, m, n]$$

from which, equating components,

$$x = x_1 + dl; \quad y = y_1 + dm; \quad z = z_1 + dn \tag{1.10}$$

The Cartesian forms (1.10) give the *parametric equations* of the line when we think of the distance d as varying. The justification for equating components in this manner is that the vector **r** has a unique set of Cartesian components $[x, y, z]$ associated with the directions $\vec{OX}, \vec{OY}, \vec{OZ}$.

Elimination of d between the three equations (1.10) gives

$$\frac{x - x_1}{l} = \frac{y - y_1}{m} = \frac{z - z_1}{n}. \tag{1.11}$$

Here there are two distinct equations each of which represents a plane. Thus (1.11) defines the line through $P_1(x_1, y_1, z_1)$ with direction cosines $[l, m, n]$ as the intersection of the planes. Were we given only the direction ratios $[\lambda, \mu, \nu]$ of the line, the corresponding direction cosines $[l, m, n]$ would be calculable from

$$[l, m, n] = \left[\frac{\lambda, \mu, \nu}{\sqrt{(\lambda^2 + \mu^2 + \nu^2)}}\right],$$

so that the direction cosines are in fact proportional to $[\lambda, \mu, \nu]$ and an alternative representation to (1.11) would then be

$$(x - x_1)/\lambda = (y - y_1)/\mu = (z - z_1)/\nu. \tag{1.12}$$

A fuller account of coordinates and lines and planes is found in Section 1.3 of Ref. 3.

1.4 SCALAR PRODUCT OF TWO VECTORS

In Fig. 1.4, $\vec{OA} \equiv \mathbf{a}$, $\vec{OB} \equiv \mathbf{b}$, $A\hat{O}B = \theta$. The angle θ is measured in the positive counter-clockwise sense from OA to OB. Then the *scalar product* of **a** and **b** is defined to be $ab \cos\theta$. It is denoted by **a . b** so that

$$\mathbf{a} \cdot \mathbf{b} = ab \cos\theta = |\mathbf{a}|\,|\mathbf{b}| \cos\theta. \tag{1.13}$$

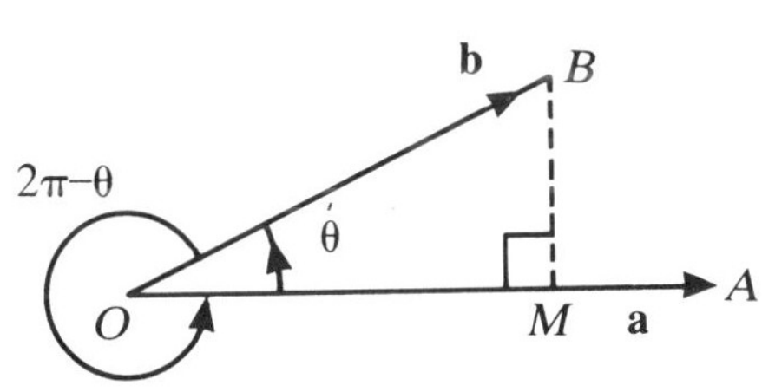

Fig. 1.4

From this definition,

$$\mathbf{b} \cdot \mathbf{a} = ba \cos(2\pi - \theta) = ab \cos\theta$$

and so we have established the *commutative law for scalar products*:

$$\mathbf{a} \cdot \mathbf{b} = \mathbf{b} \cdot \mathbf{a}. \tag{1.14}$$

It follows that

$$\hat{\mathbf{a}} \cdot \mathbf{b} = \mathbf{b} \cdot \hat{\mathbf{a}} = b \cos\theta = OM$$

where OM is the orthogonal projection of OB on OA. Further, if $\mathbf{a}$ and $\mathbf{b}$ are orthogonal, so that $\theta = \pi/2$, then $\mathbf{a} \cdot \mathbf{b} = 0$. Conversely, if we are given that $\mathbf{a} \cdot \mathbf{b} = 0$ then in the same notation as above, $ab \cos\theta = 0$ so that either $a = 0$, $b = 0$ or $\cos\theta = 0$. If we know that neither $\mathbf{a}$ nor $\mathbf{b}$ is a zero vector, the only possibility is $\cos\theta = 0$ so that the two vectors are at right-angles. Hence *a necessary and sufficient condition for two non-zero vectors of finite magnitude to be perpendicular (orthogonal) is that their scalar product should vanish.* This is a useful test of perpendicularity.

For any vector $\mathbf{a}$ if we define $\mathbf{a}^2$ to mean $\mathbf{a} \cdot \mathbf{a}$, then

$$\mathbf{a}^2 = a \times a \times \cos 0 = a^2 = |\mathbf{a}|^2.$$

For the unit vectors $\mathbf{i}, \mathbf{j}, \mathbf{k}$ in $\overrightarrow{OX}, \overrightarrow{OY}, \overrightarrow{OZ}$ (Fig. 1.3),

$$\begin{aligned} \mathbf{i}^2 &= 1, \quad \mathbf{j}^2 = 1, \quad \mathbf{k}^2 = 1, \\ \mathbf{i} \cdot \mathbf{j} &= 0, \quad \mathbf{j} \cdot \mathbf{k} = 0, \quad \mathbf{k} \cdot \mathbf{i} = 0. \end{aligned} \tag{1.15}$$

It is easy to establish the *distributive law for scalar products* in the form

$$\mathbf{a} \cdot (\mathbf{b} + \mathbf{c}) = \mathbf{a} \cdot \mathbf{b} + \mathbf{a} \cdot \mathbf{c}. \tag{1.16}$$

A proof of this is given in Ref. 3, pp. 24, 25.

If two vectors $\mathbf{a} = [a_x, a_y, a_z]$; $\mathbf{b} = [b_x, b_y, b_z]$ are expressed in such Cartesian component forms, then the above distributive law gives

$$\mathbf{a} \cdot \mathbf{b} = a_x b_x + a_y b_y + a_z b_z \tag{1.17}$$

on using the relations (1.15).

The form (1.17) is easily remembered and it affords an easy means for finding the cosine of the angle θ between two lines whose direction cosines are known in the forms $[l_1, m_1, n_1]$, $[l_2, m_2, n_2]$. These direction cosines are the components of unit vectors, say $\hat{\mathbf{a}}_1$ and $\hat{\mathbf{a}}_2$. Their scalar product is by definition

$$\hat{\mathbf{a}}_1 \cdot \hat{\mathbf{a}}_2 = 1 \times 1 \times \cos\theta$$

and in virtue of (1.17),

$$\hat{\mathbf{a}} \cdot \hat{\mathbf{a}} = l_1 l_2 + m_1 m_2 + n_1 n_2$$

and so

$$\cos\theta = l_1 l_2 + m_1 m_2 + n_1 n_2.$$

As a further illustration any vector **F** may be expressed in the form

$$\mathbf{F} = (\mathbf{F}\,.\,\mathbf{i})\mathbf{i} + (\mathbf{F}\,.\,\mathbf{j})\mathbf{j} + (\mathbf{F}\,.\,\mathbf{k})\mathbf{k}, \tag{1.18}$$

since $(\mathbf{F}\,.\,\mathbf{i})$ is the component of **F** in the x-direction, etc.

1.5 VECTOR PRODUCT OF TWO VECTORS

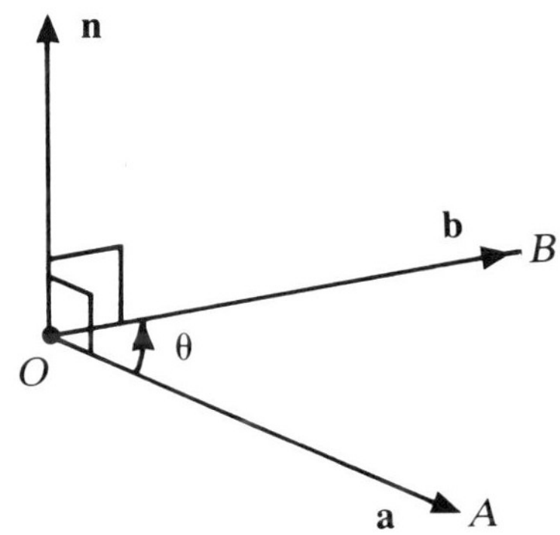

Fig. 1.5

In Fig. 1.5, $\vec{OA} \equiv \mathbf{a}$, $\vec{OB} \equiv \mathbf{b}$ and $A\hat{O}B = \theta$. The vectors **a** and **b** are assumed to be in different directions so that AOB determines a plane. The direction of the unit vector **n** of the normal to this plane is the direction in which a right-handed screw advances when given a positive rotation from OA to OB. Its negative $-\mathbf{n}$ is in the sense of a negative rotation from **a** to **b** or, equivalently, a positive rotation from **b** to **a**. The *vector product* of **a** and **b** is defined to be a third vector whose magnitude is $ab \sin \theta$ and whose direction is **n**. This vector product is denoted by either $\mathbf{a} \times \mathbf{b}$ or by $\mathbf{a} \wedge \mathbf{b}$, the former notation being nowadays the more universal. Thus

$$\mathbf{a} \times \mathbf{b} = (ab \sin \theta)\,\mathbf{n}. \tag{1.19}$$

From the diagram and the definition embodied in (1.19)

$$\mathbf{b} \times \mathbf{a} = ba \sin (2\pi - \theta)\mathbf{n}$$

which shows that

$$\mathbf{b} \times \mathbf{a} = -\mathbf{a} \times \mathbf{b}. \tag{1.20}$$

(1.20) expresses a *non-commutative law for vector multiplication*.

Accepting the universality of the form (1.19), on taking $\mathbf{b} = \mathbf{a}$, we obtain $\theta = 0$ so that

$$|\mathbf{a} \times \mathbf{a}| = a \times a \times \sin 0 = 0$$

and so

$$\mathbf{a} \times \mathbf{a} = \mathbf{0}. \tag{1.21}$$

If for two vectors **a** and **b** we are given

$$\mathbf{a} \times \mathbf{b} = \mathbf{0} \tag{1.22}$$

then either $\mathbf{a} = \mathbf{0}$, $\mathbf{b} = \mathbf{0}$ or **a** and **b** are in the same or in opposite directions ($\theta = 0$ or π). Thus, for two given non-zero vectors **a** and **b**, (1.22) expresses the condition of parallelism when **a** and **b** are localised at different points and collinearity when they are localised at the same point.

For the unit vectors **i**, **j**, **k** in the axes $\vec{OX}$, $\vec{OY}$, $\vec{OZ}$ of a tri-rectangular Cartesian coordinate frame, the reader will easily establish, using the basic form (1.19), the important results:

$$\left.\begin{array}{lll} \mathbf{i} \times \mathbf{i} = \mathbf{0}, & \mathbf{j} \times \mathbf{j} = \mathbf{0}, & \mathbf{k} \times \mathbf{k} = \mathbf{0} \\ \mathbf{j} \times \mathbf{k} = \mathbf{i}, & \mathbf{k} \times \mathbf{i} = \mathbf{j}, & \mathbf{i} \times \mathbf{j} = \mathbf{k} \\ \mathbf{k} \times \mathbf{j} = -\mathbf{i}, & \mathbf{i} \times \mathbf{k} = -\mathbf{j}, & \mathbf{j} \times \mathbf{i} = -\mathbf{k} \end{array}\right\}. \tag{1.23}$$

The distributive law for vector products holds in the form

$$\mathbf{a} \times (\mathbf{b} + \mathbf{c}) = \mathbf{a} \times \mathbf{b} + \mathbf{a} \times \mathbf{c}. \tag{1.24}$$

The proof of this is established in Ref. 3, p. 31.

Using this rule and the relations (1.23) it is easy to show that, given two vectors $\mathbf{a} = [a_x, a_y, a_z]$; $\mathbf{b} = [b_x, b_y, b_z]$ in component forms, their vector product is

$$\mathbf{a} \times \mathbf{b} = (a_y b_z - a_z b_y)\mathbf{i} + (a_z b_x - a_x b_z)\mathbf{j} + (a_x b_y - a_y b_x)\mathbf{k}$$

$$= \begin{vmatrix} \mathbf{i} & \mathbf{j} & \mathbf{k} \\ a_x & a_y & a_z \\ b_x & b_y & b_z \end{vmatrix}, \tag{1.25}$$

the determinantal form being easily remembered.

The last result enables one to find the sine of the angle between two lines having known direction cosines $[l_1, m_1, n_1]$; $[l_2, m_2, n_2]$. These direction cosines are the Cartesian components of unit vectors $\hat{\mathbf{a}}_1$ and $\hat{\mathbf{a}}_2$, respectively. If θ is the angle between them, such that $0 \leqslant \theta \leqslant \pi$,

$$\sin\theta = |\hat{\mathbf{a}}_1 \times \hat{\mathbf{a}}_2|.$$

From

$$\hat{\mathbf{a}}_1 \times \hat{\mathbf{a}}_2 = \begin{vmatrix} \mathbf{i} & \mathbf{j} & \mathbf{k} \\ l_1 & m_1 & n_1 \\ l_2 & m_2 & n_2 \end{vmatrix}$$

it is easy to show

$$\sin\theta = \sqrt{\{(m_1 n_2 - m_2 n_1)^2 + (n_1 l_2 - n_2 l_1)^2 + (l_1 m_2 - l_2 m_1)^2\}}.$$

1.6 TRIPLE PRODUCTS OF VECTORS

Given three vectors in the Cartesian component forms

$$\mathbf{a} = [a_x, a_y, a_z]\,;\ \ \mathbf{b} = [b_x, b_y, b_z]\,;\ \ \mathbf{c} = [c_x, c_y, c_z]\,,$$

we can form the *scalar triple product* $\mathbf{a} \,.\, (\mathbf{b} \times \mathbf{c})$, meaning the scalar product of the vector $\mathbf{a}$ with the vector $(\mathbf{b} \times \mathbf{c})$. This scalar triple product will be denoted by $[\mathbf{a}, \mathbf{b}, \mathbf{c}]$. Since $\mathbf{b} \times \mathbf{c} = \Sigma \left\{\mathbf{i}(b_y c_z - b_z c_y)\right\}$, it follows that

$$[\mathbf{a}, \mathbf{b}, \mathbf{c}] = \mathbf{a} \,.\, (\mathbf{b} \times \mathbf{c}) = \begin{vmatrix} a_x & a_y & a_z \\ b_x & b_y & b_z \\ c_x & c_y & c_z \end{vmatrix}. \qquad (1.26)$$

From elementary determinantal properties, the last result shows

$$[\mathbf{a}, \mathbf{b}, \mathbf{c}] = [\mathbf{b}, \mathbf{c}, \mathbf{a}] = [\mathbf{c}, \mathbf{a}, \mathbf{b}]\,. \qquad (1.27)$$

But

$$[\mathbf{b}, \mathbf{a}, \mathbf{c}] = -[\mathbf{a}, \mathbf{b}, \mathbf{c}]\,, \text{etc.} \qquad (1.28)$$

Thus *cyclic permutation of the elements in a given scalar triple product* $[\mathbf{a}, \mathbf{b}, \mathbf{c}]$ *leave the value of it unchanged, but an acyclic permutation will change its sign leaving the magnitude unaltered.*

However, if we pre-vector multiply $(\mathbf{b} \times \mathbf{c})$ by $\mathbf{a}$ we obtain the vectorial quantity $\mathbf{a} \times (\mathbf{b} \times \mathbf{c})$, known as the *vector triple product* of the elements $\mathbf{a}, \mathbf{b}, \mathbf{c}$. It is shown in Ref. 3, pp. 39, 40 that

$$\mathbf{a} \times (\mathbf{b} \times \mathbf{c}) = (\mathbf{a} \,.\, \mathbf{c})\mathbf{b} - (\mathbf{a} \,.\, \mathbf{b})\mathbf{c}. \qquad (1.29)$$

This is the formula for the *vector triple product expansion.*

1.7 VECTOR MOMENT ABOUT A POINT: SCALAR MOMENT ABOUT AN AXIS

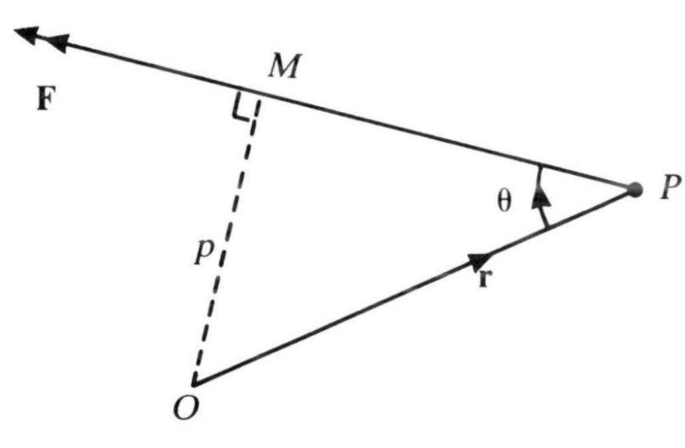

Fig. 1.6

Let $\mathbf{F}$ be a vector quantity acting at a point P whose position vector with respect to a chosen origin O is given by $\overrightarrow{OP} = \mathbf{r}$. The *vector moment* $\mathbf{G}$ of $\mathbf{F}$ about 0 is defined to be

$$\mathbf{G} = \mathbf{r} \times \mathbf{F}.$$

If $p = OM$ is the perpendicular from O to $\mathbf{F}$ and $O\hat{P}M = \theta$ then

$$\mathbf{G} = (rF \sin \theta)\mathbf{n} = pF\mathbf{n},$$

where $\mathbf{n}$ is the unit normal to the plane OPM in the sense of a positive rotation from $\mathbf{r}$ to $\mathbf{F}$.

When $\mathbf{F}$ is identified specifically with a force, the magnitude $G = pF$ of the vector moment $\mathbf{G}$ about O is clearly the physical moment of the force about the axis $\mathbf{n}$ through O. If P were a point in a rigid body which was free to rotate about this axis, the effect of the force would be to produce positive rotation of the body about the axis specified by $\mathbf{n}$ through O. (The reader will recall that a rigid body is a collection of particles such that the distance between every pair remains invariant whatever the motion of the collection.) Thus there is a specific direction associated with the moment about O: the moment is vectorial in kind and justifies the vectorial form $\mathbf{r} \times \mathbf{F}$.

Since moments are vectorial in kind, they may be combined as vectors. Hence if we have n forces $\mathbf{F}_k$ acting at n points P_k in 3-space where $\overline{OP}_k \equiv \mathbf{r}_k$ $(k = 1, 2, \ldots, n)$, then the total vector moment about O is

$$\mathbf{G} = \sum_{k=1}^{n} \mathbf{r}_k \times \mathbf{F}_k.$$

Let us now find the physical moment of a force about any axis. Suppose that $\mathbf{F}$ is a force acting at a point P distant p from an axis specified by a unit vector $\hat{\mathbf{a}}$ (Fig. 1.7).

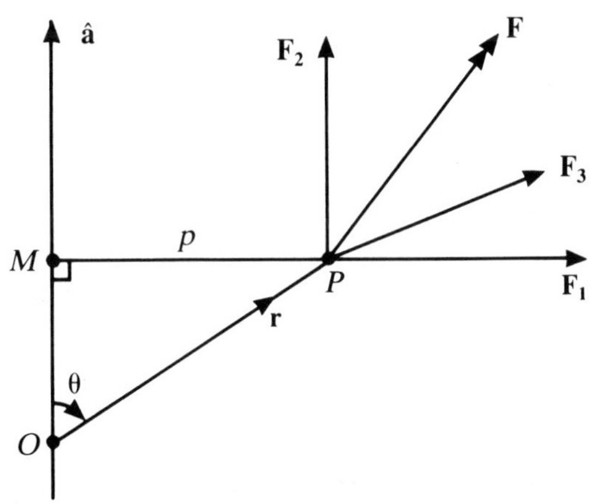

Fig. 1.7

Choosing an origin O in $\hat{\mathbf{a}}$, we let $\overrightarrow{OP} \equiv \mathbf{r}$. Let $\mathbf{F}$ have components F_1 along $\overrightarrow{MP}$. F_2 parallel to $\hat{\mathbf{a}}$ and F_3 in the sense $\mathbf{r} \times \mathbf{F}$, respectively. The three directions are mutually perpendicular. Since F_1 and F_2 produce no turning effect about $\hat{\mathbf{a}}$, the physical moment of $\mathbf{F}$ about $\hat{\mathbf{a}}$ is simply pF_3. Now $p = r \sin\theta = |\hat{\mathbf{a}} \times \mathbf{r}|$, and the vector $\hat{\mathbf{a}} \times \mathbf{r}$ has the direction of the component F_3 of $\mathbf{F}$. Thus

$$pF_3 = |\hat{\mathbf{a}} \times \mathbf{r}|F_3 = (\hat{\mathbf{a}} \times \mathbf{r}) \,.\, \mathbf{F} = [\hat{\mathbf{a}}, \mathbf{r}, \mathbf{F}] \,.$$

Thus the physical moment of $\mathbf{F}$ about the axis $\hat{\mathbf{a}}$ is given by the scalar triple product $[\hat{\mathbf{a}}, \mathbf{r}, \mathbf{F}]$ or by $\hat{\mathbf{a}} \,.\, \mathbf{G}$, where $\mathbf{G} = \mathbf{r} \times \mathbf{F}$ is the vector moment of $\mathbf{F}$ about the point O.

Two equal and opposite vectors $\pm\,\mathbf{F}$ *not* acting in the same straight line are said to constitute a *couple*. Let P_1, P_2 be any points chosen on $\mathbf{F}$ and $-\mathbf{F}$ respectively and let $\overrightarrow{OP_1} \equiv \mathbf{r}$, $\overrightarrow{OP_2} \equiv \mathbf{r}_2$, $\overrightarrow{P_2P_1} \equiv \mathbf{s}$ for a chosen origin O (Fig. 1.8). Then $\mathbf{s} = \mathbf{r}_1 - \mathbf{r}_2$.

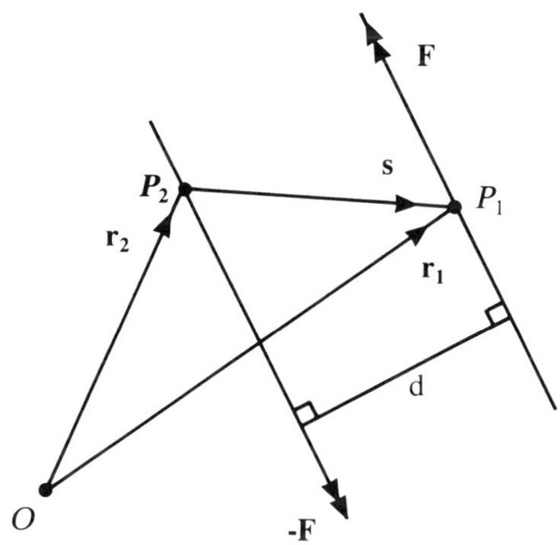

Fig. 1.8

The total vector moment about O is

$$\begin{aligned} \mathbf{G} &= \mathbf{r}_1 \times \mathbf{F} + \mathbf{r}_2 \times (-\mathbf{F}) \\ &= \mathbf{s} \times \mathbf{F}. \end{aligned}$$

This is seen to be invariant for any choice of O. It is also independent of $\mathbf{s}$ and hence of the positions of P_1 and P_2 since $|\mathbf{s} \times \mathbf{F}| = dF$, where d is the perpendicular distance between $\pm\,\mathbf{F}$. Also the direction of $\mathbf{G}$ is perpendicular to the plane $f \pm \mathbf{F}$ and in the sense of a positive rotation from $\mathbf{s}$ to $+\,\mathbf{F}$.

1.8 VECTOR FUNCTION OF A SCALAR VARIABLE: DERIVATIVE OF THE VECTOR FUNCTION

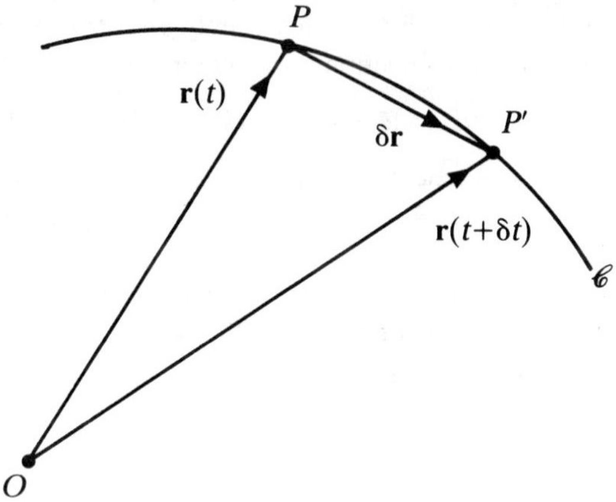

Fig. 1.9

In Fig. 1.9, O is a fixed origin, $\overrightarrow{OP} \equiv \mathbf{r} = [x, y, z]$, the components being referred to a tri-rectangular Cartesian frame through O. Suppose x, y, z are continuous and differentiable functions of a scalar t so that the derivatives $\mathrm{d}x/\mathrm{d}t$, $\mathrm{d}y/\mathrm{d}t$, $\mathrm{d}z/\mathrm{d}t$ exist uniquely for each t. The locus of P as t varies is a curve in 3-space. Since $x = x(t)$, etc., we may write

$$\mathbf{r} = \mathbf{r}(t) = x(t)\mathbf{i} + y(t)\mathbf{j} + z(t)\mathbf{k}.$$

Here $x(t)$, $y(t)$, $z(t)$ are scalar functions of t and $\mathbf{r}(t)$ is a *vector function of the scalar* t. In many problems of mechanics, t often specifies time so that the point P moves along $\mathscr{C}$ in such a way that its position vector $\mathbf{r} = \mathbf{r}(t)$ is known at each instant of time t. When the scalar t is increased to $t + \delta t$, the position vector changes from $\overrightarrow{OP}$ to $\overrightarrow{OP}'$, where

$$\overrightarrow{OP}' \equiv \mathbf{r}(t + \delta t) = \mathbf{r} + \delta\mathbf{r}.$$

Then

$$\overrightarrow{PP}' \equiv \overrightarrow{OP}' - \overrightarrow{OP} \equiv \delta\mathbf{r}.$$

The mean rate of change of the position vector as the scalar parameter specifying $\mathscr{C}$ changes from t to $(t + \delta t)$ is $\delta\mathbf{r}/\delta t$. Following conventional scalar differential calculus, we define the derivative of $\mathbf{r}(t)$ at t to be

$$\frac{\mathrm{d}\mathbf{r}}{\mathrm{d}t} = \lim_{\delta t \to 0} \left(\frac{\delta\mathbf{r}}{\delta t}\right),$$

assuming this limit to exist uniquely at t. Clearly $\mathrm{d}\mathbf{r}/\mathrm{d}t$ is a vector along the tangent at P to $\mathscr{C}$. When t specifies time, this limit signifies the instantaneous velocity at P on $\mathscr{C}$. We also note from the Cartesian form of $\mathbf{r}$ that

$$\frac{d\mathbf{r}}{dt} = \frac{dx}{dt}\mathbf{i} + \frac{dy}{dt}\mathbf{j} + \frac{dz}{dt}\mathbf{k}.$$

The components $[dx/dt, dy/dt, dz/dt]$ are the velocity components of P along the x-, y-, z-directions when t specifies time.

If $\mathbf{r} = \mathbf{r}(t)$, $\mathbf{s} = \mathbf{s}(t)$ are two vector functions of a scalar t, then

$$\begin{aligned}\frac{d(\mathbf{r}\,.\,\mathbf{s})}{dt} &= \lim_{\delta t \to 0}\left\{\frac{(\mathbf{r}+\delta\mathbf{r})\,.\,(\mathbf{s}+\delta\mathbf{s}) - \mathbf{r}\,.\,\mathbf{s}}{\delta t}\right\} \\ &= \lim_{\delta t \to 0}\left\{\frac{\mathbf{r}\,.\,\delta\mathbf{s} + \mathbf{s}\,.\,\delta\mathbf{r} + \delta\mathbf{r}\,.\,\delta\mathbf{s}}{\delta t}\right\} \\ &= \mathbf{r}\cdot\frac{d\mathbf{s}}{dt} + \mathbf{s}\cdot\frac{d\mathbf{r}}{dt}. \qquad (1.30)\end{aligned}$$

Similarly we can show

$$\frac{d(\mathbf{r}\times\mathbf{s})}{dt} = \mathbf{r}\times\frac{d\mathbf{s}}{dt} + \frac{d\mathbf{r}}{dt}\times\mathbf{s}. \qquad (1.30')$$

Here it is essential to preserve the correct order of the vector products on the right-hand side.

Further details of this topic are contained in Ref. 3, Ch. 3.

1.9 NOTION OF SCALAR AND VECTOR FIELDS

Suppose $P(x, y, z)$ is any point of a region R_3 of 3-space, which may be either finite or infinite, and that at each point P a scalar quantity ϕ is defined. Then the function ϕ is a *scalar function* of the position of P and it may be specified as either $\phi(x, y, z)$ or as $\phi(\mathbf{r})$, where $\overrightarrow{OP} \equiv \mathbf{r}$ and O is the chosen fixed origin. The aggregate of all values ϕ in R_3 constitutes a *scalar field*. We suppose ϕ to be continuous and single-valued throughout R_3, in which case it is also a *uniform scalar function* in R_3. An example of a uniform scalar field is provided by the temperature inside a kettle of water heated by a gas flame. Before boiling takes place the temperature distribution throughout the water is non-uniform but single-valued at each point of the water. In the case of a uniform scalar function ϕ in R_3, we can construct the surfaces $\phi(x, y, z)$ = constant in R_3. These are *level surfaces* or *iso-ϕ surfaces*. No two surfaces can intersect at any P in R_3 since the value of ϕ is unique at each point P. The level surfaces laminate the region R_3. In the case when R_3 is the volume of heated water in the kettle, the level surfaces are the isothermals throughout R_3.

Now suppose that at each point $P(x, y, z)$ of R_3 a unique vector function $\mathbf{F}$ is defined. The totality of all values of $\mathbf{F}$ throughout R_3 constitutes a *vector field*. An example is afforded by a fish-aquarium with water entering through one tube and leaving by another. This gives the stored water varying but unique velocities at different points P of R_3, the volume of water in the aquarium. The totality of values of the velocity at all points of the aquarium constitutes a vector field. With a uniform vector function **F** defined at each $P(x, y, z)$ of R_3, we may write $\mathbf{F} = \mathbf{F}(x, y, z)$ or $\mathbf{F} = \mathbf{F}(\mathbf{r})$, where $\overrightarrow{OP} \equiv \mathbf{r}$. Such a vector function may be expressed in terms of its scalar components:

$$\mathbf{F} = [F_x, F_y, F_z] = F_x\mathbf{i} + F_y\mathbf{j} + F_z\mathbf{k},$$

where $F_x = F_x(x, y, z)$, etc. At each point P of R_3 a line may be drawn with direction ratios $[F_x, F_y, F_z]$: it will be tangential to a certain curve $\mathscr{C}$ in R_3 and the differential equations of $\mathscr{C}$ may be specified by

$$\mathrm{d}x/F_x = \mathrm{d}y/F_y = \mathrm{d}z/F_z. \tag{1.31}$$

The solution of these equations involves two arbitrary constants corresponding to a double family of surfaces. The intersection of a pair of such surfaces determines a curve called a *field-line* of **F** in R_3. Examples of field-lines are afforded by the streamlines of the flow of air over an aerofoil and the lines of magnetic field intensity emanating from the poles of a bar magnet.

We note that if a family of surfaces is specified by the differential equation

$$F_x\mathrm{d}x + F_y\mathrm{d}y + F_z\mathrm{d}z = 0$$

then such a family is cut orthogonally by the field-lines

$$\mathrm{d}x/F_x = \mathrm{d}y/F_y = \mathrm{d}z/F_z.$$

This follows because the first equation shows that if the vectors $[F_x, F_y, F_z]$ and $[\mathrm{d}x, \mathrm{d}y, \mathrm{d}z]$ are non-zero, then they are orthogonal. The second vector is the displacement from a point $P(x, y, z)$ on a surface to a neighbouring point $P'(x + \mathrm{d}x, y + \mathrm{d}y, z + \mathrm{d}z)$ on the same surface. As $\overrightarrow{PP'}$ is tangential to the surface it follows that the first vector is normal to it and hence in the direction of the field line through P.

1.10 VECTOR GRADIENT

Suppose $\phi = \phi(x, y, z)$ is a uniform differentiable function in R_3. Throughout R_3 the level surfaces are drawn: S_P is the level surface through $P(x, y, z)$ on which the scalar function takes the constant value $\phi(x, y, z)$ and S_Q that through $Q(x + \delta x, y + \delta y, z + \delta z)$ on which it takes the constant value $\phi(x + \delta x, y + \delta y, z + \delta z) = \phi + \delta\phi$. Then $\phi_N = \phi_{P'} = \phi + \delta\phi$. Let PN be drawn normally at P to cut S_Q in N so that PN is approximately normal to S_P when we take Q very close to P. Then

$$\frac{\phi_Q - \phi_P}{\delta s} = \frac{\delta\phi}{\delta s} = \frac{\delta\phi}{\delta n} \times \frac{\delta n}{\delta s},$$

where $PQ = \delta s$, $PN = \delta n$. Let **s** and **n** be unit vectors along $\overrightarrow{PQ}$, $\overrightarrow{PN}$, respectively.

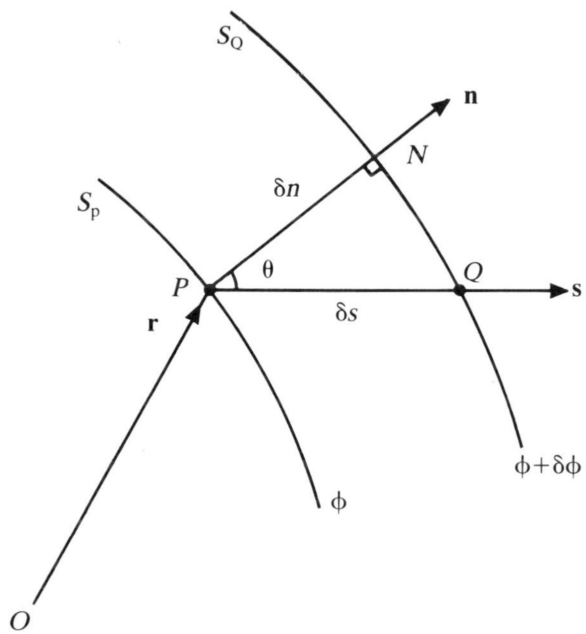

Fig. 1.10

As $S_Q \to S_P$ so that $Q \to P$ along the direction $\overrightarrow{QP}$, θ being kept constant, the last relation gives in the limit

$$\frac{\partial \phi}{\partial s} = \frac{\partial \phi}{\partial n} \cos \theta . \tag{1.32}$$

The left-hand side measures the rate of change of ϕ at P in the direction of s. Thus if we allow θ to vary but fix P on S_P,

$$\max |\partial\phi/\partial s| = |\partial\phi/\partial n|,$$

showing that the maximum rate of change of ϕ with distance at P occurs along the normal to S_P at P. We associate **n** with $\partial\phi/\partial n$ to define the uniform vector function $(\partial\phi/\partial n)\mathbf{n}$ at each $P(x, y, z)$ of S_P and hence throughout R_3 since S_P is any arbitrary level surface in R_3. This vector function is denoted either by grad ϕ or by ϕ and it is called the *(vector) gradient* of the scalar function ϕ in R_3. Thus

$$\text{grad}\, \phi = \nabla\phi = \frac{\partial \phi}{\partial n}\, \mathbf{n}. \tag{1.33}$$

Since $\cos\theta = \mathbf{n} \,.\, \mathbf{s}$, equation (1.32) may be written

$$\frac{\partial \phi}{\partial s} = \mathbf{s} \,.\, \mathbf{n}\, \frac{\partial \phi}{\partial n} = \mathbf{s} \,.\, \nabla\phi, \tag{1.34}$$

on using (1.33). As this relation is true for any uniform differentiable scalar function $\phi(x, y, z)$ in R_3, we may abstract from it the operational equivalence

$$\frac{\partial}{\partial s} \equiv \mathbf{s} \,.\, \nabla. \tag{1.34'}$$

In (1.34), $\partial/\partial s$ or $\mathbf{s}\,.\,\nabla$ denotes the *directional derivative* since when applied to ϕ, it differentiates ϕ in the direction s.

The Cartesian components of $\nabla\phi$ are $[\mathbf{i}\,.\,\nabla\phi, \mathbf{j}\,.\,\nabla\phi, \mathbf{k}\,.\,\nabla\phi]$. Taking $\mathbf{s} = \mathbf{i}$ in (1.34), we have $\partial\phi/\partial x = \mathbf{i}\,.\,\nabla\phi$. Similarly $\mathbf{j}\,.\,\nabla\phi = \partial\phi/\partial y$ and $\mathbf{k}\,.\,\nabla\phi = \partial\phi/\partial z$. Thus we have found the Cartesian components of ϕ and we may write

$$\nabla\phi = \text{grad}\,\phi = \frac{\partial\phi}{\partial x}\mathbf{i} + \frac{\partial\phi}{\partial y}\mathbf{j} + \frac{\partial\phi}{\partial z}\mathbf{k}, \tag{1.35}$$

a form useful for evaluating $\nabla\phi$ when ϕ is specified in Cartesians, so that $\phi = \phi(x, y, z)$. Since $\phi(x, y, z)$ is an arbitrary uniform, continuous and differentiable function, we may abstract from (1.35) the Cartesian form of the operator ∇, i.e.,

$$\nabla \equiv \mathbf{i}\,\frac{\partial}{\partial x} + \mathbf{j}\,\frac{\partial}{\partial y} + \mathbf{k}\,\frac{\partial}{\partial z}. \tag{1.35'}$$

The total differential of $\phi(x, y, z)$ is

$$\begin{aligned} \mathrm{d}\phi &= \frac{\partial\phi}{\partial x}\mathrm{d}x + \frac{\partial\phi}{\partial y}\mathrm{d}y + \frac{\partial\phi}{\partial z}\mathrm{d}z \\ &= \left(\frac{\partial\phi}{\partial x}\mathbf{i} + \frac{\partial\phi}{\partial y}\mathbf{j} + \frac{\partial\phi}{\partial z}\mathbf{k}\right).(\mathrm{d}x\,\mathbf{i} + \mathrm{d}y\,\mathbf{j} + \mathrm{d}z\,\mathbf{k}). \end{aligned}$$

In vector form, this gives

$$\mathrm{d}\phi = \mathrm{d}\mathbf{r}\,.\,\nabla\phi. \tag{1.36}$$

Here we use $\overrightarrow{OP} \equiv \mathbf{r} = x\mathbf{i} + y\mathbf{j} + z\mathbf{k}$, so that $\mathrm{d}\mathbf{r} = \mathrm{d}x\mathbf{i} + \mathrm{d}y\mathbf{j} + \mathrm{d}z\mathbf{k}$. To the first order of approximation we may replace (1.36) by the approximation

$$\delta\phi \simeq \delta\mathbf{r}\,.\,\nabla\phi,$$

for sufficiently small $|\delta\mathbf{r}|$.

Since ϕ is a uniform differentiable scalar function of position it may be expressed either as $\phi(x, y, z)$, as above, or as $\phi(\mathbf{r})$. This follows since the position of a general point P in R_3 is specified either by the Cartesian coordinates (x, y, z) or by the position vector $\overrightarrow{OP} \equiv \mathbf{r}$. At a neighbouring point $P'(x + \delta x, y + \delta y, z + \delta z)$, ϕ would be specified as either $\phi(x + \delta x, y + \delta y, z + \delta z)$ or as $\phi(r + \delta\mathbf{r})$. When $|\delta\mathbf{r}|$ is sufficiently small, Taylor's theorem may be expressed in the form

$$\phi(x+\delta x, y+\delta y, z+\delta z) = \phi(x,y,z) + $$

$$+\left(\delta x\frac{\partial}{\partial x}+\delta y\frac{\partial}{\partial y}+\delta z\frac{\partial}{\partial z}\right)\phi(x,y,z)$$

$$+\frac{1}{2!}\left(\delta x\frac{\partial}{\partial x}+\delta y\frac{\partial}{\partial y}+\delta z\frac{\partial}{\partial z}\right)^2\phi(x,y,z)+\ldots$$

Since $\delta x(\partial/\partial x)+\delta y(\partial/\partial y)+\delta z(\partial/\partial z)\equiv\delta\mathbf{r}\,.\,\nabla$, this has the vectorial form

$$\phi(\mathbf{r}+\delta\mathbf{r})=\phi(\mathbf{r})+(\delta\mathbf{r}\,.\,\nabla)\,\phi(\mathbf{r})+\frac{1}{2!}\,(\delta\mathbf{r}\,.\,\nabla)^2\,\phi(\mathbf{r})+\ldots \tag{1.37}$$

When $\phi=\phi(r)$, where $r=|\mathbf{r}|$, the scalar distance from O to P, grad ϕ is easily found from the definition (1.33). The level surfaces of $\phi(r)$ are given by $\phi(r) = \text{const.}$, i.e. $r = \text{const.}$ These surfaces are the family of concentric spheres centred on O. Thus in the notation of Fig. 1.10, $\delta n=\delta r$ and $\hat{\mathbf{n}}=\hat{\mathbf{r}}=\mathbf{r}/r$, since the radius of any sphere is normal to its surface. Thus

$$\nabla\phi(r)=\frac{\partial\phi}{\partial n}\,\mathbf{n}=\frac{\mathrm{d}\phi(\mathrm{r})}{\mathrm{d}r}\,\hat{\mathbf{r}}=\phi'(r)\,\hat{\mathbf{r}}.$$

This form may also be derived by conversion to Cartesian coordinates. For with ϕ expressed as a function of x, y, z,

$$\frac{\partial\phi}{\partial x}=\frac{\mathrm{d}\phi(\mathrm{r})}{\mathrm{d}r}\times\frac{\partial r}{\partial x}=\frac{x}{r}\,\phi'(r),$$

since $r^2=x^2+y^2+z^2$ gives $\partial r/\partial x=x/r$, etc. Similar forms are obtained for $\partial\phi/\partial y$ and $\partial\phi/\partial z$ and so (1.35) gives

$$\nabla\phi(r)=\frac{1}{r}\,\phi'(r)\,(x\mathbf{i}+y\mathbf{j}+z\mathbf{k})=\phi'(r)\hat{\mathbf{r}},$$

as before.

From the Cartesian form (1.35) it is easy to establish the important results

$$\nabla(c_1\phi_1+c_2\phi_2)=c_1\,\nabla\phi_1+c_2\,\nabla\phi_2, \tag{1.38}$$

$$\nabla(\phi_1\phi_2)=\phi_1\,\nabla\phi_2+\phi_2\,\nabla\phi_1, \tag{1,39}$$

where ϕ_1, ϕ_2 are each uniform differentiable scalar functions of (x, y, z) and c_1, c_2 are constants. These equations follow similar patterns obtained in elementary scalar calculus. For if $D\equiv\mathrm{d}/\mathrm{d}x$, then

$$D[c_1u_1(x)+c_2u_2(x)]=c_1Du_1+c_2Du_2;$$

$$D[u_1(x)u_2(x)]=u_1Du_2+u_2Du_1,$$

for continuous differentiable functions $u_1(x)$ and $u_2(x)$. For fuller details, see Ref. 3, p. 85.

1.11 THE OPERATOR ∇

The Cartesian form (1.35′) of ∇ shows its likeness to a vectorial quantity: it is a *vector operator*. Consequently we expect it to behave, in some respects at least, in the manner of a vector quantity. In fact, following the forms for $\mathbf{a} . \mathbf{b}$ and $\mathbf{a} \times \mathbf{b}$ we can define $\nabla . \mathbf{F}$ and $\nabla \times \mathbf{F}$ for a given uniform differentiable vector function $\mathbf{F} = [F_x, F_y, F_z]$, where $F_x = F_x(x, y, z)$, etc. Thus we define the divergence of $\mathbf{F}$ or div $\mathbf{F}$ to be $\nabla . \mathbf{F}$ where

$$\operatorname{div} \mathbf{F} = \nabla . \mathbf{F} = \frac{\partial F_x}{\partial x} + \frac{\partial F_y}{\partial y} + \frac{\partial F_z}{\partial z} . \tag{1.40}$$

Essentially this is a scalar quantity and div can act only on a vector function. Likewise we can define the *curl* of $\mathbf{F}$ or $\nabla \times \mathbf{F}$ to be

$$\operatorname{curl} \mathbf{F} = \nabla \times \mathbf{F} = \begin{vmatrix} \mathbf{i} & \mathbf{j} & \mathbf{k} \\ \partial/\partial x & \partial/\partial y & \partial/\partial z \\ F_x & F_y & F_z \end{vmatrix} . \tag{1.41}$$

The expansion of the determinant in (1.41) is

$$\mathbf{i}(\partial F_z/\partial y - \partial F_y/\partial z) + \mathbf{j}(\partial F_x/\partial z - \partial F_z/\partial x) + \mathbf{k}(\partial F_y/\partial x - \partial F_x/\partial y).$$

The operator curl can act only on a vector function to produce another vector function.

Further we may define the vector operator $\mathbf{F} . \nabla$ or $\mathbf{F}$. grad to be

$$\mathbf{F} . \nabla \equiv F_x \frac{\partial}{\partial x} + F_y \frac{\partial}{\partial y} + F_z \frac{\partial}{\partial z} . \tag{1.42}$$

This operator is applicable to both scalar and vectorial operands. Thus if $\phi = \phi(x, y, z)$ is a uniform differentiable scalar function in R_3, then

$$\mathbf{F} . \nabla\phi = F_x\, \partial\phi/\partial x + F_y\, \partial\phi/\partial y + F_z\, \partial\phi/\partial z. \tag{1.43}$$

For a uniform differentiable vector function $\mathbf{V} = [V_x, V_y, V_z]$, however

$$\mathbf{F} . \nabla\mathbf{V} = F_x\, \partial\mathbf{V}/\partial x + F_y\, \partial\mathbf{V}/\partial y + F_z\, \partial\mathbf{V}/\partial z, \tag{1.44}$$

where $\partial\mathbf{V}/\partial x = (\partial V_x/\partial x)\mathbf{i} + (\partial V_y/\partial x)\mathbf{j} + (\partial V_z/\partial x)\mathbf{k}$, etc., since $\mathbf{i}, \mathbf{j}, \mathbf{k}$ are constant vectors. Full particulars feature in Ref. 3, pp. 85, 86.

From the Cartesian form for divergence it is easy to show that

$$\nabla . (c_1 \mathbf{F}_1 + c_2 \mathbf{F}_2) = c_1 \nabla . \mathbf{F}_1 + c_2 \nabla . \mathbf{F}_2 \tag{1.45}$$

where c_1, c_2 are arbitrary scalar constants and $\mathbf{F}_1$, $\mathbf{F}_2$ are uniform differentiable vector functions in R_3.

Another important result is

$$\nabla.(\phi\mathbf{F}) = \phi\ \nabla.\mathbf{F} + (\nabla\phi).\mathbf{F}, \tag{1.46}$$

where ϕ is a uniform differentiable scalar function and $\mathbf{F}$ a uniform differentiable vector function in R_3. Equations (1.45) and (1.46) are obtained in Ref. 3, pp. 94, 95.

For a uniform twice-differentiable function in R_3, we have grad $\phi = \nabla\phi = [\partial\phi/\partial x,\ \partial\phi/\partial y,\ \partial\phi/\partial z]$, a vector function. This vector function has a divergence $\nabla.(\nabla\phi)$ or $\nabla^2\ \phi$ so that

$$\nabla^2\phi = \nabla.(\nabla\phi) = \frac{\partial^2\phi}{\partial x^2} + \frac{\partial^2\phi}{\partial y^2} + \frac{\partial^2\phi}{\partial z^2}. \tag{1.47}$$

The operator $\nabla^2 \equiv \Sigma\ \partial^2/\partial x^2$ is called the *Laplacian*. If in particular ϕ satisfies the partial differential equation

$$\nabla^2\ \phi = 0 \tag{1.48}$$

then $\phi(x, y, z)$ is said to be a *harmonic function* in R_3 and (1.48) is called *Laplace's equation* and it is frequently met in many branches of classical and modern mathematical physics. A simple example of a harmonic function is

$$\phi(x, y, z) = \exp\ [\surd(m^2 + n^2)x]\ \sin my\ \cos nz.$$

For

$$\partial^2\phi/\partial x^2 = (m^2 + n^2)\phi,\quad \partial^2\phi/\partial y^2 = -m^2\phi,\quad \partial^2\phi/\partial z^2 = -n^2\phi,$$

so that $\nabla^2\ \phi = 0$.

An example of a very important harmonic function is $\phi(r) = 1/r$. We have seen in Section 1.10 that

$$\nabla\phi(r) = -\frac{1}{r^2}\,\hat{\mathbf{r}} = -\frac{1}{r^3}\,\mathbf{r}.$$

Thus

$$\nabla^2\phi(r) = -\frac{1}{r^3}\ \nabla.\mathbf{r} - \mathbf{r}.\nabla\left(\frac{1}{r^3}\right).$$

As $\mathbf{r} = [x, y, z]$, $\nabla.\mathbf{r} = 3$. The level surfaces of $1/r^3$ are the concentric spheres $r =$ const. so that

$$\nabla\left(\frac{1}{r^3}\right) = \hat{\mathbf{r}}\ \frac{\mathrm{d}}{\mathrm{d}r}\left(\frac{1}{r^3}\right) = -\frac{3}{r^4}\,\hat{\mathbf{r}}.$$

Hence

$$\nabla^2\phi(r) = -\frac{3}{r^3} + \mathbf{r}.\hat{\mathbf{r}}\left(\frac{3}{r^4}\right) = 0,$$

since $\mathbf{r}\,.\,\hat{\mathbf{r}} = r$. Alternatively, the Cartesian form of the Laplacian, namely $\partial^2/\partial x^2 + \partial^2/\partial y^2 + \partial^2/\partial z^2$, may be used on $1/r = (x^2 + y^2 + z^2)^{-1/2}$.

1.12 FURTHER IDENTITIES INVOLVING ∇

In R_3, uniform differentiable scalar functions ϕ and ψ and uniform differentiable vector functions $\mathbf{F}$ and $\mathbf{G}$ are defined, the functions being differentiable as many times as are required to validate the identities listed. The proofs of these identities are given in Ref. 3, Section 6.5.

First we restate the results

$$\nabla(\phi\psi) = \phi\nabla\psi + \psi\nabla\phi, \tag{1.49}$$

$$\nabla\,.\,(\phi\mathbf{F}) = \phi\nabla\,.\,\mathbf{F} + \mathbf{F}\,.\,\nabla\phi \tag{1.50}$$

and observe how closely they follow the rule for differentiating the product of two functions of the same single variable in elementary scalar calculus.

Next, we observe that

$$\text{curl grad}\,\phi = \nabla\times\nabla\phi = \mathbf{0}, \tag{1.51}$$

$$\text{div curl}\,\mathbf{F} = \nabla.\nabla\times\mathbf{F} = [\nabla, \nabla, \mathbf{F}] = 0. \tag{1.52}$$

In (1.51) the form $\nabla\times\nabla$ suggests that a zero vector may ensue by analogy with the form $\mathbf{a}\times\mathbf{a} = \mathbf{0}$ in vector algebra. Also the equal elements in the scalar triple product form of (1.52) suggest the scalar zero here. However, these are only useful mnemonics for recapitulation. Full proofs are found in Ref. 3, Section 6.5.

The form

$$\text{curl curl}\,\mathbf{F} = \text{grad div}\,\mathbf{F} - \nabla^2\mathbf{F}$$

or, symbolically,

$$\nabla\times(\nabla\times\mathbf{F}) = \quad(\nabla\,.\,\mathbf{F}) - \nabla^2\mathbf{F} \tag{1.53}$$

obeys the expansion rule of the vector triple product when taken in the form

$$\mathbf{a}\times(\mathbf{b}\times\mathbf{c}) = \mathbf{b}(\mathbf{a}\,.\,\mathbf{c}) - (\mathbf{a}\,.\,\mathbf{b})\,\mathbf{c},$$

with **a, b** and **c** formally replaced by ∇, ∇ and $\mathbf{F}$ respectively. It is noteworthy that ∇ must act on an operand: in this case on $\nabla\,.\,\mathbf{F}$.

Following the list in Ref. 3 in increasing order of complexity, we have

$$\text{div}\,(\mathbf{F}\times\mathbf{G}) = \mathbf{G}\,.\,\text{curl}\,\mathbf{F} - \mathbf{F}\,.\,\text{curl}\,\mathbf{G}$$

or

$$\nabla\,.\,(\mathbf{F}\times\mathbf{G}) = \mathbf{G}\,.\,(\nabla\times\mathbf{F}) - \mathbf{F}\,.\,(\nabla\times\mathbf{G}). \tag{1.54}$$

Contrast this with the result in vector algebra

$$\mathbf{a}\,.\,(\mathbf{b}\times\mathbf{c}) = \mathbf{c}\,.\,(\mathbf{a}\times\mathbf{b})$$

where the cyclic order is preserved. In (1.54) an additional term features on the right-hand side. ∇ is a vector operator and not a vector. We recall that in vector algebra

$$k(uv) = v(ku).$$

but in elementary scalar calculus, if $u = u(x)$, $v = v(x)$ $\mathrm{D} \equiv \mathrm{d}/\mathrm{d}x$, then

$$\mathrm{D}(uv) = u\mathrm{D}v + v\mathrm{D}u,$$

showing that more replacement of algebraic k by the calculus operator D will yield an incorrect result. This is why naive replacement of **a** by ∇, **b** by **F**, **c** by **G** yields an incorrect result.

Other identities listed include:

$$\operatorname{curl}(\mathbf{F} \times \mathbf{G}) = (\mathbf{G} \,.\, \nabla)\mathbf{F} - (\mathbf{F} \,.\, \nabla)\mathbf{G} + \mathbf{F} \operatorname{div} \mathbf{G} - \mathbf{G} \operatorname{div} \mathbf{F}$$

or

$$\nabla \times (\mathbf{F} \times \mathbf{G}) = (\mathbf{G} \,.\, \nabla)\mathbf{F} - (\mathbf{F} \,.\, \nabla)\mathbf{G} + \mathbf{F}(\nabla \,.\, \mathbf{G}) - \mathbf{G}(\nabla \,.\mathbf{F}); \tag{1.55}$$

$$\nabla(\mathbf{F} \,.\, \mathbf{G}) = (\mathbf{G} \,.\, \nabla)\mathbf{F} + (\mathbf{F} \,.\, \nabla)\mathbf{G} + \mathbf{G} \times (\nabla \times \mathbf{G}) + \mathbf{F} \times (\nabla \times \mathbf{G}). \tag{1.56}$$

In (1.56) on taking $\mathbf{F} = \mathbf{G} = \mathbf{V}$ we obtain the result

$$(\mathbf{V} \,.\, \nabla)\mathbf{V} = \nabla(\tfrac{1}{2}\mathbf{V}^2) - \mathbf{V} \times \operatorname{curl} \mathbf{V}, \tag{1.57}$$

important in hydrodynamics.

1.13 DEFINITIONS OF LINE, SURFACE AND VOLUME INTEGRALS

Further development of the vector calculus requires the extension of ordinary integration techniques to special cases in which the integration is restricted to R_1 in the case of a curve, R_2 in the case of a surface and R_3 in the case of a volume. Here we treat the more elementary aspects of such integrations. Their further developments will be treated heuristically as and when required within the vector calculus proper.

(i) First, we consider the *line integral of a scalar function* $\phi(t)$ along an arc of a curve $\mathscr{C}$ specified parametrically by t such that along the arc of $\mathscr{C}$,

$$x = x(t), \quad y = y(t), \quad z = z(t) \qquad \text{where } t_1 \leqslant t \leqslant t_2 .$$

Then the *line integral of $\phi(t)$ with respect to t along $\mathscr{C}$ from t_1 to t_2 is* defined to be

$$\int_{t_1}^{t_2} \phi(x(t), y(t), z(t))\, \mathrm{d}t.$$

In the particular case when t is identified with the arc length s measured from a fixed point on $\mathscr{C}$ and such that $a \leqslant s \leqslant b$ defines the extent of the arc of $\mathscr{C}$, the line integral becomes simply

$$\int_a^b \phi(x(s), y(s), z(s))\, \mathrm{d}s, \qquad a \leqslant s \leqslant b.$$

In this form, a physical meaning can be assigned to the line integral. If at the location on $\mathscr{C}$ specified by s the line density of the curve is $\phi(s) = \phi(x/s), y(s), z(s))$, then $\phi(s)\,\delta s$ is the mass of a further arc element located at distance s and so the total mass of the entire arc $a \leqslant s \leqslant b$ of $\mathscr{C}$ is

$$\int_a^b \phi(x(s), y(s), z(s))\,\mathrm{d}s = \lim_{\substack{N\to\infty \\ \max \delta_{si}\to 0}} \sum_{i=0}^{N-1} \phi(x(s_i), y(s_i)\, z(s_i))\,\delta s_i,$$

where $a = s_0 < s_1 < \ldots < s_N = b$ and $\delta s_i = s_{i+1} - s_i$.

Example [1.1]
A wire of variable line density and finite length is in the form of the circular helix

$$x = a\cos\theta,\quad y = a\sin\theta,\quad z = a\theta\tan\alpha,\qquad 0 \leqslant \theta \leqslant \pi/2,$$

where a and α are constants. At location θ the mass per unit length of the wire is $m_0\cos^2\theta$ (m_0 = const.). Find the total mass of the wire.

From the parametric equations,

$$\begin{aligned} \mathrm{d}s(\theta) &= \sqrt{\{(\mathrm{d}x)^2 + (\mathrm{d}y)^2 + (\mathrm{d}z)^2\}} \\ &= a\sec\alpha\, d\theta \end{aligned}$$

and so the mass of the arc element δs is $m_0 a\sec\alpha\cos^2\theta\,\delta\theta$. Hence the total mass of the wire is

$$m_0 a\sec\alpha\int_0^{\pi/2}\cos^2\theta\,\mathrm{d}\theta = \frac{\pi m_0 a\sec\alpha}{4}.$$

The types of line integral considered above are scalar. The *vector line integral* $\int\phi(x, y, z)\mathrm{d}\mathbf{s}$ or $\int\phi(x, y, z)\mathrm{d}\mathbf{r}$ (since $\mathrm{d}\mathbf{s} = \mathrm{d}\mathbf{r}$) means

$$\int\phi(x, y, z)\,(\mathbf{i}\,\mathrm{d}x + \mathbf{j}\,\mathrm{d}y + \mathbf{k}\,\mathrm{d}z),$$

where all three integrals are taken between appropriately specified limits.

Example [1.2]
If $\phi = xy$, evaluate the vector line integral for the helix arc.

As $\phi(\theta) = a^2\sin\theta\cos\theta$, the required line integral is

$$\int_0^{\pi/2}\phi(\theta)\left(\mathbf{i}\,\frac{\mathrm{d}x}{\mathrm{d}\theta} + \mathbf{j}\,\frac{\mathrm{d}y}{\mathrm{d}\theta} + \mathbf{k}\,\frac{\mathrm{d}z}{\mathrm{d}\theta}\right)\mathrm{d}\theta$$

$$= a^3 \int_0^{\pi/2} (-\sin^2\theta\cos\theta\,\mathbf{i} + \sin\theta\cos^2\theta\,\mathbf{j} + \tan\alpha\sin\theta\cos\theta\,\mathbf{k})\,d\theta$$

$$= a^3\,(-\tfrac{1}{3}\mathbf{i} + \tfrac{1}{3}\mathbf{j} + \tfrac{1}{2}\tan\alpha\,\mathbf{k}).$$

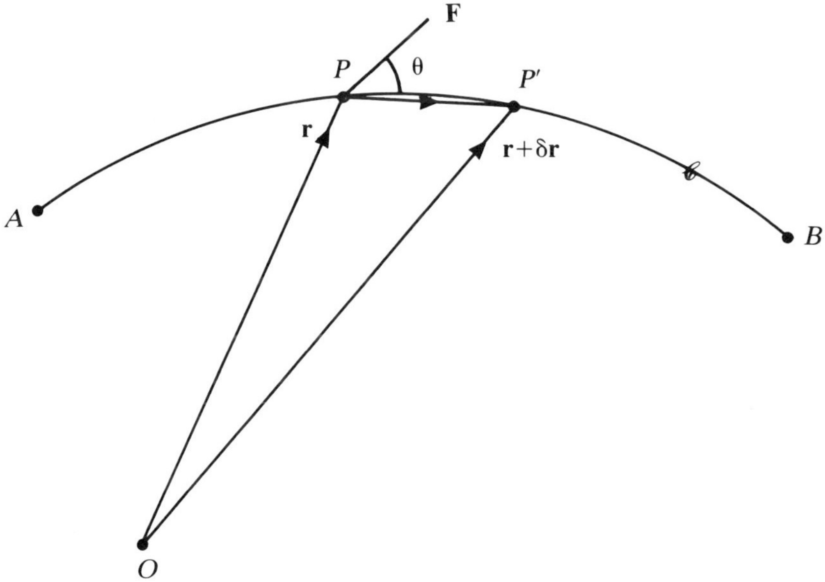

Fig. 1.11

Now suppose a vector function $\mathbf{F} = [F_x, F_y, F_z]$ is defined uniquely at each $P(x, y, z)$ of an arc AB of a curve $\mathscr{C}$. At P in Fig. 1.11, $\mathbf{F}$ makes an angle θ with the tangent at P to $\mathscr{C}$. If P' is the neighbouring point $(x + \delta x, y + \delta y, z + \delta z)$ on $\mathscr{C}$, then

$$F\cos\theta\,\delta s = \mathbf{F}\,.\,\delta\mathbf{s} = \mathbf{F}\,.\,\delta\mathbf{r} = F_x\,\delta x + F_y\delta y + F_z\delta z,$$

where arc $PP' = \delta s$ and $\overrightarrow{PP'} \equiv \delta\mathbf{r} = [\delta x, \delta y, \delta z]$. If $\mathbf{F}$ were specifically identified with a force applied to a ring free to move along a rigid wire bent into the form of the arc of $\mathscr{C}$ then each of the above four expressions would represent the work done by the force $\mathbf{F}$ in moving the ring from P to P' along the wire. A line integral of $\mathbf{F}$ along $\mathscr{C}$ from A to B is obtained from the above appropriate limiting summation forms analogous to the earlier model. It is represented by any of the four forms

$$\int_A^B F\cos\theta\,ds = \int_A^B \mathbf{F}\,.\,d\mathbf{s} = \int_A^B \mathbf{F}\,.\,d\mathbf{r}$$

$$= \int_A^B \{F_x(x, y, z)dx + F_y(x, y, z)dy + F_z(x, y, z)dz\}\,.$$

With **F** identified specifically as a force applied to a movable ring, as above, each of these forms represents the total work done by the force in moving the ring from A to B along the wire.

(ii) We next pass on to the notion of *surface integral.* In R_3 a surface S may be specified by a Cartesian equation of the form

$$\phi(x, y, z) = 0, \tag{1.58}$$

or, equivalently,

$$z = f(x, y). \tag{1.58$'$}$$

The Monge form (1.58$'$) is obtained from the more symmetrical form (1.58) by solving for z in terms of x, y. Alternatively, the surface may be expressed by a two-parameter set of equations of the form

$$x = x(y, v), \quad y = y(u, v), \quad z = z(u, v) \tag{1.59}$$

where u, v are the parameters. If $\psi = \psi(x, y, z)$ is a scalar function which can be associated with each $P(x, y, z)$ of S, then the *surface integral of* ψ *taken over* S is

$$\int_S \psi \, \mathrm{d}S = \lim_{\substack{N\to\infty \\ \max \delta S_i \\ \to 0}} \sum_{i=1}^{N} \psi_i \, \delta S_i, \tag{1.60}$$

where S has been divided into N small elements of area δS_i on which ψ assumes a mean value $\psi_i (i = 1, \ldots, N)$. When $\psi = 1$, we find

$$\int_S \mathrm{d}S = \text{area of surface.}$$

However, for evaluation purposes one usually needs to replace the integral form in (1.60) by an appropriate double integral. To this end suppose S is bounded by a closed curve $\mathscr{C}$ whose orthogonal projection on the plane $z = 0$ is $\mathscr{C}_0$, enclosing an area S_0 (Fig. 1.12). At $P_0(x, y, 0)$ in S_0, the element $\delta x \times \delta y$ is taken: it is the orthogonal projection on $z = 0$ of the surface element δS, so that $\delta x \delta y = \delta S \cos \theta$, where θ is the acute angle between the normal to δS and the unit vector **k** in $\overrightarrow{OZ}$. Thus $\delta x \delta y = \delta S |\mathbf{n} \,.\, \mathbf{k}|$, where **n** is the unit vector in one of the two chosen normal directions to δS. Hence $\delta S = \delta x \delta y / |\mathbf{n} \,.\, \mathbf{k}|$ and so

$$\int_S \psi \, \mathrm{d}S = \int_{S_0}\!\!\int \frac{\psi(x, y)}{|\mathbf{n} \,.\, \mathbf{k}|} \, \mathrm{d}x \, \mathrm{d}y, \tag{1.61}$$

it being assumed that $\mathbf{n} \,.\, \mathbf{k} \neq 0$ over S.

When S is specified in the Monge form (1.58$'$), a vector in the normal to (x, y, z) on S is given by

$$\text{grad } [f(x, y) - z] = [f_x, f_y, -1] \quad (f_x = \partial f/\partial x, \text{etc.}),$$

since this vector is normal to the level surfaces $f - z = \text{const.}$ Hence

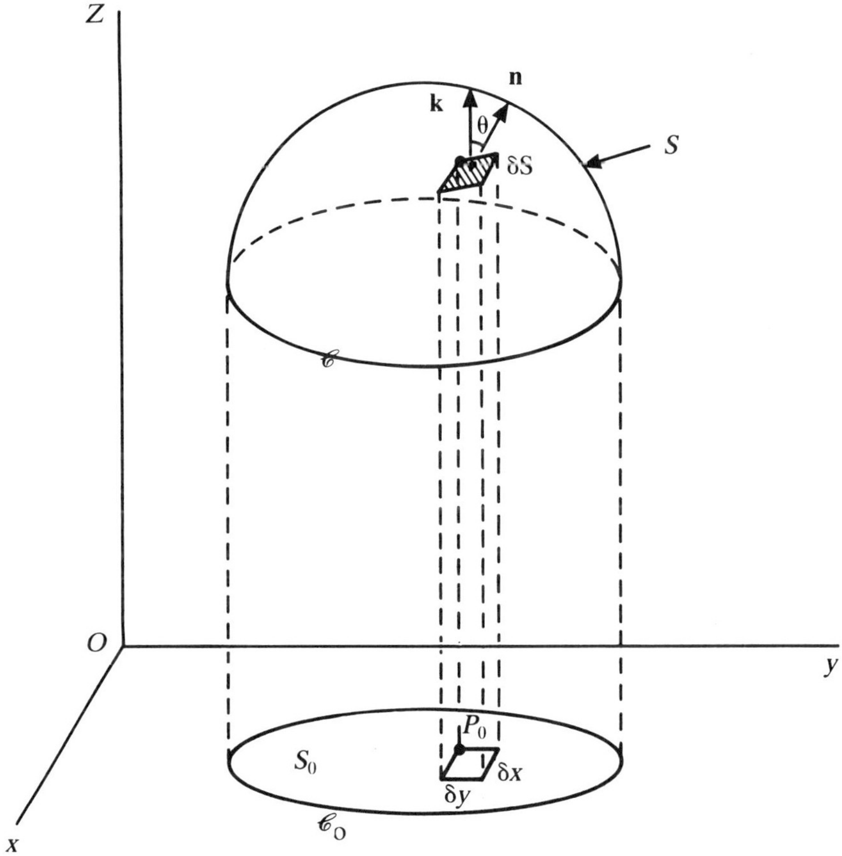

Fig. 1.12

$$\mathbf{n} = \pm [f_x, f_y, -1]/\sqrt{(f_x^2 + f_y^2 + 1)}.$$

Thus the form (1.61) becomes

$$\int_S \psi \, dS = \int_{S_0} \int \psi(x, y) \sqrt{(f_x^2 + f_y^2 + 1)} dx \, dy. \tag{1.62}$$

This is then evaluated as a double integral.

Example [1.3]

A metal bowl is in the form of the section of the paraboloid of revolution $z = x^2 + y^2$, $0 \leqslant z \leqslant h^2$. The mass per unit area of the metal at (x, y, z) of the surface is $k(x^2 + y^2)$. Find the total mass of the bowl.

Since $k(x^2 + y^2)\delta S$ is the mass of a surface element δS at (x, y, z) of the bowl, the total mass M is given by

$$M = k \int (x^2 + y^2)\, \mathrm{d}S.$$

where δS is related to $\delta x \delta y$ by

$$\begin{aligned}\delta x \delta y &= \text{projection of } \delta S \text{ on } z = 0 \\ &= \frac{\delta S}{\sqrt{(1 + z_x^2 + z_y^2)}} = \frac{\delta S}{\sqrt{(1 + 4x^2 + 4y^2)}}.\end{aligned}$$

Hence

$$M = k \int_S \int (x^2 + y^2) \sqrt{(1 + 4x^2 + 4y^2)}\, \mathrm{d}x\mathrm{d}y$$

where S is the circle $x^2 + y^2 = h^2$, $z = 0$. Since the integrand is a function of $x^2 + y^2 = r^2$, it is evaluated more expeditiously by converting from plane Cartesian coordinates (x, y) to plane polar coordinates (r, θ). Remembering that the area element $\delta x \delta y$ is now replaced by the polar form $\delta r \times r\delta\theta$, we obtain

$$\begin{aligned}M &= k \int_S \int r^2 \sqrt{(1 + 4r^2)}\,.\, r\, \mathrm{d}r\mathrm{d}\theta \\ &= k \int_0^{2\pi} \mathrm{d}\theta \int_0^h \sqrt{(1 + 4r^2)}\,.\, r^3\, \mathrm{d}r \\ &= \tfrac{1}{2} k\,.\, 2\pi\,.\, \int_0^{h^2} \sqrt{(1 + 4t)}\,.\, t\, \mathrm{d}t \qquad (t = r^2) \\ &= \underline{\pi k \left\{10b^2 (4h^2 + 1) - (4h^2 + 1)^{5/2} + 1\right\}/60.}\end{aligned}$$

Crucial to the vector calculus is the notion of the *normal flux of a vector field across a given surface.* We first define the normal flux of a uniform differentiable vector function **F** across a surface element δS which is appreciably plane and drawn through a point P. In Fig. 1.13, **n** is a unit vector specifying one of the two possible normal directions at P to δS. For this chosen **n**, if **F** is the value of

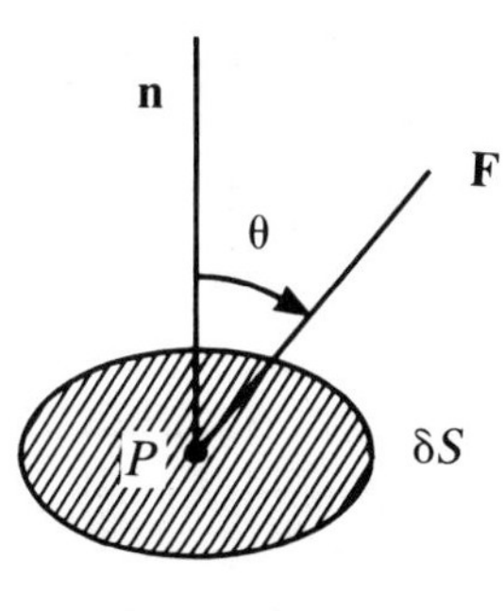

Fig. 1.13

the vector function at P and θ is the angle between the two vectors, then the **n**-component of **F** at P is $F_n = F \cos\theta = \mathbf{n} \cdot \mathbf{F}$. The normal flux of **F** across δS associated with the direction **n** is defined to be

$$F_n \, \delta S = \mathbf{n} . \mathbf{F} \, \delta S = \mathbf{F} \, . \, \delta \mathbf{S},$$

where the *vector area element* $\delta\mathbf{S}$ is defined to be $\mathbf{n} \, \delta S$. For a finite surface S the total normal flux of **F** is defined to be any of the equivalent forms

$$\int_S F_n \, \mathrm{d}S = \int_S \mathbf{n} \, . \, \mathbf{F} \, \mathrm{d}S = \int_S \mathbf{F} \, . \, \mathrm{d}\mathbf{S}, \tag{1.63}$$

where the integrals are found as limiting summations such as

$$\int_S F_n \, \mathrm{d}S = \lim_{N\to\infty} \sum_{i=1}^{N} (F_n)_i \, \delta S_i, \text{ etc.}$$

Throughout, **n** always lies on the same side of each element δS. The reader conversant with fundamental electromagnetic field theory will recall that if S is a closed surface in R_3 in which there exists a magnetic field whose local intensity at any element δS of S is **H**, then the total normal flux of **H** across S when measured in the direction outwards from the interior of S is

$$\int_S H_n \, \mathrm{d}S = \int_S \mathbf{n} \, . \, \mathbf{H} \, \mathrm{d}S = \int_S \mathbf{H} \, . \, \mathrm{d}\mathbf{S},$$

where the unit normal vector **n** to each δS is taken to be directed outwards from the interior of S.

Example [1.4]

Find the total outward-directed normal flux of **H** across the surfaces of the finite cylinder

$$x^2 + y^2 = a^2, \quad 0 \leqslant z \leqslant h$$

when at each $P(x, y, z)$ of R_3,

$$\mathbf{H} = 2x\mathbf{i} - 3y\mathbf{j} + 4z\mathbf{k}.$$

On the plane surface $S_1 : z = 0$, the total outward normal component of flux, is since $\hat{\mathbf{n}} = -\mathbf{k}$ and $-\mathbf{k} \, . \, \mathbf{H} = 0$,

$$\int_{S_1} (-\mathbf{k}) \, . \, \mathbf{H} \, \mathrm{d}S = 0.$$

On the plane surface $S_2 : z = h$, the total outward normal component of flux is, since $\mathbf{n} = \mathbf{k}$ and $\mathbf{k} \, . \, \mathbf{H} = 4h$,

$$\int_{S_2} \mathbf{k} \, . \, \mathbf{H} \, \mathrm{d}S = 4h \int_{S_2} \mathrm{d}S = 4\pi a^2 h.$$

Any point on the curved surface of the cylinder may be represented by

$$x = a\cos\theta, \quad y = a\sin\theta, \quad z = z,$$

where $0 \leqslant \theta \leqslant 2\pi$, $0 \leqslant z \leqslant h$. At such a point,

$$\mathbf{H} = 2a\cos\theta\,\mathbf{i} - 3a\sin\theta\,\mathbf{j} + 4z\,\mathbf{k}$$

and the appropriate outward unit normal vector is

$$\mathbf{n} = \cos\theta\,\mathbf{i} + \sin\theta\,\mathbf{j},$$

so that

$$\mathbf{n}\,.\,\mathbf{H} = a(2\cos^2\theta - 3\sin^2\theta).$$

A suitable surface element at $(a\cos\theta, a\sin\theta, z)$ is

$$\delta S = a\delta\theta \times \delta z.$$

Hence the total outward normal flux across the curved surface S_3 is

$$\int_{S_3} \mathbf{n}\,.\,\mathbf{H}\,\mathrm{d}S = a^2 \int_0^h \mathrm{d}z \int_0^{2\pi} (2\cos^2\theta - 3\sin^2\theta)\,\mathrm{d}\theta$$
$$= -\pi a^2 h.$$

Hence the total normal flux across all surfaces of the cylinder is $3\pi a^2 h$.

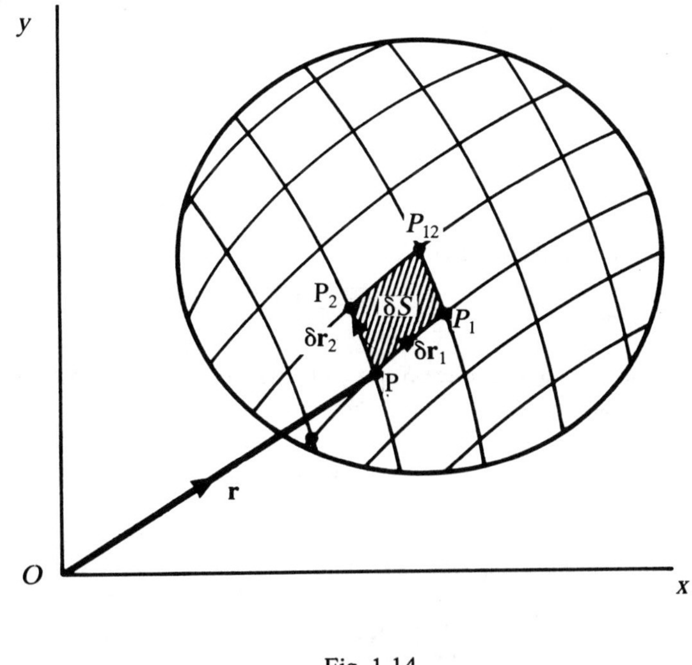

Fig. 1.14

We have seen that the integral of $f(x, y)$ over a surface S in the (x, y)-plane and enclosed by a closed curve S is

$$I = \int_S f(x, y)\, \mathrm{d}S = \int_S \int f(x, y)\, \mathrm{d}x\mathrm{d}y.$$

Often the double integral is difficult or impossible to evaluate and it may be necessary to change the coordinates from (x, y) to new ones (u, v), which are not necessarily rectangular. To this end, suppose we know x, y as functions of u, v:

$$x = x(u, v), \quad y = y(u, v).$$

In the (x, y)-plane (Fig. 1.14), draw the two families of curves

$$u(x, y) = \text{const.}, \quad v(x, y) = \text{const.}$$

The curves $u(x, y) = \text{const.}$, $v(x, y) = \text{const.}$ pass through $P(x, y)$ and $u + \delta u = \text{const.}$, $v + \delta v = \text{const.}$ are neighbouring curves. A cell $P P_1 P_{12} P_2 P$ is formed of area δS. We let

$$\overrightarrow{OP} \equiv \mathbf{r} = x\mathbf{i} + y\mathbf{j},$$

$$\overrightarrow{OP}_1 \equiv \mathbf{r} + \delta\mathbf{r}_1, \qquad \overrightarrow{OP}_2 \equiv \mathbf{r} + \delta\mathbf{r}_2.$$

Then $\qquad \overrightarrow{PP}_1 \equiv \delta\mathbf{r}_1, \qquad \overrightarrow{PP}_1 \equiv \delta\mathbf{r}_2.$

Along PP_1, $v = \text{const.}$ and u varies from u at P to $u + \delta u$ at P. Hence, to the first order,

$$\delta\mathbf{r}_1 = \frac{\partial \mathbf{r}}{\partial u}\delta u = \left(\frac{\partial x}{\partial u}\,\mathbf{i} + \frac{\partial y}{\partial u}\,\mathbf{j}\right)\delta u,$$

and similarly,

$$\delta\mathbf{r}_2 = \frac{\partial \mathbf{r}}{\partial v}\,\delta v = \left(\frac{\partial x}{\partial v}\,\mathbf{i} + \frac{\partial y}{\partial v}\,\mathbf{j}\right)\delta v.$$

Now

$$\delta S = |\delta\mathbf{r}_1 \times \delta\mathbf{r}_2| = \delta u \delta v \left|\frac{\partial(x, y)}{\partial(u, v)}\right|,$$

where the Jacobian form is

$$\frac{\partial(x, y)}{\partial(u, v)} = \frac{\partial x}{\partial u}\frac{\partial y}{\partial v} - \frac{\partial x}{\partial v}\frac{\partial y}{\partial u}.$$

Hence the integral I has the alternative representations

$$I = \int_S f(x, y)\, \mathrm{d}S$$

$$= \int_S\int f(x, y)\,\mathrm{d}x\mathrm{d}y$$

$$= \int_S\int f(x(u,v),\, y(u,v))\left|\frac{\partial(x,y)}{\partial(u,v)}\right| \mathrm{d}u\mathrm{d}v. \qquad (1.64)$$

Example [1.5]

Evaluate the double integral

$$I = \int_0^\infty \int_0^\infty \exp[-(x+y)]\,\sqrt{(xy)}\,\mathrm{d}x\mathrm{d}y$$

using the substitutions

$$x = \tfrac{1}{2}u(1+v);\; y = \tfrac{1}{2}u(1-v).$$

The domain of integration is the positive quadrant $x \geqslant 0, y \geqslant 0$. Also

$$\frac{\partial(x,y)}{\partial(u,v)} = \frac{\partial x}{\partial u}\frac{\partial y}{\partial v} - \frac{\partial x}{\partial v}\frac{\partial y}{\partial u} = \tfrac{1}{2}(1+v)(-\tfrac{1}{2}u) - \tfrac{1}{2}u\,.\,\tfrac{1}{2}(1-v)$$

$$= -\tfrac{1}{2}u.$$

The u-curves (i.e. u = const.) are the family of parallel lines

$$x + y = u = \text{const.}$$

Clearly $0 \leqslant u \leqslant \infty$. The v-curves are the lines

$$y/x = (1-v)/(1+v)$$

through the origin. The x-axis is given by $v = 1$, the y-axis by $v = -1$, so that $-1 \leqslant v \leqslant 1$. The integral reduces to the form

$$I = \tfrac{1}{4}\int_0^\infty u^2 \exp(-u)\,\mathrm{d}u \int_{-1}^{2} (1-v^2)^{1/2}\,\mathrm{d}v.$$

Both integrals are elementary and separated. Evaluation gives

$$\underline{I = \pi/4.}$$

(iii) Finally we examine the meaning of *volume integral*. To this end, suppose S is a closed surface in R_3 containing a volume v. Let the interior of v be dissected into N elemental volumes of δv_i and let P_i (x_i, y_i, z_i) be a point within δv_i $(i = 1, 2, \ldots, N)$. Then

$$\int_v \phi\,\mathrm{d}v = \lim_{N\to\infty} \sum_{i=1}^{N} \phi(x_i, y_i, z_i)\,\delta v_i,$$

the limit being taken such that each $\delta v_i \to 0$.

When we choose the volume elements δv_i to be rectantular boxes $\delta x_i \times \delta y_i \times \delta z_i$ as in Fig. 1.15, the last form gives

$$\int_v \phi \, dv = \lim_{N \to 0} \sum_{i=1}^{N} \phi(x_i, y_i, z_i) \, \delta x_i \, \delta y_i \, \delta z_i .$$

The latter form is usually written as a *triple integral* so that

$$\int_v \phi \, dv = \iiint \phi(x, y, z) \, dx dy dz, \tag{1.65}$$

the triple integral on the right of (1.65) being evaluated throughout v, so that appropriate limits for x, y, z attach to this form. These limits depend very much on the shape of S. The triple integral form is often used for evaluation.

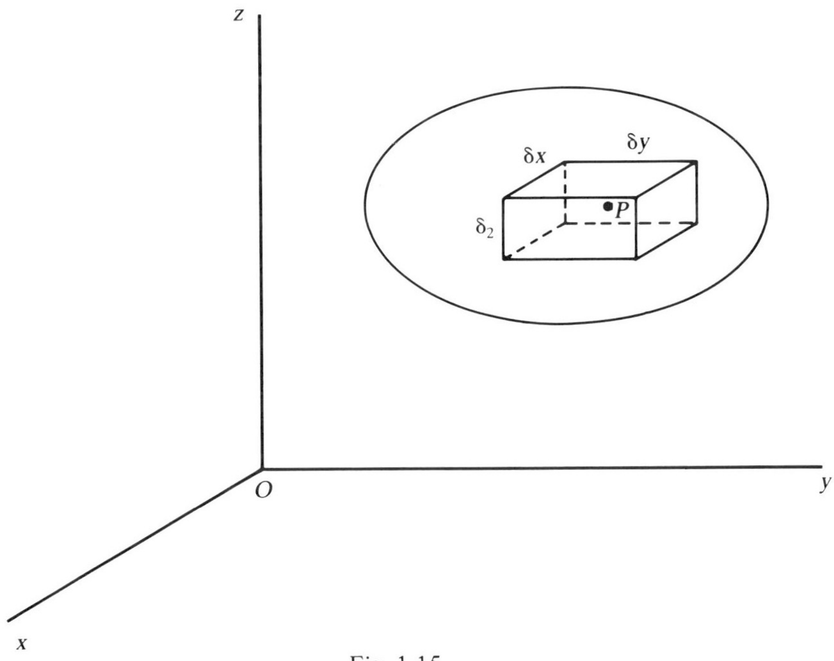

Fig. 1.15

A physical meaning of the above is readily imparted if we think of the interior v of S as matter (solid, liquid or gas) not necessarily uniformly distributed. If we assign to $\phi(x, y, z)$ the specific meaning of density of the matter (i.e. mass per unit volume) at $P(x, y, z)$, then $\phi \delta v$ is the mass of the box element at P and so either of the forms (1.64) represent the mass of the material occupying the interior v of S.

The above ideas are readily extensible to the case of a uniform vector function $\mathbf{F}(x, y, z)$ defined at each $P(x, y, z)$ of v. Throughout scalar ϕ is replaced by vectorial $\mathbf{F}$.

Example [1.6]
Find the mass of the spherical octant

$$x^2 + y^2 + z^2 \leqslant a^2, \qquad 0 \leqslant x, y, z \leqslant a$$

if the density at an internal point $P(x, y, z)$ is $kxyz$.

The mass of the element $\delta v = \delta x \times \delta y \times \delta z$ is $kxyz\ \delta v = kxyz\ \delta x\delta y\delta z$. Hence the total mass M of the octant (Fig. 1.16) is

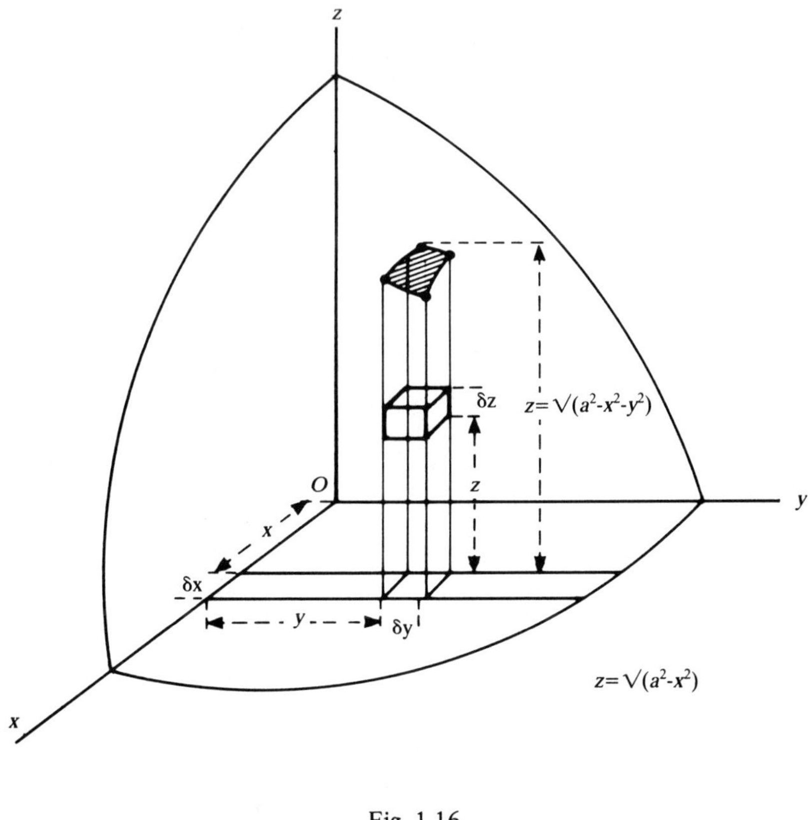

Fig. 1.16

$$M = k\int xyz\ \mathrm{d}v$$

$$= k\int_0^a x\ \mathrm{d}x \int_0^{\sqrt{(a^2-x^2)}} y\ \mathrm{d}y \int_0^{\sqrt{(a^2-x^2-y^2)}} z\ \mathrm{d}z$$

$$= k\int_0^a x\ \mathrm{d}x \int_0^{\sqrt{(a^2-x^2)}} \tfrac{1}{2}y\,(a^2 - x^2 - y^2)\ \mathrm{d}y$$

$$= \tfrac{1}{4}k\int_0^a x\left[(a^2 - x^2 - y^2)\right]_{y=\sqrt{(a^2-x^2)}}^{y=0} \mathrm{d}x$$

$$= \tfrac{1}{4}k\int_0^a x(a^2 - x^2)^2 \,\mathrm{d}x$$

$$= \frac{1}{24}k\left[(a^2 - x^2)^3\right]_{x=a}^{x=0}$$

$$= \frac{1}{24}ka^6.$$

Example [1.7]
If ϕ is a continuous function in R_3 and if $\int_v \phi \,\mathrm{d}v = 0$ for all subregions v of R_3, then $\phi = 0$ throughout R_3.

Let us suppose that $\phi \neq 0$ at some point P of R_3. Since ϕ is continuous in R_3, there exists a neighbourhood τ of P throughout which ϕ has the same sign as ϕ_P. Then $\int_\tau \phi \,\mathrm{d}v$ has the same sign as ϕ_P, i.e.

$$\int_\tau \phi \,\mathrm{d}v \neq 0.$$

This contradicts the given information so that $\phi_P = 0$ at all P of R_3.

1.14 GREEN'S THEOREM AND THE GAUSS DIVERGENCE THEOREM

At this stage, we relate volume and surface integrals through one of the many theorems due to Green. The enunciation of the theorem required here is:

If in R_3, specified by rectangular Cartesian coordinates (x, y, z), uniform differentiable scalar functions $P(x, y, z)$, $Q(x, y, z)$, $R(x, y, z)$ are defined in a volume v of R_3 bounded by a closed surface S, then provided the first partial derivatives of P, Q, R are continuous,

$$\int_S (lP + mQ + nR)\,\mathrm{d}S = \int_v \left(\frac{\partial P}{\partial x} + \frac{\partial Q}{\partial y} + \frac{\partial R}{\partial z}\right)\mathrm{d}v \tag{1.66}$$

where $[l, m, n]$ are the direction cosines of the outward normal to δS of S.

Suppose that any parallel to Oz meets S in at most two points, as in Fig. 1.17. Then a curve $\mathscr{C}$ can be drawn on S dividing it into a lower portion S_1 and an upper portion S_2. $\mathscr{C}$ is projected orthogonally on to the plane $z = 0$ to form another closed curve $\mathscr{C}_0$ enclosing an area A. An element δA of A is the ortho-

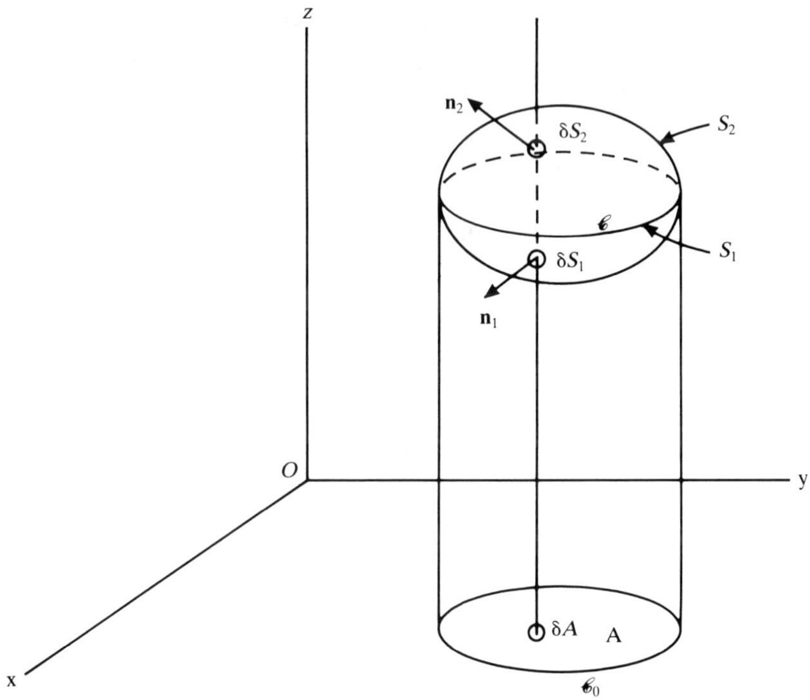

Fig. 1.17

gonal projection of both δS_1 and δS_2. Let S_1, S_2 have the equations $z = z_1(x, y)$, $z = z_2(x, y)$, respectively. Then for the function $R(x, y, z)$,

$$\int_v \frac{\partial R}{\partial z}\, \mathrm{d}v = \int_A \mathrm{d}A \int_{z_1}^{z_2} \frac{\partial R}{\partial z}\, \mathrm{d}z$$

$$= \int_A [R(x, y, z_2) - R(x, y, z_1]\, \mathrm{d}A,$$

since $\partial R/\partial z$ is continuous. Now $\delta A = -\delta S_1 n_1 = +\delta S_2 n_2$ since the normal $\mathbf{n}_1$ to δS_1 makes an obtuse angle with $\mathbf{k}$ and $\mathbf{n}_2$ makes an acute angle with it. Consequently

$$\int_A [R(x, y, z_2) - R(x, y, z_1)]\, \mathrm{d}A$$

$$= \int_{S_2} R(x, y, z_2)\mathbf{n}_2\, \mathrm{d}S + \int_{S_1} R(x, y, z_1)\mathbf{n}_1\, \mathrm{d}S$$

$$= \int_S nR(x, y, z)\, \mathrm{d}S.$$

Similarly,

$$\int_v \frac{\partial P}{\partial x}\,\mathrm{d}v = \int_S lP\,\mathrm{d}S;$$

$$\int_v \frac{\partial Q}{\partial y}\,\mathrm{d}v = \int_S mQ\,\mathrm{d}S.$$

By addition, then, (1.65) follows at once.

In the case when S is intersected more than once by the line through δA perpendicular to $z = 0$, we can divide v into a number of sub-regions, each bounded by a surface satisfying the same conditions as in Fig. 1.16. On applying the result

$$\int_S nR(x, y, z)\,\mathrm{d}S = \int_A [R(x, y, z_2) - R(x, y, z_1)]\,\mathrm{d}A$$

to each sub-region and adding, we can show that the result is true generally. Rigorous treatment may be found in pure mathematical texts.

The *Gauss divergence theorem* states that:

If in R_3 the vector function $\mathbf{F} = \mathbf{F}(x, y, z)$ is uniform, differentiable and has continuous first-order partial derivatives, then

$$\int_S \mathbf{n}\,.\,\mathbf{F}\,\mathrm{d}S = \int_v \nabla\,.\,\mathbf{F}\,\mathrm{d}v, \tag{1.67}$$

in the notation of the enunciation of the previous theorem.

On taking $\mathbf{F} = [P, Q, R]$, $\mathbf{n} = [l, m, n]$ in Green's theorem, the result follows at once.

From (1.67) the mean divergence per unit volume throughout v is $\langle$ div F $\rangle$ where

$$\langle \operatorname{div} \mathbf{F} \rangle = \frac{1}{v}\int_v \nabla\,.\,\mathbf{F}\,\mathrm{d}v = \frac{1}{v}\int_S \hat{\mathbf{n}}\,.\,\mathbf{F}\,\mathrm{d}S. \tag{1.68}$$

Suppose $P(x, y, z)$ is any point in v and, keeping P fixed, let $v \to 0$ at P. Then the equation (1.67) gives

$$(\operatorname{div} \mathbf{F})_P = \lim_{v\to 0}\left\{\frac{1}{v}\int_S \mathbf{n}\,.\,\mathbf{F}\,\mathrm{d}S\right\}. \tag{1.69}$$

Equation (1.69) tells us that the local divergence of $\mathbf{F}$ at a point P in R_3 is the limiting total outward normal flux per unit volume of $\mathbf{F}$ across a surface containing P as we make the surface collapse on to the point P. This affords an alternative definition of the divergence and from it the appropriate Cartesian form (1.40) is derivable. For particulars the reader should consult Ref. 3, pp. 90–92. As an illustration of the method let us revert to the Example [1.4] of

Section 1.13. We showed there that the total outward normal flux of $\mathbf{H} = 2x\mathbf{i} - 3y\mathbf{j} + 4z\mathbf{k}$ across the surfaces of the finite cylinder $x^2 + y^2 = a^2, 0 \leqslant z \leqslant h$ is $3\pi a^2 h$. Thus the mean flux per unit volume throughout the cylinder is 3. As we let the volume $v = \pi a^2 h \to 0$ this remains equal to 3 which is also the value of $\text{div}(2x\mathbf{i} - 3y\mathbf{j} + 4x\mathbf{k})$.

Two extensions immediately stem from the Gauss divergence theorem (1.67), namely

$$\int_S \mathbf{n}\,\phi\,\mathrm{d}S = \int_v \nabla\phi \mathrm{d}v \tag{1.70}$$

for a uniform differentiable scalar function ϕ, and

$$\int_S \mathbf{n} \times \mathbf{F}\,\mathrm{d}S = \int_v \nabla \times \mathbf{F}\,\mathrm{d}v. \tag{1.71}$$

To derive (1.70), we put $\mathbf{F} = \phi\mathbf{a}$ in (1.67), where $\mathbf{a}$ is a constant non-zero vector. Then

$$\mathbf{n}\,.\,(\phi\,\mathbf{a}) = \mathbf{a}\,.\,(\mathbf{n}\phi)$$

and

$$\nabla.\,(\phi\,\mathbf{a}) = \mathbf{a}\,.\,(\nabla\phi),$$

using (1.46) and recalling the constancy of $\mathbf{a}$. Hence (1.67) becomes on simplification,

$$\mathbf{a}\,.\left\{\int_S \mathbf{n}\,\phi\,\mathrm{d}S - \int_v \nabla\phi\,\mathrm{d}v\right\} = 0.$$

Since $\mathbf{a} \neq \mathbf{0}$ and its direction is arbitrary so that it need not be perpendicular to the second vector, this second vector must vanish, leading to (1.70).

For (1.71), put $\mathbf{G} = \mathbf{F} \times \mathbf{a}$ in

$$\int_S \mathbf{n}\,.\,\mathbf{G}\,\mathrm{d}S = \int_v \nabla\,.\,\mathbf{G}\,\mathrm{d}v,$$

where $\mathbf{a}$ is an arbitrary non-zero constant vector. Now use the scalar triple product property

$$\mathbf{n}\,.\,(\mathbf{F} \times \mathbf{a}) = \mathbf{a}\,.\,(\mathbf{n} \times \mathbf{F})$$

together with

$$\nabla.\,(\mathbf{F} \times \mathbf{a}) = \mathbf{a}\,.\,(\nabla \times \mathbf{F}),$$

from (1.54). An argument similar to that of the previous case then leads to (1.71).

Example [1.8]

If u, v, w are harmonic uniform differentiable scalar functions in R_3 and the level surfaces $u = \text{const.}$, $v = \text{const.}$, $w = \text{const.}$, intersect mutually orthogonally, prove that

$$\int_S \frac{\partial}{\partial n}(uvw)\,\mathrm{d}S = 0,$$

where S is a closed surface in R_3 and $\partial/\partial n$ signifies differentiation along the outward-directed normal to any element δS of S.

If $\mathbf{n}$ is the unit normal vector to δS drawn outwards from the interior V of S, then $\partial/\partial n \equiv \mathbf{n}\,.\,\nabla$, the direction differentiator. Thus

$$\int_S \frac{\partial}{\partial n}(uvw)\,\mathrm{d}S = \int_S \mathbf{n}\,.\,\nabla(uvw)\,\mathrm{d}S$$

$$= \int_V \nabla.\nabla\,(uvw)\,\mathrm{d}\tau$$

where $\delta\tau$ is a volume-element of V (and replaces δv to avoid confusion).

A double application of (1.39) gives

$$\nabla(uvw) = (\nabla u)vw + u(\nabla v)w + uv\nabla w.$$

Taking the divergence of this gives

$$\begin{aligned}\nabla^2(uvw) &= (\nabla^2 u)vw + (\nabla^2 v)uw + (\nabla^2 w)uv \\ &\quad + 2u\nabla v.\,\nabla w + 2v\nabla w.\ \ u + 2w\nabla u\,.\,\nabla v \\ &= 2(u\nabla v\,.\,\nabla w + v\nabla w\,.\,\nabla u + w\nabla u\,.\,\nabla v),\end{aligned}$$

since u, v, w are harmonic. As the level surfaces $u = \text{const.}$, $v = \text{const.}$, $w = \text{const.}$, intersect orthogonally, so do their normals. Thus $\nabla v\,.\,\nabla w = 0$, etc., since ∇v is a vector in the normal to $v = \text{const.}$, etc. Thus all scalar products in the last expression vanish so that

$$\nabla^2(uvw) = 0 \text{ in } V.$$

Thus

$$\int_S \frac{\partial}{\partial n}(uvw)\,\mathrm{d}S = \int_V 0\,\mathrm{d}\tau = 0.$$

1.15 STOKES'S THEOREM

(i) Case of a simple plane closed curve

In Fig. 1.18, $\mathscr{C}$ is a closed curve in the (x, y)-plane containing a plane area S. $\mathscr{C}$ lies within the rectangle $a \leqslant x \leqslant b$; $c \leqslant y \leqslant d$. If $P(x, y)$ is a continuous function possessing the derivative $\partial P/\partial y$, then

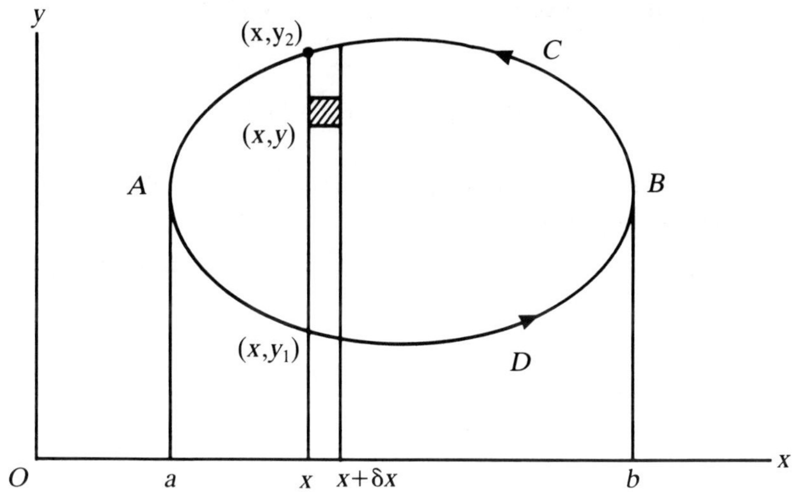

Fig. 1.18

$$\int_S\int \frac{\partial P}{\partial y}\,\mathrm{d}x\mathrm{d}y = \int_a^b \mathrm{d}x \int_{y_1}^{y_2} \frac{\partial P}{\partial y}\,\mathrm{d}y$$

$$= \int_a^b [P(x, y_2) - P(x, y_1)]\,\mathrm{d}x$$

$$= \int_{ACB} P(x, y_2)\,\mathrm{d}x - \int_{ADB} P(x, y_1)\,\mathrm{d}x$$

$$= -\int_{ADB} P(x, y)\,\mathrm{d}x - \int_{BCA} P(x, y)\,\mathrm{d}x$$

$$= -\oint_{\mathscr{C}} P(x, y)\,\mathrm{d}x.$$

Similarly it is easy to show that if $Q(x, y)$ is a continuous function with derivative $\partial Q/\partial x$, then

$$\int_S\int \frac{\partial Q}{\partial x}\,\mathrm{d}x\mathrm{d}y = \oint_{\mathscr{C}} Q(x, y)\,\mathrm{d}y.$$

From these two results, each of which connects a double integral over an area with a single one around its perimeter, we obtain at once a third such relation, namely

$$\int_S\int \left(\frac{\partial Q}{\partial x} - \frac{\partial P}{\partial y}\right)\mathrm{d}x\mathrm{d}y = \oint_{\mathscr{C}} (P\,\mathrm{d}x + Q\,\mathrm{d}y). \tag{1.72}$$

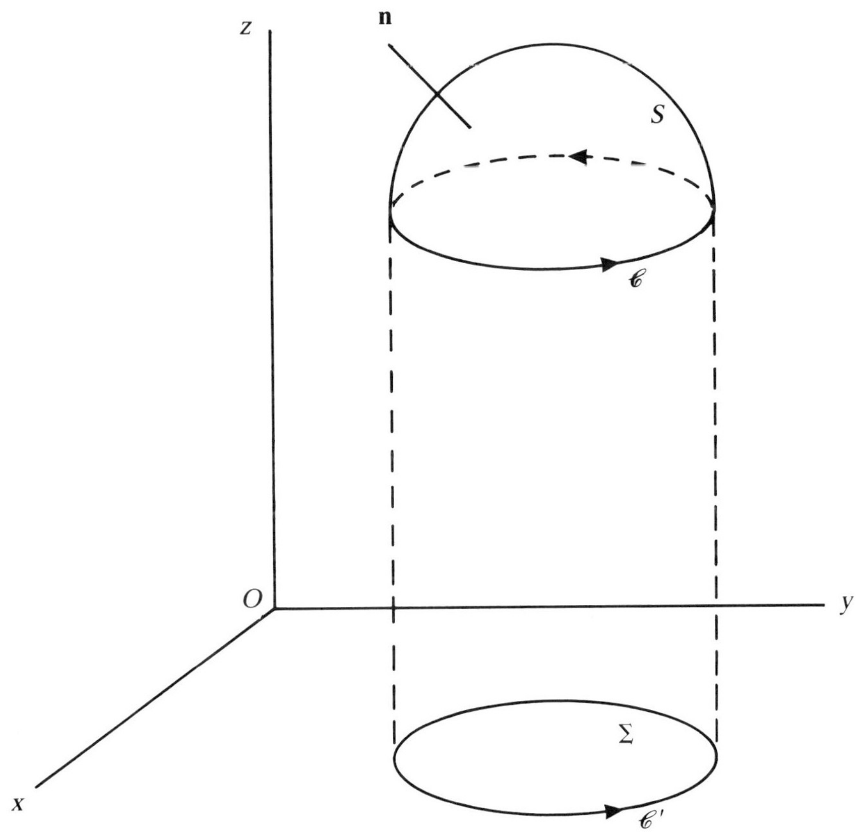

Fig. 1.19

This form is *Stokes's theorem for two dimensions* (and known in some texts as Green's theorem). The reader will note that the form (1.72) still holds when the coordinate axes are not rectangular.

The above theorem is also true for a non-simple closed plane curve, such as a figure of eight.

(ii) Case of a closed twisted curve

Let S be an open surface rimmed by a simple twisted closed curve $\mathscr{C}$ (which is therefore not plane) and suppose that $\mathbf{F}$ is a uniform differentiable vector function in R_3 and has continuous partial derivatives of the first order. Then we prove *Stokes's theorem for three dimensions*, namely

$$\oint_{\mathscr{C}} \mathbf{F} \, . \, \mathrm{d}\mathbf{r} = \int_S \operatorname{curl} \mathbf{F} \, . \, \mathrm{d}\mathbf{S}, \tag{1.73}$$

where the directions of $\mathrm{d}\mathbf{S}$ and the sense of description of $\mathscr{C}$ are right-handedly related, i.e. the direction of $\mathbf{n}$, the unit vector normal to $\mathrm{d}S$ for which $\mathrm{d}\mathbf{S} = \mathrm{d}S\mathbf{n}$, is taken in the direction of movement of a right-handed screw when rotated in the sense of description of $\mathscr{C}$.

With $\mathbf{F} = [F_x, F_y, F_z]$, where $F_x = F_x(x, y, z)$, etc., we first observe that

$$\operatorname{curl}(F_x\mathbf{i}) = (\partial F_x/\partial z)\mathbf{j} - (\partial F_x/\partial y)\mathbf{k}$$

and so

$$\int_S \operatorname{curl}(F_x\,\mathbf{i}) \,.\, \mathrm{d}\mathbf{S} = \int_S \mathbf{n} \,.\, \operatorname{curl}(F_x\,\mathbf{i})\,\mathrm{d}S$$

$$= \int_S \left(\frac{\partial F_x}{\partial z}\mathbf{j} \,.\, \mathbf{n} - \frac{\partial F_x}{\partial y}\mathbf{k} \,.\, \mathbf{n}\right)\mathrm{d}S.$$

Suppose the equation of S is $z = f(x, y)$ (Monge form). Then grad $(z - f)$ is a vector normal to the surface S, i.e. the vector $[-\partial f/\partial x, -\partial f/\partial y, 1]$ is normal to S and so these direction ratios are proportional to those of $\hat{\mathbf{n}}$, which are $[\hat{\mathbf{n}} \,.\, \mathbf{V}$, $\mathbf{n} \,.\, \mathbf{j}, \mathbf{n} \,.\, \mathbf{k}]$. Thus on S,

$$\mathbf{j} \,.\, \mathbf{n} = (-\partial f/\partial y)\mathbf{k} \,.\, \mathbf{n} = (-\partial z/\partial y)\mathbf{k} \,.\, \mathbf{n},$$

so that

$$\int_S \operatorname{curl}(F_x\,\mathbf{i}) \,.\, \mathrm{d}\mathbf{S} = -\int_S \left\{\frac{\partial F_x}{\partial y} + \frac{\partial F_x}{\partial z}\frac{\partial z}{\partial y}\right\}(\mathbf{k} \,.\, \mathbf{n})\,\mathrm{d}S$$

$$= -\int_S \frac{\partial}{\partial y} F_x(x, y, f)\,(\mathbf{k} \,.\, \mathbf{n})\,\mathrm{d}S$$

$$= -\iint_\Sigma \frac{\partial}{\partial y} F_x(x, y, f)\,\mathrm{d}x\mathrm{d}y,$$

where Σ is the orthogonal projection of S on to the coordinate plane $z = 0$. By Stokes's theorem (i) above, the last integral transforms into a line integral as

$$-\iint_\Sigma \frac{\partial}{\partial y} F_x(x, y, f)\,\mathrm{d}x\mathrm{d}y = +\oint_{\mathscr{C}'} F_x(x, y, f)\,\mathrm{d}x$$

where $\mathscr{C}'$ is the closed boundary of Σ. In the single integral $\mathscr{C}'$ may be replaced by $\mathscr{C}$ without altering its value. Thus

$$\int_S \operatorname{curl}(F_x\,\mathbf{i}) \,.\, \mathrm{d}\mathbf{S} = \oint_{\mathscr{C}} F_x\,\mathrm{d}x.$$

Similarly,

$$\int_S \operatorname{curl}(F_y\,\mathbf{j}) \,.\, \mathrm{d}\mathbf{S} = \oint_{\mathscr{C}} F_y\,\mathrm{d}y,$$

$$\int_S \operatorname{curl}(F_z\,\mathbf{k}) \,.\, \mathrm{d}\mathbf{S} = \oint_{\mathscr{C}} f_z\,\mathrm{d}z$$

Adding together the last three relations gives at once the desired form (1.73).

The theorem is true for any open surface S that is rimmed by the same $\mathscr{C}$. For if S_1, S_2 are two surfaces rimmed by $\mathscr{C}$, then the combined surfaces $S_1 + S_2$ is closed and it contains a volume v, say. Suppose for this combined surface $d\mathbf{S} = \mathbf{n}\, dS$, where $\mathbf{n}$ is measured outwards from v. Then

$$\int_{S_1+S_2} \text{curl } \mathbf{F} \,.\, d\mathbf{S} = \int_v \text{div curl } \mathbf{F}\, dv = 0.$$

Thus

$$\int_{S_1} \text{curl } \mathbf{F} \,.\, d\mathbf{S} = -\int_{S_2} \text{curl } \mathbf{F} \,.\, d\mathbf{S}$$

where $\mathbf{n}$ satisfies this newly defined convention. On reverting to the former convention of taking $\mathbf{n}$ in the direction of motion of a right-handed screw rotated in the sense of description of $\mathscr{C}$, the last relation becomes

$$\int_{S_1} \text{curl } \mathbf{F} \,.\, d\mathbf{S} = +\int_{S_2} \text{curl } \mathbf{F}.d\mathbf{S} = \oint_{\mathscr{C}} \mathbf{F}.d\mathbf{r},$$

as required.

If in (1.73) we take $\mathbf{F} = [P, Q, R]$, where $P = P(x, y, z)$, etc., and $\mathbf{n} = [l, m, n]$, we obtain the corresponding form in Cartesian coordinates, namely

$$\oint_{\mathscr{C}} (P\, dx + Q\, dy + R\, dz) = \int_S \left\{ l\left(\frac{\partial R}{\partial y} - \frac{\partial Q}{\partial y}\right) + m\left(\frac{\partial P}{\partial z} - \frac{\partial R}{\partial x}\right) + n\left(\frac{\partial Q}{\partial x} - \frac{\partial P}{\partial y}\right)\right\} dS. \tag{1.74}$$

Reverting to the form (1.73), the mean value of $\mathbf{n}$.curl $\mathbf{F}$ taken over the area S is

$$\frac{1}{S}\int_S \mathbf{n} \,.\, \text{curl } \mathbf{F}\, dS = \frac{1}{S}\oint_{\mathscr{C}} \mathbf{F} \,.\, d\mathbf{r}.$$

As we let $S \to 0$ at a point $P(x, y, z)$ of S this gives

$$(\mathbf{n} \,.\, \text{curl } \mathbf{F})_P = \lim_{S\to 0} \left\{\frac{1}{S}\oint_{\mathscr{C}} \mathbf{F} \,.\, d\mathbf{r}\right\}. \tag{1.75}$$

For a small *plane* element δS bounded by $\mathscr{C}$ this gives the approximation

$$\oint_{\mathscr{C}} \mathbf{F} \,.\, d\mathbf{r} = (\mathbf{n} \,.\, \text{curl } \mathbf{F})\, \delta S. \tag{1.76}$$

If δS is taken through $P(x, y, z)$ and the orientation of $\mathbf{n}$ is varied, (1.76) show that *the total circulation or line integral around the boundary is a maximum when the vectors* $\mathbf{n}$ *and curl* $\mathbf{F}$ *are in alignment.* Thus at a point P of a vector field **F**, the vector curl **F** has the direction such that the circulation of **F** around the boundary of an elemental surface ΔS through P is a maximum when curl **F** is normal to ΔS.

In (1.74), let us write $\mathbf{F} = \phi\mathbf{a}$, where **a** is an arbitrary non-zero constant vector and ϕ is a uniform differentiable scalar function. Since

$$\nabla \times (\phi\,\mathbf{a}) = (\nabla\phi) \times \mathbf{a}$$

and

$$\mathbf{n}\,.\,[\nabla \times (\mathbf{a}\phi)] = (\mathbf{n} \times \nabla\phi)\,.\,\mathbf{a},$$

$$\mathbf{a}\,.\oint_{\mathscr{C}} \phi\,\mathrm{d}\mathbf{s} = \mathbf{a}\,.\int_S \mathbf{n} \times \nabla\phi\,\mathrm{d}S,$$

leading to

$$\oint_{\mathscr{C}} \phi\,\mathrm{d}\mathbf{s} = \int_S \mathbf{n} \times \nabla\phi\,\mathrm{d}S. \tag{1.77}$$

1.16 CONSERVATIVE VECTOR FIELDS

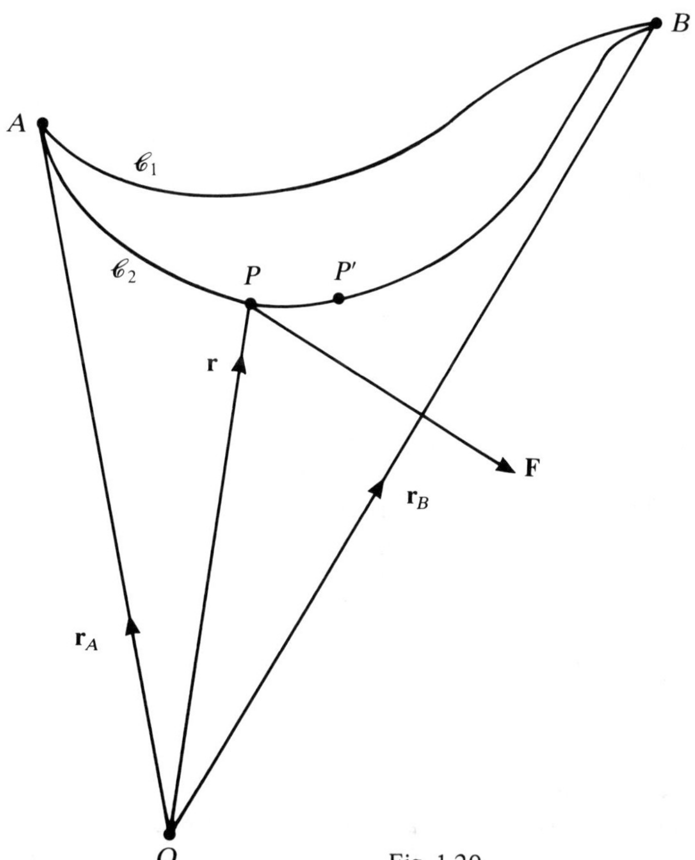

Fig. 1.20

Let A and B be any two points in R_3 and suppose $\mathscr{C}_1$ and $\mathscr{C}_2$ are any two arcs joining A, B and lying entirely in R_3. Let us further suppose that for all pairs of such arcs lying entirely in R_3 either may be deformed into the other without passing outside R_3. Then the region R_3 is said to be *simply connected.* Examples of simply connected regions include: the interior of a spherical surface; the exterior of a spherical surface; the region between two concentric spherical surfaces; unbounded space. If, however, R_3 is taken to be the region between two coaxial infinite cylinders then such a region is not simply connected. For if we take a closed curve entirely in R_3 encircling the inner cylinder and passing through any two chosen points A, B of R_3 then the closed curve will be divided into two arcs $\mathscr{C}_1$, $\mathscr{C}_2$ each in R_3. It is not possible, however, to deform either arc to produce the other without going outside the region between the cylindrical surfaces: such a region is said to be *multiply connected.*

Now suppose $\mathbf{F}$ is a uniform differentiable vector function defined everywhere in a simply connected region R_3 (Fig. 1.20). The line integral of $\mathbf{F}$ along the arc $\mathscr{C}_1$ joining A, B in R_3 is

$$I(\mathscr{C}_1) = \int_A^B \mathbf{F} \,.\, \mathbf{dr}.$$

With A, B fixed this will in general depend on the shape of the arc $\mathscr{C}_1$ joining them so that $I(\mathscr{C}_2)$, the line integral taken along another path $\mathscr{C}_2$ in R_3 and joining A to B, will in general differ from $I(\mathscr{C}_1)$. However, in certain important cases, the value of I along all paths $\mathscr{C}$ joining A to B and lying entirely in R_3 will be quite independent of the shape of $\mathscr{C}$. In these circumstances the field $\mathbf{F}$ in R_3 is said to be *conservative*. The following theorem establishes necessary and sufficient conditions for the field $\mathbf{F}$ to be conservative.

Theorem *A necessary and sufficient condition for the field* $\mathbf{F}$ *in a simply connected region* R_3 *to be conservative is that there exists a uniform differentiable scalar function* ϕ *such that* $\mathbf{F} = \text{grad}\,\phi$ *at each* P *of* R_3.

Proof (i) First suppose that such a $\phi = \phi(x, y, z)$ exists at each $P(x, y, z)$ of R_3. Then $\mathbf{F} = \nabla\phi$ and $\mathbf{F} \,.\, \mathbf{dr} = \mathbf{dr} \,.\, \nabla\phi = \mathrm{d}\phi$ and so

$$\int_A^B \mathbf{F} . \mathbf{dr} = \int_A^B \mathrm{d}\phi = \phi_B - \phi_A.$$

This last form depends only on the coordinates of A, B and not on the path $\mathscr{C}$ joining them and so we have shown that the field $\mathbf{F}$ is conservative in R_3.

(ii) Now suppose that $\mathbf{F}$ is given to be conservative in R_3. Then we have to show there exists a uniform scalar field ϕ in R_3 such that $\mathbf{F} = \nabla\phi$.

Let $P(x, y, z)$; $P'(x + \delta x, y + \delta y, z + \delta z)$ be neighbouring points on any curve $\mathscr{C}$ (Fig. 1.20). Let

$$I = \int_A^P \mathbf{F} \,.\, \mathbf{dr}.$$

This is independent of the particular curve joining A to P and since A is fixed and $P(x, y, z)$ is varying, it follows that $I = I(x, y, z)$. The value of the line integral from A to P' is therefore $I(x + \delta x, y + \delta y, z + \delta z)$ and so the increase from P to P' is

$$\delta I = I(x + \delta x, \delta y + y, \delta z + z) - I(x, y, z)$$

$$= \int_P^{P'} \mathbf{F} \,.\, \mathrm{d}\mathbf{r} \triangleq \mathbf{F} \,.\, \delta\mathbf{r} \text{ to the first order } (\overrightarrow{PP'} \equiv \delta\mathbf{r}).$$

Since $I = I(x, y, z)$, $\delta I = \nabla I \,.\, \delta\mathbf{r}$ and so

$$\mathbf{F} \,.\, \delta\mathbf{r} = \nabla I \,.\, \delta\mathbf{r}$$

This is generally true for all non-zero displacements $\delta\mathbf{r}$ and so

$$\mathbf{F} = \nabla I.$$

Thus $\phi = I(x, y, z)$.

The reader will easily establish the corollary to this theorem that *a necessary and sufficient condition for the field* $\mathbf{F}$ *to be conservative in a simply connected* R_3 *is* curl $\mathbf{F} = \mathbf{0}$.

1.17 SOME FURTHER THEOREMS DUE TO GREEN

Let us refer back to the Gauss Divergence Theorem, equation (1.67) in the form

$$\int_S \mathbf{n} \,.\, \mathbf{F}\, \mathrm{d}S = \int \nabla \,.\, \mathbf{F}\, \mathrm{d}v.$$

Let ϕ and ψ be uniform differentiable scalar functions in a sub-region v of R_3. If we choose $\mathbf{F} = \phi\nabla\psi$, then

$$\mathbf{n} \,.\, \mathbf{F} = \phi\, \mathbf{n} \,.\, \nabla\psi = \phi \partial\psi/\partial n;$$

$$\nabla \,.\, \mathbf{F} = \phi\, \nabla^2 \psi + \nabla\phi \,.\, \nabla\psi,$$

and the Divergence Theorem gives at once

$$\int_S \phi \frac{\partial\psi}{\partial n}\, \mathrm{d}S = \int (\phi\nabla^2 \psi + \nabla\phi \,.\, \nabla\psi)\, \mathrm{d}v. \tag{1.78}$$

As ϕ and ψ are arbitrary differentiable scalar functions, they may be interchanged to give

$$\int_S \psi \frac{\partial\phi}{\partial n}\, \mathrm{d}S = \int (\psi\, \nabla^2\phi + \nabla\phi \,.\, \nabla\psi)\, \mathrm{d}v. \tag{1.79}$$

Subtracting (1.79) from (1.78),

$$\int_S \left(\phi \frac{\partial\psi}{\partial n} - \psi \frac{\partial\phi}{\partial n}\right) \mathrm{d}S = \int (\phi\, \nabla^2 \psi - \psi\, \nabla^2 \phi)\, \mathrm{d}v. \tag{1.80}$$

The results (1.78), (1.79), (1.80) are all due to Green: the last is Green's theorem in symmetrical form. [(1.80) may be compared with the rule for integration by parts in elementary calculus in the form

$$\int [u(x)\,v''(x) - v(x)u''(x)]\,\mathrm{d}x = u(x)v'(x) - v(x)u'(x).]$$

In (1.80) we take for ψ the harmonic function $1/r$, where r is the distance from a fixed point P outside S to a variable point in v or on S.

$$\int_S \left\{\phi\,\frac{\partial}{\partial n}\left(\frac{1}{r}\right) - \frac{1}{r}\frac{\partial\phi}{\partial n}\right\}\mathrm{d}S + \int_v \frac{1}{r}\,\nabla^2\phi\,\mathrm{d}v = 0. \tag{1.81}$$

Now suppose P is taken within v so that $1/r$ is infinite at P. With centre P and radius ϵ, construct a small sphere Σ within v and let v' be the region between the surfaces S and Σ (Fig. 1.21). Then P is external to the total surface $S + \Sigma$ of v'.

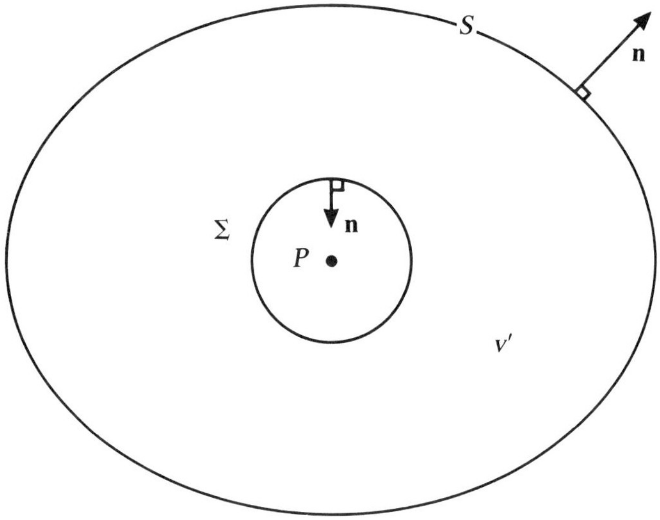

Fig. 1.21

Application of (1.81) to the region v' and its total surface gives

$$\left(\int_S + \int_\Sigma\right)\left\{\phi\,\frac{\partial}{\partial n}\left(\frac{1}{r}\right) - \frac{1}{r}\frac{\partial\phi}{\partial n}\right\}\mathrm{d}S + \int_{v'} \frac{1}{r}\nabla^2\phi\,\mathrm{d}v = 0. \tag{1.82}$$

On Σ, the unit normal vector outwards from v' enters the interior of Σ and so $\mathrm{d}n = -\,\mathrm{d}r$ and

$$\left\{\frac{\partial}{\partial n}\left(\frac{1}{r}\right)\right\}_{r=\epsilon} = \left\{-\frac{\mathrm{d}}{\mathrm{d}r}\left(\frac{1}{r}\right)\right\}_{r=\epsilon} = \frac{1}{\epsilon^2}.$$

Thus

$$\int_{\Sigma} \phi \frac{\partial}{\partial n}\left(\frac{1}{r}\right) \mathrm{d}S = \frac{1}{\epsilon^2} \int_{\Sigma} \phi \, \mathrm{d}S = \frac{1}{\epsilon^2} \times \bar{\phi} \times 4\pi\epsilon^2,$$

where $\bar{\phi}$ is the mean surface value of the continuous function ϕ on Σ. As we let $\epsilon \to 0$, the last form $\to 4\pi\phi_P$. Further,

$$\int_{\Sigma} \frac{1}{r} \frac{\partial \phi}{\partial n} \mathrm{d}S = \frac{1}{\epsilon} \int_{\Sigma} \frac{\partial \phi}{\partial n} \mathrm{d}S = \frac{1}{\epsilon}\left(\overline{\frac{\partial \phi}{\partial n}}\right) \times 4\pi\epsilon^2,$$

where $(\overline{\partial\phi/\partial n})$ signifies an appropriate mean value of $\partial\phi/\partial n$ (assumed to be continuous) is taken on S. This mean value is presumed to be finite: then the last expression $\to 0$ as $\epsilon \to 0$. Thus letting $\epsilon \to 0$ in (1.82) gives

$$4\pi\phi_P + \int_S \left\{\phi \frac{\partial}{\partial n}\left(\frac{1}{r}\right) - \frac{1}{r}\frac{\partial \phi}{\partial n}\right\} \mathrm{d}S + \int_v \frac{1}{r} \nabla^2 \phi \, \mathrm{d}v = 0. \tag{1.83}$$

(1.83) is known as *Green's formula*, giving the value of ϕ at any P in v where r is the distance of P from a volume element $\mathrm{d}v$ or a surface element $\mathrm{d}S$.

When ϕ is a harmonic function the volume integral in (1.83) vanishes and if the surface distributions of ϕ and $\partial\phi/\partial n$ are known on S, we can evaluate ϕ at an internal point P of v.

When $\phi = 1$, equation (1.83) gives

$$4\pi = -\int_S \frac{\partial}{\partial n}\left(\frac{1}{r}\right) \mathrm{d}S.$$

Now

$$\frac{\partial}{\partial n}\left(\frac{1}{r}\right) = \mathbf{n} \, . \nabla \left(\frac{1}{r}\right);$$

$$\nabla\left(\frac{1}{r}\right) = \hat{\mathbf{r}} \frac{\mathrm{d}}{\mathrm{d}r}\left(\frac{1}{r}\right) = -\frac{1}{r^2} \hat{\mathbf{r}}.$$

Thus

$$4\pi = \int_S \frac{\mathbf{n} \, . \, \hat{\mathbf{r}}}{r^2} \mathrm{d}S = \int_S \frac{\cos\theta}{r^2} \mathrm{d}S,$$

where θ is the angle between $\hat{\mathbf{r}}$ and $\mathbf{n}$. The expression on the right-hand side is the total solid angle subtended by S at an internal point P (Ref. 3, Section 5.7).

1.18 GENERAL ORTHOGONAL CURVILINEAR COORDINATES

Suppose that at each point $P(x, y, z)$ of R_3 there exist three uniform differentiable scalar functions $u_i(x, y, z)$ $(i = 1, 2, 3)$ having as level surfaces

$$u_i(x, y, z) = c_i \qquad (i = 1, 2, 3), \tag{1.84}$$

where each c_i is independent of the rectangular coordinates (x, y, z). If in the three equations (1.78) the constants c_i are prescribed and if these equations can be solved uniquely for (x, y, z) to obtain a corresponding unique P in R_3, then clearly the system (u_1, u_2, u_3) affords an alternative specification to the Cartesian coordinates (x, y, z) of P. In general two surfaces u_2 = const., u_3 = const., will intersect along some curve in space along which only u_1 varies and such a curve is called the *u_1-line*. The set (u_1, u_2, u_3) are called the *curvilinear coordinates of P* and the three surfaces defined by (1.84) are called the *curvilinear coordinate surfaces for P*. If the three level surfaces (1.84) meet at no point other than P, then the normals to these three surfaces are not coplanar at P and so the scalar triple product of the gradients of the $u_i(x, y, z)$ $(i = 1, 2, 3)$ is non-zero, i.e.

$$[\nabla u_1, \nabla u_2, \nabla u_3] = \begin{vmatrix} \partial u_1/\partial x & \partial u_1/\partial y & \partial u_1/\partial z \\ \partial u_2/\partial x & \partial u_2/\partial y & \partial u_2/\partial z \\ \partial u_3/\partial x & \partial u_3/\partial y & \partial u_3/\partial z \end{vmatrix} \neq 0, \tag{1.85}$$

or, in Jacobian notation

$$\frac{\partial(u_1, u_2, u_3)}{\partial(x, y, z)} \neq 0. \tag{1.85$'$}$$

This is a well-known condition for the equations (1.84) to possess a unique solution in the form

$$x = x(u_1, u_2, u_3), y = y(u_1, u_2, u_3), z = z(u_1, u_2, u_3). \tag{1.86}$$

If for an origin O in R_3, $\overrightarrow{OP} = \mathbf{r} \equiv [x, y, z]$, then we can express $\mathbf{r}$ in the form $\mathbf{r} = \mathbf{r}(u_1, u_2, u_3)$, a vector function of the curvilinear coordinates. For this vector function, we can form $\partial\mathbf{r}/\partial u_1$, another vector function which, when localised at P, is directed along the tangent at P to the u_1-line, since u_2 and u_3 are both kept constant in forming the partial derivative. If $\hat{\mathbf{a}}_i$ denote the unit vectors in the three u_i-lines through P, taken in the sense of $u_i (i = 1, 2, 3)$ increasing, then

$$\partial\mathbf{r}/\partial u_i = h_i\,\hat{\mathbf{a}}_i \qquad (i = 1, 2, 3), \tag{1.87}$$

where $h_i = h_i(u_1, u_2, u_3)$ are uniform scalar functions of the u_i $(i = 1, 2, 3)$. With $\mathbf{r} = \mathbf{r}(u_1, u_2, u_3)$, its total differential is

$$\mathrm{d}\mathbf{r} = \mathrm{d}\mathbf{s} = (\partial\mathbf{r}/\mathrm{d}u_1)\,\mathrm{d}u_1 + (\partial\mathbf{r}/\partial u_2)\,\mathrm{d}u_2 + (\partial\mathbf{r}/\partial u_3)\,\mathrm{d}u_3. \tag{1.88}$$

From (1.88), (1.87), we obtain the *vector arc element*

$$\mathrm{d}\mathbf{r} = \mathrm{d}\mathbf{s} = (h_1\,\mathrm{d}u_1)\hat{\mathbf{a}}_1 + (h_2\,\mathrm{d}u_2)\hat{\mathbf{a}}_2 + (h_3\,\mathrm{d}u_3)\hat{\mathbf{a}}_3. \tag{1.89}$$

Let P, P' be two neighbouring points in R_3 specified by rectangular Cartesian coordinates (x, y, z); $(x + \delta x, y + \delta y, z + \delta z)$ and by the respective curvilinear coordinates (u_1, u_2, u_3) ; $(u_1 + \delta u_1, u_2 + \delta u_2, u_3 + \delta u_3)$. Throughout each of these points the three coordinate surfaces u_i = const. $(i = 1, 2, 3)$ are constructed to form a volume element approximating to a parallelepiped (Fig. 1.22).

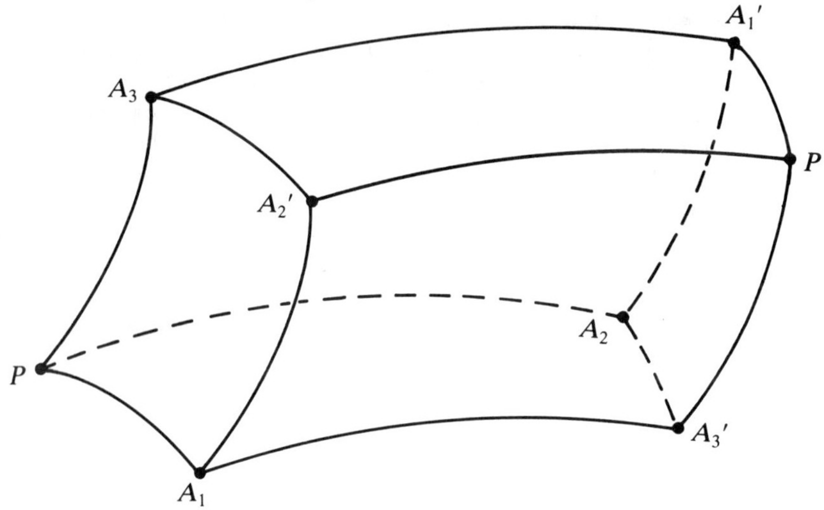

Fig. 1.22

In the diagram the primed vertices are diagonally opposite to the corresponding unprimed ones. The curved line PA_1 is the u_1-line through P along which only u_1 varies, so that A_1 has curvilinear coordinates $(u_1 + \delta u_1, u_2, u_3)$. The point A_2 has the coordinates $(u_1, u_2 + \delta u_2, u_3)$ and A_3 is $(u_1, u_2, u_3 + \delta u_3)$. Since A_3' is accessible from P via the route $\overrightarrow{PA_1} + \overrightarrow{A_1A_3}'$, i.e. via a u_1- and a u_2-line, so that u_3 is constant. Hence A_3' has the coordinates $(u_1 + \delta u_1, u_2 + \delta u_2, u_3)$. The coordinates of all six vertices are easily obtained. As

$$\overrightarrow{PP'} = \overrightarrow{PA}_1 + \overrightarrow{PA}_2 + \overrightarrow{PA}_3$$

and
$$\delta \mathbf{r} = h_1 \, \delta u_1 \, \hat{\mathbf{a}}_1 + h_2 \, \delta u_2 \, \hat{\mathbf{a}}_2 + h_3 \, \delta u_3 \, \hat{\mathbf{a}}_3 ,$$

since the unit vector $\hat{\mathbf{a}}_i$ is tangential to the u_i curve at P, it follows that

$$\overrightarrow{PA}_i \equiv h_i \, \delta u_i \, \hat{\mathbf{a}}_i \qquad (i = 1, 2, 3)$$

and
$$PA_i = h_i \, \delta u_i \qquad (i = 1, 2, 3).$$

Now suppose the curvilinear system (u_1, u_2, u_3) is so chosen that at each P of R_3 the curves of intersection of the three families of coordinate surfaces (1.78) are mutually orthogonal, so that $\hat{\mathbf{a}}_2 \cdot \hat{\mathbf{a}}_3 = \hat{\mathbf{a}}_3 \cdot \hat{\mathbf{a}}_1 = \hat{\mathbf{a}}_1 \cdot \hat{\mathbf{a}}_2 = 0$. Then (u_1, u_2, u_3) are called the *general orthogonal curvilinear coordinates* of P. For convenience we further suppose the coordinates to be so labelled that

$\hat{\mathbf{a}}_1 \times \hat{\mathbf{a}}_2 = \hat{\mathbf{a}}_3$ and not $-\hat{\mathbf{a}}_3$: then the unit vectors $\hat{\mathbf{a}}_i$ $(i = 1, 2, 3)$ form a right-handed system. For an orthogonal system, (1.89) gives the arc element

$$ds = |d\mathbf{r}| = \sqrt{\{(h_1\, du_1)^2 + (h_2\, du_2)^2 + (h_3\, du_3)^2\}}. \tag{1.90}$$

The rectangular Cartesian coordinate system (x, y, z) is orthogonal since the three planes are mutually at right-angles. Hence we may take $u_1 = x$, $u_2 = y$, $u_3 = z$ and $\hat{\mathbf{a}}_1 = \mathbf{i}, \hat{\mathbf{a}}_2 = \mathbf{j}, \hat{\mathbf{a}}_3 = \mathbf{k}$. Also,

$$\mathbf{r} = x\,\mathbf{i} + y\,\mathbf{j} + z\mathbf{k},$$

$$d\mathbf{r} = dx\,\mathbf{i} + dy\,\mathbf{j} + dz\,\mathbf{k}.$$

But
$$d\mathbf{r} = h_1\, du_1\, \hat{\mathbf{a}}_1 + h_2\, du_2\, \hat{\mathbf{a}}_2 + h_3\, du_3\, \hat{\mathbf{a}}_3$$

and so, for the Cartesian system,

$$h_1 = h_2 = h_3 = 1.$$

However, it is often convenient to employ systems of orthogonal coordinates other than the Cartesian. Two of these we now consider.

(i) Cylindrical Polar Coordinates

In Fig. (1.23), P is the point having rectangular Cartesian coordinates (x, y, z) in the right-handed frame OX, OY, OZ. PM is drawn perpendicularly from P on to the plane XOY and MN is perpendicular to OX. Then $PM = z$. We write $OM = R$, $X\hat{O}M = \phi$. Then (R, ϕ, z) are called the *cylindrical polar coordinates* of P. For a given P, these coordinates are unique, or, conversely, given (R, ϕ, z) it is easy to see that P is unique. The admissible ranges of variation of the coordinates (R, ϕ, z) are

$$R \geqslant 0; \quad 0 \leqslant \phi \leqslant 2\pi; \quad -\infty < z < \infty.$$

From the diagram it is easy to obtain the following relations between the Cartesian and cylindrical polar coordinates:

$$x = R\cos\phi, \quad y = R\sin\phi, \quad z = z;$$

$$R = \sqrt{(x^2 + y^2)}, \quad \tan\phi = y/x, \quad z = z.$$

The orthogonal nature of the cylindrical polar coordinate system is established by considering in turn the three surfaces

$$R = \text{const.}, \quad \phi = \text{const.}; \quad z = \text{const.}$$

The first is a right-circular cylinder having OZ as axis. The second is a half-plane through OZ at an angle ϕ with the coordinate plane XOZ. The third is the plane through P parallel to the coordinate plane XOY. Clearly the three intersect mutually orthogonally at P. Thus (R, ϕ, z) is an orthogonal curvilinear coordinate system, and in applying the general theory we may take $u_1 = R$, $u_2 = \phi$, $u_3 = z$. It is also necessary to evaluate the h_i in the (R, ϕ, z) system.

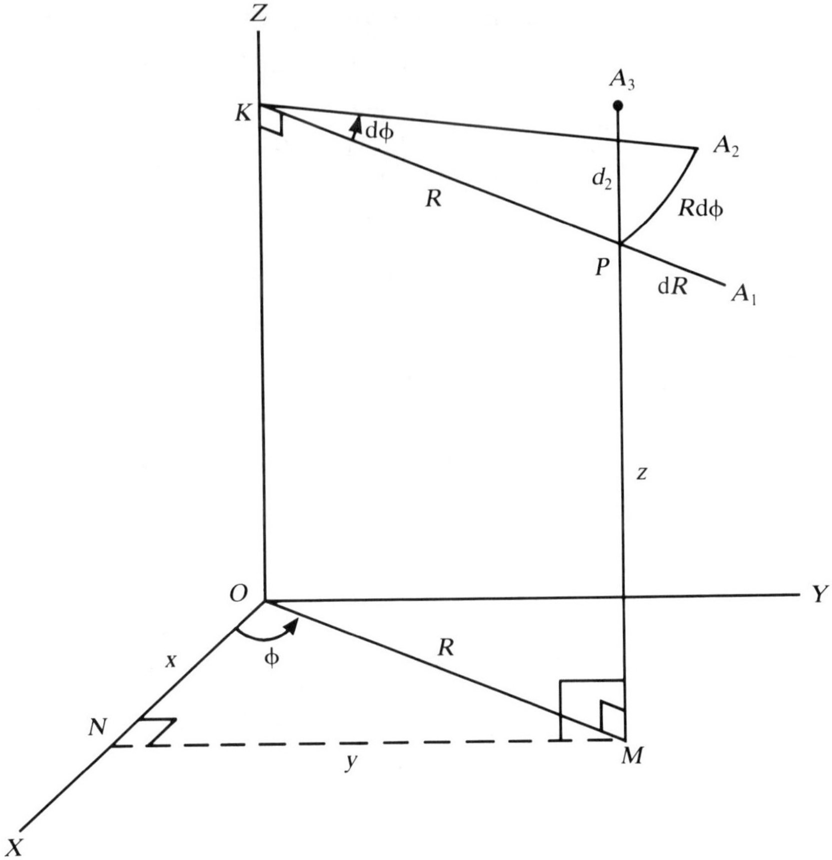

Fig. 1.23

This is undertaken by finding the displacement of P when each coordinate is increased by an infinitesimal amount.

We start at $P(R, \phi, z)$. First, we keep ϕ, z constant and increase R to $R + \mathrm{d}R$ when the displacement of P is along $\overrightarrow{KP}$ to A, where $PA_1 = \mathrm{d}R$. We write

$$\overrightarrow{PA}_1 \equiv \mathrm{d}R\ \hat{\mathbf{R}}$$

to define the unit vector $\hat{\mathbf{R}}$ at P in the direction of R increasing. Now revert to P and increase ϕ to $\phi + \mathrm{d}\phi$, keeping R and z fixed. Then P moves to A_2 along the circular arc PA_2 parallel to the plane XOY, the arc's centre being K (on OZ), where $P\hat{K}A_2 = \mathrm{d}\phi$ so that arc $PA_2 = R\ \mathrm{d}\phi$. Write;

$$\overrightarrow{PA}_2 \equiv R\ \mathrm{d}\phi\ \hat{\phi}$$

to define the unit vector $\hat{\phi}$ at P in the direction of ϕ increasing. Reverting again to P, we increase z to $z + \mathrm{d}z$, keeping R, ϕ both fixed. P moves to A_3 along $\overrightarrow{MP}$ where $PA_3 = \delta z$ and

$$\overrightarrow{PA}_3 \equiv \mathrm{d}z\, \mathbf{k}.$$

Thus we have obtained the components of $\mathrm{d}\mathbf{r}$ and so

$$\mathrm{d}\mathbf{r} = \mathrm{d}R\, \hat{\mathbf{R}} + R\, \mathrm{d}\phi\, \hat{\boldsymbol{\phi}} + \mathrm{d}z\, \mathbf{k}. \tag{1.91}$$

$\hat{\mathbf{R}}$ and $\hat{\boldsymbol{\phi}}$ are variable unit vectors, but $\mathbf{k}$ is fixed which is why quite different notations are employed. In the general system (u_1, u_2, u_3) equation (1.89) holds and comparison with (1.91) shows

$$h_1 = 1, \quad h_2 = R, \quad h_3 = 1. \tag{1.92}$$

Also in (R, θ, z) coordinates the arc element is

$$\mathrm{d}s = \sqrt{\{(\mathrm{d}R)^2 + (R\, \mathrm{d}\phi)^2 + (\mathrm{d}z)^2\}}.$$

(ii) Spherical Polar Coordinates

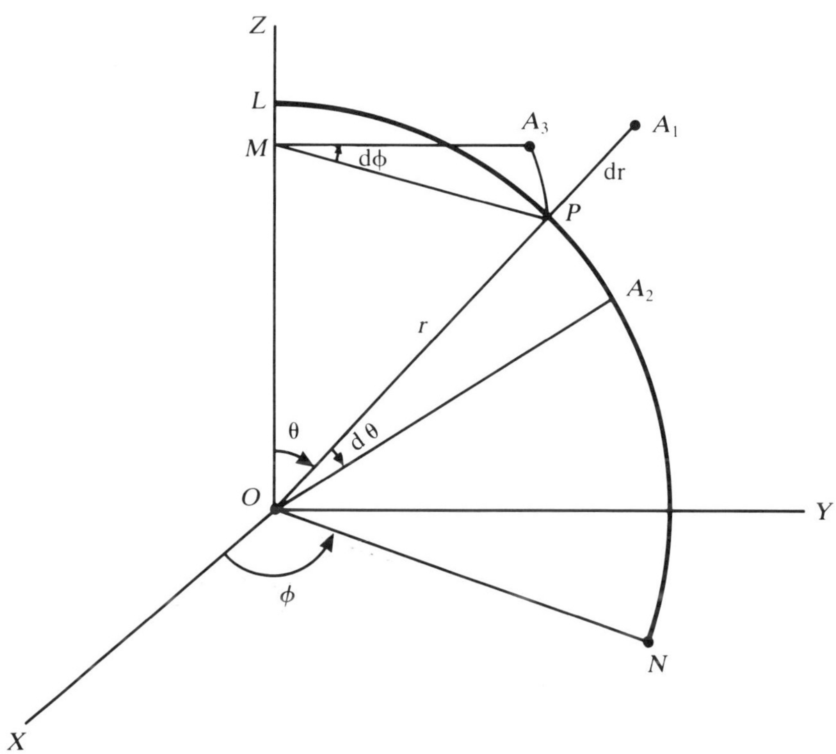

Fig. 1.24

In Fig. 1.24, P has the Cartesian coordinates (x, y, z). We let $OP = r$, $P\hat{O}Z = \theta$, $X\hat{O}N = \phi$, the angle between the planes XOZ, ZOP. Then (r, θ, ϕ) are called the *spherical polar coordinates of P*. They determine P uniquely for prescribed values and it is easy to see that, with P fixed in the frame, (r, θ, ϕ) are unique. In the diagram the circle LPN is drawn in the plane POZ to cut OZ in L and the plane XOY in N. PM is perpendicular to OZ. The ranges of the spherical polar coordinates are

$$r \geqslant 0; \quad 0 \leqslant \theta \leqslant \pi; \quad 0 \leqslant \phi \leqslant 2\pi.$$

It is easy to see that

$$x = r \sin\theta \cos\phi, \qquad y = r \sin\theta \sin\phi, \qquad z = r\cos\theta;$$

$$r = \sqrt{(x^2 + y^2 + z^2)}, \qquad \theta = \cos^{-1}(z/r), \qquad \tan\phi = y/x.$$

To establish the orthogonality of the (r, θ, ϕ) system, one needs to consider in turn the nature of the following three surfaces through P:

$$r = \text{const.}, \qquad \theta = \text{const.}, \qquad \phi = \text{const.}$$

The first is a sphere centred on O, the second the right-circular cone having OZ as axis and θ as semi-vertical angle, and the final surface is the plane through OZ set at angle ϕ to the plane XOZ. Clearly these three surfaces intersect along curves which are mutually orthogonal, so that (r, θ, ϕ) is an orthogonal system and we may take $(u_1, u_2, u_3) = (r, \theta, \phi)$. It now remains to find the scale factors h_1, h_2, h_3.

If we start at $P(r, \theta, \phi)$ in Fig. 1.22 and advance r to $r + \mathrm{d}r$ keeping both θ, ϕ constant, then P moves outwards along $\overline{OP}$ to A_1 where $PA_1 = \mathrm{d}r$. We may now introduce the unit vector $\hat{\mathbf{r}}$ along $\overline{OP}$ and write

$$\overrightarrow{PA}_1 \equiv \mathrm{d}r\, \hat{\mathbf{r}}.$$

If we revert to P, keep r and ϕ constant but advance θ to $\theta + \mathrm{d}\theta$, then P moves along the circle LPN to A_2 where $P\hat{O}A_2 = \mathrm{d}\theta$ so that the arc $PA_2 = r\,\mathrm{d}\theta$. A unit vector $\hat{\theta}$ may now be introduced at P, in the direction tangential to the circle at P, through the relation

$$\overrightarrow{PA} \equiv r\,\delta\theta\, \hat{\boldsymbol{\theta}}.$$

Reverting again to P and keeping both r and θ constant, we increase ϕ to $\phi + \mathrm{d}\phi$. The movement of P is along the circular arc PA_3 to A_3. M is the centre of the arc, $\mathrm{d}\phi$ the angle PMA_3 so that arc $PA_3 = r\sin\theta\,\mathrm{d}\phi$. The unit vector $\hat{\phi}$ in the tangent at P to this circular arc is defined through

$$\overrightarrow{PA} \equiv r \sin\theta\, \mathrm{d}\phi\, \hat{\boldsymbol{\phi}}.$$

Hence

$$\mathrm{d}\mathbf{r} = \mathrm{d}r\, \hat{\mathbf{r}} + r\,\mathrm{d}\theta\, \hat{\boldsymbol{\theta}} + r\sin\theta\,\mathrm{d}\phi\, \hat{\boldsymbol{\phi}}. \tag{1.94}$$

(Note that $|\mathrm{d}\mathbf{r}| \neq \mathrm{d}r$ here.) Comparison of (1.89), (1.94) shows that for the (r, θ, ϕ) system

$$h_1 = 1, \quad h_2 = r, \quad h_3 = r \sin\theta \tag{1.95}$$

when we take $(u_1, u_2, u_3) = (r, \theta, \phi)$ with $(\hat{\mathbf{a}}_1, \hat{\mathbf{a}}_2, \hat{\mathbf{a}}_3) = (\hat{\mathbf{r}}, \hat{\theta}, \hat{\phi})$. Each of the unit vectors $\hat{\mathbf{r}}, \hat{\theta}, \hat{\phi}$ at P is of variable direction.

1.19 EVALUATION OF THE FIELD FUNCTIONS grad, div, curl IN ORTHOGONAL COORDINATES

In the following, (u_1, u_2, u_3) are general orthogonal curvilinear coordinates, $U(u_1, u_2, u_3)$ is a uniform differentiable scalar function in R_3 and $\mathbf{F} = F_1\,\hat{\mathbf{a}}_1 + F_2\,\hat{\mathbf{a}}_2 + F_3\,\hat{\mathbf{a}}_3$ is a uniform differentiable vector function with $F_i = F_i(u_1, u_2, u_3)$ $(i = 1, 2, 3)$ specifying its scalar components.

The components of ∇U in the directions of u_1, u_2, u_3 increasing are $\hat{\mathbf{a}}_i \,.\, \nabla U$ $(i = 1, 2, 3)$. When u_i is increased to $u_i + \mathrm{d}u_i$ keeping the other two coordinates constant, the arc element through P has length h_i, $\mathrm{d}u_i$. Hence, if in (1.34) we take $\mathrm{s} = \hat{\mathbf{a}}_i$, $\mathrm{d}s = h_i\,\mathrm{d}u_i$, then

$$\hat{\mathbf{a}}_i \,.\, \nabla U = \frac{\partial U}{h_i\,\partial u_i} \qquad (i = 1, 2, 3).$$

Thus it follows that in orthogonal coordinates (u_1, u_2, u_3),

$$\nabla U = \frac{1}{h_1}\frac{\partial U}{\partial u_1}\,\hat{\mathbf{a}}_1 + \frac{1}{h_2}\frac{\partial U}{\partial u_2}\,\hat{\mathbf{a}}_2 + \frac{1}{h_3}\frac{\partial U}{\partial u_3}\,\hat{\mathbf{a}}_3. \tag{1.96}$$

In (1.96), if we take $U = u_i$, then for $i = 1, 2, 3$,

$$\hat{\mathbf{a}}_i = h_i\,\nabla u_i. \tag{1.97}$$

We note further from this form that

$$h_i = 1/|\nabla u_i|, \tag{1.98}$$

and so

$$\hat{\mathbf{a}}_i = \nabla u_i/|\nabla u_i|. \tag{1.99}$$

A further consequence of the relations (1.97) is the following. For a right-handed system, $\hat{\mathbf{a}}_1 = \hat{\mathbf{a}}_2 \times \hat{\mathbf{a}}_3 = h_2 h_3\,\nabla u_2 \times \nabla u_3$. Hence

$$\nabla.\,\hat{\mathbf{a}}_1 = h_2 h_3\,\nabla.(\nabla u_2 \times \nabla u_3) + \nabla(h_1 h_3)\,.\,(\nabla u_2 \times \nabla u_3).$$

It is easy to prove that $\nabla.(\nabla u_2 \times \nabla u_3) = 0$. Then using (1.97)

$$\nabla.\,\hat{\mathbf{a}}_1 = \frac{1}{h_2 h_3}\,\hat{\mathbf{a}}_1\,.\,\nabla(h_2 h_3)$$

or, using $\hat{\mathbf{a}}_1\,.\,\nabla \equiv (1/h_1)\,\partial/\partial u_1$, (1.100)

$$\nabla.\,\hat{\mathbf{a}}_1 = \frac{1}{h_1 h_2 h_3}\frac{\partial}{\partial u_1}(h_2 h_3),$$

with two similar results obtaining when the subscripts 1, 2, 3 are permuted cyclically. These results enable us to evaluate $\nabla . \mathbf{F}$ when

$$\mathbf{F} = F_1\,\hat{\mathbf{a}}_1 + F_2\,\hat{\mathbf{a}}_2 + F_3\,\hat{\mathbf{a}}_3, \qquad F_i = F_i(u_1, u_2, u_3)\,(i = 1, 2, 3).$$

For

$$\begin{aligned}\nabla . (F_1\,\hat{\mathbf{a}}_1) &= F_1\,\nabla . \hat{\mathbf{a}}_1 + \hat{\mathbf{a}}_1 . \nabla F_1 \\ &= \frac{F_1}{h_1 h_2 h_3}\,\frac{\partial}{\partial u_1}\,(h_2 h_3) + \frac{1}{h_1}\,\frac{\partial}{\partial u_1}\,(F_1) \\ &= \frac{1}{h_1 h_2 h_3}\,\frac{\partial}{\partial u_1}\,(h_2 h_3 F_1), \text{ etc.}\end{aligned}$$

Hence we obtain the *divergence in orthogonal coordinates*:

$$\nabla . \mathbf{F} = \frac{1}{h_1 h_2 h_3}\left\{\frac{\partial}{\partial u_1}\,(h_2 h_3 F_1) + \frac{\partial}{\partial u_2}\,(h_3 h_1 F_2) + \frac{\partial}{\partial u_3}\,(h_1 h_2 F_3)\right\}. \tag{1.101}$$

In (1.95), if we take

$$\mathbf{F} = \nabla U, \qquad U = U(u_1, u_2, u_3),$$

then $F_1 = (1/h_1)\,\partial U/\partial u_1$, etc., and so we obtain the *Laplacian in orthogonal coordinates*:

$$\begin{aligned}\nabla^2 U = \frac{1}{h_1 h_2 h_3}\Bigg\{&\frac{\partial}{\partial u_1}\left(\frac{h_2 h_3}{h_1}\,\frac{\partial U}{\partial u_1}\right) + \frac{\partial}{\partial u_2}\left(\frac{h_3 h_1}{h_2}\,\frac{\partial U}{\partial u_2}\right) \\ &+ \frac{\partial}{\partial u_3}\left(\frac{h_1 h_2}{h_3}\,\frac{\partial U}{\partial u_3}\right)\Bigg\}.\end{aligned} \tag{1.102}$$

For the expression of $\nabla \times \mathbf{F}$ in orthogonal coordinates, the procedure is as follows. Using (1.91),

$$\begin{aligned}\nabla \times (F_1 \hat{\mathbf{a}}) &= \nabla \times (h_1 F_1\,\nabla u_1) \\ &= h_1 F_1\,\nabla \times (\nabla u_1) + \nabla(h_1 F_1) \times \nabla u_1 \\ &= \mathbf{0} + \left[\frac{\hat{\mathbf{a}}_1}{h_1}\,\frac{\partial(h_1 F_1)}{\partial u_1} + \frac{\hat{\mathbf{a}}_2}{h_2}\,\frac{\partial(h_1 F_1)}{\partial u_2} + \frac{\hat{\mathbf{a}}_3}{h_3}\,\frac{\partial(h_1 F_1)}{\partial u_2}\right] \times \frac{\hat{\mathbf{a}}_1}{h_1} \\ &= \frac{\hat{\mathbf{a}}_2}{h_3 h_1}\,\frac{\partial}{\partial u_3}\,(h_1 F_1) - \frac{\hat{\mathbf{a}}_3}{h_1 h_2}\,\frac{\partial}{\partial u_2}\,(h_1 F_1).\end{aligned}$$

Similarly,

$$\nabla \times (F_2 \hat{\mathbf{a}}_2) = \frac{\hat{\mathbf{a}}_3}{h_1 h_2} \frac{\partial}{\partial u_1} (h_2 F_2) - \frac{\hat{\mathbf{a}}_1}{h_2 h_3} \frac{\partial}{\partial u_3} (h_2 F_2);$$

$$\nabla \times (F_3 \hat{\mathbf{a}}_3) = \frac{\hat{\mathbf{a}}_1}{h_2 h_3} \frac{\partial}{\partial u_2} (h_3 F_3) - \frac{\hat{\mathbf{a}}_2}{h_3 h_1} \frac{\partial}{\partial u_1} (h_3 F_3).$$

Thus as $\mathbf{F} = \Sigma\, (F_1 \hat{\mathbf{a}}_1)$,

$$\nabla \times \mathbf{F} = \frac{1}{h_1 h_2 h_3} \sum h_1\, \hat{\mathbf{a}}_1 \left\{ \frac{\partial}{\partial u_2} (h_3 F_3) - \frac{\partial}{\partial u_3} (h_2 F_2) \right\},$$

or, in a more easily remembered determinantal form,

$$\nabla \times \mathbf{F} = \frac{1}{h_1 h_2 h_3} \begin{vmatrix} h_1 \hat{\mathbf{a}} & h_2 \hat{\mathbf{a}}_2 & h_3 \hat{\mathbf{a}}_3 \\ \partial/\partial u_1 & \partial/\partial u_2 & \partial/\partial u_3 \\ h_1 F_1 & h_2 F_2 & h_3 F_3 \end{vmatrix}. \tag{1.103}$$

The reader should now verify that the rectangular Cartesian system (x, y, z) is orthogonal with $(u_1, u_2, u_3) = (x, y, z)$ and $h_1 = h_2 = h_3 = 1$. Using these results and the equations (1.96), (1.101), (1.102), (1.103) it is easy to recover the Cartesian forms for ∇U, $\nabla \cdot \mathbf{F}$, $\nabla^2 U$, $\nabla \times \mathbf{F}$.

1.20 EXPRESSION OF THE FIELD FUNCTIONS IN CYLINDRICAL AND IN SPHERICAL POLARS

The reader should now use the general orthogonal forms alluded to above for finding the following field functions in (i) cylindrical polars and (ii) spherical polars.

(i) Cylindrical polars (R, ϕ, z)

$$(u_1, u_2, u_3) = (R, \phi, z), \quad h_1 = 1, \quad h_2 = R, \quad h_3 = 1$$

$$U = U(R, \phi, z),$$

$$\mathbf{F} = F_R\, \hat{\mathbf{R}} + F_\phi\, \hat{\phi} + F_z \mathbf{k}, \text{ with } F_R = F_R(R, \phi, z), \text{ etc.}$$

Then

$$\nabla U = \frac{\partial U}{\partial R} \hat{\mathbf{R}} + \frac{1}{R} \frac{\partial U}{\partial \phi} \hat{\phi} + \frac{\partial U}{\partial z} \mathbf{k}; \tag{1.104}$$

$$\nabla . \mathbf{F} = \frac{1}{R}\left\{\frac{\partial}{\partial R}(RF_R) + \frac{\partial F_\phi}{\partial \phi} + \frac{\partial}{\partial z}(RF_z)\right\}; \tag{1.105}$$

$$\nabla^2 U = \frac{1}{R}\left\{\frac{\partial}{\partial R}\left(R\frac{\partial U}{\partial R}\right) + \frac{\partial}{\partial \phi}\left(\frac{1}{R}\frac{\partial U}{\partial \phi}\right) + \frac{\partial}{\partial z}\left(R\frac{\partial U}{\partial z}\right)\right\}; \tag{1.106}$$

$$\nabla \times \mathbf{F} = \frac{1}{R}\begin{vmatrix} \hat{\mathbf{R}} & R\hat{\boldsymbol{\phi}} & \mathbf{k} \\ \partial/\partial R & \partial/\partial \phi & \partial/\partial z \\ F_R & RF_\phi & F_z \end{vmatrix}. \tag{1.107}$$

(ii) Spherical polars (r, θ, ϕ)

$(u_1, u_2, u_3) = (r, \theta, \phi)$; $\quad h_1 = 1, \quad h_2 = r, \quad h_3 = r \sin\theta$

$U = U(r, \theta, \phi)$,

$\mathbf{F} = F_r\,\hat{\mathbf{r}} + F_\theta\,\hat{\boldsymbol{\theta}} + F_\phi\,\hat{\boldsymbol{\phi}}$, where $F_r = F_r(r, \theta, \phi)$, etc.

Then

$$\nabla U = \frac{\partial U}{\partial r}\hat{\mathbf{r}} + \frac{1}{r}\frac{\partial U}{\partial \theta}\hat{\boldsymbol{\theta}} + \frac{1}{r\sin\theta}\frac{\partial U}{\partial \phi}\hat{\boldsymbol{\phi}}, \tag{1.108}$$

$$\nabla . \mathbf{F} = \frac{1}{r^2\sin\theta}\left\{\frac{\partial(r^2\sin\theta\, F_r)}{\partial r} + \frac{\partial(r\sin\theta\, F_\theta)}{\partial \theta} + \frac{\partial(rF_\phi)}{\partial \phi}\right\}, \tag{1.109}$$

$$\nabla^2 U = \frac{1}{r^2\sin\theta}\left\{\frac{\partial}{\partial r}\left(r^2\sin\theta\frac{\partial U}{\partial r}\right) + \frac{\partial}{\partial \theta}\left(\sin\theta\frac{\partial U}{\partial \theta}\right) + \operatorname{cosec}\theta\frac{\partial^2 U}{\partial \phi^2}\right\}, \tag{1.110}$$

$$\nabla \times \mathbf{F} = \frac{1}{r^2\sin\theta}\begin{vmatrix} \hat{\mathbf{r}} & r\hat{\boldsymbol{\theta}} & r\sin\theta\hat{\boldsymbol{\phi}} \\ \partial/\partial r & \partial/\partial \theta & \partial/\partial \phi \\ F_r & rF_\theta & r\sin\theta F_\phi \end{vmatrix}. \tag{1.111}$$

Example [1.9]

Toroidal coordinates (r, θ, ϕ) are related to rectangular Cartesian coordinates (x, y, z) through

$$x = (c + r\sin\phi)\cos\theta, \quad y = (c + r\sin\phi)\sin\theta, \quad z = r\cos\phi,$$

where c is a constant. Prove that the toroidal system is an orthogonal curvilinear one and express $\nabla^2 U(r, \theta, \phi)$ in the coordinates.

The vector arc element ds at P may be expressed as

$$\mathrm{d}\mathbf{s} = h_r\, \mathrm{d}r\, \hat{\mathbf{r}} + h_\theta\, \mathrm{d}\theta\, \hat{\boldsymbol{\theta}} + h_\phi\, \mathrm{d}\phi\, \hat{\boldsymbol{\phi}},$$

where $h_r = h_r(r, \theta, \phi)$, etc., and $\hat{\mathbf{r}}, \hat{\boldsymbol{\theta}}, \hat{\boldsymbol{\phi}}$ are the unit vectors at P in the directions of r only increasing, θ only increasing and ϕ only increasing. Then

$$(\mathrm{d}\mathbf{s})^2 = (h_r\, \mathrm{d}r)^2 + (h_\theta\, \mathrm{d}\theta)^2 + (h_\phi\, \mathrm{d}\theta)^2 + 2h_\theta h_\phi\, \mathrm{d}\theta \mathrm{d}\phi\, \hat{\boldsymbol{\theta}} \,.\, \hat{\boldsymbol{\phi}} + \text{etc.}$$

Now

$$(\mathrm{d}s)^2 = (\mathrm{d}x)^2 + (\mathrm{d}y)^2 + (\mathrm{d}z)^2,$$

since (x, y, z) form an orthogonal system.
Now

$$\mathrm{d}x = -(c + r\sin\phi)\sin\theta\, \mathrm{d}\theta + \cos\theta\, \mathrm{d}(c + r\sin\phi),$$

$$\mathrm{d}y = (c + r\sin\phi)\cos\theta\, \mathrm{d}\theta + \sin\theta\, \mathrm{d}(c + r\sin\phi)$$

and so

$$\begin{aligned}(\mathrm{d}x)^2 + (\mathrm{d}y)^2 &= (c + r\sin\phi)^2 (\mathrm{d}\theta)^2 + [\mathrm{d}(c + r\sin\phi)]^2 \\ &= (c + r\sin\phi)^2 (\mathrm{d}\theta)^2 + (r\cos\phi\, \mathrm{d}\phi + \mathrm{d}r \sin\phi)^2, \\ \mathrm{d}z &= \mathrm{d}r\cos\phi - r\sin\phi\, \mathrm{d}\phi.\end{aligned}$$

Hence

$$(\mathrm{d}s)^2 = (c + r\sin\theta)^2 (\mathrm{d}\theta)^2 + (\mathrm{d}r)^2 + r^2 (\mathrm{d}\phi)^2.$$

There are no terms in $\mathrm{d}\theta\mathrm{d}\phi$, $\mathrm{d}\phi\mathrm{d}r$, $\mathrm{d}\theta\mathrm{d}r$ and as $(\mathrm{d}\mathbf{s})^2 = (\mathrm{d}s)^2$ comparison of the two forms shows

$$h_r = 1, \quad h_\theta = c + r\sin\phi, \quad h_\phi = r;$$

$$h_\theta h_\phi\, \mathrm{d}\theta\mathrm{d}\phi\, \hat{\boldsymbol{\theta}} \,.\, \hat{\boldsymbol{\phi}} = 0, \text{ etc.}$$

We thus obtain

$$\hat{\boldsymbol{\theta}} \,.\, \hat{\boldsymbol{\phi}} = 0, \quad \hat{\boldsymbol{\phi}} \,.\, \hat{\mathbf{r}} = 0, \quad \hat{\mathbf{r}} \,.\, \hat{\boldsymbol{\theta}} = 0,$$

as the scale factors h_r, h_θ, h_ϕ are all generally non-zero. These relations establish the orthogonality of the system.

The Laplacian in the coordinates (r, θ, ϕ) has the form

$$\nabla^2 U(r, \theta, \phi) = \frac{1}{h_r h_\theta h_\phi}\left\{\frac{\partial}{\partial r}\left(\frac{h_\theta h_\phi}{h_r}\frac{\partial U}{\partial r}\right) + \frac{\partial}{\partial \theta}\left(\frac{h_\phi h_r}{h_\theta}\frac{\partial U}{\partial \theta}\right) + \frac{\partial}{\partial \phi}\left(\frac{h_r h_\theta}{h_\phi}\frac{\partial U}{\partial \phi}\right)\right\}$$

$$\nabla^2 U(r, \theta, \phi) = \frac{1}{r(c + r \sin \phi)} \left\{ \frac{\partial}{\partial r} \left[r(c + r \sin \theta) \frac{\partial U}{\partial r} \right] \right.$$

$$\left. + \frac{r}{c + r \sin \phi} \frac{\partial^2 U}{\partial \theta^2} + \frac{1}{r} \frac{\partial}{\partial \phi} \left[(c + r \sin \phi) \frac{\partial U}{\partial \phi} \right] \right\}.$$

(N.B. An alternative treatment for establishing the orthogonality of the torus system is found in Ref. 4. Here the scale factors are derived geometrically and used to investigate the geometry of the torus.)

The above example illustrates the more general result that, if for a curvilinear system (u_1, u_2, u_3), $\mathrm{d}s = h_1 \mathrm{d}u_1 \hat{\mathbf{a}}_1 + h_2 \mathrm{d}u_2 \hat{\mathbf{a}}_2 + h_3 \mathrm{d}u_3 \hat{\mathbf{a}}_3$, and if we can prove that $(\mathrm{d}s)^2 = (h_1 \ \mathrm{d}u_1)^2 + (h_2 \ \mathrm{d}u_2)^2 + (h_3 \ \mathrm{d}u_3)^2$, then $\hat{\mathbf{a}}_2 \cdot \hat{\mathbf{a}}_3 = 0$, etc., thereby establishing orthogonality. In fact the absence of terms in $\mathrm{d}u_2 \mathrm{d}u_3$, etc., from the quadratic form for $(\mathrm{d}s)^2$ is a necessary and sufficient condition for (u_1, u_2, u_3) to be an orthogonal system.

1.21 BRIEF OUTLINE OF CARTESIAN TENSOR ANALYSIS

Resolution of vectorial formulae obtained in various branches of mechanics and theoretical physics into scalar component forms corresponding to the (x, y, z) rectangular Cartesian coordinates often leads to protracted forms which are tedious to read and write. An abridged notation is therefore often employed to obviate these difficulties. This notation has the additional advantage of permitting certain results to be obtained more simply: it is especially advantageous in the theoretical treatment of viscous fluids in motion.

In the abridged system, rectangular Cartesian coordinates in R_3 are specified by x_i $(i = 1, 2, 3)$ which are related to the usual unabridged coordinates (x, y, z) by $(x_1, x_2, x_3) = (x, y, z)$. The coordinate axes for the abridged system are specified by OX_i and the unit vectors in their positive directions by $\hat{\mathbf{e}}_i$ $(i = 1, 2, 3)$. The system is so labelled as to form a right-handed frame so that

$$\begin{aligned} &\hat{\mathbf{e}}_i \cdot \hat{\mathbf{e}}_j = \delta_{ij} \ (i, j = 1, 2, 3) \\ &\hat{\mathbf{e}}_1 \times \hat{\mathbf{e}}_2 = \hat{\mathbf{e}}_3, \text{etc.} \end{aligned} \quad , \qquad (1.112)$$

where δ_{ij}, known as the *Kronecker delta*, is unity when $j = i$ and is zero when $j \neq i$ (e.g. $\delta_{11} = 1, \delta_{23} = 0$).

The scalar components $[a_1, a_2, a_3]$ of a vector $\mathbf{a}$ are often denoted by a_i or by a_j. The i and j are *free subscripts* and each may assume the values 1, 2, 3. In fact it is quite common to regard a_i as the vector $\mathbf{a}$ itself in the abridged system. If b_i denote the scalar components of another vector $\mathbf{b}$ then

$$\mathbf{a} \cdot \mathbf{b} = a_1 b_1 + a_2 b_2 + a_3 b_3 = a_i b_i.$$

On the far right the once-repeated literal subscript i appears and the *summation convention* is used. According to this convention a *literal subscript may be either free or repeated once only: in the latter case the repeated letter sums the quantity with respect to itself over the values 1, 2, 3*. The quantity $\mathbf{a} \cdot \mathbf{b}$ is quite independent of i and so i is a dummy subscript which may be replaced by any other letter so that, for example, $\mathbf{a} \cdot \mathbf{b} = a_j b_j = a_k b_k$.

Extension of the convention takes place readily. For example in the expression $a_{ij}x_j$, the unrepeated subscript i is therefore free, but j is a repeated dummy subscript and is therefore summed. We first sum with respect to j from $j = 1$ to 3 to give $a_{i1}x_1 + a_{i2}x_2 + a_{i3}x_3$. Then we allow i to range in turn through the set of values 1, 2, 3 thereby generating the three sums

$$\left.\begin{aligned} &a_{11}x_1 + a_{12}x_2 + a_{13}x_3, \\ &a_{21}x_1 + a_{22}x_2 + a_{23}x_3, \\ &a_{31}x_1 + a_{32}x_2 + a_{33}x_3. \end{aligned}\right\}$$

Already the advantage of writing these sums tersely as $a_{ij}x_j$ is realised. Since j is dummy, the form may also be written as $a_{ik}x_k$ but the free i must be maintained and cannot also be used as the dummy.

Further examples illustrating the summation convention are:

(i) $$a_{ij}x_ix_j = \sum_{i=1}^{3}\sum_{j=1}^{3} a_{ij}x_ix_j \qquad \text{(9 terms)};$$

(ii) $$a_{ijk}x_ix_jx_k = \sum_{i=1}^{3}\sum_{j=1}^{3}\sum_{k=1}^{3} a_{ijk}x_ix_jx_k \qquad \text{(27 terms)}.$$

For the Kronecker delta defined above, we note that

$$\delta_{ii} = \delta_{11} + \delta_{22} + \delta_{33} = 3,$$

since $\delta_{11} = \delta_{22} = \delta_{33} = 1$. Also,

$$\delta_{ij}x_j = \delta_{i1}x_1 + \delta_{i2}x_2 + \delta_{i3}x_3 \qquad (i = 1, 2, 3).$$

When $i = 1$, the right-hand side is $\delta_{11}x_1 + \delta_{12}x_2 + \delta_{13}x_3 = x_1$. Similarly $i = 2$ gives the value x_2 to the right-handed side and $i = 3$ gives the value x_3. Thus we obtain

$$\delta_{ij}x_j = x_i, \tag{1.113}$$

a result which is easily remembered: the action of the δ_{ij} on x_j is merely to substitute the free i for the dummy j in the x_j. This is part of the reason why δ_{ij} is often called the 'substitution tensor'.

So far we have deliberated on the summation convention affording a terse form for protracted notations involving the unabridged (x, y, z) system of rectangular coordinates. An important conception which we now develop is that of *orthogonal rotation of coordinate axes*.

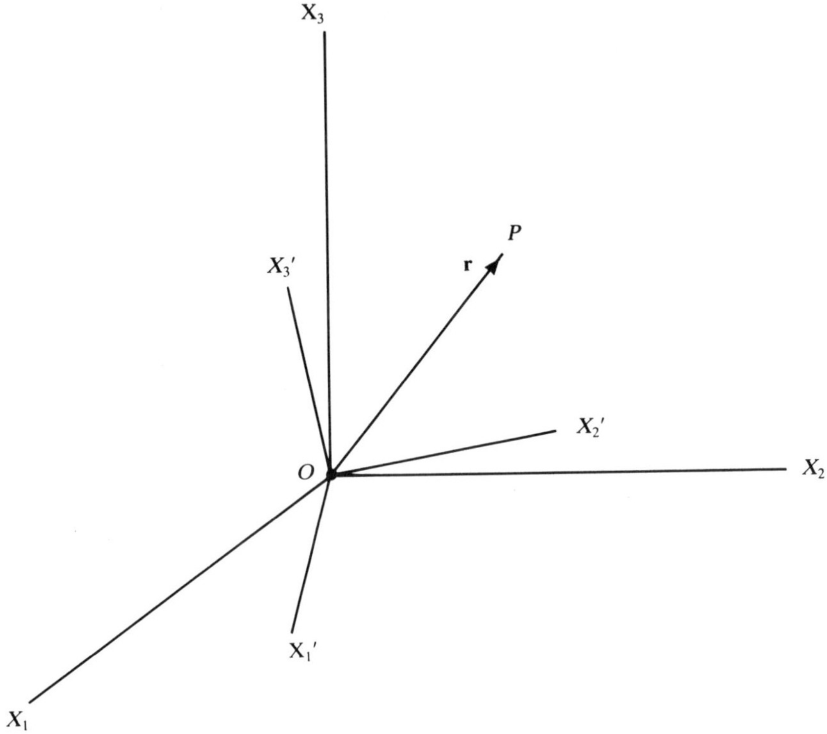

Fig. 1.25

In Fig. 1.25, the axes OX_i $(i = 1, 2, 3)$ form a right-handed orthogonal coordinate frame. This frame is rotated about O, preserving right-handed and orthogonal features, to the new position OX'_i $(i = 1, 2, 3)$. We specify the direction cosines of the new axes in terms of the old ones as exhibited in the following table.

	OX_1	OX_2	OX_3
OX'_1	l_{11}	l_{12}	l_{13}
OX'_2	l_{21}	l_{22}	l_{23}
OX'_3	l_{31}	l_{32}	l_{33}

The table shows that, for example, OX'_2 has direction cosines $[l_{21}, l_{22}, l_{23}]$ with respect to the former frame of unprimed axes whilst OX_3 has direction cosines $[l_{13}, l_{23}, l_{33}]$ with respect to the new frame of primed axes. We observe that from the entries in the horizontal rows we obtain six equations of the form

$$l_{11}^2 + l_{12}^2 + l_{13}^2 = 1, \text{etc.};$$

$$l_{11}l_{21} + l_{12}l_{22} + l_{13}l_{23} = 0, \text{etc.}$$

These six equations may be tersely abbreviated by

$$l_{ij}\, l_{kj} = \delta_{ik}.$$

Similarly the entries in the vertical columns give the six equations

$$l_{11}^2 + l_{21}^2 + l_{31}^2 = 1, \text{etc.};$$

$$l_{11}l_{12} + l_{21}l_{22} + l_{31}l_{32} = 0, \text{etc.}$$

In abridged notation, these become simply

$$l_{ij}\, l_{ik} = \delta_{jk}.$$

The *orthonormal properties* of the direction cosines in the table may thus be expressed by the equations

$$l_{ij}\, l_{kj} = \delta_{ik} = l_{ji}\, l_{jk}. \tag{1.114}$$

The elements exhibited in the table of direction cosines form a 3×3 matrix **T** called the *transformation matrix*. It specifies the nature of the orthogonal rotation and it may be tersely written as

$$\mathbf{T} = [l_{ij}]. \tag{1.115}$$

The transpose of this matrix is

$$\mathbf{T}' = [l_{ji}]. \tag{1.116}$$

The inner product of the ith row of **T** and the jth column of $\mathbf{T}'$ is

$$l_{i1}l_{j1} + l_{i2}l_{j2} + l_{i3}l_{j3} = l_{ir}\, l_{jr} = \delta_{ij},$$

using the orthonormal properties listed. Thus

$$\mathbf{TT}' = [\delta_{ij}] = \mathbf{I}, \tag{1.117}$$

where **I** is the unit 3×3 matrix having each element in the principal diagonal unity and all other elements zero. Since $\det \mathbf{T}' = \det \mathbf{T}$,

$$(\det \mathbf{T})^2 = \det \mathbf{I} = 1$$

i.e. $$\det \mathbf{T} = \pm 1.$$

As $\det \mathbf{T} \neq 0$, we see that $\mathbf{T}^{-1}$, the inverse of **T**, exists and hence

$$\mathbf{T}' = \mathbf{T}^{-1}. \tag{1.118}$$

Thus the transpose of the transformation matrix is equal to the inverse of the matrix: such a matrix is said to be *orthogonal*.

We observe that

$$\det \mathbf{T} = [\hat{\mathbf{e}}_1', \hat{\mathbf{e}}_2', \hat{\mathbf{e}}_3'].$$

This scalar triple product is + 1 for a right-handed system and − 1 for a left-handed one. As we have stipulated a right-handed orthogonal transformation,

$$\det \mathbf{T} = +1. \tag{1.119}$$

In the diagram P is x_i or x_i' $(i = 1, 2, 3)$ and $\overline{OP} \equiv \mathbf{r}$. Then

$$\mathbf{r} = x_i\,\hat{\mathbf{e}}_i = x_i'\,\hat{\mathbf{e}}_i'. \tag{1.120}$$

$$x_i' = \mathbf{r}\,.\,\hat{\mathbf{e}}_i' = x_j\,\hat{\mathbf{e}}_j\,.\,\hat{\mathbf{e}}_i' = l_{ij}x_j,$$

since $\hat{\mathbf{e}}_i'\,.\,\hat{\mathbf{e}}_j$ is the cosine of the angle between OX_i' and OX_j and from the pattern in the table this is l_{ij}. Likewise

$$x_i = \mathbf{r}\,.\,\hat{\mathbf{e}}_i = x_j'\,\hat{\mathbf{e}}_j'\,.\,\hat{\mathbf{e}}_i = l_{ji}\,x_j'.$$

Thus we have found that the unprimed coordinates x_j under orthogonal transformation become the primed ones x_i' where

$$x_i' = l_{ij}\,x_j \tag{1.121}$$

and the primed ones transform into the unprimed ones according to

$$x_i = l_{ji}\,x_j'. \tag{1.122}$$

Mathematically these transformations are subject to (1.114) which expresses the orthogonality of the transformation.

From (1.121), we have

$$x_i'x_i' = (l_{ij}\,x_j)\,(l_{ik}\,x_k),$$

where a dummy subscript has been replaced by another letter to prevent that first letter from being repeated more than once. Thus

$$\begin{aligned} x_i'x_i' &= (l_{ij}\,l_{ik})\,(x_jx_k) \\ &= \delta_{jk}(x_jx_k) \qquad \text{(using orthogonality conditions)} \\ &= x_kx_k \qquad \text{(using } \delta\text{-substitution properties).} \end{aligned} \tag{1.114}$$

In full this means

$$x_1'^2 + x_2'^2 + x_3'^2 = x_1^2 + x_2^2 + x_3^2. \tag{1.123}$$

Either side of (1.123) expresses OP^2; this obvious result says that OP^2, and hence also OP, are invariant under the orthogonal transformation. Although the result obtained seems nugatory, it nevertheless illustrates very well at an elementary level the methods of manipulating systems which we shall define to be tensors.

The above models permit us to define Cartesian tensors of orders unity and zero.

The quantities u_i $(i = 1, 2, 3)$ are said to form the components of a *first-order tensor (or vector)* if under a right-handed orthogonal transformation these transform into quantities u_i' where the u_i' satisfy the *law of transformation*

$$u_i' = l_{ir}\, u_r \tag{1.124}$$

subject to the orthonormal relations

$$l_{ri} l_{si} = \delta_{rs} = l_{ir} l_{is}. \tag{1.125}$$

Essentially the components x_i of **r** which are considered satisfy this law of transformation of a first-order tensor.

If, however, a quantity remains invariant under the orthogonal transformation (1.10) then it is called a *scalar* or a *tensor of order zero*. Such cases are afforded by the magnitude of a distance between two fixed points, mass, energy, etc. As an exercise let us show that the scalar product of two vectors $\mathbf{a} = a_i \hat{\mathbf{e}}_i$ and $\mathbf{b} = b_i \hat{\mathbf{e}}_i$ is a tensor of zero order. First, under orthogonal transformation, a_i transforms into

$$a_i' = l_{ir}\, a_r,$$

subject to (1.125). Similarly the b_i transform into

$$b_i' = l_{ir}\, b_r.$$

Avoiding the repetition of a subscript more than once,

$$\begin{aligned} a_i' b_i' &= (l_{ir}\, a_r)\,(l_{is}\, b_s), \\ &= (l_{ir}\, l_{is})\,(a_r b_s) \\ &= \delta_{rs} a_r b_s \\ &= a_s b_s \end{aligned}$$

i.e. $a_1' b_1' + a_2' b_2' + a_3' b_3' = a_1 b_1 + a_2 b_2 + a_3 b_3$, as is required.

At this stage it is convenient to define a *tensor of order (or rank) n*. Let $u_{ijk} \ldots$ (n subscripts) be a quantity which under orthogonal transformation (right-handed) becomes $u'_{pqr} \ldots$ (n subscripts) in accordance with the law of transformation

$$u'_{pqr} \ldots = l_{pi}\, l_{qj}\, l_{rk} \ldots u_{ijk} \ldots \tag{1.126}$$

(involving n of the l's on the right) subject to the orthogonality condition

$$l_{ri}\, l_{si} = \delta_{rs} = l_{ir}\, l_{is}.$$

The $u_{ijk} \ldots$ is said to be a *tensor of order n*. If (1.126) and (1.125) hold, then we can prove

$$u_{pqr} \ldots = l_{ip}\, l_{jq}\, l_{kr} \ldots u'_{ijk} \ldots \tag{1.127}$$

For taking up the right-hand side of (1.127),

$$\begin{aligned} & l_{ip}\, l_{jq}\, l_{kr} \ldots u'_{ijk} \ldots \\ &= l_{ip}\, l_{jq}\, l_{kr} \ldots (l_{i\alpha}\, l_{j\beta}\, l_{k\gamma} \ldots u_{\alpha\beta\gamma} \ldots) \\ &= (l_{ip}\, l_{i\alpha})\,(l_{jq}\, l_{j\beta})\,(l_{kr}\, l_{k\gamma}) \ldots u_{\alpha\beta\gamma} \ldots \end{aligned}$$

$$= \delta_{p\alpha}\,\delta_{q\beta}\,\delta_{r\gamma}\,\ldots u_{\alpha\beta\gamma}\,\ldots$$

$$= u_{pqr}\,\ldots, \text{ the left-hand side of (1.127)}$$

In the form (1.126) there are n free subscripts on either side, namely p, q, r, ... The right-hand side contains additionally n dummy subscripts, namely i, j, k, ... Any tensor equation should always be checked for such consistencies. If to the nth-order tensor $u_{ijk}\ldots$ we apply δ_{ij} then

$$\delta_{ij}u_{ijk}\,\ldots = u_{iik}\,\ldots$$

using the substitution properties of the Kronecker delta. In $u_{iik}\ldots$ there are now only $(n-2)$ free subscripts and in fact

$$u_{iik}\,\ldots = u_{11k\ldots} + u_{22k\ldots} + u_{33k\ldots}\,.$$

We prove that this is in fact a tensor of order $(n-2)$. The proof requires showing that it satisfies the law of transformation of a Cartesian tensor of order $(n-2)$. The proof is established by putting $q = p$ in (1.126) to give

$$u'_{ppr}\,\ldots = (l_{pi}\,l_{pj})\,l_{rk}\,\ldots u_{ijk}\,\ldots$$

$$= \delta_{ij}\,l_{rk}\,\ldots u_{ijk}\,\ldots$$

$$= l_{rk}\,\ldots u_{iik}\,\ldots$$

Thus we obtain

$$u'_{ppr}\,\ldots = l_{rk}\,\ldots u_{iik}\,\ldots$$

which involves the $(n-2)$ free subscripts $r, \ldots$ on either side and is subject to the orthonormal relations (1.125). The combination of the two forms constitutes the law of transformation of a tensor of order $(n-2)$ and so $u_{iik}\ldots$ is such a tensor. This process of producing from the nth order tensor $u_{ijk}\ldots$ a tensor of order $(n-2)$ by forming $\delta_{ij}u_{ijk}\ldots$ or $u_{iik}\ldots$ is called *contraction*. Such a process applied to a second-order tensor u_{ij} gives $\delta_{ij}u_{ij} = u_{ii}$, which is a scalar or tensor of order zero.

Example [1.10]

Associated with any deformable (elastic) medium are two second-order tensors: e_{ij} the strain tensor and τ_{ij} the stress tensor. These tensors are known to be related through

$$\tau_{ij} = c_{ijrs}\,e_{rs} \qquad (i, j, r, s = 1, 2, 3),$$

which is a generalised form of Hooke's law. It is required to prove that c_{ijrs} is a Cartesian tensor of order four.

We have to show that under an orthogonal transformation specified by

$$l_{ri}\,l_{si} = \delta_{rs} = l_{ir}\,l_{is},$$

the c_{ijrs} will obey the law of transformation of a fourth-order tensor.

Since the stress and strain tensors are both of the second order, their laws of transformation are expressed by

$$\tau'_{pq} = l_{pi}\, l_{qj}\, \tau_{ij},$$

$$e'_{\alpha\beta} - l_{\alpha r}\, l_{\beta s}\, e_{rs},$$

both being subject to the above orthogonal conditions. From the last result,

$$e_{rs} = l_{\alpha r}\, l_{\beta s}\, e'_{\alpha\beta},$$

in virtue of equations (1.125), (1.126), (1.127) above. Invoking the given form of Hooke's law

$$\begin{aligned} \tau'_{pq} &= l_{pi}\, l_{qj}\, (c_{ijrs}\, e_{rs}) \\ &= l_{pi}\, l_{qj}\, (c_{ijrs}\, l_{\alpha r} l_{\beta s} e'_{\alpha\beta}) \\ &= (l_{pi}\, l_{qj}\, l_{\alpha r}\, l_{\beta s}\, c_{ijrs}) e'_{\alpha\beta} \\ &= c'_{pq\alpha\beta}\, e'_{\alpha\beta}, \end{aligned}$$

where

$$c'_{pq\alpha\beta} = l_{pi}\, l_{qj}\, l_{\alpha r}\, l_{\beta s}\, c_{ijrs}.$$

This, subject to the orthonormal relations, expresses the law of transformation of a fourth-order tensor and so c_{ijrs} is such a tensor.

1.22 PROPERTIES OF SECOND-ORDER TENSORS

We now establish some of the more important properties of second-order tensors. In the following a_{ij} $(i, j = 1, 2, 3)$ denote the components of a second-order tensor, i.e. they obey the law of transformation,

$$a'_{rs} = l_{ri}\, l_{sj}\, a_{ij} \tag{1.128}$$

subject to the orthogonal relations

$$l_{ri}\, l_{si} = \delta_{rs} = l_{ir}\, l_{is} \tag{1.129}$$

which hold under an orthogonal transformation. Then, as we have already seen for tensors of general order n, from (1.128) and (1.129) we can easily show

$$a_{rs} = l_{ir}\, l_{js}\, a'_{ij}. \tag{1.130}$$

Of some importance is the fact that the nine components a_{ij} can be arranged as the 3×3 matrix $[a_{ij}]$ where

$$[a_{ij}] = \begin{bmatrix} a_{11} & a_{12} & a_{13} \\ a_{21} & a_{22} & a_{23} \\ a_{31} & a_{32} & a_{33} \end{bmatrix}. \tag{1.131}$$

Since a_{ij} is a second-order tensor contraction shows that a_{ii} is a scalar (tensor of zero order). This means that under orthogonal transformation the *trace or sum of the principal diagonal elements of the matrix (1.131) is constant.* For the specific case of a stress matrix (with $a_{ij} = \tau_{ij}$), this leads to the important result that the sum of the direct stresses $\tau_{11} + \tau_{22} + \tau_{33}$ at any point P of a deformable medium is invariant for any system of orthogonal axes OX_i ($i = 1, 2, 3$).

If it happens that $a_{ij} = a_{ji} (i, j = 1, 2, 3)$, then a_{ij} is called a *symmetric tensor of the second order*. The corresponding matrix will be diagonally symmetric. If, however, $a_{ij} = -a_{ji}$, then a_{ij} is called a *skew symmetric tensor of the second order*. In general a second-order tensor will be neither symmetric nor skew symmetric, but it can always be expressed as the sum of a symmetric and a skew symmetric tensor of the second order simply by putting

$$a_{ij} = e_{ij} + \omega_{ij}, \tag{1.132}$$

where

$$e_{ij} = \tfrac{1}{2}(a_{ij} + a_{ji}), \tag{1.133}$$

$$\omega_{ij} = \tfrac{1}{2}(a_{ij} - a_{ji}) \tag{1.134}$$

and so $e_{ij} = e_{ji}$, but $\omega_{ij} = -\omega_{ji}$. We need, however, to show that e_{ij} and ω_{ij} both obey the transformation laws of second-order tensors. This is achieved simply by showing that $\lambda a_{ij} + \mu a_{ji}$ is a second-order tensor. Writing

$$b_{ij} = \lambda a_{ij} + \mu a_{ji},$$

we have

$$\begin{aligned} l_{ri}\, l_{sj}\, b_{ij} &= \lambda l_{ri}\, l_{sj}\, a_{ij} + \mu l_{sj}\, l_{ri}\, a_{ji} \\ &= \lambda a'_{rs} + \mu a'_{sr}. \end{aligned}$$

This last expression depends solely on the free subscripts r and s and it may be designated by b'_{rs}. Then

$$b'_{rs} = l_{ri}\, l_{sj}\, b_{ij}.$$

This is the law of transformation of a second-order tensor and so b_{ij} is such a tensor for all λ, μ. The case $\lambda = \mu = 1$ establishes e_{ij} as a second-order tensor. This case $\lambda = 1, \mu = -1$ establishes the tensorial nature of ω_{ij}. Hence e_{ij} is a symmetric second-order tensor and ω_{ij} is a skew-symmetric second-order tensor.

Associated with a symmetric second-order tensor e_{ij} is its quadric sruface

$$e_{ij} x_i x_j = K. \tag{1.135}$$

Essentially the quantity on the left-hand side is a scalar. Using the symmetry property $e_{ij} = e_{ji}$, the long-hand form of (1.135) is

$$\begin{aligned} &e_{11} x_1^2 + e_{22} x_2^2 + e_{33} x_3^2 \\ &+ 2e_{23} x_2 x_3 + 2e_{31} x_3 x_1 + 2e_{12} x_1 x_2 = K. \end{aligned}$$

This is a central quadric having the origin as its centre, since the equation is quite devoid of first degree terms in the coordinates. The lengths of the semi-axes of the quadric are found from the roots of the determinantal equation

$$\det(e_{ij} - \lambda\delta_{ij}) = 0 \tag{1.136}$$

or

$$\begin{vmatrix} e_{11} - \lambda & e_{12} & e_{13} \\ e_{21} & e_{22} - \lambda & e_{23} \\ e_{31} & e_{32} & e_{33} - \lambda \end{vmatrix} = 0,$$

which is a cubic equation in λ. The directions of the three principal axes are given by the ratios $X : Y : Z$ where

$$\mathbf{E}\,\mathbf{V} = \lambda\mathbf{V}, \tag{1.137}$$

$\mathbf{E}$ being the matrix $[e_{ij}]$ and $\mathbf{V}$ the column vector $\{X, Y, Z\}$. To each of the three values of λ found from (1.136), a unique ratio $X : Y : Z$ is obtained, i.e. the direction of each principal axis is found. Two such directions will, in general, be orthogonal. For suppose two distinct roots of (1.136) are $\lambda = \lambda_1, \lambda_2$ and that the corresponding column vectors satisfying (1.137) are the column vectors $\mathbf{V}_1 = \{X_1, Y_1, Z_1\}, \mathbf{V}_2 = \{X_2, Y_2, Z_2\}$. Then

$$\mathbf{E}\,\mathbf{V}_1 = \lambda_1\mathbf{V}_1,$$
$$\mathbf{E}\,\mathbf{V}_2 = \lambda_2\mathbf{V}_2.$$

Denote by $\mathbf{V}_1' = [X_1, Y_1, Z_1], \mathbf{V}_2' = [X_2, Y_2, Z_2]$, the row vectors coresponding to the column vectors $\mathbf{V}_1, \mathbf{V}_2$ respectively. Then

$$\mathbf{V}_2'\,\mathbf{E}\,\mathbf{V}_1 = \lambda_1\mathbf{V}_2'\mathbf{V}_1,$$
$$\mathbf{V}_1'\,\mathbf{E}\,\mathbf{V}_2 = \lambda_2\mathbf{V}_1'\mathbf{V}_2.$$

If we take the transpose of the second equation, we obtain

$$\mathbf{V}_2'\,\mathbf{E}'\mathbf{V}_1'' = \lambda_2\mathbf{V}_2'\mathbf{V}_1''$$

or

$$\mathbf{V}_2'\,\mathbf{E}\mathbf{V}_1 = \lambda_2\mathbf{V}_2'\mathbf{V}_1.$$

For $\mathbf{E}$ is a symmetric $3 \times$ matrix for which the transpose $\mathbf{E}' = \mathbf{E}$. Also $\mathbf{V}_1'' = \mathbf{V}_1$, etc. Moreover

$$\mathbf{V}_1'\mathbf{V}_2 = \mathbf{V}_2'\mathbf{V}_1 = X_1X_2 + Y_1Y_2 + Z_1Z_2 = \mathbf{V}_1 \,.\, \mathbf{V}_2,$$

where $\mathbf{V}_1 = X_1\mathbf{i} + Y_1\mathbf{j} + Z_1\mathbf{k}$, etc. The two expressions for $\mathbf{V}_1'\mathbf{E}\,\mathbf{V}_1'$ show that

$$\lambda_1\mathbf{V}_1 \,.\, \mathbf{V}_2 = \lambda_2\mathbf{V}_1 \,.\, \mathbf{V}_2,$$

i.e. if $\lambda_1 \neq \lambda_2$, then

$$\mathbf{V}_1 \,.\, \mathbf{V}_2 = 0,$$

which confirms that two principal axes are at right angles. If $\lambda_1, \lambda_2, \lambda_3$ are all unequal, then the corresponding three principal axes will be mutually perpendicular.

The quantities λ_1, λ_2, λ_3 are called the *eigenvalues* and the corresponding vectors are the *eigenvectors* of the 3×3 symmetric matrix $[e_{ij}] = E$.

Example [1.11] Distinct eigenvalues

Find the eigenvalues and eigenvectors of the 3×3 symmetric matrix

$$\mathbf{E} = \begin{bmatrix} -2 & 5 & 4 \\ 5 & 7 & 5 \\ 4 & 5 & 2 \end{bmatrix}.$$

The eigenvalues are simply found as the roots of the determinantal equation in λ:

$$\det(\mathbf{E} - \lambda \mathbf{I}) = 0.$$

This is a cubic equation in λ given by

$$\begin{vmatrix} -2-\lambda & 5 & 4 \\ 5 & 7-\lambda & 5 \\ 4 & 5 & -2-\lambda \end{vmatrix} = 0.$$

This reduces to

$$\lambda^3 - 3\lambda^2 - 90\lambda - 216 = 0.$$

Inspection shows that a root of this is $\lambda = -3$ and so $\lambda + 3$ is a linear factor and we obtain

$$(\lambda + 3)(\lambda^2 - 6\lambda - 72) = 0$$

or

$$(\lambda + 3)(\lambda + 6)(\lambda - 12) = 0.$$

Thus the eigenvalues are

$$\underline{\lambda = -6, -3, +12.}$$

The corresponding eigenvectors are found from

$$\begin{bmatrix} -2-\lambda & 5 & 4 \\ 5 & 7-\lambda & 5 \\ 4 & 5 & -2-\lambda \end{bmatrix} \begin{bmatrix} X \\ Y \\ Z \end{bmatrix} = \begin{bmatrix} 0 \\ 0 \\ 0 \end{bmatrix},$$

where λ has to be taken (in turn) to have each of the above values. We have the three homogeneous equations

$$\begin{aligned} -(2+\lambda)X + \quad 5Y \quad + \quad 4Z \quad &= 0, \\ 5X \quad + (7-\lambda)Y + \quad 5Z \quad &= 0, \\ 4X \quad + \quad 5Y \quad - (2+\lambda)Z &= 0. \end{aligned}$$

When we expand the above determinant, which as we have seen has zero value, across its top row we obtain

$$-(2+\lambda) \times \text{its cofactor} + 5 \times \text{its cofactor} + 4 \times \text{its cofactor} = 0.$$

From the determinant, the cofactors of $-(2+\lambda)$; 5; 4 are respectively

$$\lambda^2 - 5\lambda - 39; \quad 5\lambda + 30; \quad 4\lambda - 3.$$

Since

$$-(2+\lambda)X + 5Y + 4Z = 0,$$

and it follows that

$$\frac{X}{\lambda^2 - 5\lambda - 39} = \frac{Y}{5\lambda + 30} = \frac{Z}{4\lambda - 3}.$$

Case I: $\lambda = -6$.

$$X/27 = Y/0 = Z/(-27)$$

or

$$X/1 = Y/0 = Z/(-1).$$

We can only find the ratio $X : Y : Z$ and not absolute values of X, Y, Z. However, this suffices for fixing the direction of the eigenvector. The eigenvector, then, corresponding to $\lambda = -6$ may be taken as the column vector $\{1, 0, -1\}$, or any arbitrary scalar multiple of this.

Case II: $\lambda = -3$.

$$X/(-15) = Y/(15) = Z(-15).$$

The simplest specification of the eigenvector is the column vector $\{1, -1, 1\}$.

Case III: $\lambda = 12$.

$$X/45 = Y/90 = Z/45.$$

The eigenvector may be taken as $\{1, 2, 1\}$.
If we write

$$\mathbf{V}_1 = \mathbf{i} - \mathbf{k}, \qquad \mathbf{V}_2 = \mathbf{i} - \mathbf{j} + \mathbf{k}, \qquad \mathbf{V}_3 = \mathbf{i} + 2\mathbf{j} + \mathbf{k},$$

then we find $\mathbf{V}_1 . \mathbf{V}_2 = \mathbf{V}_2 . \mathbf{V}_3 = \mathbf{V}_3 . \mathbf{V}_1 = 0$ and as each vector is non-zero we have established mutual orthogonality. This is a check on the above.

The above evaluations would, of course, determine the lengths and directions of the principal axes of the quadric

$$\mathbf{V}'\,\mathbf{E}\,\mathbf{V} = K$$

or

$$-2X^2 + 7Y^2 - 2Z^2 + 10YZ + 8ZX + 10XZ = K.$$

Example [1.12] Two equal eigenvalues.

Let us consider

$$\mathbf{E} = \begin{bmatrix} 2 & -1 & 1 \\ -1 & 2 & -1 \\ 1 & -1 & 2 \end{bmatrix}$$

The routine gives the values

$$\lambda = 1,\ 1,\ \text{or } 4.$$

The cofactors in $\det(E - \lambda I)$ are found to be

$$(\lambda - 1)(\lambda - 3); \qquad -(\lambda - 1); \qquad (\lambda - 1).$$

These are in the proportion $X : Y : Z$ and so

$$\frac{X}{\lambda - 3} = \frac{Y}{-1} = \frac{Z}{1}.$$

When $\lambda = 1$,

$$X/(-2) = Y/(-1) = Z/(1).$$

When $\lambda = 4$

$$X/1 = Y/(-1) = Z/1.$$

Thus appropriate eigenvectors are $\{2, 1, -1\}$ and $\{1, -1, 1\}$.
A third eigenvector is found from

$$(2\mathbf{i} + \mathbf{j} - \mathbf{k}) \times (\mathbf{i} - \mathbf{j} + \mathbf{k}) = -3(\mathbf{j} + \mathbf{k}).$$

Thus a third eigenvector is $\{0, 1, 1\}$.

As a check on this third eigenvector, we have

$$\begin{bmatrix} 2 & -1 & 1 \\ -1 & 2 & -1 \\ 1 & -1 & 2 \end{bmatrix} \begin{bmatrix} 0 \\ 1 \\ 1 \end{bmatrix} = \begin{bmatrix} 0 \\ 1 \\ 1 \end{bmatrix}$$

which also confirms that the corresponding $\lambda = 1$.

The reader can easily confirm the mutual orthogonality of all three eigenvectors.

The two equal eigenvectors means that the conicoid is a surface of revolution with two equal principal axes ($\lambda = 1, 1$).

PROBLEMS 1

1. Two adjacent sides of a parallelogram represent the vectors **a**, **b**. If **p** = **a** + **b**, **q** = **a** − **b**, use a geometrical construction to show that **p** + **q** = 2**a**, **p** − **q** = 2**b**.

2. P_1 is $(-3, -1, 0)$ and P_2 is $(-1, -4, 6)$. Find (i) the distance P_1P_2; (ii) the direction cosines of $\overrightarrow{P_1P_2}$; (iii) the coordinates of the points dividing P_1P_2 internally and externally in the ration 2:1; (iv) the coordinates of the point of intersection of P_1P_2 with the coordinate plane YOZ.

3. Show that the three points having position vectors **i** + 2**j** − 2**k**, 2**i** + **j** + 2**k**, 4**i** − **j** + 10**k** referred to an origin O are collinear.

4. Given the points $A(2, 4, 3)$; $B(4, 1, 19)$, $C(10, -1, 6)$, show that the triangle ABC is isosceles and right-angled.

5. Evaluate (**a** − **b**) . (**b** − **c**) where **a**, **b**, **c** are unit vectors each making 60° with one another.

6. **a** and **b** are constant vectors and **a** cos θ + **b** sin θ is a unit vector for all values of θ. Show that **a** and **b** are perpendicular unit vectors.

7. Find the direction cosines of the line of intersection of the planes

$$3x - y + z + 1 = 0; \quad 5x + y + 3z = 0$$

and obtain the equation of the plane perpendicular to this line and passing through (2, 1, 4).

8. Prove that the line

$$(x-3)/2 = (y-4)/3 = (z-5)/4$$

is parallel to the plane $4x + 4y - 5z = 14$.

9. If $(\mathbf{c} - \frac{1}{2}\mathbf{a}) . \mathbf{a} = 0 = (\mathbf{c} - \frac{1}{2}\mathbf{b}) . \mathbf{b}$, prove that in general $\mathbf{c} - \frac{1}{2}(\mathbf{a} + \mathbf{b})$ is perpendicular to $\mathbf{a} - \mathbf{b}$, the exceptional cases being (i) $\mathbf{a} = \mathbf{b}$; (ii) $\mathbf{c} = \frac{1}{2}(\mathbf{a} + \mathbf{b})$.

10. Prove that the centroid of equal point-masses at the vertices of a triangle is at the point of intersection of its medians.

11. If $\mathbf{a} + \mathbf{b} + \mathbf{c} = \mathbf{0}$, prove $\mathbf{b} \times \mathbf{c} = \mathbf{c} \times \mathbf{a} = \mathbf{a} \times \mathbf{b}$ and hence deduce the sine rule for a triangle. Establish that, for a triangle ABC, the given equation also embodies the cosine rule and such results as $a = b \cos C + c \cos B$, etc.

12. If $\mathbf{n}$ is the unit vector along the internal bisector of the angle between two unit vectors $\mathbf{d}$ and $\mathbf{d}'$ localized at a point, prove that $2(\mathbf{n}\,.\,\mathbf{d})\mathbf{n} = \mathbf{d} + \mathbf{d}'$.

Hence prove that a ray of light emerges parallel to itself and of reversed direction after successive reflection in each of three mutually perpendicular plane mirrors.

Establish the identities in Problems 13–19 inclusive.

13. $\mathbf{a} \times (\mathbf{b} \times \mathbf{c}) + \mathbf{b} \times (\mathbf{c} \times \mathbf{a}) + \mathbf{c} \times (\mathbf{a} \times \mathbf{b}) = \mathbf{0}$.

14. $\mathbf{c} \times \{(\mathbf{b} \times \mathbf{c}) \times (\mathbf{c} \times \mathbf{a})\} = \mathbf{0}$.

15. $(\mathbf{a} \times \mathbf{b})^2 = a^2 b^2 - (\mathbf{a}\,.\,\mathbf{b})^2$.

16. $(\mathbf{b} \times \mathbf{c}) \times (\mathbf{c} \times \mathbf{d}) = [\mathbf{b}, \mathbf{c}, \mathbf{d}]\,\mathbf{c}$.

17. $[(\mathbf{a} \times \mathbf{b}), (\mathbf{b} \times \mathbf{c}), (\mathbf{c} \times \mathbf{a})] = [\mathbf{a}, \mathbf{b}, \mathbf{c}]^2$.

18. $[(\mathbf{a} - \mathbf{b}), (\mathbf{b} - \mathbf{c}), (\mathbf{c} - \mathbf{a})] = 0$.

19. $[(\mathbf{a} \times \mathbf{b}), (\mathbf{b} \times \mathbf{c}), (\mathbf{c} \times \mathbf{d})] = [\mathbf{b}, \mathbf{c}, \mathbf{d}]\ [\mathbf{a}, \mathbf{b}, \mathbf{c}]$.

20. If $\mathbf{V}(t)$ is a non-zero differentiable vector function of the scalar parameter t and if it is of constant magnitude show that $\mathbf{V}$ and $\mathrm{d}\mathbf{V}/\mathrm{d}t$ are perpendicular. (Differentiate the relation $\mathbf{V}^2 = $ const. w.r.t. t.)

21. Given that $\mathbf{r} = \mathbf{a} \cos nt + \mathbf{b} \sin nt$, where $\mathbf{a}$ and $\mathbf{b}$ are constant vectors, n a constant scalar, t a variable scalar, prove that

(i) $\ddot{\mathbf{r}} + n^2 \mathbf{r} = \mathbf{0}$; (ii) $\mathbf{r} \times \dot{\mathbf{r}} = n\,\mathbf{a} \times \mathbf{b}$.

22. Given that $\mathbf{x} + \mathbf{y} + \mathbf{z} = \mathbf{d}$, where $\mathbf{x}, \mathbf{y}, \mathbf{z}$ are each differentiable functions of a scalar t and $\mathbf{d}$ is a constant vector, prove

(i) $[\dot{\mathbf{x}}, \dot{\mathbf{y}}, \dot{\mathbf{z}}] = 0$; (ii) $(\dot{\mathbf{x}} \times \dot{\mathbf{y}}) \times (\dot{\mathbf{y}} \times \dot{\mathbf{z}}) = \mathbf{0}$.

23. Given $\mathbf{a}$, $\mathbf{b}$, $\mathbf{c}$ in that order form a right-handed set of mutually perpendicular unit vectors which are differentiable functions of a scalar t, prove $[\dot{\mathbf{a}}, \dot{\mathbf{b}}, \dot{\mathbf{c}}] = 0$.

24. $\boldsymbol{\omega}, \mathbf{r}_1, \mathbf{r}_2$ are non-zero differentiable vector functions of a scalar t and

$$\boldsymbol{\omega} \times \mathbf{r}_1 = \dot{\mathbf{r}}_1, \quad \boldsymbol{\omega} \times \mathbf{r}_2 = \dot{\mathbf{r}}_2.$$

Show that $\mathbf{r}_1$ and $\mathbf{r}_2$ have constant magnitudes and that the angle $(\mathbf{r}_1 ; \mathbf{r}_2)$ is also constant. Prove that if $\dot{\mathbf{r}}_1$ is perpendicular to $\mathbf{r}_2$ then $\dot{\mathbf{r}}_2$ is also perpendicular to $\mathbf{r}_1$ and show that in this case $\boldsymbol{\omega}$ may be expressed in the form $\boldsymbol{\omega} = \lambda \mathbf{r}_1 + \mu \mathbf{r}_2$, where λ, μ are given by $\lambda \mathbf{r}_1 \times \mathbf{r}_2 = \dot{\mathbf{r}}_2$, $\mu \mathbf{r}_2 \times \mathbf{r}_1 = \dot{\mathbf{r}}_1$ provided $\mathbf{r}_2 \neq \mathbf{r}_1$.

25. Given $\phi = xy + yz + zx$, find grad ϕ at the point $(1, 2, 3)$. Determine also the directional derivative of ϕ at the same point in the direction towards the point $(3, 4, 4)$.

26. At a point P having position vector $\mathbf{r}$ referred to the origin O a scalar ϕ is defined by $\phi = f(r)$ where $r = |\mathbf{r}|$. Evaluate $\nabla\phi$ at points where both $r = 1$ and $f'(r) = 1$. Obtain the directional derivative of ϕ at the same points inclined at $120°$ to $\nabla\phi$.

27. A uniform differentiable scalar function ϕ is of the form $\phi(R)$ where R is the distance of the field point from a fixed line. Find grad ϕ in simplest form.

28. Find $\phi(r)$ such that $\nabla\phi = \mathbf{r}/r^3$ where $r = |\mathbf{r}| \neq 0$ and $\phi = \phi_0$ when $\dot{\mathbf{r}} = r_0$.

29. Evaluate grad $\left(\dfrac{\mathbf{a}\,.\,\mathbf{r}}{r}\right)$ in terms of $\mathbf{a}$, $\mathbf{r}$, r where $\mathbf{a}$ is a constant vector, $\mathbf{r} = [x, y, z]$, $r = |\mathbf{r}|$.

30. Given $\phi = x^2 + y^2$, $\psi = z^2$, prove that $\nabla\phi . \nabla\psi = 0$. Comment on the nature of the relationship of the level surfaces $\phi =$ const., $\psi =$ const. Prove that if at a point $P(x, y, z)$ the directional derivatives of ϕ and ψ in the direction of the unit vector $[l, m, n]$ are equal, then the locus of P is a plane through the origin.

31. It is known that the gravitational potential ϕ at a point P due to a uniform rod AB, of mass m per unit length and of length $2l$, is given by

$$\phi = -\gamma m \log\left(\frac{r_1 + r_2 + l}{r_1 + r_2 - l}\right) \quad (\gamma,\ m,\ l \text{ constants})$$

where $AP = r_1$, $BP = r_2$. What are the equipotentials, i.e. level surfaces, for ϕ? Show that at P

$$\nabla\phi = \frac{2\gamma m l \cos\alpha}{a^2 - l^2}\,\mathbf{n}$$

where $r_1 + r_1 = 2a$, $A\hat{P}B = 2\alpha$ and $\mathbf{n}$ is the unit vector along the normal to the equipotential through P.

32. If $\mathbf{F}$ and ϕ are uniform differentiable vector and scalar point functions, prove that the components of $\mathbf{F}$ normal and tangential to a level surface of ϕ are

$$\frac{(\mathbf{F}\,.\,\nabla\phi)\nabla\phi}{(\nabla\phi)^2} \quad \text{and} \quad \frac{\nabla\phi \times (\mathbf{F} \times \nabla\phi)}{(\nabla\phi)^2}.$$

33. **F**, **G** are uniform differentiable vector functions and ϕ is a uniform differentiable scalar function in a region $\mathcal{R}$ given by $\nabla\phi = \mathbf{F} \times \mathbf{G}$. Prove that $d\phi = [d\mathbf{r}, \mathbf{F}, \mathbf{G}]$.

If the field-lines of **F** and those of **G** coincide in $\mathcal{R}$, show that ϕ = constant throughout $\mathcal{R}$.

If $\mathbf{F} \times \mathbf{G} \neq \mathbf{0}$, show that when the variations $d\mathbf{r}$ of the position vector **r** of a field-point P in $\mathcal{R}$ are taken in the surface through P containing the field-lines of both **F** and **G**, then ϕ is constant over any one such surface, but the constant differs in value from one such surface to another.

34. A right-circular cylinder of radius a has its axis along Oz and is bounded by the planes $z = c \pm h$. Evaluate the total flux of **F** over the surface of the cylinder including the ends when (i) $\mathbf{F} = x\mathbf{i} + y\mathbf{j}$; (ii) $\mathbf{F} = x\mathbf{i} + y\mathbf{j} + z\mathbf{k}$. In each case use the basic definition of divergence (limiting outward normal flux per unit volume) to find div **F** at $(0, 0, c)$.

35. Two concentric spherical surfaces of radii $a \pm h$ have their centres at the origin O and a pair of planes intersecting in a line through O and terminating at this line are inclined at an acute angle α to each other. The smaller volume V bounded by the spherical surfaces and the planes contains the point A where $\overrightarrow{OA} \equiv \mathbf{a}$. If there exists a vector field $\mathbf{F} = r^2\,\hat{\mathbf{r}}$ in V where $r = |\mathbf{r}|$ and **r** is the position vector of a field point P in V, show that the total outward normal flux of **F** across the boundary of V is $16\alpha ah(a^2 + h^2)$ and hence, deduce from the basic definition, that div $\mathbf{F} = 4a$ at A.

In the case when $\mathbf{F} = r^{-2}\,\hat{\mathbf{r}}$, show the total outward normal flux over the boundary of V is zero and hence div $\mathbf{F} = 0$ at A.

36. Two concentric spherical surfaces of radii $a \pm h$ have their centres at the origin O and a single cone whose solid angle is ω has its vertex at O. The volume V bounded by the two spherical surfaces and the surface of the cone contains the point P where $\overrightarrow{OP} \equiv \mathbf{a}$ and $|\mathbf{a}| = a$. If there is a vector field $\mathbf{F} = r^2\,\hat{\mathbf{r}}$ in V show that the total outward flux of **F** across its boundary is $8\omega ah(a^2 + h^2)$ and deduce that at P, div $\mathbf{F} = 4a$.

37. The surface of a sphere of radius a is cut into two parts by a plane at a distance $a/2$ from the centre. Determine the area of the smaller portion of the surface and find the solid angle it subtends at the centre of the sphere.

38. In plane polar coordinates (r, θ) the equation of a curve is $r = a \tan\theta$ and S is the area between the straight line $\theta = \pi/6$ and the arc of the curve from $\theta = 0$ to $\theta = \pi/6$. O is the pole and $OP\,(=a)$ is perpendicular to the plane of S. Find the solid angle subtended at P by S.

39. In polar coordinates the equation of a circle is $r = 3a/4$. O is the pole and $OP\ (=a)$ is perpendicular to the plane of the circle. Show that the area of the circle subtends a solid angle at P of $2\pi/5$.

40. Use the divergence theorem to find $\int_S \operatorname{grad}\phi \,.\, d\mathbf{S}$ where $\phi = x^4 + y^4 + z^4$ and S is the surface of the sphere $x^2 + y^2 + z^2 = a^2$.

41. Prove (i) $\nabla^2 e^r = e^r(1 + 2/r)$; (ii) $\nabla^4 e^r = \nabla^2(\nabla^2 e^r) = e^r(1 + 4/r)$.

42. Prove that a closed surface S contains a volume $\frac{1}{3}\int_S \mathbf{r} \,.\, d\mathbf{S}$ and use this formula to find the volume of a cone.

43. Prove that the function r^n is harmonic only for $n = 0, n = -1$.

44. Solve the differential equation $\operatorname{div}(\phi\mathbf{r}) = 0$ for ϕ if $\phi = \phi(r)$.

45. If $\phi = \phi(r)$, solve the differential equation $\nabla^2(\phi r) = 0$.

46. Given the vector field $\mathbf{F} = r\mathbf{i}$ where $r = |\mathbf{r}|$ and $\mathbf{r}$ is the position vector of (x, y, z) referred to the origin O, evaluate $\int_0^B \mathbf{F} \,.\, d\mathbf{r}$ from O to $B\,(1, 1, 0)$.

(a) along the path formed by the two straight line segments OA, OB;
(b) along the straight line OB,

where A is $(1, 0, 0)$.

47. If $\mathbf{F} = [yz, zx, xy]$, evaluate $\int \mathbf{F} \,.\, d\mathbf{r}$ along the straight line from $(0, 0, 0)$ to $(1, 1, 1)$.

48. In the sense of θ increasing evaluate the circulation of $\mathbf{F} = x^3\mathbf{j}$ round the ellipse having parametric equations $x = a\cos\theta, y = b\sin\theta, z = 0$.

49. Referred to right-handed orthogonal axes, Ox, Oy, Oz, the parametric equations of a circle are

$$x = a + c\cos\theta, \quad y = b + c\sin\theta, \quad z = 0; \quad a > b > c > 0.$$

Evaluate the circulation γ of $\mathbf{F} = -y^3\mathbf{i}$ round the circle in the sense of θ increasing and evaluate $\lim_{S\to 0}(\gamma/S)$ where S is the area of the circle.

A square in the plane xOy has sides of length $2c$ parallel to Ox, Oy and its centre is $(a, b, 0)$. Evaluate γ' the circulation of $\mathbf{F} = -y^3\mathbf{i}$ round the square in the same sense as the circle and evaluate $\lim_{S' \to 0} (\gamma'/S')$ where S' is the area of the square.

Deduce the value of curl **F** at $(a, b, 0)$ and explain how its direction is determined.

50. The parametric equations of a curve are

$$x = -a(m \cos\theta + nl \sin\theta),$$
$$y = a(l \cos\theta - mn \sin\theta),$$
$$z = a(l^2 + m^2) \sin\theta$$

where a, l, m, n are constants, $l^2 + m^2 + n^2 = 1$ and $a > 0$. A is the point where $\theta = 0$ and P is any other point on the curve. By considering $\overrightarrow{OA} \times \overrightarrow{OP}$ and $|OP|$, or otherwise, show that the curve is a circle with centre O. Evaluate the circulation γ of $\mathbf{F} = -z\mathbf{j}$ round the circle in the sense of θ increasing. If S is the area of the circle and the orientation of its plane may now vary, determine the greatest value of $\lim(\gamma/S)$ as $S \to 0$ and hence find curl **F** at O, explaining how to determine its direction.

51. If $\mathbf{F} = (2y^2 + 3z^2 - x^2)\mathbf{i} + (2z^2 + 3x^2 - y^2)\mathbf{j} + (2x^2 + 3y^2 - z^2)\mathbf{k}$ and S is the surface

$$x^2 + y^2 - 2ax + az = 0, \quad z \geqslant 0,$$

use Stokes's circulation theorem to show that

$$\int_S \mathbf{n} \,.\, \text{curl}\, \mathbf{F}\, dS = 6\pi a^3.$$

52. Use Stokes's theorem to show that

$$\oint (4x^3 y^3 z dx + 3x^4 y^2 z dy + x^4 y^3 dz) = 0$$

where the integral is taken round any closed curve. Establish the result in another way by expressing the integrand as an exact differential.

53. For a constant vector **a**, prove $\text{div}\{(\mathbf{a} \,.\, \mathbf{r})\mathbf{r}\} = 4\mathbf{a} \,.\, \mathbf{r}$.

54. Prove $\text{div}(\text{grad}\, \phi \times \text{grad}\, \psi) = 0$.

55. If **F** is a solution of $\mathbf{X} \,.\, (\nabla \times \mathbf{X}) = 0$. show that $\phi\mathbf{F}$ is also a solution where ϕ is any uniform differentiable scalar function.

56. If c, ρ, k are uniform differentiable scalar functions of space coordinates and V is a uniform differentiable scalar function of space coordinates and time t show that if within a closed volume v

$$c\rho \frac{\partial V}{\partial t} = \operatorname{div}(k\nabla V)$$

and if on S either $V = 0$ or $\partial V/\partial n = 0$, then

$$\int_v c\rho \frac{\partial V}{\partial t} \, \mathrm{d}v = - \int_v k(\nabla V)^2 \, \mathrm{d}v.$$

The partial differential equation of heat conduction in a body is

$$c\rho \frac{\partial \theta}{\partial t} = \operatorname{div}(k\nabla\theta),$$

where θ is the temperature, c the specific heat, ρ the density and k the thermal conductivity. To establish *uniqueness of solution*, let $\theta = \theta_1$ and $\theta = \theta_2$ be two supposedly different solutions in v but having the same value on S and the same values in v at $t = 0$. Write $V = \theta_1 - \theta_2$ and show that

$$c\rho \frac{\partial V}{\partial t} = \operatorname{div}(k\nabla V) \text{ in } v; \quad V = 0 \text{ on } S.$$

Write $J = \int_v c\rho V^2 \, \mathrm{d}v$ and show that:

(i) $\dfrac{\partial J}{\partial t} = \displaystyle\int_v c\rho \frac{\partial V}{\partial t} \, \mathrm{d}v = - \int_v k(\nabla V)^2 \, \mathrm{d}v.$

(ii) If $k > 0$, $c > 0$, $\rho > 0$, and if $V \neq 0$ and $\nabla V \neq 0$, then $J > 0$ and $\partial J/\partial t = 0$.

(iii) At $t = 0$, $J = 0$ and so $J < 0$ for $t > 0$ if $\partial J/\partial t = 0$.

(iv) From the contradictions $J > 0$, $J < 0$, in v for $t > 0$ infer that $\theta_1 = \theta_2$ in v for all $t \geqslant 0$.

57. The scalar functions u, v, w are harmonic and at each point of a region their level surfaces intersect mutually orthogonally. If S is a closed surface within the region and $\partial/\partial n$ denotes differentiation along the outward normal to a surface element $\mathrm{d}S$, prove

$$\int_S \frac{\partial}{\partial n} (uvw) \, \mathrm{d}S = 0.$$

58. If f, g are two harmonic functions defined in a volume v and on its boundary S, prove

(i) $\displaystyle\int_v \nabla f . \nabla g \, \mathrm{d}v = \int_S f \frac{\partial g}{\partial n} \, \mathrm{d}S = \int_S g \frac{\partial f}{\partial n} \, \mathrm{d}S;$

(ii) $\displaystyle\int_v \{\nabla(f - g)\}^2 \, \mathrm{d}v = 0.$

59. If $\mathbf{r}_0$ is the position vector of a point P_0 not on S and $\mathbf{r}$ that of an element $\mathrm{d}\mathbf{S}$, evaluate

$$\text{(i)} \quad \int_S \mathrm{d}\mathbf{S} \times (\mathbf{r} - \mathbf{r}_0); \qquad \text{(ii)} \quad \int_S \frac{\mathbf{r} - \mathbf{r}_0}{|\mathbf{r} - \mathbf{r}_0|^3} \cdot \mathrm{d}\mathbf{S}.$$

60. If $\mathbf{r}$ is the position vector of an element of $\mathrm{d}\mathbf{S}$ of a closed surface S containing a volume v and $\mathbf{a}$ is constant vector, prove

$$\int_S (\mathbf{r} \times \mathbf{a}) \times \mathrm{d}\mathbf{S} = 2v\mathbf{a}.$$

61. Show that the position vector of the centroid of a uniform solid of volume v and bounded by a closed surface S is given by $(1/2v) \int r^2 \, \mathrm{d}\mathbf{S}$ and use the formula to show that the centroid of the uniform solid hemisphere $x^2 + y^2 + z^2 = a^2, x \geqslant 0$ is $(\frac{3}{8}a, 0, 0)$.

62. Parabolic coordinates (ξ, η, ϕ) are related to rectangular Cartesian coordinates (x, y, z) by

$$x = \tfrac{1}{2}(\xi - \eta), \quad y = (\xi\eta)^{1/2} \cos\phi, \quad z = (\xi\eta)^{1/2} \sin\phi.$$

Prove that the arc element $\mathrm{d}s$ of a curve in such a system is given by

$$(\mathrm{d}s)^2 = \frac{\xi + \eta}{4} \left[\frac{(\mathrm{d}\xi)^2}{\xi} + \frac{(\mathrm{d}\eta)^2}{\eta} \right] + \xi\eta(\mathrm{d}\phi)^2$$

and say why the system (ξ, η, ϕ) is an orthogonal coordinate one. Show that for a uniform differentiable scalar function $\psi(\xi, \eta, \phi)$,

$$\nabla^2 \psi = \frac{4}{\xi + \eta} \left[\frac{\partial}{\partial\xi} \left(\xi \frac{\partial\eta}{\partial\xi} \right) + \frac{\partial}{\partial\eta} \left(\eta \frac{\partial\phi}{\partial\eta} \right) \right] + \frac{1}{\xi\eta} \frac{\partial^2 \psi}{\partial\phi^2}.$$

If $\psi = \psi(\xi, \eta)$ is harmonic, show that there exists $F(\xi, \eta)$ such that

$$\xi \frac{\partial\psi}{\partial\xi} = -\frac{\partial F}{\partial\eta}, \quad \eta \frac{\partial\psi}{\partial\psi} = \frac{\partial F}{\partial\xi}$$

and show

$$\xi \frac{\partial^2 F}{\partial\xi^2} + \eta \frac{\partial^2 F}{\partial\eta^2} = 0.$$

63. Oblate spheroidal coordinates (ξ, η, ζ) are defined by

$$x = c\cosh\xi\cos\eta\cos\zeta, \quad y = c\cosh\xi\cos\eta\sin\zeta, \quad z = c\sinh\xi\sin\eta;$$

where (x, y, z) are the rectangular Cartesian coordinates of any point in space.

Show that the system is an orthogonal curvilinear coordinate one and that

$$h_\xi = h_\eta = c(\cosh^2\xi - \cos^2\eta)^{1/2}, \quad h_\zeta = c\cosh\xi\cos\eta,$$

and also show that

$$\nabla^2\phi = \frac{1}{c^2(\cosh^2\xi - \cos^2\eta)}\left[\frac{\partial^2\phi}{\partial\xi^2} + \frac{\partial^2\phi}{\partial\eta^2} + (\sec^2\eta - \operatorname{sech}^2\xi)\frac{\partial^2\phi}{\partial\zeta^2} + \tanh\xi\frac{\partial\phi}{\partial\xi} - \tan\eta\frac{\partial\phi}{\partial\eta}\right].$$

64. A system of coordinates (ρ_1, ρ_2, ϕ) is related to rectangular Cartesian coordinates by

$$x = \rho_1\rho_2^2\,\sigma\cos\phi, \quad y = \rho_1\rho_2^2\,\sigma\sin\phi, \quad z = \rho_1^2\rho_2\,\sigma,$$

where $1/\sigma = \rho_1^2 + \rho_2^2$. Show that the surfaces ρ_1 = const., ρ_2 = const., ϕ = const. form an orthogonal coordinate system and that Laplace's equation in this system of coordinates is

$$\frac{\partial}{\partial\rho_1}\left[\frac{\rho_1^3}{\rho_1^2+\rho_2^2}\frac{\partial V}{\partial\rho_1}\right] + \frac{\partial}{\partial\rho_2}\left[\frac{\rho_2^4}{\rho_1(\rho_1^2+\rho_2^2)}\frac{\partial V}{\partial\rho_2}\right] + \frac{\rho_1}{\rho_1^2+\rho_2^2}\frac{\partial^2 V}{\partial\phi^2} = 0.$$

65. Show that in spherical polar coordinates (r, θ, ϕ),

$$\operatorname{curl}(\cos\theta\,\operatorname{grad}\phi) = \operatorname{grad}(1/r).$$

66. In spherical polars (r, θ, ϕ), $\mathbf{F} = [u, v, w]$. Find the components of curl $\mathbf{F}$ in this system and show that if $u = v = \partial w/\partial\phi = 0$ and curl curl $\mathbf{F} = \mathbf{0}$, then

$$r^2\frac{\partial^2 w}{\partial r^2} + 2r\frac{\partial w}{\partial r} + \frac{\partial}{\partial\theta}\left(\frac{\partial w}{\partial\theta} w\cot\theta\right) = 0.$$

Find a solution in the form $w(r, \theta) = f(r)\sin\theta$.

67. *Alternative treatment of orthogonal coordinate formulae*
In the following make use of the vector identities (1)–(3):

$$\nabla\times(\phi\mathbf{F}) = \nabla\phi\times\mathbf{F} + \phi\nabla\times\mathbf{F}, \tag{1}$$

$$\operatorname{div}(\mathbf{F}\times\mathbf{G}) = \mathbf{G}\,.\operatorname{curl}\mathbf{F} - \mathbf{F}\,.\operatorname{curl}\mathbf{G}, \tag{2}$$

$$\operatorname{div}(\operatorname{grad}\phi\times\operatorname{grad}\psi) = 0. \tag{3}$$

In the notation of the text, prove

$$\nabla u_i = \hat{\mathbf{a}}_i/h_i, \quad \operatorname{curl}(\hat{\mathbf{a}}_i/h_i) = \mathbf{0} \quad (i = 1, 2, 3). \tag{4}$$

Now take $\mathbf{F} = \Sigma\{(h_1 F_1)(\hat{\mathbf{a}}_1/h_1)\}$ and use (1), (4) to show

$$\text{curl } \mathbf{F} = \Sigma\,\{\nabla(h_1 F_1) \times (\hat{\mathbf{a}}_1/h_1)\} \tag{5}$$

and hence develop the determinantal form for curl **F**.
Next, taking $\mathbf{F} = \Sigma\{(h_2 h_3 F_1)(\hat{\mathbf{a}}_1/h_2 h_3)\}$, show that

$$\text{div } \mathbf{F} = \Sigma\{(h_2 h_3 F_1)\nabla.(\hat{\mathbf{a}}_1/h_2 h_3) + (\hat{\mathbf{a}}_1/h_2 h_3).\nabla(h_2 h_3 F_1)\}.$$

Using $\hat{\mathbf{a}}_1 = \hat{\mathbf{a}}_2 \times \hat{\mathbf{a}}_3$, etc., and (3), (4), show that

$$\nabla.(\hat{\mathbf{a}}_1/h_2 h_3) = 0, \text{ etc.}$$

and hence obtain div **F** in standard form.

68. If (u_1, u_2, u_3) form an orthogonal curvilinear system of coordinates with $\text{ds} = \Sigma(h_1\, du_1\, \hat{\mathbf{a}}_1)$, then the wave equation for the scalar function $\phi(u_1, u_2, u_3\; t)$ is

$$\nabla^2\phi = \frac{1}{c^2}\frac{\partial^2\phi}{\partial t^2}\quad (c = \text{const.})$$

in terms of the orthogonal coordinates and t.

If $\phi = \phi(u_1)$, $h_1 = K(= \text{const.})$, $h_2 h_3 = u_1^2$, obtain the general solution of the wave equation in the form

$$\phi = [F(u_1 - ct/K) + G(u_1 + ct/K)]/u_1.$$

69. Show that Poisson's equation $\nabla^2 V = -4\pi\rho$, on transformation to the coordinate system ξ, η, ζ, defined by

$$\begin{cases} y^2 = 4\xi(\xi + x), \\ y^2 = 4\eta(\eta - x), \\ z = \zeta, \end{cases}$$

becomes

$$4\pi y\rho(\xi+\eta) = \sqrt{\xi}\,.\frac{\partial}{\partial\xi}\left[\sqrt{\xi}\,\frac{\partial V}{\partial\xi}\right] + \sqrt{\eta}\,.\frac{\partial}{\partial\eta}\left[\sqrt{\eta}\,\frac{\partial V}{\partial\eta}\right] + (\xi+\eta)\frac{\partial^2 V}{\partial\zeta^2}.$$

Show how this equation can be solved for the case $\rho = 0$ by the method of separation of coordinates and, assuming the solution of the equation $y'' = y(k^2 + x^2)$ to be $y = Y_k(x)$, write down a general solution.

70. (a) When a rigid body rotates about a fixed point O with vector angular velocity $\boldsymbol{\omega}$, the velocity **v** at any other point P is $\boldsymbol{\omega} \times \mathbf{r}$, where $\overrightarrow{OP} \equiv \mathbf{r}$. Prove that $\boldsymbol{\omega} = \frac{1}{2}\text{curl } \mathbf{v}$.
(b) Explain carefully what is meant by the statement that **F** is a conservative vector field and establish curl $\mathbf{F} = \mathbf{0}$ as a necessary and sufficient condition for

such a property to hold. With ω and $\mathbf{r}$ defined as in (a), prove that $(\omega\,.\,\mathbf{r})\omega$ is a conservative field and find, in simplest form, a scalar function ϕ whose gradient is equal to $(\omega\,.\,\mathbf{r})\omega$.

71. Prove that the vector field

$$\mathbf{F} = (3x^2yz + 2y^2z - 3z^2 + 2)\mathbf{i} + (x^3z + 4xyz)\mathbf{j} + (x^3y + 2xy^2 - 6xz + 5)\mathbf{k}$$

is conservative and find a suitable potential function ϕ for which $\mathbf{F} = \nabla\phi$.

72. In the usual notation a vector field $\mathbf{V}$ is defined by

$$\mathbf{V} = \mathbf{k} \times \mathbf{r}/|\mathbf{k} \times \mathbf{r}|^2.$$

Show that at points of the field where $\mathbf{k} \times \mathbf{r} \neq \mathbf{0}$, the field is conservative and find a suitable potential function ϕ for which $\mathbf{V} = \nabla\phi$.

73. If the vector field $x^m y^m z^m (x^n\mathbf{i} + y^n\mathbf{j} + z^n\mathbf{k})$ is conservative, prove that either $m = 0$ or $n = -1$.

74. Prove that the integral

$$\int_{P_1}^{P_2} (x^2 - yz)\mathrm{d}x + (y^2 - zx)\mathrm{d}y + (z^2 - xy)\mathrm{d}z$$

depends only on the coordinates of P_1, P_2 and not at all on the shape of the path joining them and evaluate it.

75. Prove that $r^{-2}\hat{\mathbf{r}}$ is solenoidal and that a suitable vector potential for it is

$$\mathbf{A} = \frac{xyz}{3r}\left[\frac{z^2 - y^2}{x(x^2 + y^2)(x^2 + z^2)}, \frac{x^2 - z^2}{y(y^2 + z^2)(y^2 + x^2)}, \frac{y^2 - x^2}{z(z^2 + x^2)(z^2 + y^2)}\right].$$

[curl $\mathbf{A} = r^{-2}\,\hat{\mathbf{r}}$ defines the vector potential.]

76. If $\mathbf{M}$ is a constant vector, prove that $(\mathbf{M} \times \mathbf{r})/r^3$ is solenoidal. Show that the field lines of the function are circles in planes perpendicular to $\mathbf{M}$ and centres on the line of $\mathbf{M}$.

77. If $\mathbf{A}\,.\,\text{curl}\,\mathbf{B} = \mathbf{B}\,.\,\text{curl}\,\mathbf{A}$, prove that $\mathbf{A} \times \mathbf{B}$ is solenoidal and that it satisfies the equation

$$\text{curl curl}\,\mathbf{V} + \nabla^2\mathbf{V} = \mathbf{0}.$$

78. Determine which, if any, of the following vector fields are conservative:

(i) $(\mathbf{a}\,.\,\mathbf{r})\mathbf{b}$; (ii) $(\mathbf{a}\,.\,\mathbf{b})\mathbf{r}$;
(iii) $(\mathbf{a}\,.\,\mathbf{r})\mathbf{b}+(\mathbf{b}\,.\,\mathbf{r})\mathbf{a}$; (iv) $(\mathbf{a}\,.\,\mathbf{r})\mathbf{b}-(\mathbf{b}\,.\,\mathbf{r})\mathbf{a}$,

where $\mathbf{a}$ and $\mathbf{b}$ are costant vectors with $\mathbf{a}\times\mathbf{b}\neq\mathbf{0}$. In the case where a vector field is conservative, find an appropriate scalar potential function whose gradient is equal to the vector.

79. If the field

$$\frac{(\mathbf{A}\,.\,\mathbf{r})}{r^5}\,\mathbf{r}+\frac{1}{r^3}\,\mathbf{B}\qquad(\mathbf{A},\mathbf{B}\text{ constant vectors})$$

is conservative, prove that $\mathbf{A}=-3\mathbf{B}$ and that a suitable potential function for its derivation is $-(\mathbf{B}\,.\,\mathbf{r})/r^3$.

80. Prove Gauss's divergence theorem and deduce Green's theorem

$$\int_v(\phi\nabla^2\psi-\psi\nabla^2\phi)\,\mathrm{d}v=\int_S(\phi\nabla\psi-\psi\nabla\phi)\,.\,\mathrm{d}\mathbf{S},$$

where ϕ, ψ are functions of position defined in a region v and on its boundary S. Show that $(\cos\lambda r)/r$ satisfies the equation

$$(\nabla^2+\lambda^2)\phi=0,$$

where λ is constant and $r=(x^2+y^2+x^2)^{1/2}$, in any region excluding the origin. If $\phi=\Phi$ is any solution of the above equation which has no singularity on or inside S, show that the value of Φ at the origin, assumed to be inside S, is given by

$$-\frac{1}{4\pi}\int_S\left(\Phi\,\frac{\partial}{\partial n}\,\frac{\cos\lambda r}{r}-\frac{\cos\lambda r}{r}\,\frac{\partial\Phi}{\partial n}\right)\mathrm{d}S,$$

where $\partial/\partial n$ denotes differentiation along the outward normal to S.

81. Verify that the transformation

$$\begin{cases} x_1'=\tfrac{1}{3}(2x_1+2x_2-x_3),\\ x_2'=\tfrac{1}{3}(2x_1-x_2+2x_3),\\ x_3'=\tfrac{1}{3}(-x_1+2x_2+2x_3)\end{cases}$$

satisfies the orthogonality conditions. Write out at length the transformation equations for a second-order tensor A_{ij} w.r.t. this transformation and verify that $A_{ii}'=A_{ii}$.

82. Verify that the transformation

$$\begin{cases} x_1' = \frac{1}{15}(5x_1 - 14x_2 + 2x_3), \\ x_2' = -\frac{1}{3}(2x_1 + x_2 + 2x_3), \\ x_3' = \frac{1}{15}(10x_1 + 2x_2 - 11x_3) \end{cases}$$

is orthogonal. A vector field $\mathbf{A}$ is defined in the x-frame by

$$A_1 = x_1^2,\ A_2 = x_2^2,\ A_3 = x_3^2.$$

Evaluate the field in the x'-frame and verify that div $\mathbf{A}$ is an invariant. (Same origin for both frames.)

83. If A_i is a vector field in the x-frame where

$$A_1 = x_2 x_3,\ A_2 = x_3 x_1,\ A_3 = x_1 x_2$$

and the transformation to the x'-frame (same origin) is

$$x_1' = \tfrac{1}{13}(5x_1 + 12x_2),\quad x_2' = \tfrac{1}{13}(12x_1 - 5x_2),\quad x_3' = x_3.$$

Show the transformation is orthogonal. Evaluate the components A_i' of the field in the new frame in terms of the coordinates x_i. Verify that div $\mathbf{A}$ is invariant under the transformation.

84. A_{ij} is a tensor field defined in the x-frame by $A_{ij} = x_i x_j$. Evaluate its components at the point P where $x_1 = 0$, $x_2 = x_3 = 1$. The coordinates x_i of a point in the x-frame are related to those in the x'-frame thus:

$$\begin{bmatrix} x_1' \\ x_2' \\ x_3' \end{bmatrix} = \frac{1}{7}\begin{bmatrix} -3 & -6 & -2 \\ -2 & 3 & -6 \\ 6 & -2 & -3 \end{bmatrix}\begin{bmatrix} x_1 \\ x_2 \\ x_3 \end{bmatrix}.$$

Show that the transformation is orthogonal and evaluate the component A_{11}' of the tensor field at P.

85. Prove that the necessary and sufficient conditions for the square matrix $[a_{ij}]$ to be orthogonal are $a_{ir}a_{jr} = \delta_{ij} = a_{ri}a_{rj}$.

86. *Affine transformations*
The transformation from the right-handed orthogonal frame x to the right-handed orthogonal frame x' described by

$$x_i' = \alpha_{i0} + (\delta_{ij} + \alpha_{ij})x_j \quad (i, j = 1, 2, 3)$$

is called an affine transformation. Suppose such a transformation is followed by another affine transformation from the x'-frame to another right-handed orthogonal frame x'' in accordance with the law

$$x_i'' = \beta_{io} + (\delta_{ij} + \beta_{ij})x_j' \quad (i, j = 1, 2, 3).$$

Show that this is equivalent to a single transformation from the x- to the x''-frame by an affine transformation of the form

$$x_i'' = \gamma_{io} + (\delta_{ij} + \gamma_{ij})x_j \quad (i, j = 1, 2, 3)$$

giving the appropriate values of γ_{io}, γ_{ij}.

87. In the theory of dynamics of a rigid body in three dimensions, the inertia matrix $\mathbf{M}$ of a body with respect to a right-handed tri-rectangular frame OX, OY, OZ is defined to be

$$\mathbf{M} = \begin{bmatrix} A & -D & -E \\ -D & B & -F \\ -E & -F & C \end{bmatrix},$$

where A, B, C are the moments of inertia of the body about OX, OY, OZ, and D, E, F are its products of inertia with respect to the pairs of axes (OY, OZ); (OZ, OX); (OX, OY). The principal moments of inertia are the eigenvalues of $\mathbf{M}$ and the principal axes of inertia are its eigenvectors.

If, for a given distribution,

$$A = B = C = 8\ \text{ma}^2; \quad D = E = F = 2\ \text{ma}^2$$

show that the principal moments of inertia are

$$4\ \text{ma}^2, \quad 10\ \text{ma}^2, \quad 10\ \text{ma}^2$$

and that unit vectors in the three principal axes are represented by the three column vectors in the matrix

$$\begin{bmatrix} \frac{1}{\sqrt{3}} & \frac{2}{\sqrt{6}} & 0 \\ \frac{1}{\sqrt{3}} & -\frac{1}{\sqrt{6}} & \frac{1}{\sqrt{2}} \\ \frac{1}{\sqrt{3}} & -\frac{1}{\sqrt{6}} & -\frac{1}{\sqrt{2}} \end{bmatrix}$$

88. The stress matrix for an elastic medium is

$$\mathbf{A} = \begin{bmatrix} \sigma_{xx} & \sigma_{xy} & 0 \\ \sigma_{xy} & \sigma_{yy} & 0 \\ 0 & 0 & \sigma_{zz} \end{bmatrix}.$$

The principal stresses of the medium are the eigenvalues of the stress matrix. Show that these principal stresses are

$$\tfrac{1}{2}\{\sigma_{xx} + \sigma_{yy} \pm [(\sigma_{xx} - \sigma_{yy})^2 + 4\sigma_{xy}^2]^{1/2}\}; \quad \sigma_{zz}.$$

If, corresponding to eigenvalue λ of $[A]$, there is an eigenvector $\mathbf{X}$ having transpose $\widetilde{\mathbf{X}} = [x, y, z]$, show that

$$\begin{cases} (\sigma_{xx} - \lambda)x + \sigma_{xy}y & = 0, \\ \sigma_{xy}x + (\sigma_{yy} - \lambda)y & = 0, \\ \qquad\qquad (\sigma_{zz} - \lambda)z & = 0, \end{cases}$$

and prove or verify that

$$\frac{x}{\sigma_{xy}} = \frac{y}{\tfrac{1}{2}\{\sigma_{yy} - \sigma_{xx} + [(\sigma_{xx} - \sigma_{yy})^2 + 4\sigma_{xy}^2]^{1/2}\}} = \frac{z}{0};$$

$$\frac{x}{\sigma_{xy}} = \frac{y}{\tfrac{1}{2}\{\sigma_{yy} - \sigma_{xx} - [(\sigma_{xx} - \sigma_{yy})^2 + 4\sigma_{xy}^2]^{1/2}\}} = \frac{z}{0};$$

$$\frac{x}{0} = \frac{y}{0} = \frac{z}{1},$$

corresponding to each eigenvalue. Confirm that the three eigenvectors are mutally perpendicular.

2

Kinematics of fluids in motion

2.1 SOLIDS, LIQUIDS AND GASES

By a *fluid* we mean a substance which *flows*. Material which does not have this property is a *solid*. Although this definition is unequivocal, in real life, however, the distinction between a solid and fluid can often be dubious since there are many materials, particularly certain products of the petrochemical industry, which exhibit properties suggestive in some respects of behaviour as solids and in other respects as fluids. An everyday example is a non-drip paint which is a *thixotropic* substance and behaves as an elastic solid after it has been allowed to stand at rest for a time, but when shaken or brushed, it loses its elasticity and solid-like character and behaves as a fluid. Most common materials can, however, be broadly categorised according to the simple criterion as to whether or not they flow.

Fluids in turn can be divided into *liquids* and *gases*. Liquids are essentially *incompressible* so their volumes do not change with variations in applied pressure. For instance, the *density*, i.e. mass per unit volume, of water increases by no more than 0.5% when the pressure applied is increased by 100 atmospheres for constant temperatures. An *atmosphere* is a measure of pressure equal to the weight of a column of air of unit cross-section taken from sea level to the edge of the air layer enveloping the earth. This great resistance to compression is the important characteristic of a liquid, and allows us to regard it for most purposes as incompressible to a very high degree of accuracy. Of course, pressures found in the depths of the oceans can be of the order of several hundreds of atmospheres and for investigating the flow properties in such great depths the compressibility of water could be significant. The term *hydrodynamics* is often used to describe the science of incompressible fluids in motion. A gas on the other hand is compressible, and its volume can be readily changed by variation in pressure.

To understand why a liquid and a gas have such different properties, we have to take a *microscopic* view of the fluids. Matter is composed of molecules or groups of atoms which are in random relative motion under the action of intermolecular forces. In *solids*, the spacings of the molecules are small compared with a typical molecular dimension and the molecules are held in recognisably ordered arrangements under very strong intermolecular forces. In *liquids*, the spacings of the molecules are greater resulting in less well-ordered arrangements of molecules and weaker intermolecular forces, and in the case of *gases*, intermolecular forces are very much weaker still and the molecules appear in disordered arrays in which very large intermolecular spacings may occur.

If we imagine that our microscope, with which we have observed the molecular structure of matter, has a variable focal length, we could change our observation of matter from the fine detailed microscopic viewpoint to a longer range *macroscopic* viewpoint in which we would not see the gaps between the molecules and the matter would appear to be continuously distributed. Throughout this book we shall take this macroscopic view of fluids in which physical quantities associated with the fluids within a given volume are assumed to be distributed continously and, within a sufficiently small volume, uniformly. This is the *continuum hypothesis* which is in complete accord with our everyday observation of such fluids as water and air where the idea of regarding their structure as anything other than continuous would be unnatural.

The continuum hypothesis inevitably implies that it is possible to attach a definite meaning to the notion of the value of some property of the fluid *at a point*, which seems to be inconsistent with the supposition that scale and dimension do not become comparable with molecular size. However, when a measurement of some fluid property is made by inserting a probe, what is actually obtained is some kind of average of that property over the volume occupied by the probe, which would normally be constructed so that the displaced volume is small enough for the measurement to be *local*, and further reduction in the size of the probe would not then significantly change the reading. At the same time it is assumed that the displaced volume of fluid is not of molecular dimensions. In this way, the notion of a small volume of fluid is both reasonable and acceptable, and if the dimensions of this small volume of fluid are small compared with say those of the vessel containing the fluid, it is then entirely reasonable to regard the small volume of fluid as a *fluid particle* occupying a geometrical point of the fluid, since the error in ignoring the size of the small volume will be insignificant.

2.2 VELOCITY AT A POINT OF A FLUID

A fluid in motion occupies a specific region of three-dimensional space and it is possible to describe the motion of a fluid in two ways. One way is to identify each of the *particles* of the fluid and put tags on them so to speak by labelling them $P_1, P_2, \ldots$. Then, if we know the velocity of *each particle* $P_i (i = 1, 2, \ldots)$ at a particular time t, we have a picture of the motion of the whole fluid at that time t. This would be achieved if we determined the position vector $\mathbf{r}_i$ of each particle P_i relative to a fixed origin O. The velocity $\mathbf{q}_i$ of the particle P_i is thus given by $\mathbf{q}_i = \mathrm{d}\mathbf{r}_i/\mathrm{d}t$, and we then have a complete description of the motion of the fluid at time t. This is the *Lagrangian* or *historical* view of the motion of a fluid. The reader already familiar with solving problems in the dynamics of a particle or rigid body will at once recognise that this is precisely the way in which the motion of a rigid body is described. But a rigid body has the great simplification, compared with a fluid, that the particles are rigidly attached to one another and this means that to determine the velocity at *every* point of a rigid body at a given instant, it is only necessary to find the translational velocity of the centre of mass and the instantaneous angular velocity of the body. Although the Lagrangian view of fluid motion may seem both intuitive and sensible, it is found in practice that the determination of $\mathbf{r}_i(t)$ for all particles P_i and all times t is such a formidable task, except for very simple flows, that an alternative approach to describing a flow is desirable, and this is the *Eulerian* view of fluid motion.

In the Eulerian view, the region occupied by the fluid is a geometrical entity made up of a collection of *points* P_i, each with a position vector $\mathbf{r}_i$ relative to a fixed origin O, so that now $\mathbf{r}_i$ is *independent* of the time t. We have a description of the state of motion of the fluid *at time* t if we know the velocity of the particle of fluid which at time t instantaneously coincides with the point P_i $(i = 1, 2, \ldots)$. Evidently in the Eulerian view, with changing t, different fluid particles coincide with P_i.

The velocity at a point P of a fluid at time t is defined to be the velocity of the fluid particle which at time t coincides with the point P. We denote this by $\mathbf{q}$ and clearly

$$\mathbf{q} = \mathbf{q}(\mathbf{r}, t), \tag{2.1}$$

with $\mathbf{r}$ and t *independent*. The Eulerian description of the flow can be thought of as providing the spatial distribution of fluid velocity at any instant during the motion. It is just like taking a snapshot, for the state of motion within the whole fluid is recorded at the particular instant of time considered. If P has Cartesian coordinates (x, y, z) relative to a fixed frame passing through a fixed origin O, then clearly

$$\mathbf{q} = \mathbf{q}(x, y, z, t), \tag{2.2}$$

and if (u, v, w) are the Cartesian resolutes of $\mathbf{q}$ in this frame, then

$$\mathbf{q} = u\mathbf{i} + v\mathbf{j} + w\mathbf{k}. \tag{2.3}$$

A point where $\mathbf{q} = \mathbf{0}$ is called a *stagnation* point.

2.3 STREAMLINES, PATHLINES AND STREAKLINES

A curve $\mathscr{C}$ is drawn in the fluid such that at each point of it, the local fluid velocity is along the direction of the tangent to $\mathscr{C}$ at that point. Such a curve $\mathscr{C}$ is called a *streamline* and is analogous to a line of force in electromagnetic theory. The fluid velocity vector $\mathbf{q}$ is a vector field, so it follows from (1.31) that the differential equations of the streamlines are

$$\frac{\mathrm{d}x}{u} = \frac{\mathrm{d}y}{v} = \frac{\mathrm{d}z}{w} \tag{2.4}$$

in Cartesian coordinates. A knowledge of $\mathbf{q}$ at neighbouring points $P_1, P_2, \ldots$ of a streamline enables an approximate form of the streamline to be obtained by drawing the line segments $P_1P_2, P_2P_3, \ldots$ since P_iP_{i+1} and $\mathbf{q}_i$ are approximately parallel with $\mathbf{q}_i$ the value of $\mathbf{q}$ at P_i $(i = 1, 2, \ldots)$. Such a picture is shown in Fig. 2.1 and it resembles the plot of the line of force due to a magnet using a small compass when the direction of the compass needle is marked on a sheet of drawing paper.

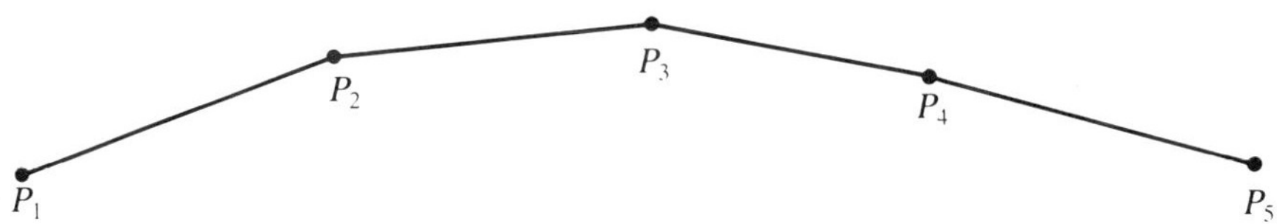

Fig. 2.1

In general the fluid velocity at a given point of a fluid depends not only on the position of the point but also on the time. The streamlines will accordingly alter from instant to instant. Thus snapshot type photographs of the fluid taken at different instants will reveal different systems of streamlines. The aggregate of all streamlines in any one such photograph constitutes the flow pattern at that instant.

When the flow is such that the velocity at each point is independent of time, the flow pattern is the same at each instant and the motion is described as *steady*. This would be the type of motion observed when a uniform stream flows past a fixed obstacle such as a bridge support in a river. There is a class of flows which may be called *relatively steady*. Such motions arise when the flow can be regarded as steady if a constant velocity is superposed on the whole system, including the observer. Thus when a ship steers a straight course with constant speed on an otherwise undisturbed sea, to an observer on the ship, the flow pattern in the sea appears to be steady and could in fact be made so by superimposing the reversed velocity of the ship on the whole system of ship and sea. The motion would then be equivalent to a uniform stream flowing past a ship at rest.

If we draw the streamline through each point of a closed curve drawn in the fluid, the resulting surface is a *stream tube*. A *stream filament* is a stream tube of very small (infinitesimal) cross-section, so that the fluid velocity may be regarded as effectively constant over the cross-section of the filament. In a steady flow, a stream tube (or filament) behaves like an actual tube through which the fluid is flowing, for there can be no flow into or out of the tube across the 'walls' since the flow is, by definition, always tangential to the walls. Moreover, the walls are fixed in space when the motion is steady, and therefore the motion of the fluid within the walls would be unaltered if we replaced the walls of the stream tube by rigid material, assuming of course that the fluid may flow freely over a rigid surface.

An important result follows on considering a stream filament in a steady flow. Let σ_1 and σ_2 be two cross-sections of a stream filament which may be taken as perpendicular to the velocity over the section, as shown in Fig. 2.2.

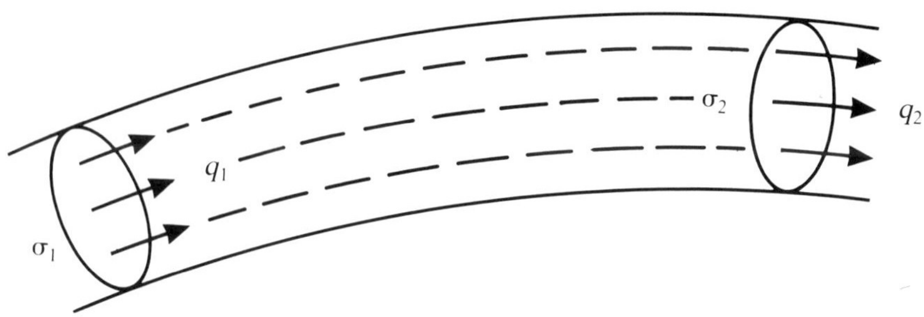

Fig. 2.2

Let q_1 and q_2 be the speeds of the flow over the sections σ_1 and σ_2 respectively. If the fluid is a liquid and therefore *incompressible*, in a steady flow, the volume of fluid flowing across σ_1 in a given time t must be exactly the same as the amount of fluid flowing across σ_2 in the same time. Thus

$$\sigma_1 q_1 t = \sigma_2 q_2 t$$

and it therefore follows that

$$q_1 \sigma_1 = q_2 \sigma_2 . \tag{2.5}$$

This is the simplest example of the equation of conservation of mass or the *equation of continuity* which asserts in its general form that the rate of generation of mass within a given volume must be balanced by an equal net outflow of mass from the volume. We consider the implications of the general form in Section 2.6.

A consequence of (2.5) is that a stream filament is widest at places where the fluid speed is least and vice versa. Thus a stream filament cannot terminate at a point within the liquid unless the velocity becomes infinite at that point. Leaving aside such exceptional *singular* points which occur for instance when sources and sinks are present, as discussed in Chapter 4, it follows that in general, stream filaments are either closed or terminate at the boundary of a liquid. The same is true of streamlines, since the cross-section of the filament may be considered as small as we please.

From the definition of velocity at a point of a fluid, it is evident that the paths of the particles called the *pathlines*, generally do *not* coincide with the streamlines. The direction of motion of a fluid particle must necessarily be tangential to the pathline of that particle and the pathline touches the streamline which passes through the instantaneous position of the particle as it describes its path. Thus the streamlines show how *all* particles are moving at a given instant while the pathlines show how a *given* particle is moving at each instant. When the motion is steady, the pathlines coincide with the streamlines.

The differential equations of the pathlines are

$$\frac{dx}{dt} = u, \quad \frac{dy}{dt} = v, \quad \frac{dz}{dt} = w \tag{2.6}$$

where now (x, y, z) are the Cartesian coordinates of the fluid particle and not a fixed point of space. The equation of the pathline which passes through the point (x_0, y_0, z_0), which is fixed in space, at time $t = 0$ say, is accordingly the solution of (2.6) which satisfies the initial condition that $x = x_0$, $y = y_0$, $z = z_0$ when $t = 0$. The solution will yield a set of equations of the form

$$\left.\begin{aligned} x &= x(x_0, y_0, z_0, t) \\ y &= y(x_0, y_0, z_0, t) \\ z &= z(x_0, y_0, z_0, t) \end{aligned}\right\}, \tag{2.7a}$$

which, as t takes all values greater than zero, will trace out the required pathline.

In addition to streamlines and pathlines, it is useful for observational purposes to define a *streakline*. This is the curve of *all* fluid particles which at some time have coincided with a particular fixed point of space. The streakline is observed when a neutrally buoyant marker fluid is continously injected into the flow at a fixed point of space from time $\tau = -\infty$. The marker fluid may be smoke if the main flow involves a gas such as air, or a dye such as potassium permanganate if the main flow involves a liquid such as water.

If the coordinates of a particle of marker fluid are (x, y, z) at time t and the particle coincided with the injection point (x_0, y_0, z_0) at some time τ, where $\tau \leqslant t$, then the time-history of this particle is obtained by solving the equations for a pathline, given by (2.6), subject to the initial condition that $x = x_0$, $y = y_0$, $z = z_0$ at $t = \tau$. As τ takes on all possible values in the range $-\infty \leqslant \tau \leqslant t$, the locations of all fluid particles on the streakline through (x_0, y_0, z_0) are obtained. Thus the equation of the streakline at time t is given by

$$\left.\begin{aligned} x &= x(x_0, y_0, z_0, t, \tau) \\ y &= y(x_0, y_0, z_0, t, \tau) \\ z &= z(x_0, y_0, z_0, t, \tau) \end{aligned}\right\} (-\infty \leqslant \tau \leqslant t). \tag{2.7b}$$

When the flow is steady, the streaklines also coincide with the streamlines and pathlines, thus a time exposure may be used to photograph the streamlines in a steady flow.

Example [2.1]

Determine the equation of the streamlines when

$$\mathbf{q} = 2mr^{-3} \cos\theta\, \hat{\mathbf{r}} + mr^{-3} \sin\theta\, \hat{\boldsymbol{\theta}},$$

where (r, θ, ϕ) denote spherical polar coordinates and m is a constant.

In solving this problem, we could obtain the Cartesian resolutes of $\mathbf{q}$ and substitute into equation (2.4). However, it is simpler to note that if $\mathrm{d}\mathbf{r} = \mathrm{d}x\mathbf{i} + \mathrm{d}y\mathbf{i} + \mathrm{d}z\mathbf{k}$, equation (2.4) is equivalent to $\mathrm{d}\mathbf{r} = \lambda\mathbf{q}$ for some scalar λ which is not necessarily a constant. In spherical polar coordinates,

$$\mathrm{d}\mathbf{r} = \mathrm{d}r\, \hat{\mathbf{r}} + r\, \mathrm{d}\theta\, \hat{\boldsymbol{\theta}} + r \sin\theta\, \mathrm{d}\phi\, \hat{\boldsymbol{\phi}},$$

$$\mathbf{q} = q_r\, \hat{\mathbf{r}} + q_\theta\, \hat{\boldsymbol{\theta}} + q_\phi\, \hat{\boldsymbol{\phi}},$$

and (2.4) is evidently equivalent to

$$\frac{\mathrm{d}r}{q_r} = \frac{r\, \mathrm{d}\theta}{q_\theta} = \frac{r \sin\theta\, \mathrm{d}\phi}{q_\phi}. \qquad (1)$$

For the example under consideration, $q_\phi = 0$ and (1) gives

$$\phi = \text{constant} = c,$$

$$2 \cot\theta\, \mathrm{d}\theta = \mathrm{d}r/r,$$

which on integration gives $r = A \sin^2\theta$ with A a constant. The streamlines are therefore the intersections of the planes $\phi = c$ and the surfaces of revolution $r = A \sin^2\theta, (A > 0)$.

The motion is *axisymmetric* about the axis $\theta = 0, \pi$, and a trace of the streamlines in a typical plane $\phi = c$ is indicated in Fig. 2.3.

Notice that for $A > 0$, the surface $r = A \sin^2\theta$ lies within the sphere $r = A$, and when $A \to \infty$, the streamline is the axis of symmetry $\theta = 0, \pi$.

The flow field considered in this example is that due to a *dipole* or *doublet* placed at the origin $r = 0$. We discuss this type of singularity (note that $\mathbf{q}$ is infinite at $r = 0$) in Chapter 4. It is worth noting that it is the fluid mechanical analogue of a magnetic dipole or small bar magnet in electromagnetism. The lines of force of a small bar magnet placed at the origin with axis along $\theta = 0, \pi$ would look like the streamlines depicted in Fig. 2.3.

Since the flow is steady, the pathlines and streaklines coincide with the streamlines.

Example [2.2]

Obtain the equations of the streamline, pathline and streakline which pass through $(l, l, 0)$ at $t = 0$ for the two-dimensional flow

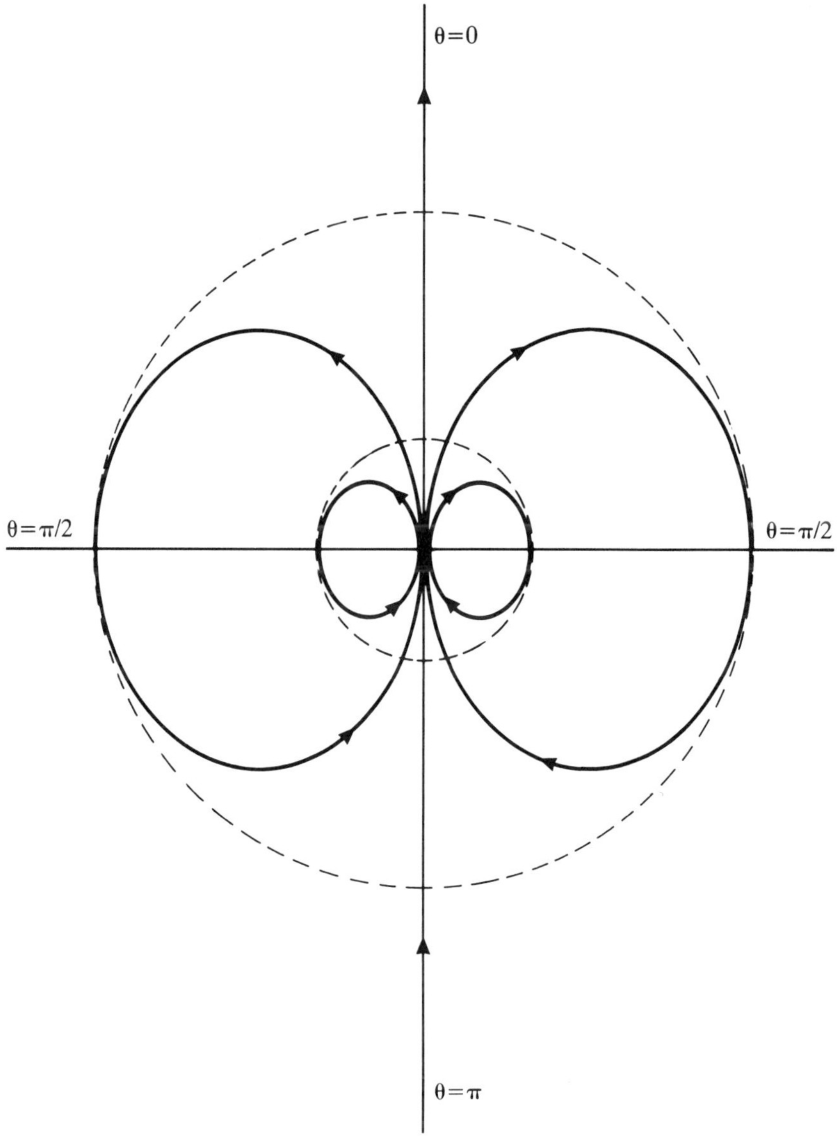

Fig. 2.3

$$u = \frac{x}{t_0}\left(1 + \frac{t}{t_0}\right), \quad v = \frac{y}{t_0}, \quad w = 0,$$

where l and t_0 are constants having respectively the dimensions of length and time.

In this problem, it is convenient to introduce *dimensionless* coordinates X, Y, Z and time T by writing

$$X = x/l, \quad Y = y/l, \quad Z = z/l, \quad T = t/t_0.$$

Streamlines

Equation (2.4), in terms of the dimensionless variables, gives

$$\frac{\mathrm{d}X}{X(1+T)} = \frac{\mathrm{d}Y}{Y} = \frac{\mathrm{d}Z}{0}.$$

The variables X, Y, Z and T are independent in these equations and on integration we obtain

$$X = c_1\, Y^{(1+T)}, \quad Z = c_2$$

with c_1 and c_2 constants. Equations (1) give the complete family of streamlines at all times $t = t_0 T$. There is, however, only one streamline which passes through $X = Y = 1, Z = 0$ when $T = 0$ and this streamline has the equation $Y = X$.

Pathlines

In dimensionless form, equation (2.6) gives

$$\frac{\mathrm{d}X}{\mathrm{d}T} = X(1+T), \quad \frac{\mathrm{d}Y}{\mathrm{d}T} = Y, \quad \frac{\mathrm{d}Z}{\mathrm{d}T} = 0. \tag{2}$$

Now X, Y, Z are the dimensionless coordinates of a fluid particle and are therefore functions of T. Integration of (2) leads to

$$X = K_1\, \mathrm{e}^{T + T^2/2}, \qquad Y = K_2\, \mathrm{e}^{T}, \qquad Z = K_3 \tag{3}$$

with K_1, K_2, K_3 constants. These are parametric equations of the pathlines and the particular pathline which passes through $X = Y = 1, Z = 0$ when $T = 0$ is given by (3) when $K_1 = K_2 = 1, K_3 = 0$. Eliminating T leads to

$$X = Y^{[1 + \frac{1}{2}(\ln Y)^2]}, \quad Z = 0$$

Streaklines

The pathline which passes through $X = Y = 1, Z = 0$ when $T = \tau$ is

$$X = \exp[T + \tfrac{1}{2}T^2 - \tau - \tfrac{1}{2}\tau^2], \quad Y = \exp[T - \tau], \quad Z = 0.$$

These are the parametric equations of the streaklines and are valid for all T. At $T = 0$, the equations give

$$X = \exp[-\tau - \tfrac{1}{2}\tau^2], \quad Y = \exp[-\tau], \quad Z = 0,$$

and elimination of τ gives

$$X = Y^{[1 - \frac{1}{2}(\ln Y)^2]}.$$

This example gives a striking illustration of how the three types of curve associated with a fluid motion can deviate when the flow is unsteady. In Fig. 2.4, the curves are sketched in the (X, Y) plane.

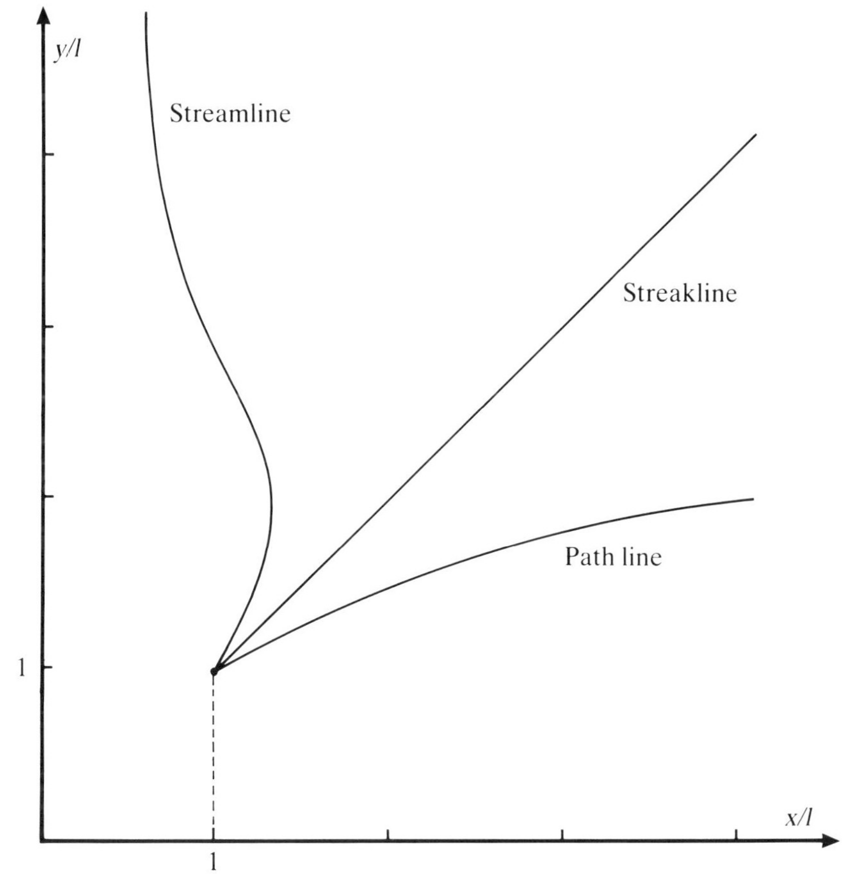

Fig. 2.4

2.4 VORTICITY AND CIRCULATION

The *vorticity vector* $\boldsymbol{\zeta}$ is defined by

$$\boldsymbol{\zeta} = \text{curl } \mathbf{q} = \nabla \times \mathbf{q}. \tag{2.8a}$$

A curve drawn in the fluid such that at each point of it, the local vorticity is along the direction of the tangent to the curve at that point, is a *vortex line*. In Cartesian form, with $\boldsymbol{\zeta} = \zeta_1 \mathbf{i} + \zeta_2 \mathbf{j} + \zeta_3 \mathbf{k}$, the differential equations of the vortex lines are

$$\frac{dx}{\zeta_1} = \frac{dy}{\zeta_2} = \frac{dz}{\zeta_3}. \tag{2.8b}$$

In general, the vorticity is not parallel to the velocity, so vortex lines and streamlines do not coincide.

Analogous to stream tubes, vortex tubes are defined as the surfaces formed when the vortex lines are drawn which pass through all points on a closed curve drawn within the fluid. A typical vortex tube is shown in Fig. 2.5 where S_1 and S_2 are any two sections of the tube, not necessarily normal. In $\mathbf{n}_1$ and $\mathbf{n}_2$ are the unit vectors normal to these sections and in each case drawn *out* of the fluid within the tube and S_3 is the curved surface of the tube, the Divergence Theorem (see Section 1.14) applied to the volume τ bounded by S_1, S_2 and S_3 gives

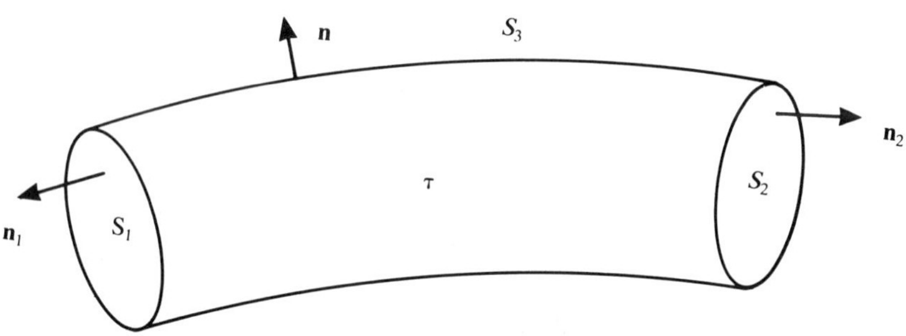

Fig. 2.5

$$\int_{S_1} \mathbf{n} \cdot \boldsymbol{\zeta}\, dS + \int_{S_2} \mathbf{n} \cdot \boldsymbol{\zeta}\, dS + \int_{S_3} \mathbf{n} \cdot \boldsymbol{\zeta}\, dS = \int_{\tau} \operatorname{div} \boldsymbol{\zeta}\, d\tau. \tag{2.9}$$

But div $\boldsymbol{\zeta} \equiv$ div curl $\mathbf{q} \equiv 0$ and $\mathbf{n} \cdot \boldsymbol{\zeta} = 0$ over S_3 by definition of a vortex tube. Thus

$$\int_{S_1} \mathbf{n} \cdot \boldsymbol{\zeta}\, dS + \int_{S_2} \mathbf{n} \cdot \boldsymbol{\zeta}\, dS = 0. \tag{2.10}$$

A *vortex filament* is a vortex tube whose cross section is of infinitesimal dimensions. The vorticity is therefore effectively constant over any cross-section. Thus if ζ_1 and ζ_2 represent the vorticities over S_1 and S_2 respectively and choosing S_1 and S_2 to be normal cross-sections, we see that for a vortex filament, (2.10) reduces to

$$\zeta_1 S_1 = \zeta_2 S_2.$$

Thus the magnitude of the product of vorticity and cross-sectional area is a constant along a vortex filament. This shows that a vortex filament cannot terminate at a point within the fluid. Vortex filaments must either be closed, i.e. vortex rings or terminate at boundaries as in whirlpools. The analogy between vortex filaments and stream filaments is immediately apparent. However, in the case of vorticity, it is not necessary to assume incompressibility of the fluid.

Let $\mathscr{C}$ be a closed curve drawn in the fluid. The *circulation* Γ of the fluid velocity $\mathbf{q}$ about $\mathscr{C}$ is defined by

$$\Gamma = \oint_{\mathscr{C}} \mathbf{q} \cdot \mathrm{d}\mathbf{r} \tag{2.11}$$

Thus Stokes's Theorm (Section 1.15) gives

$$\Gamma = \oint_{\mathscr{C}} \mathbf{q} \cdot \mathrm{d}\mathbf{r} = \int_S \boldsymbol{\zeta} \cdot \mathrm{d}\mathbf{S}, \tag{2.12}$$

with S any surface rimmed by $\mathscr{C}$.

An interesting and important class of flows occurs when $\boldsymbol{\zeta} = \text{curl}\,\mathbf{q} = \mathbf{0}$. The flow is then said to be *irrotational*. A consequence of irrotational flow is that the circulation about any closed curve drawn in the fluid is zero. This in turn implies (see Section 1.16) that $\mathbf{q}$ is expressible in terms of a scalar function $\Phi(x, y, z, t)$ by the equation

$$\mathbf{q} = \nabla\Phi. \tag{2.13}$$

The scalar function Φ is called the *velocity potential*. Irrotational flows are therefore also known as *potential flows*.

The algebraic sign of the velocity potential may be chosen to be of the opposite sign if preferred, and this convention is adopted by many authors. The choice of sign then follows the generally accepted convention used for the electrostatic and magnetostatic potentials, in which case lines of force are directed from points of high to points of low potential. The choice of sign adoped in (2.13) follows the convention used in defining gravitational potential, but there is no physical reason why either convention should not be adopted for the velocity potential. In this book we shall use that defined by (2.13).

The reader should note that in an irrotational flow, the streamlines are orthogonal to the equi-Φ surfaces.

Example [2.3]

Show that $\mathbf{q} = (x^2 + y^2)^{-1}\,(x\mathbf{j} - y\mathbf{i})$ is a possible fluid velocity for irrotational flow of an incompressible fluid. Determine the velocity potential and obtain the equations of the streamlines and equipotential surfaces.

For the given $\mathbf{q}$ to be the velocity vector of an incompressible fluid in motion, $\text{div}\,\mathbf{q} = 0$.

Now

$$\text{div}\,\mathbf{q} = -\frac{\partial}{\partial x}\left(\frac{y}{x^2+y^2}\right) + \frac{\partial}{\partial y}\left(\frac{x}{x^2+y^2}\right) = 0.$$

For irrotational flow, curl $\mathbf{q} = \mathbf{0}$.
But

$$\text{curl}\,\mathbf{q} = \begin{vmatrix} \mathbf{i} & \mathbf{j} & \mathbf{k} \\ \partial/\partial x & \partial/\partial y & \partial/\partial z \\ \dfrac{-y}{x^2+y^2} & \dfrac{x}{x^2+y^2} & 0 \end{vmatrix}$$

$$= \mathbf{k}\left\{\frac{y^2-x^2}{(x^2+y^2)^2} + \frac{x^2-y^2}{(x^2+y^2)^2}\right\} = \mathbf{0}.$$

The velocity potential $\Phi(x, y, z)$ must satisfy $\mathbf{q} = \nabla\Phi$, in which case

$$\frac{\partial\Phi}{\partial x} = \frac{-y}{x^2+y^2}, \quad \frac{\partial\Phi}{\partial y} = \frac{x}{x^2+y^2}, \quad \frac{\partial\Phi}{\partial z} = 0.$$

The third equation gives $\Phi = \Phi\,(x, y)$ while the first and second give respectively on integration

$$\Phi = -\tan^{-1}\,(x/y) + f_1(y),$$

and

$$\Phi = -\tan^{-1}\,(x/y) + f_2(x).$$

Thus $f_1(y) = f_2(x) =$ constant, which without loss of generality may be chosen to be zero since a constant potential gives rise to no velocity.

The equipotentials are the planes $y = cx$ with c any constant. The differential equations of the streamlines are

$$\frac{-\mathrm{d}x(x^2+y^2)}{y} = \frac{\mathrm{d}y(x^2+y^2)}{x} = \frac{\mathrm{d}z}{0},$$

which on integration give

$$x^2+y^2 = a^2, \quad z = b,$$

for constants a, b.

The streamlines are therefore circles whose centres lie on the z-axis and whose planes are perpendicular to this axis. The streamlines are accordingly orthogonal to the equipotential surfaces.

Example [2.4]

Describe the fluid motion with

$$\mathbf{q} = -\omega(y\mathbf{i} - x\mathbf{j}),$$

where ω is a constant.

We first note that div $\mathbf{q} = 0$. This means that $\mathbf{q}$ is a possible velocity vector for motion of an *incompressible* fluid. Furthermore,

$$\boldsymbol{\zeta} = \text{curl } \mathbf{q} = 2\omega\mathbf{k}.$$

This type of motion is an example of a 'rigid-body' motion of the fluid, i.e. the fluid moves as if it were a rigid body rotating about the z-axis with constant angular velocity $\omega\mathbf{k}$. In fact, if the fluid were suddenly frozen solid, it would continue to move in this manner. It is no coincidence that $\frac{1}{2}\boldsymbol{\zeta}$ is the angular velocity of the fluid in this example, and we shall show in Section 2.9 that this is true at a general point of any flow. The reader will therefore appreciate why a flow in which curl $\mathbf{q} = \mathbf{0}$ is referred to as *irrotational*.

The form of $\mathbf{q}$ in the example shows that for this type of flow, the circulation about the circle $x^2 + y^2 = a^2$, $z =$ constant, when a is a constant, is $2\pi\omega a^2$.

2.5 ACCELERATION AT A POINT OF A FLUID

In Section 2.2, we defined the velocity at a point P of the fluid to be that of the fluid particle which coincides with the geometrical point P at the instant in time under consideration. It is therefore natural to define the acceleration at P at time t to be the acceleration of the same fluid particle which coincides with P.

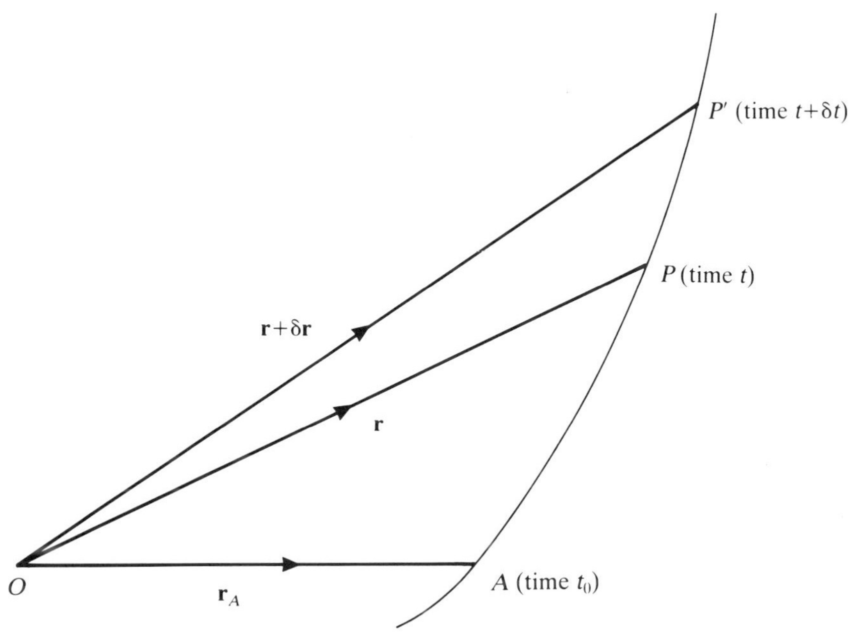

Fig. 2.6

To determine the particle acceleration at time t, it is necessary to follow the motion of the particle during a small interval from time t to time $t + \delta t$, since the acceleration of the particle is by definition the rate of change of its velocity at time t. Fig. 2.6 shows the pathline of a particle of fluid which at time t coincides with the geometrical point P. Its position vector relative to a fixed origin O is $\mathbf{r}$ at time t. At some earlier time t_0, which could of course be zero, this particle was at A with position vector $\mathbf{r}_A$, and at time $t + \delta t$ it has moved to coincide with the point P' with position vector $r + \delta\mathbf{r}$, so clearly $\delta\mathbf{r} \equiv \overrightarrow{PP'}$. The reader should bear in mind that A, P, P' are geometrical points of the region occupied by the fluid and they coincide with the locations of the *same* fluid particle at the times t_0, t, $t + \delta t$ respectively. Let $\mathbf{f}$ be the acceleration of the particle at time t when it coincides with P. From definition it is given by

$$\mathbf{f} = \lim_{\delta t \to 0} \frac{(\text{change in particle velocity in time } \delta t)}{\delta t}. \tag{2.14}$$

But the particle velocity at time t is $\mathbf{q}(\mathbf{r}, t)$ while the particle velocity at time $t + \delta t$ is $\mathbf{q}(\mathbf{r} + \delta\mathbf{r}, t + \delta t)$. Thus (2.14) gives

$$\mathbf{f} = \lim_{\delta t \to 0} \frac{[\mathbf{q}(\mathbf{r} + \delta\mathbf{r}, t + \delta t) - \mathbf{q}(\mathbf{r}, t)]}{\delta t} \tag{2.15}$$

Now

$$\frac{[\mathbf{q}(\mathbf{r} + \delta\mathbf{r}, t + \delta t) - \mathbf{q}(\mathbf{r}, t)]}{\delta t} = \frac{[\mathbf{q}(\mathbf{r} + \delta\mathbf{r}, t + \delta t) - \mathbf{q}(\mathbf{r}, t + \delta t)]}{\delta t} + \frac{[\mathbf{q}(\mathbf{r}, t + \delta t) - \mathbf{q}(\mathbf{r}, t)]}{\delta t}. \tag{2.16}$$

Since $\mathbf{r}$ is independent of t, it follows that

$$\lim_{\delta t \to 0} \frac{[\mathbf{q}(\mathbf{r}, t + \delta t) - \mathbf{q}(\mathbf{r}, t)]}{\delta t} = \frac{\partial \mathbf{q}}{\partial t}. \tag{2.17}$$

Furthermore, using the Taylor expansion,

$$\mathbf{q}(\mathbf{r} + \delta\mathbf{r}, t + \delta t) - \mathbf{q}(\mathbf{r}, t + \delta t) = (\delta\mathbf{r} \cdot \nabla)\, \mathbf{q}\, (\mathbf{r}, t + \delta t) + \boldsymbol{\epsilon} \tag{2.18}$$

where $|\boldsymbol{\epsilon}| = O((\delta r)^2)$. But $\delta\mathbf{r}$ is merely the displacement of the fluid particle in time δt. Thus

$$\delta\mathbf{r} = \mathbf{q}(\mathbf{r}, t)\delta t \tag{2.19}$$

plus second order terms in small quantities which vanish as $\delta t \to 0$. It therefore follows that

$$\lim_{\delta t \to 0} \frac{[\mathbf{q}(\mathbf{r}+\delta\mathbf{r}, t+\delta t) - \mathbf{q}(\mathbf{r}, t+\delta t)]}{\delta t} = (\mathbf{q} \cdot \nabla)\mathbf{q} \tag{2.20}$$

with the right-hand sides of (2.20) and (2.17) evaluated at (**r**, **t**). Combining these two equations with (2.15) and (2.16) leads to the result that at time t, the acceleration of the fluid at P is given by

$$\mathbf{f} = \frac{\partial \mathbf{q}}{\partial t} + (\mathbf{q} \cdot \nabla)\,\mathbf{q}. \tag{2.21}$$

The acceleration is an example of the rate of change of a quantity, associated with the fluid, which has to be determined by *following* the fluid particle as it moves in the time-interval from t to $t + \delta t$. The same procedure can be applied to find the rate of change of any physical property associated with the fluid, such as density, and the reader may easily verify that if $F = F(\mathbf{r}, t)$ is any scalar or vector quantity associated with the fluid, its rate of change at time t is given by

$$\frac{\partial F}{\partial t} + (\mathbf{q} \cdot \nabla)F. \tag{2.22}$$

This rate of change is often referred to as the *total* rate of change since it is made up of two terms – the *local* rate of change $\partial F/\partial t$ and the *convective* rate of change $(\mathbf{q} \cdot \nabla)F$. We shall denote it by

$$\frac{\mathrm{D}F}{\mathrm{D}t} \equiv \frac{\partial F}{\partial t} + (\mathbf{q} \cdot \nabla)F. \tag{2.23}$$

The operator $\mathrm{D}/\mathrm{D}t$ is essentially Lagrangian since the motion of a specific fluid particle is followed in order to evaluate it. However, the operators on the right-hand side are Eulerian since **r** and t are independent. For a steady motion, $\partial F/\partial t = 0$. In the particular case when $F = \rho$, the *density* of the fluid, equation (2.23) gives for a general unsteady flow,

$$\frac{\mathrm{D}\rho}{\mathrm{D}t} \equiv \frac{\partial \rho}{\partial t} + (\mathbf{q} \cdot \nabla)\rho. \tag{2.24}$$

Consider a small element of fluid of volume $\delta\tau$ which encloses the point P and moves with the fluid. If $\delta\tau$ is sufficiently small for variations in ρ over $\delta\tau$ to be ignored, then mass conservation of the fluid within $\delta\tau$ requires that $\rho\delta\tau =$ constant, and therefore

$$\frac{\mathrm{D}}{\mathrm{D}t}(\rho\delta\tau) = \frac{\mathrm{D}\rho}{\mathrm{D}t}\,\delta\tau + \rho\,\frac{\mathrm{D}}{\mathrm{D}t}\,(\delta\tau) = 0. \tag{2.25}$$

If the fluid is *incompressible*, the volume $\delta\tau$ is a constant. Thus a necessary and sufficient condition for a fluid to be incompressible is that

$$\frac{\mathrm{D}\rho}{\mathrm{D}t} = 0 \tag{2.26}$$

at all points of the fluid. From (2.24) it will be clear that (2.26) does not necessarily require that ρ is a constant at all points of the fluid. However, from the definition of $\mathrm{D}\rho/\mathrm{D}t$, the density ρ is a constant along each pathline in an incompressible fluid.

Some authors use the notation $\mathrm{d}F/\mathrm{d}t$ rather than $\mathrm{D}F/\mathrm{D}t$, but this can lead to confusion since in the Eulerian view of fluid mechanics, the variables $\mathbf{r}$ and t are *independent* and consequently the operator $\mathrm{d}/\mathrm{d}t$ as a total differential operator in the usual sense of the differential calculus does not exist. If, however, $\mathbf{r}$ denotes the position vector of the fluid particle coinciding with P at time t, then $\mathbf{r} = \mathbf{r}(t)$ and therefore

$$\frac{\mathrm{d}F}{\mathrm{d}t} = \frac{\partial F}{\partial t} + \frac{\partial F}{\partial x}\frac{\mathrm{d}x}{\mathrm{d}t} + \frac{\partial F}{\partial y}\frac{\mathrm{d}y}{\mathrm{d}t} + \frac{\partial F}{\partial z}\frac{\mathrm{d}z}{\mathrm{d}t},$$

and since

$$u = \frac{\mathrm{d}x}{\mathrm{d}t}, \quad v = \frac{\mathrm{d}y}{\mathrm{d}t}, \quad w = \frac{\mathrm{d}z}{\mathrm{d}t},$$

it follows that

$$\frac{\mathrm{d}F}{\mathrm{d}t} = \frac{\partial F}{\partial t} + (\mathbf{q}\,.\,\nabla)F = \frac{\mathrm{D}F}{\mathrm{D}t}, \tag{2.27}$$

but it must be clearly understood that in equation (2.27) $\mathbf{r}$ is now the *Lagrangian* position vector of the moving particle rather than the *Eulerian* position vector of the fluid point. Thus the operators $\mathrm{d}/\mathrm{d}t$ and $\mathrm{D}/\mathrm{D}t$ are the same only in Lagrangian coordinates.

2.6 EQUATION OF CONTINUITY

Let us consider a region of space through which a fluid is flowing, perhaps from an external reservoir. If the region is devoid of *sources*, or points where the fluid actually enters from the reservoir, and also *sinks*, or points of fluid exit such as a drain, then the total mass of fluid within the region must be a constant in accordance with the principle of conservation of mass. The *density* ρ at a point P of the fluid is defined to be the mass per unit volume of fluid at that point. It is measured by taking out a small volume of fluid which includes the point P and determining its mass. The mass is then divided by the volume. If this calculation is repeated with successively smaller volumes of fluid, the values obtained for the mass/volume ratio will approach a sensibly constant quantity which we can identify as the density ρ at the point P. In practice, this sensibly constant value for ρ would be reached long before the volume of fluid became so small that

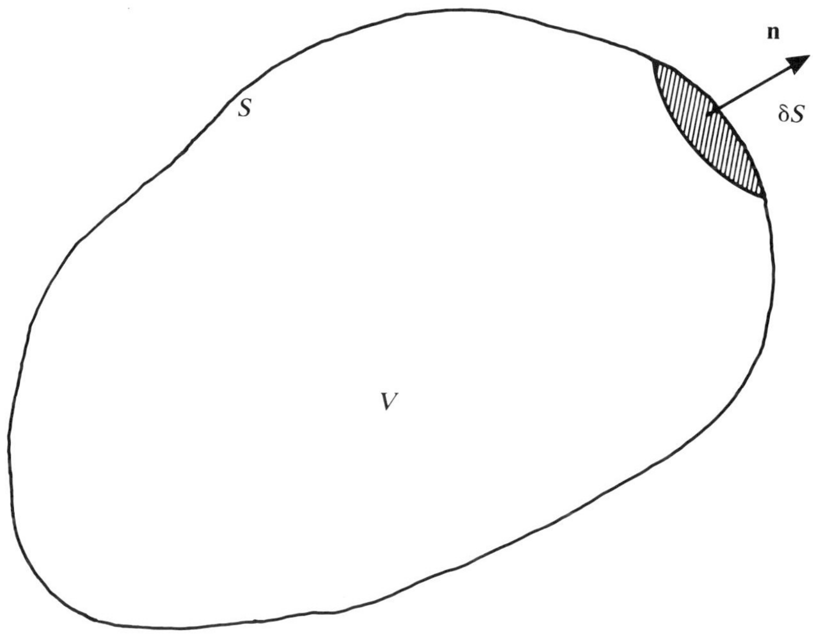

Fig. 2.7

there was any question of the assumptions of the continuum hypothesis, discussed in Section 2.1, being violated.

Now let us consider a closed geometrical surface S, *fixed in space*, which is drawn in a region through which fluid flows. The surface S may be arbitrarily chosen except that we suppose that it does not include points which are either sources or sinks. Let V be the volume enclosed by S and $\mathbf{n}$ denote the unit vector normal to a general element δS of S direction out of V, as indicated in Fig. 2.7. The mass of fluid contained within the volume V at time t is M say, where

$$M = \int_V \rho \, \mathrm{d}V.$$

The rate of *decrease* of this mass is $-\mathrm{d}M/\mathrm{d}t$ and, in the absence of sources and sinks within V, it is exactly matched by the rate at which fluid *leaves* the region V by flowing across the surface, S. To determine what this is, consider first the mass of fluid which leaves V by flowing across an element δS of S in time δt. This quantity of fluid is simply that which is contained in a small cylinder of cross-section δS and length $(\mathbf{q} \cdot \mathbf{n})\delta t = q_n \delta t$, as indicated in Fig. 2.8. The mass of this fluid is accordingly $\rho q_n \delta S \delta t$ and the *rate* at which fluid leaves V by flowing across the element δS is

$$\rho q_n \, \delta S = \rho(\mathbf{q} \cdot \mathbf{n}) \, \delta S. \tag{2.28}$$

Summing over all such elements δS of S, we obtain the rate of flow of fluid out of V across the entire surface S. It therefore follows that

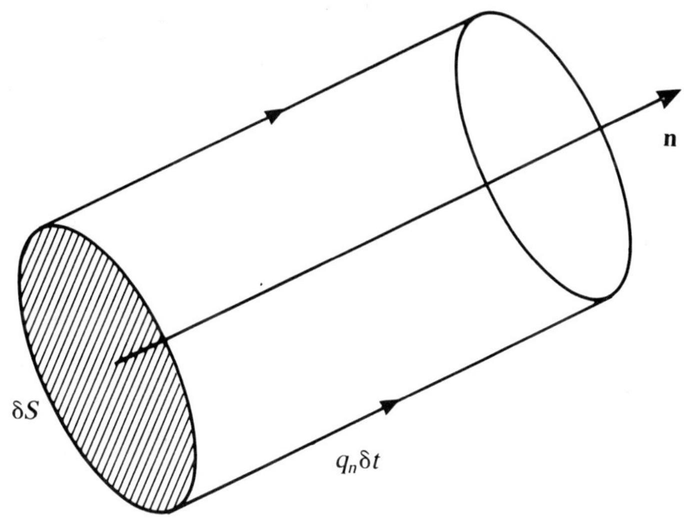

Fig. 2.8

$$-\frac{\mathrm{d}M}{\mathrm{d}t} = -\frac{\mathrm{d}}{\mathrm{d}t}\int_V \rho\,\mathrm{d}V = \int_S \rho(\mathbf{q}\,.\,\mathbf{n})\mathrm{d}S = \int_S \rho\mathbf{q}\,.\,\mathrm{d}\mathbf{S}. \tag{2.29}$$

But $\rho = \rho(x, y, z, t)$ with (x, y, z) the Cartesian coordinates of a general point of V, a *fixed* region of space. Consequently the space coordinates are independent of the time t and therefore

$$\frac{\mathrm{d}}{\mathrm{d}t}\int_V \rho\,\mathrm{d}V = \int_V \frac{\partial\rho}{\partial t}\,\mathrm{d}V. \tag{2.30}$$

If now the Divergence Theorem, proved in Section 1.14 of Chapter 1, is used to transform the surface integral in (2.29) to a volume integral over V, then (2.30) together with (2.29) yield the equation

$$\int_V \left\{\frac{\partial\rho}{\partial t} + \mathrm{div}\,(\rho\mathbf{q})\right\}\mathrm{d}V = 0. \tag{2.31}$$

But V is arbitrary so it follows from the result proved in Section 1.13 of Chapter 1 that at *any point* of the fluid which is neither a source nor a sink,

$$\frac{\partial\rho}{\partial t} + \mathrm{div}(\rho\mathbf{q}) = 0. \tag{2.32}$$

Equation (2.32) is known as the *equation of continuity* of the fluid. By use of the identity

$$\mathrm{div}(\rho\mathbf{q}) = \rho\,\mathrm{div}\,\mathbf{q} + \mathbf{q}\,.\,\nabla\rho,$$

equation (2.32) can be rewritten in the alternative forms.

$$\frac{\partial \rho}{\partial t} + \rho \operatorname{div} \mathbf{q} + \mathbf{q} \cdot \nabla \rho = 0, \tag{2.33}$$

$$\frac{\mathrm{D}\rho}{\mathrm{D}t} + \rho \operatorname{div} \mathbf{q} = 0, \tag{2.34}$$

$$\frac{\mathrm{D}}{\mathrm{D}t} (\ln \rho) + \operatorname{div} \mathbf{q} = 0. \tag{2.35}$$

Thus a necessary and sufficient condition for *incompressible* flow is

$$\frac{\mathrm{D}\rho}{\mathrm{D}t} = 0, \tag{2.36}$$

which was established in Section 2.5. Equation (2.34) shows that for such a fluid, the equation of continuity is

$$\operatorname{div} \mathbf{q} = 0. \tag{2.37}$$

Equation (2.36) is trivially satisfied by ρ = constant, in which case the fluid is *homogeneous*. There are other solutions of (2.36) with variable density which occur in spatially stratified fluids such as salt water with varying salinity. Equations (2.33) to (2.35) apply to all fluids and although (2.37) applies strictly to incompressible fluids, i.e. liquids, it also has applications in flows of gases when their compressibility is not significant. This occurs when the flow velocity is not comparable with the speed of sound propagation for the particular gas. We shall discuss this matter more fully in a subsequent volume.

Example [2.5]

If $\sigma(s)$ is the cross-sectional area of a stream filament, prove that the equation of continuity is

$$\frac{\partial}{\partial t} (\rho\sigma) + \frac{\partial}{\partial s} (\rho\sigma q) = 0,$$

where δs is an element of arc of the filament and q is the fluid speed.

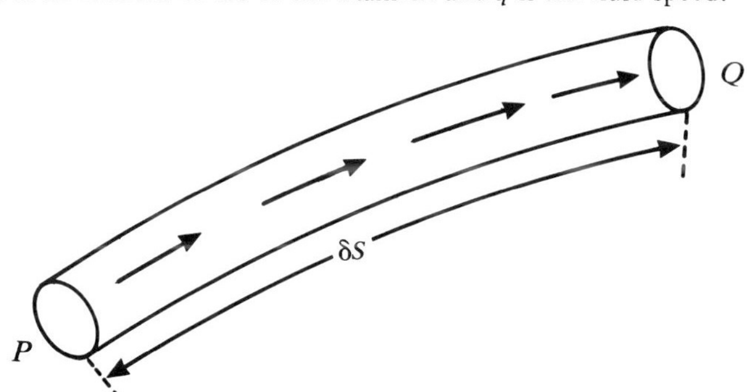

Fig. 2.9

Rate of flow of fluid out of volume of filament between ends is

$$(\rho q\sigma)_Q - (\rho q\sigma)_P = \frac{\partial}{\partial s}(\rho q\sigma)_P\delta s + O(\delta s^2) \tag{1}$$

with P, Q points on end sections.

Rate of decrease of mass of fluid within segment of filament is

$$-\frac{\partial}{\partial t}(\rho\sigma)_P\delta s + O(\delta s^2). \tag{2}$$

Equate (1) and (2), divide by δs and let $\delta s \to 0$ to establish the required result at any point P of the filament.

2.7 REYNOLDS' TRANSPORT THEOREM

In deriving the equation of continuity in the previous section, the reader will have noticed that we chose to focus our attention on a fixed region of space, so that different fluid particles occupy this region at different times. A volume V defined in this way is therefore *Eulerian*. Sometimes, however, it is convenient to consider a particular region of the fluid consisting of the same fluid particles. The region τ then moves with the fluid and is therefore *Lagrangian* and the space coordinates defining the region τ will naturally be functions of time since they are dependent on where the fluid particles have moved to as time progresses.

Let $\mathbf{r}$ be the position vector of a general point of the Lagrangian region τ which moves with the fluid. Since τ consists of the same fluid particles, then $\mathbf{r} = \mathbf{r}(t)$ defines the position vector of a typical fluid particle of the set which constitutes τ. Let $F(\mathbf{r}, t)$ denote some scalar field associated with the fluid. It could be density or a component of velocity for example. Whatever the physical quantity $F(\mathbf{r}, t)$ denotes, it is clear that the integral

$$\int_\tau F(\mathbf{r}, t)\,\mathrm{d}\tau \tag{2.38}$$

is a function of t only. Reynolds' Transport Theorem states that

$$\frac{\mathrm{d}}{\mathrm{d}t}\int_\tau F\,\mathrm{d}\tau = \int_\tau \left\{\frac{\partial F}{\partial t} + \operatorname{div}(F\mathbf{q})\right\}\mathrm{d}\tau \tag{2.39}$$

and gives us the means of evaluating the time rate of change of an integral of the type (2.38).

To prove this theorem, we first note that if t is increased by a small amount δt, then τ will change from $\tau(t)$ to $\tau(t + \delta t)$, and accordingly

$$\frac{\mathrm{d}}{\mathrm{d}t}\int_\tau F\,\mathrm{d}\tau = \lim_{\delta t \to 0}\frac{1}{\delta t}\left\{\int_{\tau(t+\delta t)} F(\mathbf{r}, t + \delta t)\,\mathrm{d}\tau - \int_{\tau(t)} F(\mathbf{r}, t)\,\mathrm{d}\tau\right\}. \tag{2.40}$$

The right-hand side of (2.40) can be written as

$$\lim_{\delta t \to 0} \left\{ \frac{1}{\delta t} \int_{\tau(t+\delta t)-\tau(t)} F(\mathbf{r}, t + \delta t)\, \mathrm{d}\tau \right\}$$
$$+ \lim_{\delta t \to 0} \left\{ \frac{1}{\delta t} \int_{\tau(t)} [F(\mathbf{r}, t + \delta t) - F(\mathbf{r}, t)]\, \mathrm{d}\tau \right\}. \tag{2.41}$$

The second limit in (2.41) is clearly

$$\int_{\tau(t)} \frac{\partial F}{\partial t}\, \mathrm{d}\tau. \tag{2.42}$$

To evaluate the first limit in (2.41), note that an element of the region $\tau(t + \delta t) - \tau(t)$ is the volume of a cylindrical region of cross-section $\delta\Sigma$ and length $(\mathbf{q} \,.\, \mathbf{n})\delta t$, where $\delta\Sigma$ is an element of the surface $\Sigma(t)$ which is the boundary of $\tau(t)$, as illustrated in Fig. 2.10(b), since the fluid particles which make up $\Sigma(t)$ move to become the surface $\Sigma(t + \delta t)$ at time $t + \delta t$. Thus

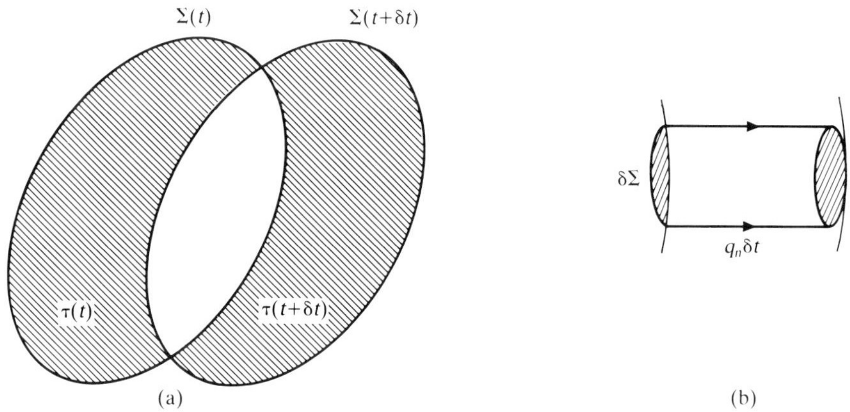

Fig. 2.10

$$\lim_{\delta t \to 0} \left\{ \frac{1}{\delta t} \int_{\tau(t+\delta t)-\tau(t)} F(\mathbf{r}, t + \delta t)\, \mathrm{d}\tau \right\}$$
$$= \lim_{\delta t \to 0} \left\{ \frac{1}{\delta t} \int_{\Sigma(t)} F(\mathbf{r}, t + \delta t)\, (\mathbf{q} \,.\, \mathbf{n})\delta t\, \mathrm{d}\Sigma \right\}$$
$$= \int_{\Sigma} (\mathbf{n} \,.\, \mathbf{q}F)\, \mathrm{d}\Sigma. \tag{2.43}$$

The surface integral appearing in (2.43) may be transformed back to a volume integral by means of the Divergence Theorem and we have the result

$$\lim_{\delta t \to 0} \left\{ \frac{1}{\delta t} \int_{\tau(t+\delta t)-\tau(t)} F(r, t+\delta t)\, \mathrm{d}\tau \right\}$$

$$= \int_\tau \operatorname{div}(\mathbf{q}F)\, \mathrm{d}\tau, \tag{2.44}$$

completing the proof of Reynolds' Transport Theorem to give

$$\frac{\mathrm{d}}{\mathrm{d}t} \int_\tau F\, \mathrm{d}\tau = \int_\tau \left\{ \frac{\partial F}{\partial t} + \operatorname{div}(F\mathbf{q}) \right\} \mathrm{d}\tau. \tag{2.45}$$

Some important results can be deduced immediately from the theorem. With $F \equiv 1$, equation (2.45) gives

$$\frac{\mathrm{d}\tau}{\mathrm{d}t} = \frac{\mathrm{d}}{\mathrm{d}t} \int_\tau \mathrm{d}\tau = \int_\tau \operatorname{div} \mathbf{q}\, \mathrm{d}\tau. \tag{2.46}$$

If the fluid is *incompressible*, the volume τ of the fluid is a constant. Thus

$$\int_\tau \operatorname{div} \mathbf{q}\, \mathrm{d}\tau = 0,$$

and since the choice of region τ is arbitrary, it follows that at any point of an incompressible fluid,

$$\operatorname{div} \mathbf{q} = 0. \tag{2.47}$$

With $F = \rho$, the density of the fluid, equation (2.45) now gives

$$\frac{\mathrm{d}}{\mathrm{d}t} \int_\tau \rho\, \mathrm{d}\tau = \int_\tau \left\{ \frac{\partial \rho}{\partial t} + \operatorname{div}(\rho \mathbf{q}) \right\} \mathrm{d}\tau. \tag{2.48}$$

However, the integral on the left-hand side is the mass of the fluid within the region τ, which, because it is made up of the same fluid particles *at all times*, is a constant in accord with the principle of conservation of matter. Therefore

$$\int_\tau \left\{ \frac{\partial \rho}{\partial t} + \operatorname{div}(\rho \mathbf{q}) \right\} \mathrm{d}\tau = 0,$$

and again since τ is arbitrary, it follows that at all points of the fluid

$$\frac{\partial \rho}{\partial t} + \operatorname{div}(\rho \mathbf{q}) = 0, \tag{2.49}$$

which is the *equation of continuity* of the fluid. This equation was obtained in the previous section by considering the rate of change of the mass of fluid within a region V fixed in space. The reader should notice that in using the Divergence

Theorem to derive the Reynolds' Transport Theorem, it is assumed that the functions F and $\mathbf{q}$ are differentiable throughout τ which, when $F = \rho$, is compatible with the conditions of no sources and sinks imposed in deriving (2.31) as the equation of continuity in the previous section.

Since (2.49) is exactly the same equation as (2.32), the variants (2.33) to (2.35) follow. In the case of an incompressible fluid, we have already shown that div $\mathbf{q} = 0$, as given by (2.46), and it follows that a further property of such a fluid is that at any point

$$\frac{\partial \rho}{\partial t} + (\mathbf{q} \,.\, \nabla)\rho = \frac{\mathrm{D}\rho}{\mathrm{D}t} = 0. \tag{2.50}$$

We have therefore been able to arrive at the equation of continuity by two routes, using both the Eulerian and Lagrangian view of fluid motion.

For an incompressible fluid (2.39) can be written in a local form by letting $\tau \to 0$ about a general point P of the fluid. We then obtain

$$\frac{\mathrm{d}F}{\mathrm{d}t} = \frac{\partial F}{\partial t} + \operatorname{div}(F\mathbf{q}) = \frac{\mathrm{D}F}{\mathrm{D}t}, \tag{2.51}$$

noting that div $\mathbf{q} = 0$ for an incompressible fluid. It should be understood that on the left-hand side of (2.51), $\mathbf{r}$ and t are Lagrangian variables so that $\mathbf{r} = \mathbf{r}(t)$ denotes the position vector of a fluid particle, while on the right-hand side of (2.51), $\mathbf{r}$ and t could be Lagrangian, in which case $\mathrm{D}/\mathrm{D} \equiv \mathrm{d}/\mathrm{d}t$, or Eulerian so that $\mathbf{r}$ and t are independent.

Equation (2.45) may be rewritten as

$$\frac{\mathrm{d}}{\mathrm{d}t}\int_\tau F\,\mathrm{d}\tau = \int_\tau \left\{\frac{\mathrm{D}F}{\mathrm{D}t} + F \operatorname{div} \mathbf{q}\right\} \mathrm{d}\tau. \tag{2.52}$$

In constructing the proof of (2.44), we assumed that $F(\mathbf{r}, t)$ is a *scalar* field over the region occupied by the fluid. However, F could be a Cartesian resolute of a vector along one of the axes of a fixed Cartesian frame of reference. It therefore follows that the Reynolds' Transport Theorem in the form given by (2.52) is valid for both *scalar* and *vector* fields $F(\mathbf{r}, t)$.

Example [2.6]

Obtain the condition for the surface $z = \eta(x, y, t)$ to be the boundary of a moving fluid

Write $\zeta(x, y, z, t) = z - \eta(x, y, t)$ and let $P(x, y, z)$ be the position of a fluid particle at time t. If its position vector is $\mathbf{r}$ at time t, after a short interval of time δt, its position vector is approximately

$$\mathbf{r} + \mathbf{q}\,\delta t,$$

where $\mathbf{q}$ is the fluid velocity. Thus

$$\zeta(\mathbf{r}, t) = \zeta, \mathbf{r} + \mathbf{q}\,\delta t, t + \delta t) = 0$$

and it follows from definition that

$$\frac{\mathrm{D}\zeta}{\mathrm{D}t} = \frac{\partial \zeta}{\partial t} + (\mathbf{q}\,.\,\nabla)\zeta = 0.$$

With (u, v, w) the Cartesian components of $\mathbf{q}$, we find after substitution of $\zeta = z - \eta(x, y, t)$ that

$$u\,\frac{\partial \eta}{\partial x} + v\,\frac{\partial \eta}{\partial y} + \frac{\partial \eta}{\partial t} - w = 0. \tag{2.53}$$

In particular, $z = \eta(x, y, t)$ could describe the elevation of a wave on the surface of a fluid above the equilibrium configuration. Equation (2.53) is then a kinematic condition which must be satisfied on the wave.

2.8 RATES OF CHANGE OF MATERIAL INTEGRALS

The volume integral (2.38) of the preceding section has the property that the region of integration τ moves with the fluid and consists of the *same* fluid particles. It is referred to as a *material integral*. We may likewise define line and surface integrals which are material integrals provided that the domain of integration in each case consists of the same fluid particles.

Consider for instance the line integral

$$\int_P^Q F\,\mathrm{d}\mathbf{r}, \tag{2.54}$$

taken along a material curve joining points P and Q, where $F = F(\mathbf{r}, t)$ is some scalar field associated with the fluid. Since the curve joining P to Q moves with the fluid and consists of the same fluid particles at all times, it is clear that as for the integral (2.38), the integral (2.54) is a function of t only. What we now wish to determine is its rate of change with time. At time $t + \delta t$, the curve PQ will have moved to become the curve $P'Q'$ as indicated in Fig. 2.11. The curve PQ may be subdivided by the points P_i, where $i = 0, 1, \ldots, n$ and $P_0 \equiv P$, $P_n \equiv Q$. The element of curve P_iP_{i+1} will move to become the element $P_i'\,P_{i+1}'$ of the curve $P'Q'$ at time $t + \delta t$.

Now

$$\frac{\mathrm{d}}{\mathrm{d}t}\int_P^Q F(\mathbf{r}, t)\,\mathrm{d}\mathbf{r} = \lim_{\delta t \to 0} \frac{1}{\delta t}\left\{\int_{P'}^{Q'} F(\mathbf{r}, t + \delta t)\,\mathrm{d}\mathbf{r} - \int_P^Q F(\mathbf{r}, t)\,\mathrm{d}\mathbf{r}\right\}. \tag{2.55}$$

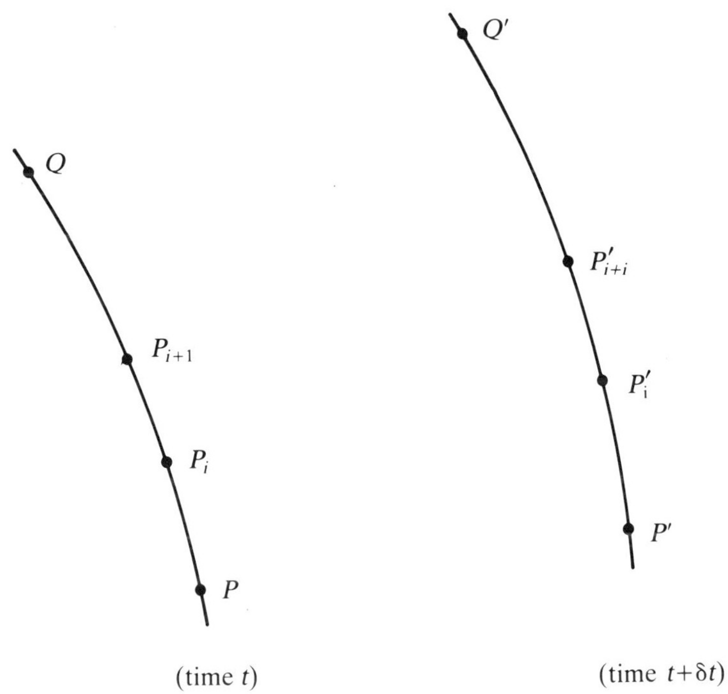

Fig. 2.11

However,

$$\int_P^Q F(\mathbf{r}, t)\, d\mathbf{r} = \lim_{\substack{n\to\infty \\ \max P_iP_{i+1}\to 0}} \sum_{i=1}^{n-1} F(\mathbf{r}_i, t)\, \overrightarrow{P_iP_{i+1}}\,, \tag{2.56}$$

$$\int_{P'}^{Q'} F(\mathbf{r}, t+\delta t)\, d\mathbf{r} = \lim_{\substack{n\to\infty \\ \max P'_iP'_{i+1}\to 0}} \sum_{i=1}^{n-1} F(\mathbf{r}'_i, t+\delta t)\, \overrightarrow{P'_iP'_{i+1}} \tag{2.57}$$

where $\mathbf{r}_i$ and $\mathbf{r}'_i$ are respectively the position vectors of P_i and P'_i. Now

$$F(\mathbf{r}'_i, t+\delta t) = F(\mathbf{r}_i, t+\delta t) + (\overrightarrow{P_iP'_i}\,.\,\nabla)\, F(\mathbf{r}_i, t+\delta t) \tag{2.58}$$

plus terms $O(|\overrightarrow{P_iP'_i}|^2)$, and since P'_i is the location at time $t + \delta t$ of the fluid particle which was at P_i at time t it follows that

$$\overrightarrow{P_iP'_i} = \mathbf{q}(\mathbf{r}_i, t)\delta t \tag{2.59}$$

plus terms of the second order in δt. Thus

$$\begin{aligned}\overrightarrow{P'_iP'_{i+1}} &= \overrightarrow{P_iP_{i+1}} + [\mathbf{q}(\mathbf{r}_{i+1}, t) - \mathbf{q}(\mathbf{r}_i, t)]\,\delta t \\ &= \overrightarrow{P_iP_{i+1}} + (\overrightarrow{P_iP_{i+1}}\,.\,\nabla)\,\mathbf{q}(\mathbf{r}_i, t)\delta t,\end{aligned} \tag{2.60}$$

ignoring terms $O(\delta t^2)$. In addition

$$F(\mathbf{r}_i, t+\delta t) = F(\mathbf{r}_i, t) + \frac{\partial F}{\partial t}(\mathbf{r}_i, t)\delta t + O(\delta t^2) \tag{2.61}$$

and it therefore follows that

$$\begin{aligned} &F(\mathbf{r}_i', t+\delta t)\,\overrightarrow{P_i'P_{i+1}'} - F(\mathbf{r}_i, t)\,\overrightarrow{P_iP_{i+1}} \\ &= \left\{\left[\frac{\partial F}{\partial t} + (\mathbf{q}\,.\,\nabla)F\right]\overrightarrow{P_iP_{i+1}} + F\,(\overrightarrow{P_iP_{i+1}}\,.\,\nabla)\mathbf{q}\right\}_{(r_i,t)}\delta t \end{aligned}$$

plus terms of order δt^2. Since $\overrightarrow{P_i'P_{i+1}'} \to 0$ as $P_iP_{i+1} \to 0$, it follows from (2.55), (2.56) and (2.57), when the appropriate limits are taken, that

$$\frac{\mathrm{d}}{\mathrm{d}t}\int_P^Q F\,\mathrm{d}\mathbf{r} = \int_P^Q \frac{\mathrm{D}F}{\mathrm{D}t}\,\mathrm{d}\mathbf{r} + \int_P^Q F\,(\mathrm{d}\mathbf{r}\,.\,\nabla)\mathbf{q}, \tag{2.62}$$

when $F(\mathbf{r}, t)$ is a scalar field. If the material curve is a closed curve $\mathscr{C}$, then

$$\frac{\mathrm{d}}{\mathrm{d}t}\oint_{\mathscr{C}} F\,\mathrm{d}\mathbf{r} = \oint \frac{\mathrm{D}F}{\mathrm{D}t}\,\mathrm{d}\mathbf{r} + \oint F\,(\mathrm{d}\mathbf{r}\,.\,\nabla)\mathbf{q}, \tag{2.63}$$

A line integral of particular importance is

$$\Gamma = \oint \mathbf{q}\,.\,\mathrm{d}\mathbf{r}$$

which represents the circulation Γ around the curve $\mathscr{C}$, as defined by (2.11). With $\mathbf{q} = u\mathbf{i} + v\mathbf{j} + w\mathbf{k}$, equation (2.63) gives

$$\begin{aligned} \frac{\mathrm{d}}{\mathrm{d}t}\oint_{\mathscr{C}} u\mathbf{i}\,.\,\mathrm{d}\mathbf{r} &= \mathbf{i}\,.\,\frac{\mathrm{d}}{\mathrm{d}t}\oint_{\mathscr{C}} u\,\mathrm{d}\mathbf{r} \\ &= \mathbf{i}\,.\left\{\oint_{\mathscr{C}} \frac{\mathrm{D}u}{\mathrm{D}t}\,\mathrm{d}\mathbf{r} + \oint_{\mathscr{C}} u\,(\mathrm{d}\mathbf{r}\,.\,\nabla)\mathbf{q}\right\}. \end{aligned} \tag{2.64}$$

Similarly

$$\frac{\mathrm{d}}{\mathrm{d}t}\oint_{\mathscr{C}} v\mathbf{j}\,.\,\mathrm{d}\mathbf{r} = \mathbf{j}\,.\left\{\oint_{\mathscr{C}} \frac{\mathrm{D}v}{\mathrm{D}t}\,\mathrm{d}r + \oint_{\mathscr{C}} v\,(\mathrm{d}\mathbf{r}\,.\,\nabla)\mathbf{q}\right\}, \tag{2.65}$$

$$\frac{\mathrm{d}}{\mathrm{d}t}\oint_{\mathscr{C}} w\mathbf{k}\,.\,\mathrm{d}\mathbf{r} = \mathbf{k}\,.\left\{\oint_{\mathscr{C}} \frac{\mathrm{D}w}{\mathrm{D}t}\,\mathrm{d}\mathbf{r} + \oint_{\mathscr{C}} w\,(\mathrm{d}\mathbf{r}\,.\,\nabla)\mathbf{q}\right\}. \tag{2.66}$$

Since the constant unit vectors may be taken under the integral signs, equations (2.64), (2.65), (2.66) give on adding

$$\frac{\mathrm{d}}{\mathrm{d}t}\oint_{\mathscr{C}}\mathbf{q}\,.\,\mathrm{d}\mathbf{r}=\oint_{\mathscr{C}}\frac{\mathrm{D}\mathbf{q}}{\mathrm{D}t}\,.\,\mathrm{d}\mathbf{r}+\oint\mathbf{q}\,.\,(\mathrm{d}\mathbf{r}\,.\,\nabla)\mathbf{q}. \tag{2.67}$$

But

$$\mathbf{q}\,.\,(\mathrm{d}\mathbf{r}\,.\,\nabla)\mathbf{q}=\left(\mathbf{q}\,.\,\frac{\partial\mathbf{q}}{\partial s}\right)\mathrm{d}s=\frac{\partial}{\partial s}\left(\tfrac{1}{2}\,\mathbf{q}^2\right)\mathrm{d}s.$$

Thus the second integral on the right-hand side of (2.67) vanishes, leaving the result that

$$\frac{\mathrm{d}}{\mathrm{d}t}\oint_{\mathscr{C}}\mathbf{q}\,.\,\mathrm{d}\mathbf{r}=\oint_{\mathscr{C}}\frac{\mathrm{D}\mathbf{q}}{\mathrm{D}t}\,.\,\mathrm{d}\mathbf{r}. \tag{2.68}$$

2.9 ANALYSIS OF LOCAL FLUID MOTION

The force exerted by a portion of fluid on an adjacent portion depends to a large extent on the way in which the fluid is deformed as it moves. Before considering the dynamical aspects of fluid motion, it is desirable to examine in detail the local character of fluid motion, which has similarities to the local deformation of an elastic solid.

Consider a line element of fluid particles at time t whose ends P, Q have position vectors, relative to a fixed origin O, given by $\mathbf{r}$ and $\mathbf{r}+\delta\mathbf{r}$ respectively, as shown in Fig. 2.12. At time $t+\delta t$, the ends P, Q will have moved to P', Q' respectively with position vectors $\mathbf{r}'$ and $\mathbf{r}'+\delta\mathbf{r}'$ relative to O. If $\mathbf{q}$ is the velocity of the fluid, to the first order in δt,

$$\mathbf{r}'=\mathbf{r}+\mathbf{q}(\mathbf{r},t)\delta t, \tag{2.69}$$

$$\mathbf{r}'+\delta\mathbf{r}'=\mathbf{r}+\delta\mathbf{r}+\mathbf{q}(\mathbf{r}+\delta\mathbf{r},t)\delta t. \tag{2.70}$$

The last term on the right-hand side of (2.70) can be expanded by Taylor's theorem to give

$$\mathbf{r}'+\delta\mathbf{r}'=\mathbf{r}+\delta\mathbf{r}+\mathbf{q}(\mathbf{r},t)\delta t+(\delta\mathbf{r}\,.\,\nabla)\mathbf{q}(\mathbf{r},t)\delta t \tag{2.71}$$

ignoring terms of $O(\delta r^2)$. We therefore see from (2.69) and (2.71) that both ends of the line element move a distance $\mathbf{q}(\mathbf{r},t)\delta t$, which represents a *rigid body translation* of the entire line element, while in addition there is a relative motion between the ends represented by the term

$$\boldsymbol{\xi}=(\delta\mathbf{r}\,.\,\nabla)\,\mathbf{q}(\mathbf{r},t)\delta t=\delta x_j\,\frac{\partial q_i}{\partial x_j}\,\hat{\mathbf{i}}_i\,\delta t=\xi_i\,\hat{\mathbf{i}}_i, \tag{2.72}$$

where the derivative is understood to be evaluated at $(\mathbf{r},t)$ with $\mathbf{r}=x_j\,\hat{\mathbf{i}}_j$ and the summation convention on repeated suffices is adopted. Defining

$$q_{i,j}=\frac{\partial q_i}{\partial x_j},$$

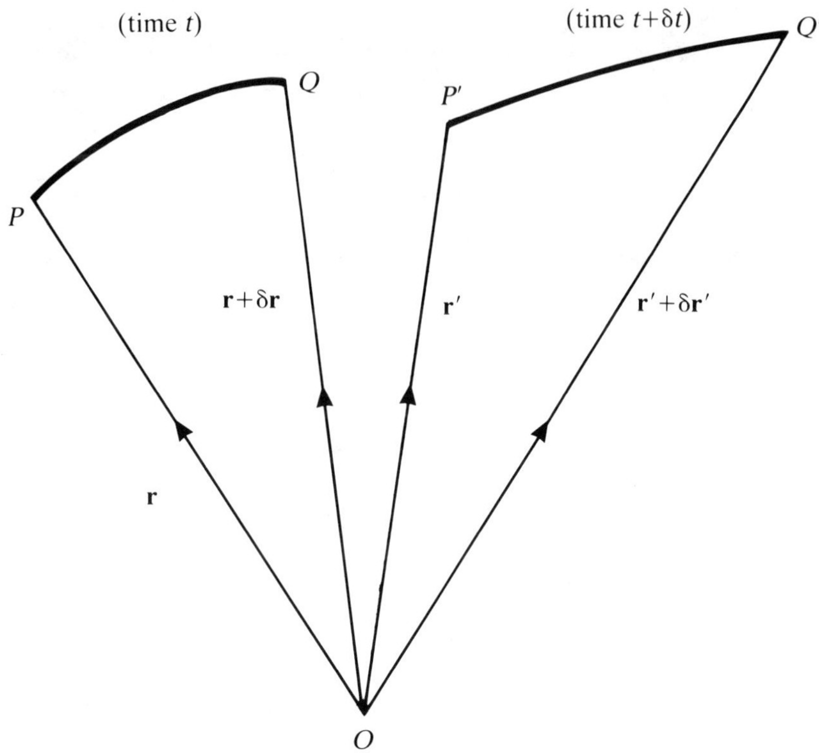

Fig. 2.12

the first thing to notice is that $q_{i,j}$ is a second-order tensor. This follows because the components q_i of the vector **q** form the components of a first-order tensor obeying the transformation law

$$q_r' = l_{ri} q_i,$$

where $l_{ri} l_{rj} = \delta_{ij} = l_{ir} l_{jr}$ as shown in Section 1.21, equation (1.124). Thus

$$\begin{aligned} q'_{r,s} &= \frac{\partial}{\partial x_s'}(l_{ri} q_i) = l_{ri}\frac{\partial q_i}{\partial x_s'} = l_{ri}\frac{\partial q_i}{\partial x_j}\frac{\partial x_j}{\partial x_s'} \\ &= l_{ri}\, q_{i,j}\, l_{sj}, \end{aligned}$$

from which it follows that

$$q'_{r,s} = l_{ri}\, l_{sj}\, q_{i,j} \tag{2.73}$$

subject to

$$l_{ri}\, l_{si} = \delta_{rs} = l_{ir}\, l_{is}.$$

Equation (2.73) shows that $q_{i,j}$ obeys the transformation law of a second-order tensor.

We may write $q_{i,j}$ in the form

$$q_{i,j} = \epsilon_{ij} + \eta_{ij} \tag{2.74}$$

where

$$\epsilon_{ij} = \frac{1}{2}(q_{i,j} + q_{j,i}), \quad \eta_{ij} = \frac{1}{2}(q_{i,j} - q_{j,i}). \tag{2.75}$$

The reader may easily verify that ϵ_{ij} and η_{ij} are respectively *symmetric* and *skew-symmetric* second-order tensors. The non-zero components of η_{ij} are related to the Cartesian components of the vorticity $\boldsymbol{\zeta}$ by the equations

$$\left.\begin{aligned} \eta_{12} &= -\eta_{21} = -\tfrac{1}{2}\zeta_3 \\ \eta_{13} &= -\eta_{31} = -\tfrac{1}{2}\zeta_2 \\ \eta_{23} &= -\eta_{32} = -\tfrac{1}{2}\zeta_1 \end{aligned}\right\}. \tag{2.76}$$

It therefore follows that the skew-symmetric part of $q_{i,j}$ results in a contribution ξ_i^A to ξ_i of (2.72) given by

$$\xi_i^A = \frac{1}{2}(\boldsymbol{\zeta} \times \delta\mathbf{r})_i \delta t \tag{2.77}$$

so that this part of the relative motion arises from a *rigid body rotation* of the entire line element with angular velocity $\frac{1}{2}\boldsymbol{\zeta}$. Since the vorticity is proportional to the local angular velocity of the fluid, the reader will appreciate how the term *irrotational* has come into use to describe a fluid motion with zero vorticity. There remains the contribution to the relative motion arising from the symmetric part of $q_{i,j}$. It has proved in Section 1.21 that any symmetric second-order tensor has a system of orthogonal axes – *the principal axes* – with respect to which it has diagonal form. Without loss of generality we may suppose that the principal axes of ϵ_{ij} have been chosen as the axes of the Cartesian coordinates (x_1, x_2, x_3). In which case the non-zero elements of ϵ_{ij} are

$$\epsilon_{11} = q_{1,1}, \quad \epsilon_{22} = q_{2,2}, \quad \epsilon_{33} = q_{3,3}. \tag{2.78}$$

It then follows that the symmetric part of $q_{i,j}$ results in a contribution ξ_i^B to ξ_i given by

$$\left.\begin{aligned} \xi_1^B &= \epsilon_{11}\,\delta x_1\,\delta t \\ \xi_2^B &= \epsilon_{22}\,\delta x_2\,\delta t \\ \xi_3^B &= \epsilon_{33}\,\delta x_3\,\delta t \end{aligned}\right\}, \tag{2.79}$$

which represents a *stretching* of the line element by the amounts $\epsilon_{11}\delta x_1\delta t$, $\epsilon_{22}\delta x_2\delta t$, $\epsilon_{33}\delta x_3\delta t$ respectively along the coordinate axes. The motion characterised by the tensor ϵ_{ij} is called a pure straining motion and the tensor is referred to as the *rate of strain* tensor.

From the foregoing analysis, it is easy to see the effect of the fluid motion on a small volume of fluid contained in a box with sides $\delta x_1, \delta x_2, \delta x_3$ at time t. In a small time interval δt, this box moves with a rigid body translational velocity $\mathbf{q}(\mathbf{r}, t)$, a rigid body rotational velocity $\frac{1}{2}\boldsymbol{\zeta}(\mathbf{r}, t)$, and a pure straining motion in which the volume of the box is increased to

$$(1 + \epsilon_{11} + \epsilon_{22} + \epsilon_{33})\delta x_1 \delta x_2 \delta x_3 \tag{2.80}$$

when terms which are second order in δt or δx_i are ignored. In Section 1.22 it was shown that the sum of the diagonal elements of a second-order tensor is invariant, and from (2.78) we see that

$$\epsilon_{11} + \epsilon_{22} + \epsilon_{33} = \frac{\partial q_1}{\partial x_1} + \frac{\partial q_2}{\partial x_2} + \frac{\partial q_3}{\partial x_3} = \operatorname{div} \mathbf{q}.$$

Hence the local rate of change in volume of a fluid element as it moves is div $\mathbf{q}$ per unit volume, which is a measure of the *divergence* of the volume as the fluid moves, and explains how the terminology for the function of velocity div $\mathbf{q}$ has come into use. Since the volume of an element of an incompressible fluid consisting of the same fluid particles does not alter, it follows that a necessary condition for incompressible flow is that div $\mathbf{q} = 0$, a condition established within Sections 2.6 and 2.7. The local motion of an element of an incompressible fluid can therefore be regarded as a superposition of a rigid body translation with velocity $\mathbf{q}$ and a rigid body rotation with angular velocity $\frac{1}{2}\boldsymbol{\zeta}$. Thus small linear elements of such a fluid would appear to move like match sticks and in the case of irrotational flow, would move so as to remain parallel to themselves over very small intervals of time as illustrated in Fig. 2.13.

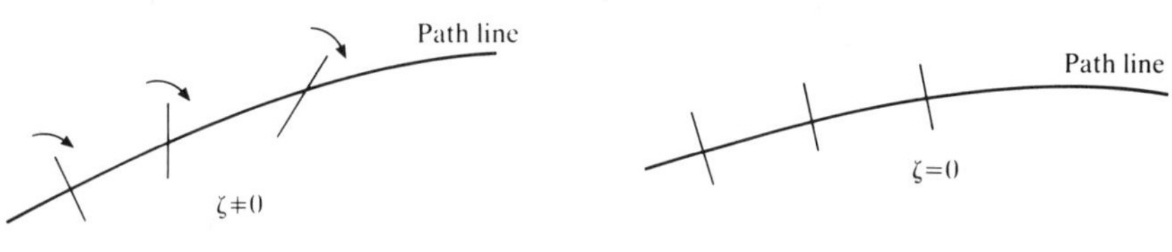

Fig. 2.13

Example [2.7]
Analyse the local flow structure for the two-dimensional flow with Cartesian components of velocity given by

$$u = \alpha y, \quad v = w = 0,$$

where α is a constant.

This is an example of a *linear shear flow* with rate of shear α. The fluid particles move in parallel planes.

Clearly,

$$\operatorname{div} \mathbf{q} = 0, \quad \text{curl} \quad = -\alpha \mathbf{k} = \boldsymbol{\zeta}.$$

The matrix array for the second-order tensor $q_{i,j}$ is

$$\begin{pmatrix} 0 & \alpha & 0 \\ 0 & 0 & 0 \\ 0 & 0 & 0 \end{pmatrix},$$

giving

$$\epsilon_{ij} = \begin{pmatrix} 0 & \frac{1}{2}\alpha & 0 \\ \frac{1}{2}\alpha & 0 & 0 \\ 0 & 0 & 0 \end{pmatrix}, \quad \eta_{ij} = \begin{pmatrix} 0 & \frac{1}{2}\alpha & 0 \\ -\frac{1}{2}\alpha & 0 & 0 \\ 0 & 0 & 0 \end{pmatrix}.$$

The tensor ϵ_{ij} has eigenvalues given by the roots of

$$\det(\epsilon_{ij} - \lambda\delta_{ij}) = 0,$$

which has solutions $\frac{1}{2}\alpha$, $-\frac{1}{2}\alpha$, 0. The principal axes are the normalised eigenvectors corresponding to these eigenvalues. These are $\frac{1}{\sqrt{2}}\{1, 1, 0\}$, $\frac{1}{\sqrt{2}}\{-1, 1, 0\}$, $\{0, 0, 1\}$ respectively. Choosing these directions as new axes, the rate of strain tensor becomes

$$\begin{pmatrix} \frac{1}{2}\alpha & 0 & 0 \\ 0 & -\frac{1}{2}\alpha & 0 \\ 0 & 0 & 0 \end{pmatrix}.$$

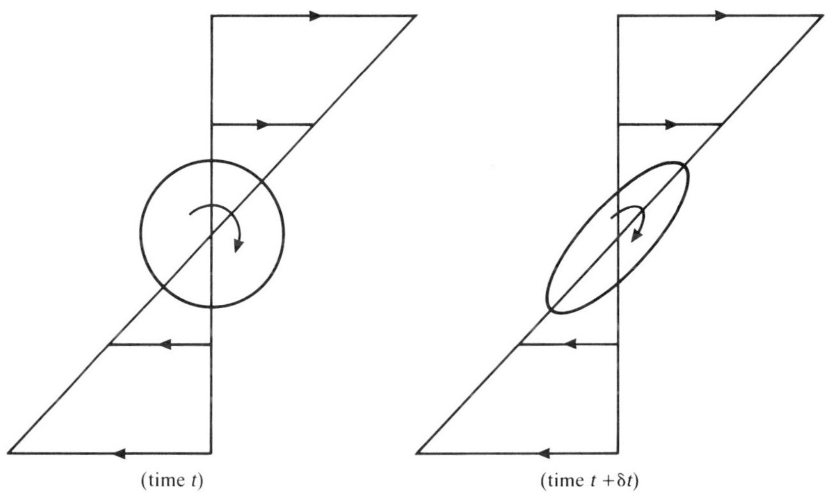

Fig. 2.14

Thus a fluid element distorts with no overall change in volume, but such that its dimension along $\mathbf{i}+\mathbf{j}$ is *extended* at the rate of $\frac{1}{2}\alpha$ and its dimension along $-\mathbf{i}+\mathbf{j}$ is *compressed* at the rate $\frac{1}{2}\alpha$. Hence a small cylinder of fluid with circular cross-section in the (x, y) plane is distorted during a small interval of time so that its cross-section is elliptical, as illustrated in Fig. 2.14. Besides distorting, the cylinder of fluid has an angular velocity $-\alpha\mathbf{k}$ and a translational velocity $\alpha y\mathbf{i}$. In the figure, the local flow is indicated in a frame of reference moving with the local translational velocity of the fluid cylinder.

PROBLEMS 2

1. Show that a fluid of constant density can have a velocity $\mathbf{q}$ given by

$$\mathbf{q}=\left[-\frac{2xyz}{(x^2+y^2)^2}, \frac{(x^2+y^2)z}{(x^2+y^2)^2}, \frac{y}{x^2+y^2}\right].$$

Find the vorticity vector.

2. For a fluid of density ρ, establish the equation of continuity,

$$\frac{D\rho}{Dt}+\rho\nabla\cdot\mathbf{q}=0.$$

Show that if $\mathbf{q}=m\mathbf{r}/r^3$, where m is a constant, the equation of continuity for an incompressible fluid is satisfied at all points other than the origin O. Show further that, if S is any closed surface not passing through O, the net volume of fluid flowing out of S per unit time is $4\pi m$ if S encloses O and zero if O is outside S.

3. A mass of fluid moves in such a way that each particle describes a circle in one plane about a fixed axis; show that the equation of continuity is

$$\frac{\partial\rho}{\partial t}+\frac{\partial(\rho\omega)}{\partial\theta}=0,$$

where ω is the angular velocity of a particle whose azimuthal angle is θ at time t.

4. A mass of fluid is in motion so that the lines of motion lie on the surface of coaxial cylinders; show that the equation of continuity is

$$\frac{\partial\rho}{\partial t}+\frac{1}{r}\frac{\partial(\rho v_\theta)}{\partial\theta}+\frac{\partial(\rho v_z)}{\partial z}=0,$$

where v_θ, v_z are the velocities perpendicular and parallel to z.

5. The particles of a fluid move symmetrically in space with regard to a fixed centre; prove that the eqaution of continuity is

$$\frac{\partial\rho}{\partial t} + u\frac{\partial\rho}{\partial r} + \frac{\rho}{r^2}\frac{\partial}{\partial r}(r^2 u) = 0,$$

where u is the velocity at distance r.

6. Each particle of a mass of liquid moves in a plane through the axis of z; find the equation of continuity.

7. If the lines of motion are curves on the surfaces of cones having their vertices at the origin and the axis of z for common axis, prove that the equation of continuity is

$$\frac{\partial\rho}{\partial t} + \frac{\partial(\rho q_r)}{\partial r} + \frac{2\rho q_r}{r} + \frac{\operatorname{cosec}\theta}{r}\frac{\partial(\rho q_\phi)}{\partial\phi} = 0.$$

8. If the lines of motion are curves on the surfaces of spheres all touching the plane of xy at the origin O, the equation of continuity is

$$r\sin\theta\,\frac{\partial\rho}{\partial t} + \frac{\partial(\rho v)}{\partial\phi} + \sin\theta\,\frac{\partial(\rho u)}{\partial\theta} + \rho u\,(1 + 2\cos\theta) = 0,$$

where r is the radius of CP of one of the spheres, θ the angle PCO, u the velocity in the plane PCO, v the perpendicular velocity, and ϕ the inclination of the plane PCO to a fixed plane through the axis of z.

9. If every particle moves on the surface of a sphere, prove that the equation of continuity is

$$\frac{\partial\rho}{\partial t}\cos\theta + \frac{\partial}{\partial\theta}(\rho\omega\cos\theta) + \frac{\partial}{\partial\phi}(\rho\omega'\cos\theta) = 0,$$

ρ being the density, θ, ϕ the latitude and longitude of any element, and ω and ω' the angular velocities of the element in latitude and longitude respectively.

10. Show that, if ξ, η, ζ be orthogonal coordinates and if U, V, W be the corresponding component velocities, the equation of continuity is

$$\frac{\partial\rho}{\partial t} + \rho(Us_1 + Vs_2 + Ws_3) + h_1\frac{\partial}{\partial\xi}(\rho U) + h_2\frac{\partial}{\partial\eta}(\rho V) + h_3\frac{\partial}{\partial\zeta}(\rho W) = 0,$$

where

$$\frac{1}{h_1^2} = \left(\frac{\partial x}{\partial \xi}\right)^2 + \left(\frac{\partial y}{\partial \xi}\right)^2 + \left(\frac{\partial z}{\partial \xi}\right)^2, \text{ etc. } \ldots\ldots$$

and s_1, s_2, s_3 are respectively the sums of the principal curvatures of the three orthogonal surfaces.

11. Show that

$$\frac{x^2}{a^2}\tan^2 t + \frac{y^2}{b^2}\cot^2 t = 1$$

is a possible form for the bounding surface of a liquid, and find an expression for the normal velocity.

12. In the steady motion of homogeneous liquid if the surfaces $f_1 = a_1$, $f_2 = a_2$ define the streamlines, prove that the most general values of the velocity components u, v, w are

$$F(f_1, f_2)\frac{\partial(f_1, f_2)}{\partial(y, z)}, \quad F(f_1, f_2)\frac{\partial(f_1, f_2)}{\partial(z, x)}, \quad F(f_1, f_2)\frac{\partial(f_1, f_2)}{\partial(x, y)},$$

where F is any arbitrary function.

13. Show that all necessary conditions can be satisfied by a velocity potential of the form

$$\phi = \alpha x^2 + \beta y^2 + \gamma z^2,$$

and a bounding surface of the form

$$F \equiv ax^4 + by^4 + cz^4 - \chi(t) = 0,$$

where $\chi(t)$ is a given function of the time, and $\alpha, \beta, \gamma, a, b, c$ suitable functions of the time.

14. If the velocity is constant in magnitude everywhere in the irrotational, two-dimensional flow of an incompressible fluid, prove that the flow must be uniform in direction. Prove also the converse of this statement.

15. Explain what is meant in hydrodynamics by a stagnation point. Referred to Cartesian axes Ox, Oy, Oz, a semi-infinite incompressible inviscid fluid occupies the region $z > 0$, and is bounded by a rigid plane at $z = 0$. The fluid is in steady irrotational motion with a stagnation point at O. If the velocity potential φ is taken as zero at O, and it is assumed valid to expand φ in a power series of the form

$$\varphi = a_1 x + a_2 y + a_3 z + a_{11} x^2 + a_{22} y^2 + a_{33} z^2 + a_{12} xy + a_{23} yz + a_{31} zx + \ldots,$$

show that, with the axes Ox, Oy suitably orientated, the series is of the form

$$\varphi = Ax^2 + By^2 - (A+B)z^2 + \text{terms of third order and higher.}$$

Discuss the form of the equipotential surfaces near the stagnation point for

(i) motion in the (x, z) plane, and
(ii) motion symmetrical about the axis Oz.

16. Show that in the general motion of a fluid, the motion of a particle relative to a neighbouring particle consists of a pure strain compounded with a rigid body rotation.

If the velocity potential at any instant be λxyz, show that the velocity at any point $P'(x+\xi, y+\eta, z+\zeta)$ relative to the fluid at $P(x, y, z)$ is normal to the quadric surface

$$x\eta\zeta + y\zeta\xi + z\xi\eta = \text{const.},$$

with centre P, when ξ, η, ζ are small.

17. If α = constant, β = constant are the equations of a curve, show that the tangent is in the direction of the vector $\nabla\alpha \times \nabla\beta$. Hence show that if the α and β surfaces are any two systems of surfaces which pass through the vortex lines, then $\boldsymbol{\zeta} = F\,\nabla\alpha \times \nabla\beta$, where F is a scalar function.

18. In problem 17 use the fact that $\nabla.\boldsymbol{\zeta} = 0$ to prove that

$$\nabla.(F\nabla\alpha \times \nabla\beta) = 0,$$

and hence show that this is equivalent to the vanishing of the Jacobian

$$\partial(F, \alpha, \beta)/\partial(x, y, z),$$

so that F is a function of α, β only.

19. Prove that $\nabla f(\alpha, \beta) = \dfrac{\partial f}{\partial \alpha}\nabla\alpha + \dfrac{\partial f}{\partial \beta}\nabla\beta$. With the notations of Problems 17, 18 show that if the scalar function $f(\alpha, \beta)$ is so chosen that $\partial f/\partial\alpha = F$, then

(i) $\mathbf{q} = f(\alpha, \beta)\,\nabla\beta$ is a solution of the equation $\boldsymbol{\zeta} = \nabla \times \mathbf{q}$.

(ii) $\boldsymbol{\zeta} = \dfrac{\partial f}{\partial \alpha}\nabla\alpha \times \nabla\beta$.

20. Use Problem 19 to prove that the general solution of $\zeta = \nabla \times \mathbf{q}$ is

$$\mathbf{q} = -\nabla\phi + f(\alpha, \beta)\ \nabla\beta,$$

where α = constant, β = constant are two systems of surfaces which pass through the vortex lines, and ϕ is a solution of Laplace's equation.

21. Obtain Clebsch's transformation that the velocity can be expressed in the form

$$\mathbf{q} = -\nabla\phi + \lambda\ \nabla\ \mu,$$

where the surfaces λ = constant, μ = constant move with the fluid, and the curves in which they intersect are vortex lines.

3

Mechanics of fluid motion

3.1 PROPERTIES OF FLUIDS

Fluid mechanics, as the terminology suggests, is the study of the application of the fundamental principles of mechanics to matter in liquid or gaseous form. The fundamental principles are those of conservation of mass, conservation of energy, Newton's laws of motion, and for compressible fluids, the laws of thermodynamics. By the application of these principles, our objective is to describe and in some cases predict the behaviour of fluids in prescribed conditions. Modern fluid mechanics finds applications in branches of engineering and the physical and medical sciences. It has evolved through the interaction of both theoretical and experimental approaches to solving problems. In the previous chapter we derived the equation of continuity for a fluid using the principle of conservation of mass. To proceed further in our study we must consider some physical properties of fluids.

Pressure (static)

Consider first a fluid at rest. As a result of innumerable random molecular collisions, any part of the fluid will experience a force exerted on it by the surrounding fluid or by the solid boundaries of the vessel containing it. This means that if a small rigid plane impermeable surface δA is inserted into the fluid at the point P, then either 'side' of δA will be bombarded by fluid molecules moving randomly in all directions. As the fluid molecules impinge on one side of δA, they transfer momentum to it in a specified time, thereby exerting a force δF on that side of δA and, in accord with Newton's third law of motion, an equal and opposite force is applied to the fluid by that side of δA. We consider the linear dimensions of δA to be large by comparison with molecular dimensions but small by comparison with macroscopic dimensions which are relevant, e.g. the dimension of the containing vessel or immersed solid. Then, δA is at rest relative to the main body of fluid since the only motions are those on a molecular scale and δF is normal to δA. The mean force per unit area exerted on one side of δA is $\delta F/\delta A$ and if δA is allowed to tend to zero about P, it is reasonable to suppose that this ratio approaches a limit. We shall suppose this to be so, for in the last resort, the apparent contradition between the continum hypothesis and the molecular model of the fluid can only be resolved by experiment. Such an experiment to measure density as referred to in Section 2.6, shows the ratio would have acquired a sensibly constant value long before the assumptions of the continuum hypothesis were violated. This limiting value of $\delta F/\delta A$ is called the *hydrostatic pressure* at P. Its value cannot be measured directly since all instruments which are used to measure pressure only measure a *difference* in pressure. This difference is frequently that between the hydrostatic pressure of the fluid and that of the atmosphere. The hydrostatic pressure of the atmosphere is therefore commonly used as a reference or datum value of hydrostatic pressure. The difference in hydrostatic pressure recorded by the measuring instrument is what engineers call the *gauge pressure*. The *absolute hydrostatic pressure* is the hydrostatic pressure difference relative to a perfect vacuum. The pressure of the atmosphere is not constant but for many engineering applications involving incompressible fluids, the variation in atmospheric pressure may be considered negligible. Pressure is measured in N/m^2 and a unit N/m^2 is called the *pascal* with the abbreviation Pa. The standard 'atmosphere' of pressure is 1.01325×10^5 Pa. A pressure of 10^5 Pa is called a *bar*, and the thousandth part of this unit is a *millibar*, which occurs commonly in meteorological measurements.

Like fluid properties such as temperature and density, the hydrostatic pressure p is a property of the fluid at each particular point. Before considering the meaning of pressure in fluids which are in motion, we shall state a few of the more important results of hydrostatics. These are to be found in texts on the subject, but can also be established from results we shall derive for fluids in motion when the fluid velocity is set equal to zero.

(i) When a fluid is at rest, the pressure at an internal point P is the same in all directions. This means that the orientation of the surface element δA at P is irrelevant.

(ii) The pressure at all points at the same depth below the free horizontal surface of a liquid at rest is the same. By a free surface we mean one which is not in contact with a rigid boundary, such as the exposed surface of a cup of tea. At a point P, a distance h below the free surface, the (hydrostatic) pressure is $\Pi + \rho gh$, with Π the atmospheric pressure at the free surface and g the acceleration due to gravity.

(iii) If a pressure is applied to a point of the free surface of a liquid, it is transmitted equally to all parts of the fluid.

These results constitute *Pascal's laws* for static fluids and find many applications in everyday life through the science of *hydraulics*, e.g. the brake systems of cars, the manoeuvring of the control surfaces and undercarriage of an aeroplane and in the designs of early lifts and elevators.

Example [3.1]
A typical lock gate in a canal is a structure which has to be built to withstand considerable force resulting from the effect of hydrostatic pressure applied on both sides of the gate where the water levels are different except when the lock is flooded to allow the passage of a boat through the lock. The lock gate is shown diagrammatically in Fig. 3.1. The pressure acting on the side of the gate where the water is deeper is $\Pi - \rho g z$ if z is measured positively *upwards* from the free surface of the deeper water and Π denotes atmospheric pressure. The force exerted by the deeper water on the gate is therefore

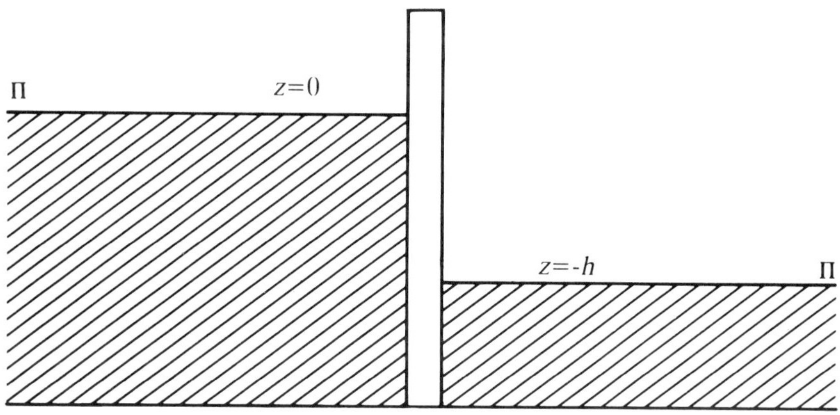

Fig. 3.1

$$l\int_{-H}^{0} (\Pi - \rho g z)\,\mathrm{d}z = \Pi\, lH + \tfrac{1}{2}\rho g l H^2,$$

where l is the width of the lock gate, in a direction from left to right in Fig. 3.1. The force acting on the other side of the gate in a direction from right to left in Fig. 3.1 is of magnitude

$$\Pi\, lH + l\int_{-H}^{-h} \{\Pi - \rho g\,(h+z)\}\,\mathrm{d}z$$

$$= \Pi\, lH + \tfrac{1}{2}\rho g l\,(H-h)^2,$$

where the first term arises from the pressure of the atmosphere on the part of the gate above the water level $z = -h$. Any change in atmospheric pressure over the range $0 \geqslant z \geqslant -h$ is of course neglected. The *net* force acting on the lock gate is accordingly

$$F = \tfrac{1}{2}\rho g l\,[H^2 - (H-h)^2]$$

in the direction from the deeper to shallower water. A couple also acts on the gate, and with moments taken about the base of the gate, this can be shown to be

$$G = \tfrac{1}{6}\rho g l\,[H^3 - (H-h)^3]$$

in the clockwise sense in Fig. 3.1. By the laws of elementary statics, this couple is equivalent to the action of a single force F acting at a height

$$\frac{1}{3}\frac{[H^3 - (H-h)^3]}{[H^2 - (H-h)^2]}$$

above the base of the gate. The point of action, which will lie on the mid-way vertical of the gate, is called the *centre of pressure*, which usually does not coincide with the centre of mass of the gate. The effect of this force must be counterbalanced in the construction of the lock gate at its pivot and where the pair of such gates in the canal meet. It is not unusual for lock gates to have to withstand forces arising from hydrostatic pressure of the order of 10^6 N.

The calculation of the force acting on immersed or floating bodies due to hydrostatic pressure can be carried out for various body shapes and the reader is referred to a text on hydrostatics if he wants further details. This force is called the *buoyancy force* and the fact that it acts at a point which does not usually coincide with the centre of mass explains why the atttitude taken up by a floating body is sometimes unexpected. This must of course be taken into account when designing a boat or a ship to ensure that in the loading and operating conditions which will be encountered, the vessel floats upright in stable equilibrium.

Pressure (dynamic)

If a fluid is in motion, we may again insert a small rigid plane impermeable surface δA into the fluid at a general point P, but instead of holding δA at rest, we shall let it move with the local fluid velocity $\mathbf{q}$ at P. As in the static case, the only momentum interchange between the fluid and the surface is on the molecular scale, but this will result in a force being applied to either side of δA. However, in contrast with the static case, this force is not necessarily in the direction of the normal to δA. Denoting the force applied to one side of δA by the vector $\delta\mathbf{F}$, we may consider the limit of the ratio $\delta\mathbf{F}/\delta A$ as the area δA shrinks to zero about P. Assuming this limit exists in the sense already alluded to in defining the hydrostatic pressure at a point, we call this limit the *stress vector*

$\mathbf{R}_n$ at P associated with the orientation $\mathbf{n}$ of the normal to δA. If $\mathbf{R}_n$ is to have a non-zero component *perpendicular* to $\mathbf{n}$, this can only arise if the surface element δA experiences a *shearing force* parallel to its plane and such a force is the result of the resistance of the fluid to motion over the element δA and more broadly the resistance of the fluid to movement of one layer over another. This 'internal friction' of the fluid is called *viscosity* and it is a property of all real fluids, but in varying degrees. We are all aware that it is easier to stir or pour water than syrup or pitch. Gases as well as liquids have viscosity but their viscosity is less obvious in everyday life. There are, however, a couple of examples which may be cited here. If a body were to fall under gravity in a vacuum, it would accelerate indefinitely. This does not happen if a body falls in a fluid, in particular air, and a constant *terminal* velocity is attained by the falling body as a result of the effect of the viscosity of the fluid. Also when a spacecraft re-enters the Earth's atmosphere, the intense heat which is generated is attributable to the viscosity of the air although it is extremely rarified in the upper levels of the atmosphere. Viscosity is an *energy dissipative* mechanism through which kinematic energy can be lost by transforming it into other energy forms, notably heat.

We shall be considering the effects of viscosity of a fluid more fully in a future volume. If the viscosity of a fluid is ignored, the fluid is said to be *inviscid* or *ideal*. In this volume we shall consider from now on the mechanics of an ideal fluid. An understanding of the mathematical description of the behaviour of flows of such fluids provides a worthwhile introduction to fluid mechanics and a useful prelude to the study of real viscous fluids.

Returning to what we mean by pressure in a moving fluid, the question becomes a lot simpler when viscosity is ignored since then the stress vector $\mathbf{R}_n$ will have no components perpendicular to $\mathbf{n}$. Thus we may write

$$\mathbf{R}_n = -p\mathbf{n}, \tag{3.1}$$

and the scalar quantity p we call the *hydrodynamic fluid pressure* at P. By convention the direction of $\mathbf{n}$ is that away from the side of δA on which the force $\delta\mathbf{F}$ acts, as illustrated in Fig. 3.2, so the pressure of the fluid impinges on the particular side of δA. For clarity δA is drawn as if it had thickness. Since we shall be concerned exclusively with fluids in motion, it is usual to drop the adjective 'hydrodynamic' and just talk about the pressure at P. Of course if the fluid is brought to rest, the (hydrodynamic) pressures reduces to the hydrostatic pressure.

Before leaving the subject of pressure, it is instructive for us to prove the dynamic analogue of Pascal's law (iii). This establishes that the pressure p at P in a moving inviscid fluid is independent of the orientation of $\mathbf{n}$. To this end, we consider the motion of a small tetrahedron $PQRS$ of fluid, as shown in Fig. 3.3, whose edges PQ, PR, PS are taken parallel to the coordinate axes OX, OY, OZ and are of length δx, δy, δz respectively at time t. The tetrahedron moves with the fluid and is therefore made up of the same fluid particles which may be assumed to have the same velocity, i.e. $\mathbf{q}$ the local velocity of the fluid at P, since

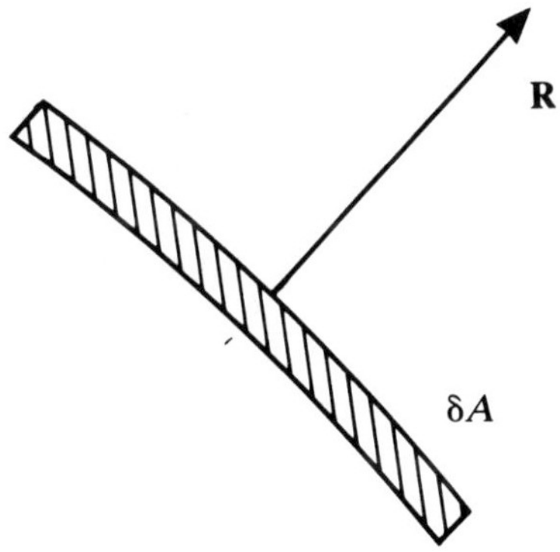

Fig. 3.2

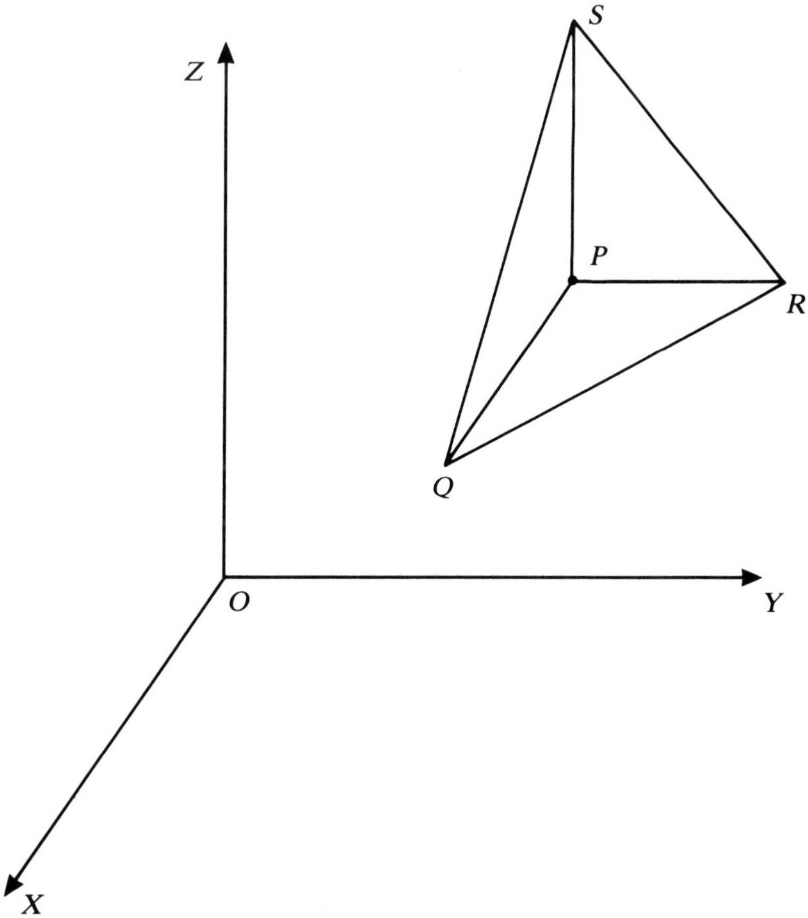

Fig. 3.3

δx, δy, δz are very small. Suppose that p_n is the value of the pressure at points on the face *SQR* while p_x, p_y, p_z are the values of the pressure over the faces of *PRS, PQS, PRQ* respectively. Then the total force exerted *on* the fluid within the tetrahedron *by* the fluid exterior to it is, using (3.1),

$$\tfrac{1}{2}\,p_x\,\delta y\delta z\mathbf{i}+\tfrac{1}{2}\,p_y\,\delta x\delta z\mathbf{j}+\tfrac{1}{2}\,p_z\,\delta x\delta y\mathbf{k}-p_n\,\delta S\,(l\mathbf{i}+m\mathbf{j}+n\mathbf{k}), \tag{3.2}$$

where δS is the area of the face *QRS*. But

$$l\delta S = \text{area of face } PRS = \tfrac{1}{2}\,\delta y\delta z,$$

$$m\delta S = \text{area of face } PQS = \tfrac{1}{2}\,\delta x\delta z,$$

$$n\delta S = \text{area of face } PQR = \tfrac{1}{2}\,\delta x\delta y.$$

There may also be a *body force* acting. This is a force field due to external causes such as gravity. If we suppose that any such body force or forces be denoted by $\mathbf{F}$ per unit mass, and note that the volume of the tetrahedron is $\frac{1}{3}h\ \delta S$, where h is the perpendicular from P to the face *QRS*, the *net* force acting on the tetrahedron is

$$l(p_x-p_n)\,\delta S\mathbf{i}+m(p_y-p_n)\delta S\mathbf{j}+n(p_z-p_n)\delta S\mathbf{k}+\tfrac{1}{3}\,\rho h\mathbf{F}\,\delta S.$$

The acceleration of the tetrahedron is $\mathrm{D}\mathbf{q}/\mathrm{D}t$ and the mass of the fluid inside it is constant. Thus the equation of motion of the fluid contained in the tetrahedron is

$$\frac{1}{3}\rho h\,\delta S\,\frac{\mathrm{D}\mathbf{q}}{\mathrm{D}t}=\frac{1}{3}\rho h\,\delta S\,\mathbf{F}$$
$$+\,l(p_x-p_n)\delta S\mathbf{i}+m(p_y-p_n)\delta S\mathbf{j}+n(p_z-p_n)\delta S\mathbf{k}. \tag{3.3}$$

On dividing by δS and letting the tetrahedron shrink to zero about P, in which case $h\to 0$, it follows that

$$p_x=p_y=p_z=p_n.$$

Since the choice of the direction $\mathbf{n}$ is quite arbitrary, as are the directions of the coordinate axes, this establishes the important result that *at any point of an ideal fluid in motion, the pressure p is the same in all directions*. This result recovers the corresponding result for fluids at rest.

Compressibility

All matter is to some extent compressible since a change in the compressive stress applied to a certain quantity of a substance always produces some change in its volume. Although the degree of compressibility of substances varies widely, the proportionate change in volume of a particular material which does not change state, e.g. from liquid to solid, during the compression is directly proportional to the compressive stress. This degree of compressibility of a substance is characterised *by the bulk modulus of elasticity* K, which is defined by

$$K = -\frac{\delta p}{\delta V/V}, \tag{3.4}$$

where δp is a small increment in the pressure applied to the material and δV is the corresponding change in the original volume V. Since an increase in pressure always produces a *decrease* in volume, δV is always negative so that K is positive. Letting $\delta p \to 0$, equation (3.4) gives

$$K = -V\frac{\partial p}{\partial V}, \tag{3.5}$$

and since the density $\rho = \text{mass}/V$, equation (3.5) can also be written as

$$K = \rho\frac{\partial p}{\partial \rho}. \tag{3.6}$$

The reciprocal of K is sometimes called the *compressibility*, so the higher value of K, the lower the compressibility, and $K = \infty$ for an incompressible fluid.

The value of K depends on the relation between pressure and density for the conditions under which the compression takes place. Two sets of conditions are particularly important. If the compression occurs while the temperature is kept constant, the value of K is the *isothermal* bulk modulus. If, on the other hand, heat is neither added to nor taken away from the fluid during the compression, the corresponding value of K is the *isentropic* bulk modulus. The ratio of isentropic to isothermal bulk moduli is γ which is also the ratio of the specific heat capacities at constant pressure and volume respectively. For liquids, the value of γ is practically unity and the two bulk moduli are almost identical. It is usual therefore not to distinguish between the bulk moduli for a liquid. The bulk modulus for a liquid is also very high and so the change in density with increase in pressure is very small even for the largest pressures encountered such as in the deep oceans. Accordingly the density of a homogeneous liquid can usually be regarded as a constant under compressible loads. But in circumstances where rapid changes in pressure can occur, as in 'water hammer', i.e. the knocking phenomenon often experienced in old pipe installations of water supply, the compressibility of liquids is then a significant effect.

As a liquid is compressed its molecules become closer together and its resistance to further compression accordingly increases. Thus K increases. For example in the case of water, the value of K effectively doubles as the pressure is raised from 1 atmostphere to 3500 atmospheres. It is also observed that K increases with temperature. In contrast, gases are easily compressed because of the relatively weak intermolecular bonds and large intermolecular spacings, but the term bulk modulus of elasticity is seldom used for a gas since pressure is usually directly proportional to $\rho\,\partial p/\partial\rho$ and therefore K cannot be regarded as even approximately constant. When gases undergo only very small variations in density, as in free flowing applications such as a ventilation system, the effects of compressibility may be ignored. The effects of compressibility come into prominence when the relative velocity between the gas and a solid body approaches the velocity at which sound is propagated through the gas. We shall defer further considerations of compressibility, like viscosity, to a later volume. From now on, unless stated otherwise, we shall assume the fluids which we are studying are both inviscid and incompressible.

3.2 BOUNDARY CONDITIONS

The simplest boundary condition occurs where a fluid (inviscid and incompressible) is in contact with a rigid impermeable boundary. This could be the wall of a container or the surface of a body which is moving through the fluid. At such a boundary, fluid cannot cross over so that the velocity of the fluid at the boundary must be such that its component normal to the boundary is exactly the same as the normal component of the velocity at the boundary. Thus at any point P of the boundary, if $\mathbf{q}$ denotes the velocity of the fluid and $\mathbf{U}$ the velocity of the boundary and $\mathbf{n}$ is a unit vector normal to the boundary,

$$\mathbf{q} \,.\, \mathbf{n} = \mathbf{U} \,.\, \mathbf{n}. \tag{3.7}$$

Notice that equation (3.7) is a purely kinematic condition. When viscosity is neglected, the fluid is free to *slip* over the boundary so no constraint is placed on the tangential component of $\mathbf{q}$ at the rigid boundary. When viscosity is taken into account, the usual boundary condition which is applied is the *non-slip* condition, in which slippage of the fluid over the boundary is not allowed and consequently both normal and tangential components of $\mathbf{q}$ must match those of the boundary.

Example [3.2]

A sphere of radius a translates with constant velocity $\mathbf{U}$ in an ideal incompressible fluid. Formulate the boundary condition on the sphere.

Referring to Fig. 3.4, let $\mathbf{U} = U\mathbf{k}$ and (r, θ, ϕ) be spherical polar coordinates. Let $\mathbf{q} = q_r \,\hat{\mathbf{r}} + q_\theta \,\hat{\boldsymbol{\theta}} + q_\phi \,\hat{\boldsymbol{\phi}}$. The boundary condition is from (3.7),

$$\mathbf{q} \,.\, \hat{\mathbf{r}} = \mathbf{U} \,.\, \mathbf{f} \qquad (r = a)$$

Thus, at a point on the surface of the sphere, the boundary condition is

$$q_r = U \cos\theta \qquad (r = a, \quad 0 \leqslant \theta \leqslant \pi).$$

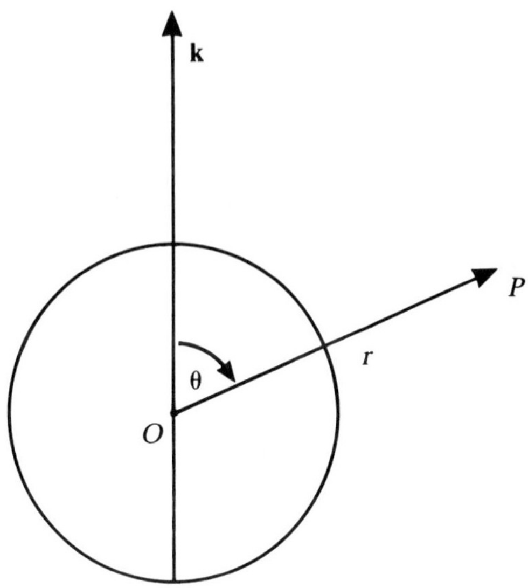

Fig. 3.4

Rigid boundaries are only one type which may be encountered in a problem. Flexible boundaries often occur. Take for example a ship's plates which deform in a heavy sea, or the walls of our arteries which expand and contract with the pumping action of our hearts, or even the case of a bubble of air in a liquid such as water. In such cases, besides equality of normal velocity components of fluid and boundary, there is also continuity of pressure. Thus at a flexible boundary

$$\mathbf{q} \,.\, \mathbf{n} = \mathbf{U} \,.\, \mathbf{n}, \tag{3.8}$$

$$p = \pi, \tag{3.9}$$

where Π is the applied pressure at the boundary. In the case of an air bubble in a liquid, it is the pressure of the air inside the bubble, and for a moving flexible boundary like an arterial wall, Π would be the loading pressure produced by the expansion or contraction of the artery.

Another type of boundary condition arises at a *free surface* S where fluid borders a vacuum. The interface between most liquids and air is usually regarded as a free surface since the ratio of densities of air and the fluid is so very small, e.g. in the case of a water/air interface, $\rho_{\text{air}}/\rho_{\text{water}} = 1.25 \times 10^{-3}$ at $10°$C. At a free surface, there is a balance of pressure. Thus

$$p = \Pi, \tag{3.10}$$

with Π denoting pressure outside the fluid. In the case of air, this would often be atmospheric pressure. Equation (3.10) is a *dynamic* boundary condition. If the equation of S in $\zeta(\mathbf{r}, t) = 0$, it follows from (2.48) that

$$\mathrm{D}\zeta/\mathrm{D}t = 0 \tag{3.11}$$

on the surface S. In terms of Cartesian coordinates, condition (3.11) is

$$u \frac{\partial \zeta}{\partial x} + v \frac{\partial \zeta}{\partial y} + w \frac{\partial \zeta}{\partial z} + \frac{\partial \zeta}{\partial t} = 0. \tag{3.11}$$

But the unit vector $\mathbf{n}$ normal to the surface $\zeta(x, y, z, t) = 0$ is given by

$$\mathbf{n} = \frac{\nabla \zeta}{|\nabla \zeta|} = \left[\left(\frac{\partial \zeta}{\partial x}\right)^2 + \left(\frac{\partial \zeta}{\partial y}\right)^2 + \left(\frac{\partial \zeta}{\partial z}\right)^2\right]^{-1/2} \left\{\mathbf{i}\frac{\partial \zeta}{\partial x} + \mathbf{j}\frac{\partial \zeta}{\partial y} + \mathbf{k}\frac{\partial \zeta}{\partial z}\right\}.$$

Accordingly (3.11) is equivalent to

$$q_n + \frac{\partial \zeta}{\partial t}\left[\left(\frac{\partial \zeta}{\partial x}\right)^2 + \left(\frac{\partial \zeta}{\partial y}\right)^2 + \left(\frac{\partial \zeta}{\partial z}\right)^2\right]^{-1/2} = 0, \tag{3.12}$$

where q_n is the normal component of the velocity $\mathbf{q}$ at the surface S.

If S is the boundary between two immiscible inviscid fluids in which the velocities are $\mathbf{q}_1$, $\mathbf{q}_2$ and the pressures are p_1, p_2 respectively, the appropriate boundary conditions are continuity of pressure and the normal component of velocity together with the kinematic condition (3.12) with $\zeta(x, y, z, t) = 0$ denoting the equation of S and q_n the common normal velocity component in the fluids at S.

The dynamic boundary condition of continuity of pressure is strictly true at a curved interface or free surface only if the *surface tension* at the surface is negligible. Surface tension arises from the intermolecular forces in the liquid and the forces between the liquid molecules and those of any adjacent substance. The result is a force which opposes any increase in the area of a liquid surface. Since an increase in the liquid surface area can only be achieved through the expenditure of mechanical energy, it follows that the existence of a free surface implies the presence of free surface energy which equals the work done in forming the surface. Any mechanical system will tend to move towards a configuration of stable equilibrium in which potential energy is a minimum. Thus, in the absence of any other constraint, a quantity of liquid adjusts its shape so that its free surface energy is a minimum. This is why a drop of liquid, free of all other forces including gravity, takes on a spherical shape since, for a given volume, the enclosing surface of minimum area is a sphere. Since the liquid surface requires mechanical energy for its formation, if it contracts it loses mechanical energy and in so doing the surface does mechanical work and so must exert on its surroundings a force in the direction in which it is moving while contracting. The surface is said to be in a state of tension. If a line is drawn in the liquid surface, then the liquid on one side of the line pulls on the liquid on the other side of the line. The magnitude of surface tension is defined as that of the tensile force acting and is perpendicular to a short straight element of the line, divided by the length of the line element. It is often denoted by γ, and from its definition has the dimensions of force/length, i.e. MT^{-2}. Water in contact with air has a value of $\gamma = 0.073$ N/m while for mercury and air the value is much larger, being of the order of 0.48 N/m as one would expect. For all liquids, surface tension decreases with rise in temperature and its value can be greatly altered by the presence of contaminants or by additives known as *surfactants*. In particular the surface tension of water is considerably reduced by the addition of small quantities of solutes such as soap and detergents, while salts in solution will raise its surface tension. Likewise a surface tension exists in the surface separating two immiscible fluids. This is referred to as *interfacial surface tension.*

If the surface of a liquid (or interface between immiscible liquids) is curved, the surface tension evidently has a resultant towards the concave side, and for equilibrium this resultant must be balanced by a greater pressure at the concave side of the surface. It can be shown that if the surface has radii of curvature R_1 and R_2 in two perpendicular planes, i.e. principal radii of curvature, the pressure at the concave side must exceed that at the convex side by the amount

$$\gamma\left[\frac{1}{R_1}+\frac{1}{R_2}\right]. \tag{3.13}$$

Inside a spherical drop, the pressure exceeds that outside by $2\gamma/R$, where R is the radius of the drop. For a spherical soap bubble the excess is doubled since the soapfilm has two surfaces, albeit very close together, in which surface tension acts.

At a free surface of a liquid in motion, or at the interface between immiscible liquids in motion, the correct dynamical boundary condition which replaces (3.10), for instance, is that the pressure excess on the concave side of the surface over that on the convex side is balanced by a surface tension stress given by (3.13). Thus for two fluids which are immiscible and in motion with an interface given by the equation $\zeta(x, y, z, t) = 0$, as shown in Fig. 3.5, the boundary conditions at the interface are

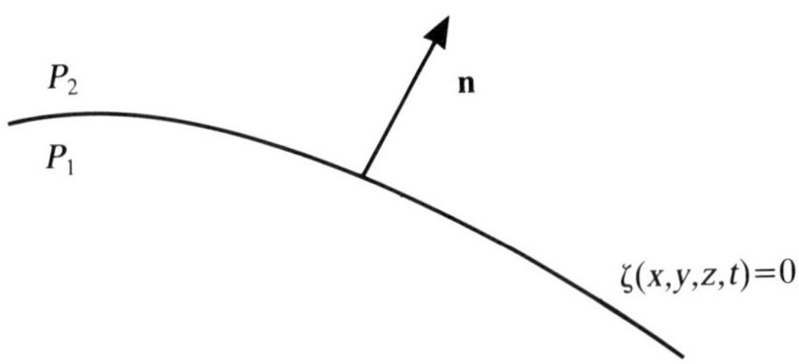

Fig. 3.5

$$\mathbf{q}_1 \cdot \mathbf{n} = \mathbf{q}_2 \cdot \mathbf{n}, \tag{3.14}$$

$$\mathbf{q}_1 \cdot \mathbf{n} + \frac{\partial \zeta}{\partial t}\left[\left(\frac{\partial \zeta}{\partial x}\right)^2 + \left(\frac{\partial \zeta}{\partial y}\right)^2 + \left(\frac{\partial \zeta}{\partial z}\right)^2\right]^{-1/2} = 0, \tag{3.15}$$

$$p_1 = p_2 + \gamma\left(\frac{1}{R_1} + \frac{1}{R_2}\right), \tag{3.16}$$

where p_1, $\mathbf{q}_1$ and p_2, $\mathbf{q}_2$ denote the pressure and velocity on either side of the interface. In the case of a free surface with the liquid having pressure and velocity p, $\mathbf{q}$ respectively the conditions are

$$\mathbf{q} \cdot \mathbf{n} + \frac{\partial \zeta}{\partial t}\left[\left(\frac{\partial \zeta}{\partial x}\right)^2 + \left(\frac{\partial \zeta}{\partial y}\right)^2 + \left(\frac{\partial \zeta}{\partial z}\right)^2\right]^{-1/2} = 0, \tag{3.17}$$

$$p = \Pi + \gamma\left(\frac{1}{R_1} + \frac{1}{R_2}\right), \tag{3.18}$$

with Π the pressure outside the free surface.

In the case of a flexible boundary when a loading pressure Π is applied, then the boundary condition (3.9) is suitably modified to take into account the effect of surface tension between the flexible boundary and the liquid. We consider the use of boundary conditions involving surface tension within our chapter on waves.

3.3 EULER'S EQUATION OF MOTION OF AN IDEAL FLUID

The equation of motion of a fluid is, in its most fundamental form, a relation equating the rate of change of momentum of a specified region of the fluid to the vector sum of all forces which act on that region of fluid, in accordance with Newton's second law of motion.

Consider a region of fluid τ bounded by the closed surface Σ, as depicted in Fig. 3.6, which consists of the same fluid particles at all times. The fluid within τ is acted on by two types of force: the *stress* acting over the surface Σ due to the rest of the fluid exterior to τ, and in the case of an ideal (inviscid) fluid, this stress is simply the pressure directed along the *inward* normal at all points Σ. The second type of force is a *body force*, such as gravity, which acts on all fluid particles. Applying Newton's second law of motion to the fluid within τ, we have

$$\frac{\mathrm{d}}{\mathrm{d}t}\int_\tau \rho\mathbf{q}\,\mathrm{d}\tau = -\int_\Sigma p\mathbf{n}\,\mathrm{d}\Sigma + \int_\tau \rho\mathbf{F}\,\mathrm{d}\tau, \tag{3.19}$$

where $\mathbf{F}$ denotes the body force per unit mass. With $\mathbf{q} = u\mathbf{i} + v\mathbf{j} + w\mathbf{k}$, it follows from (2.41) that

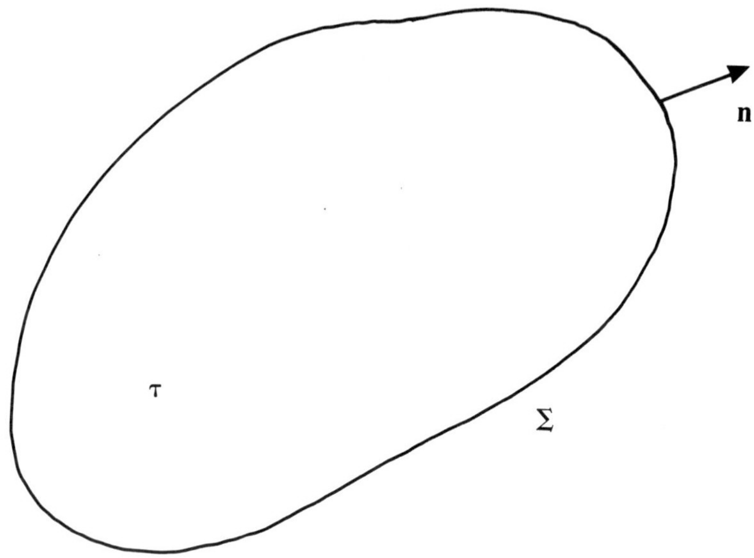

Fig. 3.6

$$\frac{\mathrm{d}}{\mathrm{d}t}\int_\tau \rho u\,\mathrm{d}\tau = \int_\tau \left\{\frac{\mathrm{D}}{\mathrm{D}t}(\rho u) + \rho u \operatorname{div}\mathbf{q}\right\}\mathrm{d}\tau$$

$$= \int_\tau \left\{\rho\frac{\mathrm{D}u}{\mathrm{D}t} + u\left[\frac{\mathrm{D}\rho}{\mathrm{D}t} + \rho \operatorname{div}\mathbf{q}\right]\right\}\mathrm{d}\tau. \tag{3.20}$$

But (2.45) shows that

$$\frac{\mathrm{D}\rho}{\mathrm{D}t} + \rho \operatorname{div}\mathbf{q} = 0, \tag{3.21}$$

and consequently

$$\frac{\mathrm{d}}{\mathrm{d}t}\int_\tau \rho u\,\mathrm{d}\tau = \int_\tau \rho\frac{\mathrm{D}u}{\mathrm{D}t}\,\mathrm{d}\tau. \tag{3.22}$$

Similar relations can be written down by replacing u by v and w in turn and combining these with (3.22) yields the vector identity

$$\frac{\mathrm{d}}{\mathrm{d}t}\int_\tau \rho\mathbf{q}\,\mathrm{d}\tau = \int_\tau \rho\frac{\mathrm{D}\mathbf{q}}{\mathrm{D}t}\,\mathrm{d}\tau. \tag{3.23}$$

Equation (3.23) expresses the rate of change of linear momentum of the fluid within τ as a volume integral over the region τ and its use in (3.19) gives

$$\int_\tau \rho\frac{\mathrm{D}\mathbf{q}}{\mathrm{D}t}\,\mathrm{d}\tau = -\int_\Sigma p\mathbf{n}\,\mathrm{d}\Sigma + \int_\tau \rho\mathbf{F}\,\mathrm{d}\tau. \tag{3.24}$$

The surface integral in (3.24) can be expressed as a volume integral over τ using the Divergence Theorem. Thus

$$\int_\tau \left\{ \rho \frac{\mathrm{D}\mathbf{q}}{\mathrm{D}t} + \nabla p - \rho \mathbf{F} \right\} \mathrm{d}\tau = 0. \qquad (3.25)$$

But τ is any arbitrarily chosen region of the moving fluid, so it follows that the integrand in (3.25) must vanish at every point of τ. Thus at any point of the fluid,

$$\frac{\mathrm{D}\mathbf{q}}{\mathrm{D}t} = \frac{\partial \mathbf{q}}{\partial t} + (\mathbf{q}\,.\nabla)\,\mathbf{q} = \mathbf{F} - \frac{1}{\rho}\nabla p. \qquad (3.26)$$

Equation (3.26) is known as *Euler's equation of motion for an ideal fluid*. Using the vector identity (1.57), the equation may be written in the alternative form

$$\frac{\partial \mathbf{q}}{\partial t} + \nabla(\tfrac{1}{2}\mathbf{q}^2) - \mathbf{q} \times (\nabla \times \mathbf{q}) = \mathbf{F} - \frac{1}{\rho}\nabla p. \qquad (3.27)$$

3.4 THE EQUATIONS OF MOTION OF AN IDEAL FLUID

Equation (3.27) is the equation which must be satisfied at every point of the fluid as a consequence of Newton's second law of motion. In Chapter 2 we saw that there is another equation – the equation of continuity – which must be satisfied also at all points of the fluid as a consequence of the conservation of matter. These two equations comprise the *equations of motion* of an ideal fluid. They are fundamental to any theoretical study of ideal fluid flow so it is worthwhile for us to restate them again here.

Euler's equation:

$$\frac{\partial \mathbf{q}}{\partial t} + (\mathbf{q}\,.\nabla)\mathbf{q} = \mathbf{F} - \frac{1}{\rho}\nabla p, \qquad (3.28)$$

Equation of continuity:

$$\frac{\partial \rho}{\partial t} + \operatorname{div}(\rho\,\mathbf{q}) = 0, \qquad (3.29)$$

and to be able to give a theoretical description of a specific flow of an ideal fluid it is necessary to solve (3.28) and (3.29) subject to the appropriate boundary and initial conditions dictated by the physical characteristics of the flow. The formulations of these boundary conditions were dealt with in Section 3.2. With the density function known, equations (3.28) and (3.29) are in fact four simultaneous differential equations for the three velocity components and pressure. It should be noted that (3.28) and (3.29) apply at each *point* of the fluid, which is a geometrical point in the space occupied by the fluid. The space variables x, y, z are therefore independent of the time variable t. The reader will realise that to solve (3.28) and (3.29) for a flow caused by the motion of some arbitrarily shaped body presents a formidable mathematical problem. Throughout this book, we shall acquaint the reader with progressively more complicated problems and their solution, but first, let us consider a couple of flows which are 'one dimensional' so that only one space variable enters into the analysis.

Example [3.3]
A homogeneous liquid occupies a length $2l$ of a straight tube of uniform small bore and is acted on by a body force which is such that the fluid is attracted to a fixed point of the tube, with a force varying as the distance from that point.

To determine the velocity and pressure within the liquid, which is of course *incompressible*, we note that the small bore of the tube allows us to ignore any variation of velocity across any cross-section of the tube and to suppose that the flow is *unidirectional*. Let u be the velocity along the tube and p be the pressure at a general point P distance x from the centre of force O. Also let h be the distance of the centre of mass G of the fluid, as indicated in Fig. 3.7.

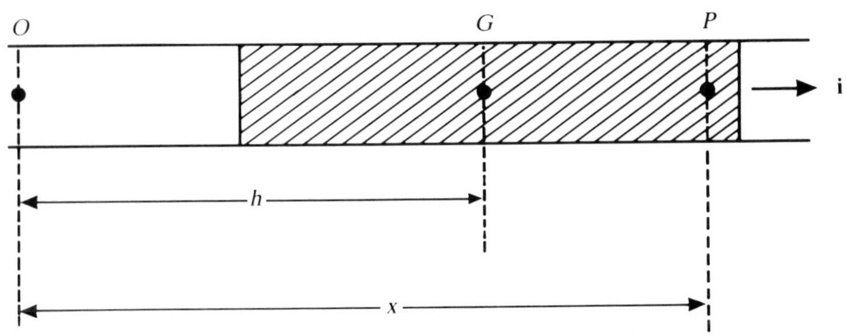

Fig. 3.7

The equations of motion of the fluid are

(1) Euler's equation

$$\frac{\partial u}{\partial t} + u\frac{\partial u}{\partial x} = -\mu x - \frac{1}{\rho}\frac{\partial p}{\partial x},$$

since the body force per unit mass is $-\mu x\mathbf{i}$, with μ a positive constant, and $u = u(x, t)$.

(2) Equation of continuity:

$$\frac{\partial u}{\partial x} = 0.$$

This equation tells us that $u = u(t)$ so Euler's equation can be integrated with respect to x to give

$$x\frac{\mathrm{d}u}{\mathrm{d}t} = -\frac{1}{2}\mu x^2 - \frac{p}{\rho} + C,$$

where C is at most a function of t only. If we suppose that the pressure is Π at the free surfaces of the liquid $x = h - l$ and $x = h + l$, it follows that

$$(h-l)\frac{\mathrm{d}u}{\mathrm{d}t} = C - \frac{1}{2}\mu(h-l)^2 - \frac{\Pi}{\rho},$$

$$(h+l)\frac{\mathrm{d}u}{\mathrm{d}t} = C - \frac{1}{2}\mu(h+l)^2 - \frac{\Pi}{\rho},$$

which on subtraction gives

$$\frac{\mathrm{d}u}{\mathrm{d}t} = -\mu h.$$

But in the fluid motion, all fluid particles move with the same velocity u and

$$u = \frac{\mathrm{d}h}{\mathrm{d}t}.$$

Thus

$$\frac{\mathrm{d}^2 h}{\mathrm{d}t^2} = -\mu h,$$

giving $h = A\cos(\sqrt{(\mu)}t + \epsilon)$, with A and ϵ constants which would be determined from initial values for the position and velocity of the liquid. We see that this type of body force results in an oscillatory motion of the fluid, which should not come as any surprise.

The pressure p at a general point P of the fluid is determined by integrating Euler's equation with respect to x from $x = h - l$ to the general value of x. This gives

$$\frac{p}{\rho} = \frac{\Pi}{\rho} - \frac{1}{2}\mu[x^2 - (h-l)^2] - (x-h+l)\frac{\mathrm{d}u}{\mathrm{d}t}$$

$$= \frac{\Pi}{\rho} - \frac{1}{2}\mu(x-h+l)(x-h-l).$$

The reader will notice that in solving this problem we first eliminated the pressure to solve for the velocity. The pressure was then determined, with the velocity now known, by integrating Euler's equation. This is the procedure which is often used in more complex problems.

A flow in which gravity is the body force is illustrated in the next example which is a model for a *water clock*.

Example [3.4]
Homogeneous liquid is in motion in a vertical plane, within a curved tube of uniform small bore, under the action of gravity. Describe the motion.

Let O be the lowest point of the tube, AB the equilibrium level of the liquid and h the height of AB above O. Let α and β be respectively the inclinations of the tube to the horizontal at A and B, and θ be the inclination of the tube at a distance s along the tube from O. Let a, b denote the arc lengths of OA, OB respectively and suppose that at time t, the liquid is displaced a small distance z along the tube from its equilibrium position.

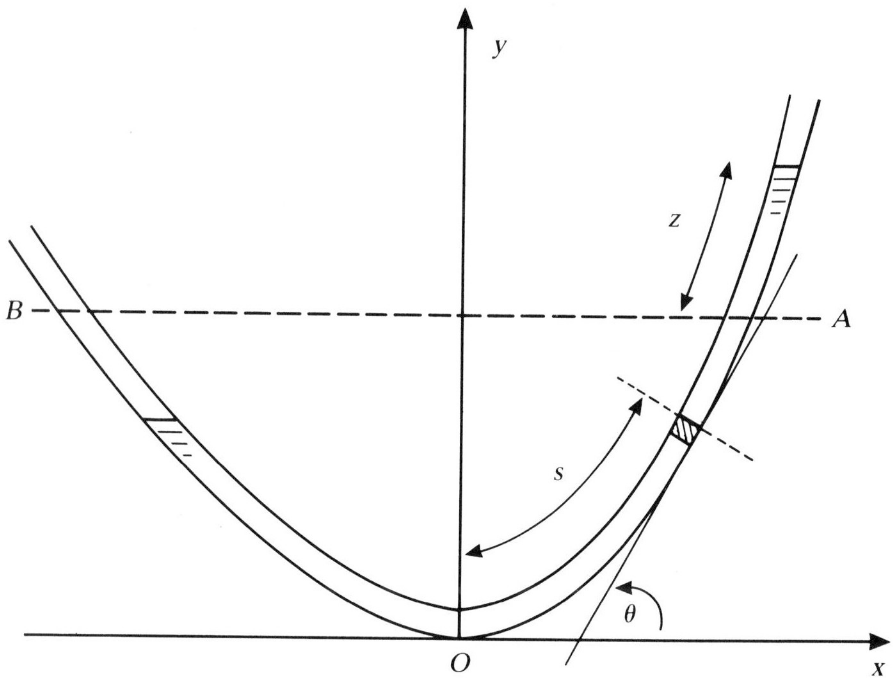

Fig. 3.8

The assumption of uniform small bore means the flow is again unidirectional along the tube. Let the velocity be $u(s, t)$. The equation of continuity gives $\partial u/\partial s = 0$, so Euler's equation of motion now reduces to

$$\frac{\mathrm{d}u}{\mathrm{d}t} = -g\sin\theta - \frac{1}{\rho}\frac{\partial p}{\partial s},$$

which may also be written as

$$\frac{\mathrm{d}u}{\mathrm{d}t} = -g\frac{\mathrm{d}y}{\mathrm{d}s} - \frac{1}{\rho}\frac{\partial p}{\partial s},$$

noting from Fig. 3.8 that $\mathrm{d}y/\mathrm{d}s = \sin\theta$. Integrating this equation with respect to s gives

$$s\,\frac{\mathrm{d}u}{\mathrm{d}t} = -gy - \frac{p}{\rho} + C(t).$$

Taking the pressure at the free surfaces of the liquid to be atmospheric pressure Π, we have the equations resulting from applying the boundary conditions at the free surfaces in the form

$$(a+z)\,\frac{\mathrm{d}u}{\mathrm{d}t} = -g\,(h + z\sin\alpha) - \frac{\Pi}{\rho} + C(t),$$

$$-(b-z)\,\frac{\mathrm{d}u}{\mathrm{d}t} = -g\,(h - z\sin\beta) - \frac{\Pi}{\rho} + C(t).$$

Subtracting these equations gives

$$(a+b)\,\frac{\mathrm{d}u}{\mathrm{d}t} = -gz\,(\sin\alpha + \sin\beta).$$

However, $u = \mathrm{d}z/\mathrm{d}t$, so this equation shows that the fluid oscillates about its equilibrium level with time period

$$2\pi\sqrt{(a+b)/g\,(\sin\alpha + \sin\beta)}.$$

The velocity u is completely determined if initial values of z and u are prescribed. Suppose $z = \epsilon$, $u = 0$ when $t = 0$, then at general time t,

$$u = -\,\epsilon n \sin nt,$$

where

$$n = g\,(\sin\alpha + \sin\beta)/\sqrt{(a+b)}.$$

The pressure p can now be determined with the function $C(t)$ found by applying the boundary conditions at the free surface where $s = a + \epsilon\cos nt$.

3.5 BERNOULLI'S EQUATION

Euler's equation (3.28) may be thought of as a differential equation for the pressure p if the velocity $\mathbf{q}$ is known. The integration of this differential equation can be accomplished very simply for certain classes of flow which cover a wide variety of situations. Usually the body force is gravitational, which is an example of a conservative force field. From Section 1.16, we see that if the body force $\mathbf{F}$ is conservative, it can be expressed in terms of a *body force potential* function Ω so that $\mathbf{F} = \nabla\Omega$.

In the case of steady motion with a conservative body force, equation (3.27) gives

$$\nabla\left\{\frac{1}{2}\mathbf{q}^2 - \Omega\right\} + \frac{1}{\rho}\nabla p = \mathbf{q} \times \boldsymbol{\zeta}, \tag{3.30}$$

where $\boldsymbol{\zeta} = \operatorname{curl} \mathbf{q}$. Let $d\mathbf{r} = dx\mathbf{i} + dy\mathbf{j} + dz\mathbf{k}$ denote the line element vector. For any function $f(x, y, z)$ with continuous derivatives of the first order, the total differential df is, by definition,

$$df = \frac{\partial f}{\partial x}\,dx + \frac{\partial f}{\partial y}\,dy + \frac{\partial f}{\partial z}\,dz = (d\mathbf{r} \,.\, \nabla)f. \tag{3.31}$$

Hence,

$$d\mathbf{r} \,.\, \left\{\frac{1}{\rho}\nabla p\right\} = \frac{1}{\rho}(d\mathbf{r} \,.\, \nabla)p = \frac{dp}{\rho}, \tag{3.32}$$

while

$$\frac{dp}{\rho} = d\int\frac{dp}{\rho} = (d\mathbf{r} \,.\, \nabla)\left\{\int\frac{dp}{\rho}\right\}. \tag{3.33}$$

It therefore follows that

$$\frac{1}{\rho}\nabla p = \nabla\left\{\int\frac{dp}{\rho}\right\}. \tag{3.34}$$

Equation (3.34) is true for any fluid whose density is variable. If the fluid is homogeneous and incompressible, the equation is trivially obvious. We can now use the general result (3.34) in (3.30) and obtain

$$\nabla\left\{\frac{1}{2}\mathbf{q}^2 - \Omega + \int\frac{dp}{\rho}\right\} = \mathbf{q} \times \boldsymbol{\zeta}. \tag{3.35}$$

If $\hat{\mathbf{s}}$ is a unit vector along the streamline through a general point of the fluid, and s measure distance along this streamline, then since $\hat{\mathbf{s}}$ is parallel to $\mathbf{q}$, equation (3.35) gives, after scalar multiplication by $\hat{\mathbf{s}}$,

$$\frac{\partial}{\partial s}\left\{\frac{1}{2}\mathbf{q}^2 - \Omega + \int\frac{dp}{\rho}\right\} = 0,$$

and consequently along any particular streamline,

$$\frac{1}{2}\mathbf{q}^2 - \Omega + \int\frac{dp}{\rho} = C, \tag{3.36}$$

where the constant C takes different values for different streamlines. This result is known as the *Bernoulli equation* and it should be noted that it applies to the steady flow of an ideal fluid in which all body forces are conservative. In many flows around finite sized bodies, the velocity at a great distance from the bodies is uniform, i.e. a uniform stream. Suppose the velocity of this stream is $\mathbf{U}$, a constant vector, then $\boldsymbol{\zeta} = \text{curl } \mathbf{U} = \mathbf{0}$ and Euler's equation shows that in the absence of a body force, the pressure p is a constant. We may also choose $\hat{\mathbf{s}}$ to be a unit vector along a vortex line, in which case equation (3.36) again follows but with the left-hand side now constant along any particular vortex line.

Further examples of the Bernoulli equation arise when $\mathbf{q} \times \boldsymbol{\zeta} = \mathbf{0}$. If $\mathbf{q}$ and $\boldsymbol{\zeta}$ are parallel, the streamlines and vortex lines coincide and $\mathbf{q}$ is said to be a Beltrami vector. If $\boldsymbol{\zeta} = \mathbf{0}$, the flow is irrotational. For both of these flow patterns,

$$\frac{1}{2}\mathbf{q}^2 - \Omega + \int \frac{\mathrm{d}p}{\rho} = C, \tag{3.37}$$

with the constant C now the same at *all* points of the fluid. In particular, for homogeneous incompressible fluids,

$$\int \frac{\mathrm{d}p}{\rho} = \frac{p}{\rho},$$

and Bernoulli's equations has the simple form

$$\frac{p}{\rho} + \frac{1}{2}\mathbf{q}^2 - \Omega = C, \tag{3.38}$$

so that when $\mathbf{q}$ is known, the pressure p can be determined immediately.

Example [3.5]

A long straight pipe of length L has a slowly tapering circular cross-section. It is inclined so that its axis makes an angle α to the horizontal with its smaller cross-section downwards. The radius of the pipe at its upper end is twice that at its lower end and water is pumped at a steady rate through the pipe to emerge at atmospheric pressure. If the pumping pressure is twice atmospheric pressure, show that the fluid leaves the pipe with speed U given by

$$U^2 = 32\,[gL \sin\alpha + \Pi/\rho]/15,$$

where Π is atmospheric pressure.

The geometry of the pipe is shown diagrammatically in Fig. 3.9.

The assumption that the pipe is *slowly* tapering means that any variation in the velocity over any cross-section can be ignored. Let the velocity at the wider end of the pipe be V and the emerging velocity by U. The only body force is that of gravity, so $\mathbf{F} = -g\mathbf{j}$ and consequently $\Omega = -gy$. Bernoulli's equation (3.38) gives

$$\frac{p}{\rho} + \frac{1}{2}\mathbf{q}^2 + gy = C.$$

Applying the equation at the two ends of the pipe gives

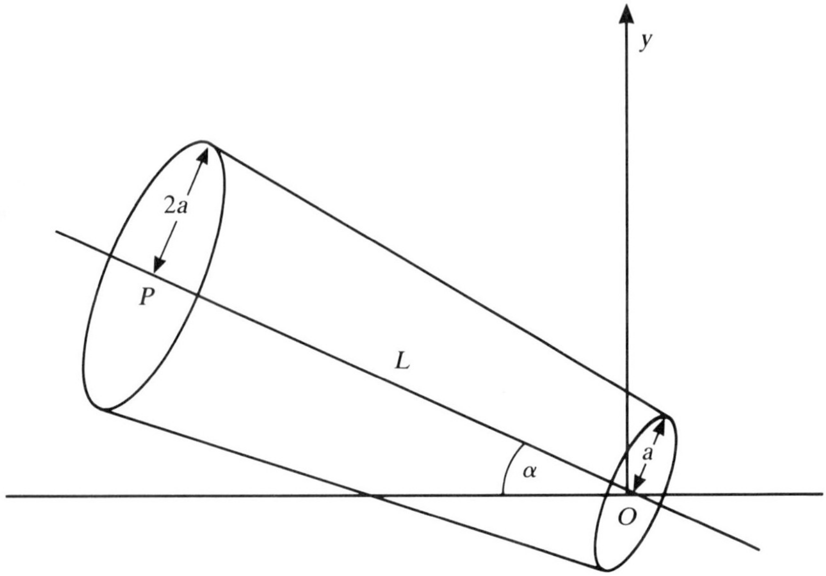

Fig. 3.9

$$\Pi + \frac{1}{2}\rho\, U^2 = 2\Pi + \frac{1}{2}\rho\, V^2 + \rho\, gL \sin\alpha,$$

while for mass conservation,

$$\pi a^2 U = 4\pi\, a^2\, V,$$

giving $V = U/4$. Using this relation,

$$\frac{1}{2}\rho\, U^2 \left(\frac{15}{16}\right) = \Pi + gL \sin\alpha,$$

yielding the required result

$$U^2 = 32\,[gL \sin\alpha + \Pi/\rho]/15.$$

Example [3.6]

Consider the flow of liquid out of a large tank of radius A through a small hole of radius a located in the bottom of the tank, as shown in Fig. 3.10.

Since the tank is assumed to be large, we may regard the flow as approximately steady. Streamlines start at the upper surface of the liquid in the tank, which we shall suppose is open to the atmosphere, and converge through the small exit hole in the base of the tank. With the y-coordinate measured upwards from the base of the tank, the only body force acting is gravity, and the body force potential $\Omega = -gy$. Applying Bernoulli's equation (3.38) along a typical

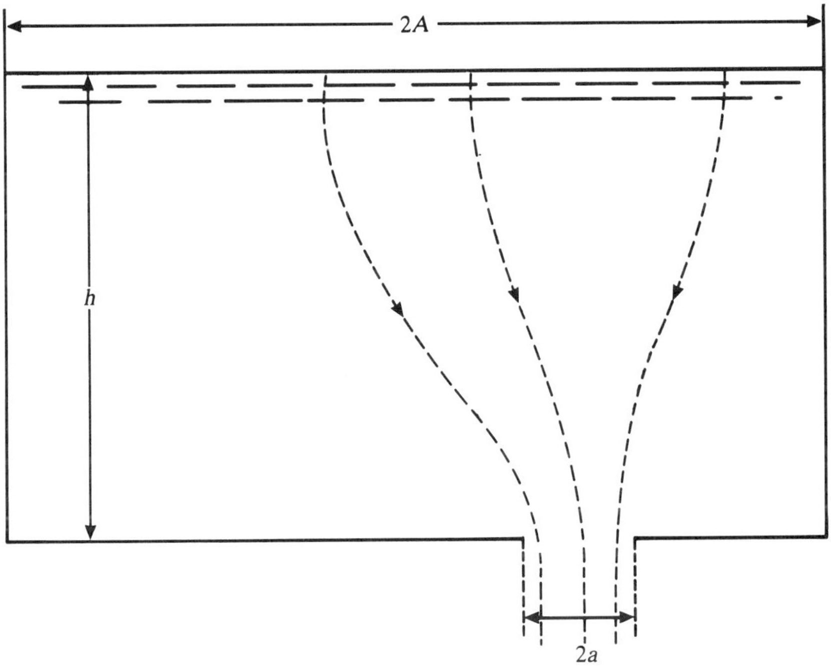

Fig. 3.10

streamline at both the upper surface of the liquid in the tank and at the exit hole, we obtain

$$\frac{1}{2}U^2 + gh + \frac{\Pi}{\rho} = \frac{1}{2}V^2 + \frac{\Pi}{\rho},$$

where U and V denote the fluid velocity at the surface and exit hole respectively and Π is the atmospheric pressure. Note that the liquid emerges from the tank at atmospheric pressure because there is no variation in velocity across the jet of liquid discharged from the tank. Consequently the pressure across a section of this jet is constant and must be the same as that of the atmosphere to satisfy the dynamic boundary conditions at the free surface of this jet of fluid. It follows that

$$V^2 = U^2 + 2gh.$$

There is a further relation between U and V to satisfy mass conservation. This is

$$\pi a^2 V = \pi A^2 U,$$

and since $U = \mathrm{d}h/\mathrm{d}t$, it follows that

$$\frac{\mathrm{d}h}{\mathrm{d}t} = -\sqrt{(2gh)}\left[\frac{A^4}{a^4} - 1\right]^{-1/2}$$

Integration of this equation yields

$$h=\left\{h_0^{1/2}-\sqrt{\left(\frac{g}{2}\right)}\left[\frac{A^4}{a^4}-1\right]^{-1/2}t\right\}^2,$$

where $h = h_0$ when $t = 0$. The time to empty the tank, i.e. when $h = 0$ is therefore $[(A^4/a^4) - 1]^{1/2}\,(2h_0/g)^{1/2}$.

The foregoing examples show how useful Bernoulli's equation is in determining certain results without the need to solve the equations of motion for the velocity and pressure at each point of the fluid, which may well be analytically intractable in certain cases; as illustrated by the second example [3.6]. The derivation of any of the forms of Bernoulli's equation given above does require that the fluid motion is steady. However, Euler's equation of motion can be integrated to give a Bernoulli equation when the flow is unsteady provided that it is irrotational and any body forces are conservative. This follows since for irrotational flow, a velocity potential Φ exists such that $\mathbf{q} = \nabla\Phi$. In which case, Euler's equation (3.28) can be integrated to give

$$\frac{\partial\Phi}{\partial t}+\frac{1}{2}\mathbf{q}^2-\Omega+\int\frac{dp}{\rho}=f(t), \tag{3.39}$$

where $f(t)$ is an arbitrary function of t arising from the integration with respect to the space variables in which t is held constant. For steady irrotational motion, $\partial\Phi/\partial t = 0$ and $f(t)$ is a constant, whereupon equation (3.37) is recovered.

3.6 FLOW MEASURING DEVICES

Since Bernoulli's equation gives a simple relation connecting pressure and velocity in liquids, devices have been constructed which can be used to measure the velocity in a particular flow. We mention two such devices here.

The Pitot tube

This device is used for measuring the speed of a stream such as a river or, in a modified form, the airspeed of an aeroplane. In one of its basic forms, the Pitot tube is illustrated in Fig. 3.11.

To measure the velocity $\mathbf{q}$ of a stream, the device is placed with the U-tube *BCD* in a vertical plane and the straight tube *AB* downstream of the flow and parallel to the local velocity $\mathbf{q}$. The tube *AB* is connected to one end of the U-tube, which contains mercury, the other end of it being open to the stream through holes *H*, as shown in the diagram. When fluid from the stream enters *AB*, it forces mercury around the U-tube until an equilibrium configuration is attained. Thus the velocity at the liquid/mercury interface *C* is zero. Let p be the pressure in the stream where the velocity is $\mathbf{q}$, then p is also the pressure at the hole *H*, it being assumed that *H* is close to *A*, and so any variation in pressure is

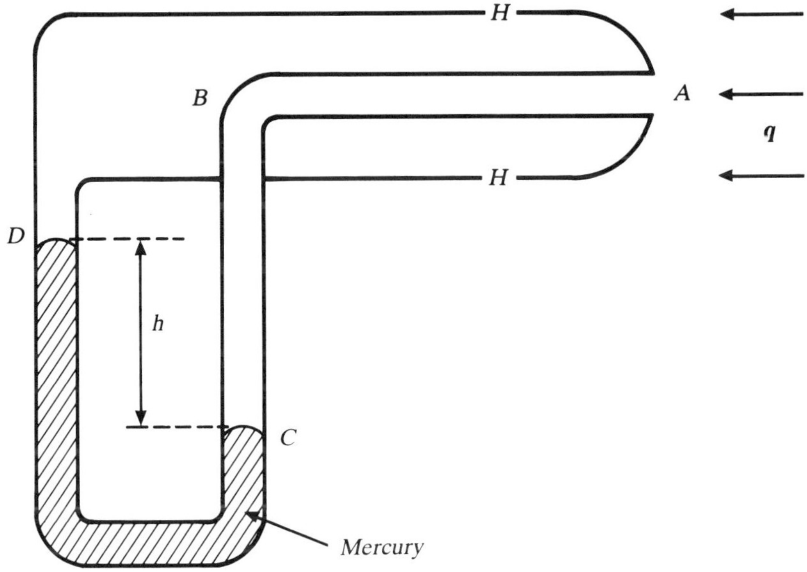

Fig. 3.11

negligible. The fluid between H and the meniscus D is at rest so that p is also the pressure at D. The stream entering the tube AB at A is brought to rest at the meniscus C. If p_0 is the pressure at C, then Bernoulli's theorem gives

$$\frac{p}{\rho} + \frac{1}{2}\mathbf{q}^2 = \frac{p_0}{\rho}$$

where ρ is the density of the liquid. This gives

$$|\mathbf{q}| = [2(p_0 - p)/\rho]^{1/2},$$

and if σ is the density of mercury, we also have

$$p_0 - p = \sigma g h.$$

Thus the local velocity of the stream can be measured. The device can be adapted to measure airspeeds of aeroplanes in subsonic flight when the compressibility of air is not significant.

The Venturi tube

This is a device for measuring flow in a pipe and is widely used in the construction of flow meters. The pipe is constricted from its normal section S_1, to a much smaller section S_2. A vertical U-tube containing mercury is inserted between broad and narrow sections A and B respectively, as shown in Fig. 3.12.

Let $\mathbf{q}_1$ and $\mathbf{q}_2$ be the velocities at A and B and p_1 and p_2 be the corresponding respective pressures. For incompressible flow, conservation of mass requires that

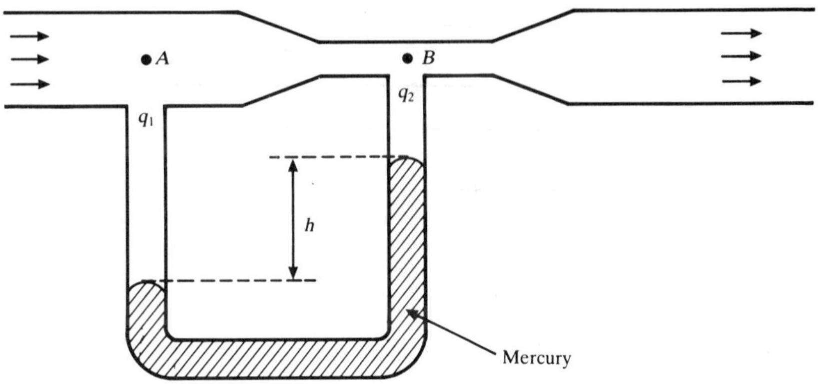

Fig. 3.12

$$q_1 S_1 = q_2 S_2,$$

and applying Bernoulli's equation along the central streamline joining A to B gives

$$\frac{p_1}{\rho} + \frac{1}{2} q_1^2 = \frac{p_2}{\rho} + \frac{1}{2} q_2^2,$$

with ρ denoting the density of the fluid. Elimination of q_2 gives the velocity q_1 along the pipe proper as

$$q_1 = \left[\frac{2(p_1 - p_2) S_2^2}{\rho(S_1^2 - S_2^2)}\right]^{1/2}.$$

This equation shows that $p_1 > p_2$ which may be surprising intuitively since it means that fluid pressure is actually *reduced* by a constriction.

If h is the difference in levels of the mercury menisci in the U-tube when equilibrium is established,

$$p_1 - p_2 = \sigma g h,$$

with σ the density of the mercury. The U-tube containing mercury when used as a device for measuring pressure difference is called a *manometer*. The mass of fluid which flows through the unconstricted part of the pipe in unit time is $\rho q_1 S_1$.

3.7 KELVIN'S CIRCULATION THEOREM

In addition to Bernoulli's equation, other results of a general nature follow from Euler's equation. We first establish Kelvin's Circulation Theorem which prescribes conditions under which the circulation around a closed circuit, moving with the fluid and consisting of the same fluid particles, is constant.

For an ideal fluid, the circulation around any closed circuit of fluid particles, which moves with the fluid, remains constant provided that the body forces are conservative and the pressure is a single valued function of density only.

Proof: In the notation previously employed, Euler's equation of motion is

$$\frac{\mathrm{D}\mathbf{q}}{\mathrm{D}t} = \mathbf{F} - \frac{1}{\rho}\nabla p = \nabla\Omega - \frac{1}{\rho}\nabla p, \tag{3.40}$$

where Ω is the potential of the conservative body force $\mathbf{F}$ per unit mass. The circulation Γ per unit mass about a closed circuit $\mathscr{C}$ of fluid particles is

$$\Gamma = \oint_{\mathscr{C}} \mathbf{q}\,.\,\mathrm{d}\mathbf{r}, \tag{3.41}$$

and since this is a material integral, (2.68) gives

$$\frac{\mathrm{d}\Gamma}{\mathrm{d}t} = \oint_{\mathscr{C}} \frac{\mathrm{D}\mathbf{q}}{\mathrm{D}t}\,.\,\mathrm{d}\mathbf{r} = \oint_{\mathscr{C}} \left\{\nabla\Omega - \frac{1}{\rho}\nabla p\right\}.\,\mathrm{d}\mathbf{r},$$

using (3.40). The first integral on the right-hand side is zero and from (3.32), the second integral is

$$-\oint_{\mathscr{C}} \frac{\mathrm{d}p}{\rho},$$

and if p is a single-valued function of ρ, this integral is also zero, showing that Γ is a constant.

The following important corollaries stem from this theorem:

1. In a closed circuit $\mathscr{C}$ of fluid particles moving under the same conditions as the theorem,

$$\int_S \operatorname{curl}\mathbf{q}\,.\,\mathrm{d}\mathbf{S} = \text{constant},$$

where S is any open surface whose rim is $\mathscr{C}$.

Proof: By Stokes's theorem (see Section 1.15),

$$\int_S \operatorname{curl}\mathbf{q}\,.\,\mathrm{d}\mathbf{S} = \oint_{\mathscr{C}} \mathbf{q}\,.\,\mathrm{d}\mathbf{r} = \Gamma,$$

and the result follows immediately.

2. Under the conditions of the theorem, vortex lines move with the fluid.

Proof: Let $\mathscr{C}$ be any closed curve drawn on the surface of a vortex tube but not embracing it. Let S be the portion of the vortex tube rimmed by $\mathscr{C}$. By definition vortex lines lie on S. Thus

$$0 = \int_S \text{curl}\,\mathbf{q}\,.\,\mathrm{d}\mathbf{S} = \oint_{\mathscr{C}} \mathbf{q}\,.\,\mathrm{d}\mathbf{r}.$$

Let $\mathscr{C}$ be a material curve and S be a material surface, then

$$\frac{\mathrm{d}}{\mathrm{d}t}\int_S (\mathbf{n}\,.\,\text{curl}\,\mathbf{q})\,\mathrm{d}S = \int_S \frac{\mathrm{D}}{\mathrm{D}t}(\mathbf{n}\,.\,\text{curl}\,\mathbf{q})\,\mathrm{d}S = 0.$$

Thus $\mathbf{n}\,.\,\text{curl}\,\mathbf{q}$ remains zero, so that S remains a surface composed of vortex lines. Consequently *vortex lines and tubes move with the fluid*. This explains why smoke rings maintain their form for long periods of time.

3. *Under the conditions of the theorem, if the flow is irrotational in a material region of the fluid at some particular time, e.g. $t = 0$, the flow is always irrotational in that material region thereafter.*

Proof: Suppose that at some instant the fluid on the material surface S is irrotational, then clearly at that instant

$$\Gamma = \oint_{\mathscr{C}} \mathbf{q}\,.\,\mathrm{d}\mathbf{r} = 0$$

for any closed curve $\mathscr{C}$ drawn on S, and by Kelvin's circulation theorem, $\Gamma = 0$ at all times. Thus

$$\int_{S'} (\mathbf{n}\,.\,\text{curl}\,\mathbf{q})\,\mathrm{d}S = 0$$

at all times, for any sub-surface S' included in S. By letting S' shrink to any point of S, it follows that curl $\mathbf{q} = \mathbf{0}$ at all points of S for all time and the motion is always irrotational.

This last result is particularly important because it shows that if a motion of ideal fluid is started from rest, e.g. by the acceleration of some body, then the fluid motion is irrotational for all time t since it is clearly irrotational when $t = 0$. The studies of irrotational flows therefore encompass a wide variety of realistic flows. We shall consider some specific flows of this type in subsequent chapters.

3.8 THE VORTICITY EQUATION

Consider Euler's equation for the motion of ideal fluid under the action of a conservative body force with potential Ω per unit mass:

$$\frac{\mathrm{D}\mathbf{q}}{\mathrm{D}t} = \frac{\partial \mathbf{q}}{\partial t} + \nabla(\tfrac{1}{2}q^2) - \mathbf{q}\times\boldsymbol{\zeta} = \nabla\Omega - \frac{1}{\rho}\nabla p, \tag{3.42}$$

where the *vorticity* $\boldsymbol{\zeta}$ = curl $\mathbf{q}$. If the fluid has constant density, on taking the curl of equation (3.42) we obtain

$$\frac{\partial \boldsymbol{\zeta}}{\partial t} = \text{curl}\,(\mathbf{q} \times \boldsymbol{\zeta})$$

$$= (\nabla \cdot \boldsymbol{\zeta})\mathbf{q} - (\mathbf{q} \cdot \nabla)\boldsymbol{\zeta} + (\boldsymbol{\zeta} \cdot \nabla)\mathbf{q} - (\nabla \cdot \mathbf{q})\boldsymbol{\zeta}.$$

But $\nabla . \boldsymbol{\zeta} = \text{div curl } \mathbf{q} = 0$ and $\nabla . \mathbf{q} = 0$ from the equation of continuity. Thus

$$\frac{\partial \boldsymbol{\zeta}}{\partial t} + (\mathbf{q} \cdot \nabla)\boldsymbol{\zeta} = (\boldsymbol{\zeta} \cdot \nabla)\mathbf{q}, \tag{3.43}$$

or equivalently

$$\frac{\mathrm{D}\boldsymbol{\zeta}}{\mathrm{D}t} = (\boldsymbol{\zeta} \cdot \nabla)\mathbf{q}. \tag{3.44}$$

For *two-dimensional* motion, the vorticity vector is perpendicular to the velocity and the right-hand side of (3.44) is identically zero. Thus for two-dimensional motion of ideal fluid, vorticity is *constant*. In the case when the body force is non-conservative, equation (3.44) becomes

$$\frac{\mathrm{D}\boldsymbol{\zeta}}{\mathrm{D}t} = (\boldsymbol{\zeta} \cdot \nabla)\mathbf{q} + \text{curl}\,\mathbf{F}, \tag{3.45}$$

where **F** is the arbitrary body force per unit mass.

Example [3.7]

A motion of inviscid incompressible fluid of uniform density is symmetric about the axis $r = 0$ where (r, θ, z) are cylindrical polar coordinates. The cylindrical polar resolutes of velocity are $[q_r(r, z), 0, q_z(r, z)]$. Show that if a fluid particle has vorticity of magnitude ζ_0 when at $r = r_0$, its vorticity when at general distance r from the axis of symmetry has magnitude

$$\zeta = (\zeta_0/r_0)r,$$

if any body forces acting are conservative.

The vorticity $\boldsymbol{\zeta}$ satisfies (3.44), so

$$\frac{\mathrm{D}\boldsymbol{\zeta}}{\mathrm{D}t} = (\boldsymbol{\zeta} \cdot \nabla)\mathbf{q}.$$

Now

$$\boldsymbol{\zeta} = \text{curl}\,\mathbf{q} = \frac{1}{r}\begin{vmatrix} \hat{\mathbf{r}} & r\hat{\boldsymbol{\theta}} & \mathbf{k} \\ \partial/\partial r & \partial/\partial\theta & \partial/\partial z \\ q_r & 0 & q_z \end{vmatrix}$$

$$= \hat{\boldsymbol{\theta}}\,[\partial q_r/\partial z - \partial q_z/\partial r],$$

and therefore

Each element is attracted towards O by a force μ/r^2 per unit mass, where μ is a constant, and r is the radial distance from O. Show that the pressure exerted on the surface of the sphere is

$$K\rho_0 \exp(\mu/KR),$$

where ρ_0 is the density at infinity.

2. A straight tube of small bore, ABC, is bent so as to make the angle ABC a right angle, and AB equal to BC. The end C is closed; and the tube is placed with the end A upwards and AB vertical, and is filled with liquid. If the end C be opened, prove that the pressure at any point of the vertical tube is instantaneously diminished one-half; and find the instantaneous change of pressure at any point of the horizontal tube, the pressure of the atmosphere being neglected.

3. Water flows steadily along a pipe of variable cross-section. If the pressure be 700 millimetres of mercury at a place where the velocity is 150 cm per second, find the pressure at a place where the cross-section of the pipe is twice as large. [Take the specific gravity of mercury as 13.6.]

4. A fine tube whose section k is a function of its length s, in the form of a closed plane curve of area A, filled with ice, is moved in any manner. When the component angular velocity of the tube about a normal to its plane is Ω the ice melts without change of volume. Prove that the velocity of the fluid relatively to the tube at a point where the section is K at any subsequent time when ω is the angular velocity is

$$2A\,(\Omega - \omega) \div K \oint \frac{ds}{k},$$

the integral being taken once round the tube.

5. A pipe of variable cross-section is given by $r = a(\cosh \alpha z)^{1/4}$, where r, θ, z are cylindrical polar coordinates with the z-axis vertical. Incompressible inviscid fluid is in steady irrotational motion along the pipe, a volume Q of fluid passing every cross-section per unit time. Assuming the vertical component of velocity to depend on z only, and neglecting the effects of the horizontal velocities, show that the pressure will not be a monotonic function of z if $\alpha Q^2 > 4\pi^2 a^4 g$.

6. A perfect incompressible fluid is moving steadily around the outside of a fixed cylinder of radius a and vertical axis Oz. The fluid particles are traversing horizontal circles with centres on Oz, the speed at distance r from Oz being a/r. Show that the motion is irrotational. If the surface of the fluid is open to the

atmosphere, and the origin O is chosen so that on the free surface $z = 0$ when $r = a$, prove, by means of Bernoulli's equation or otherwise, that the equation of the free surface is

$$2gz = 1 - a^2/r^2.$$

7. A tube AB of uniform fine bore is in the form of the arc of the cycloid $s = 4a \sin \psi$, the ends A and B being given by $\psi = 0$, $\psi = \frac{1}{2}\pi$, respectively. It is fixed in a vertical plane so that A is its lowest point and the tangent there is horizontal. The tube, which is closed at A and open at B, is full of uniform liquid. If the end A is opened, show that the tube empties in time $\pi\sqrt{(2a/g)}$.

8. Define a continuum.

The symbol $\mathbf{R}_n$ denotes the stress vector across a surface normal to the direction $\mathbf{n}$. Show that the equation of motion of a continuum is

$$\frac{\partial}{\partial x}\mathbf{R}_x + \frac{\partial}{\partial y}\mathbf{R}_y + \frac{\partial}{\partial z}\mathbf{R}_z - \rho g\mathbf{k} = \rho\,\frac{\mathrm{D}}{\mathrm{D}t}\,\mathbf{q},$$

where $\mathrm{D}/\mathrm{D}t$ denotes differentiation following a particle.

9. Explain local rate of change $\partial/\partial t$ and particle rate of change $\mathrm{D}/\mathrm{D}t$ and show that for a fluid motion with particle velocity $\mathbf{v}$

$$\frac{\mathrm{D}\mathbf{v}}{\mathrm{D}t} = \frac{\partial \mathbf{v}}{\partial t} + (\mathbf{v}\,.\,\nabla)\mathbf{v} = \frac{\partial \mathbf{v}}{\partial t} + \nabla(\tfrac{1}{2}v^2) - \mathbf{v} \times \boldsymbol{\zeta},$$

where $\boldsymbol{\zeta}$ is the vorticity vector.

The particle velocity $\mathbf{v}$ for a fluid motion referred to rectangular Cartesian axes $O(x, y, z)$ with unit vectors $\mathbf{i}, \mathbf{j}, \mathbf{k}$ is given by

$$\mathbf{v} = v_0\left(\mathbf{i}\cos\frac{\pi x}{2a}\cos\frac{\pi z}{2a} + \mathbf{k}\sin\frac{\pi x}{2a}\sin\frac{\pi z}{2a}\right),$$

where v_0 is a constant. Show that this is a possible motion of an incompressible fluid motion under no body forces in an infinite fixed rigid tube $-a \leqslant x \leqslant a$, $0 \leqslant z \leqslant 2a$ and that the pressure p is given by

$$p = \tfrac{1}{4}\rho v_0^2\left\{\cos(\pi z/a) - \cos(\pi x/a)\right\} + \text{const.}$$

10. Prove that the equation of motion of a homogeneous inviscid liquid moving under forces arising from a potential V may be written in the form

$$\frac{\partial \mathbf{q}}{\partial t} - \mathbf{q} \times \boldsymbol{\zeta} = -\,\mathrm{grad}\left(\frac{p}{\rho} + \tfrac{1}{2}q^2 + V\right),$$

where $\boldsymbol{\zeta} = \mathrm{curl}\,\mathbf{q}$ is the vorticity.

If the velocity $\mathbf{q}$ referred to cylindrical polar coordinates (r, θ, z) is given by

$$\mathbf{q} = [0, \tfrac{1}{2}\omega r, 0] \quad (0 \leqslant r \leqslant a),$$

$$\mathbf{q} = [0, \tfrac{1}{2}\omega a^2/r, 0], \quad (r \geqslant a),$$

where ω is a constant, prove that the vorticity is given by

$$\boldsymbol{\zeta} = [0, 0, \omega], \quad (0 \leqslant r \leqslant a),$$

$$\boldsymbol{\zeta} = [0, 0, 0], \quad (r \geqslant a).$$

Determine the corresponding pressure distribution if the motion takes place under no external forces, and find the smallest value of the pressure at infinity which will ensure that the pressure is everywhere positive.

11. By integrating the Euler equations of motion show that in the steady irrotational motion of an inviscid fluid of constant density, the pressure is given by

$$p = p_0 - \rho g z - \tfrac{1}{2}\rho q^2,$$

where the quantity p_0 is constant throughout the fluid.

A ship's propeller has diameter 2.4 m, and is mounted on a shaft 4 m below the surface. Show that the least pressure at blade tip occurs when the blade moves through its highest position. If the forward velocity of the ship is 8 m sec^{-1}, what is the maximum possible angular velocity of the propeller shaft before cavitation occurs?

12. Infinite inviscid fluid of constant density is attracted towards a fixed point O by a force $f(r)$ per unit mass, where r is the distance of any point from O. Initially the liquid is at rest, and there is a cavity bounded by a sphere of radius a. If there is no pressure at infinity or in the cavity, prove that the radius R of the cavity at time t is such that

$$\frac{\mathrm{d}}{\mathrm{d}t}\left[R^3\left(\frac{\mathrm{d}R}{\mathrm{d}t}\right)^2\right] + 2R^2\frac{\mathrm{d}R}{\mathrm{d}t}\int_R^\infty f(r)\,\mathrm{d}r = 0.$$

If $f(r) = kr^{-3/2}$, where k is a constant, and the cavity is filled after time T, show that $25k^2T^4 = 4a^5$.

13. A spherical globule of gas of radius a and at pressure P expands in an infinite mass of liquid of density ρ in which the pressure at infinity is zero. The gas is initially at rest and its pressure and volume are governed by the equation $pv^{4/3}$ = constant. Prove that the gas doubles its radius in time

$$(28a/15)\sqrt{(2\rho P)}.$$

14. A large mass of incompressible non-viscous fluid contains a spherical air bubble, the air inside the bubble obeying Boyle's law, pv = constant. At a great

distance from the bubble the pressure is zero. Neglecting the inertia of the system, show that the radius R of the bubble at time t satisfies an equation of the form

$$\frac{\mathrm{d}}{\mathrm{d}t}\left(R^2\,\frac{\mathrm{d}R}{\mathrm{d}t}\right) - \tfrac{1}{2}R\left(\frac{\mathrm{d}R}{\mathrm{d}t}\right)^2 - \frac{k}{R^2},$$

where k is a constant.

15. An infinite mass of ideal incompressible fluid is subjected to a force $\mu r^{-7/3}$ per unit mass directed towards the origin. If initially the fluid is at rest and there is a cavity in the form of the sphere $r = a$ in it, show that the cavity will be completely filled after an interval of time $\pi a^{5/3}\,(10\mu)^{-1/2}$.

16. A homogeneous liquid is contained between two concentric spherical surfaces, the radius of the inner being a and that of the outer indefinitely great. The fluid is attracted to the centre of these surfaces by a force $\phi(r)$, and a constant pressure Π is exerted at the outer surface.

Suppose $\int \phi(r)\,\mathrm{d}r = \psi(r)$, and that $\psi(r)$ vanishes when r is infinite. Show that if the inner surface is suddenly removed, the pressure at the distance r is suddenly diminished by

$$\Pi\,\frac{a}{r} - \frac{a\rho}{r}\,\psi(a).$$

17. The radial oscillations of a small air bubble in water induced by local pressure changes as it is swept past an obstacle in a stream, may be represented by the radial motion of a spherical bubble in an unbounded inviscid liquid of density ρ that is at rest at a great distance from the bubble with a pressure there given by $p_0\{1 + f(t)\}$, where p_0 is a constant and $f(t)$ is a small function of time such that $f(t) \to 0$ as $t \to \pm\infty$. At a great distance from the obstacle the bubble has radius R_0 and is in equilibrium with internal pressure p_0. During the subsequent motion the air is compressed adiabatically in accordance with the law $p \propto \rho^\gamma$. Neglecting the inertia of the air, and assuming that departures from equilibrium are small, show that the radius of the bubble at time t is

$$R_0 + \left(\frac{p_0}{3\gamma\rho}\right)^{1/2}\int_{-\infty}^{t} f(\tau)\,\sin\left\{\left(\frac{3p_0\gamma}{\rho R_0^2}\right)\,(\tau - t)\right\}\mathrm{d}\tau.$$

18. A stream in a horizontal pipe, after passing a contraction in the pipe at which its sectional area is A, is delivered at atmospheric pressure at a place where the sectional area is B. Show that if a side tube is connected with the pipe at the former place, water will be sucked up through it into the pipe from a reservoir at a depth $\dfrac{s^2}{2g}\left(\dfrac{1}{A^2} - \dfrac{1}{B^2}\right)$ below the pipe; s being the delivery per second.

19. A sphere whose radius at time t is $b + a \cos nt$ is surrounded by liquid extending to infinity under no forces. Prove that the pressure at distance r from the centre is less than the pressure at an infinite distance by

$$\rho \frac{n^2 a}{r}(b + a \cos nt)\left\{a(1 - 3\sin^2 nt) + b\cos nt + \frac{1}{2}\frac{a}{r^3}\sin^2 nt\,(b + a\cos nt)^3\right\}.$$

20. A sphere of radius a is alone in an unbounded liquid, which is at rest at a great distance from the sphere and is subject to no external forces. The sphere is forced to vibrate radially keeping its spherical shape, the radius r at any time being given by $r = a + b \cos nt$. Show that if Π is the pressure in the liquid at a great distance from the sphere the least pressure (assumed positive) at the surface of the sphere during the motion is $\Pi - n^2 \rho b(a + b)$.

21. Every particle of a mass of liquid is revolving uniformly about a fixed axis, the angular speed varying as the nth power of the distance from the axis. Show that the motion is irrotational only if $n + 2 = 0$.

If a very small spherical portion of the liquid be suddenly solidified, prove that it will begin to rotate about a diameter with an angular velocity $(n + 2)/2$ of that with which it was revolving about the fixed axis.

22. Investigate an expression for the change in an indefinitely short time in the mass of fluid contained within a spherical surface of small radius.

Prove that the momentum of the mass in the direction of the axis of x is greater than it would be if the whole were moving with the velocity at the centre by

$$\frac{1}{5}\frac{Ma^2}{\rho}\left\{\frac{\partial\rho}{\partial x}\frac{\partial u}{\partial x} + \frac{\partial\rho}{\partial y}\frac{\partial u}{\partial y} + \frac{\partial\rho}{\partial z}\frac{\partial u}{\partial z} + \frac{1}{2}\rho\left(\frac{\partial^2 u}{\partial x^2} + \frac{\partial^2 u}{\partial y^2} + \frac{\partial^2 u}{\partial z^2}\right)\right\}.$$

23. An infinite fluid in which is a spherical hollow of radius a is initially at rest under the action of no forces. If a constant pressure Π is applied at infinity, show that the time of filling up the cavity is

$$\pi^2 a(\rho/\Pi)^{1/2}\, 2^{5/6}\left\{\Gamma\left(\tfrac{1}{3}\right)\right\}^{-3}.$$

24. An invisicid fluid moves under a conservative system of forces. Prove that the circulation in any circuit which moves with the fluid is independent of time.

Obtain an equation for the rate of change of vorticity with respect to time and deduce the permanence of vortex rings.

Prove that in two-dimensional motion the vorticity along a streamline is constant.

25. An incompressible inviscid fluid moves under an arbitrary body force **F** per unit mass. Prove that the generation of vorticity is determined by the equation

$$\mathrm{D}\boldsymbol{\zeta}/\mathrm{D}t = (\boldsymbol{\zeta} \cdot \nabla)\mathbf{q} + \operatorname{curl} \mathbf{F},$$

where $\boldsymbol{\zeta} = \operatorname{curl} \mathbf{q}$.

4

Potential flow

4.1 EQUATIONS OF MOTION AND BOUNDARY CONDITIONS

In this chapter we shall derive properties and solutions of specific flows involving an ideal (inviscid) fluid which we shall also assume is incompressible. The equations which must be satisfied at all points of the fluid are Euler's equation of motion and the equation of continuity. For an incompressible fluid these are

$$\frac{\partial \mathbf{q}}{\partial t} + (\mathbf{q} \cdot \nabla)\mathbf{q} = -\frac{1}{\rho}\nabla p + \mathbf{F}, \tag{4.1}$$

$$\nabla \cdot \mathbf{q} = 0, \tag{4.2}$$

where **F** is the body force per unit mass. Equations (4.1) and (4.2) are sufficient to determine the velocity and pressure fields in the flow independently of any temperature distribution which may exist within the fluid. It was pointed out in Chapter 3 that compressibility is a property of the fluid which couples the equations of thermodynamics with those of mechanics while for incompressible fluids, the mechanical problem is complete and closed within itself.

At any solid boundary, the kinematic condition which must be satisfied is

$$\mathbf{q} \cdot \mathbf{n} = \mathbf{U} \cdot \mathbf{n} \tag{4.3}$$

at any point of the solid boundary, where **U** is its velocity and **n** denotes a unit vector normal to the boundary. Other dynamic and kinematic conditions must be satisfied at any free surface or interface between immiscible fluids, as set out in detail in Section 3.2, but within this chapter we shall consider only boundary conditions of the type (4.3). If the fluid is of infinite extent, then the behaviour of the velocity of the fluid at infinity must be prescribed, e.g. that the flow at infinity is that of a uniform stream.

In Section 3.7 we showed that any motion of an ideal incompressible fluid which is started from rest is *irrotational*. Thus curl $\mathbf{q} = \mathbf{0}$ at all points of the fluid and a *velocity potential* Φ exists with the property that

$$\mathbf{q} = \nabla\Phi, \tag{4.4}$$

and in order that the equation of continuity (4.2) is satisfied, it follows that

$$\nabla^2 \Phi = 0. \tag{4.5}$$

The velocity potential thus satisfies Laplace's equation and is consequently a harmonic function. At a solid boundary (4.3) gives

$$\frac{\partial\Phi}{\partial n} = \mathbf{U}\,.\,\mathbf{n}. \tag{4.6}$$

Thus for irrotational or potential flows, the equations which determine the velocity, by means of the velocity potential, are independent of the pressure, and if the body force is conservative, Euler's equation (4.1) can be integrated to give the Bernoulli equation

$$\frac{\partial\Phi}{\partial t} + \frac{1}{2}(\nabla\Phi)^2 - \Omega + \int \frac{dp}{\rho} = f(t), \tag{4.7}$$

where Ω is the body force potential per unit mass and $f(t)$ is an arbitrary function of t. For the case of constant density, equation (4.7) determines the pressure since

$$\frac{p}{\rho} + \frac{\partial\Phi}{\partial t} + \frac{1}{2}(\nabla\Phi)^2 - \Omega = f(t). \tag{4.8}$$

4.2 ACYCLIC AND CYCLIC IRROTATIONAL MOTION

When the region occupied by fluid in irrotational motion is *simply connected*, the velocity potential is single-valued and its value at a point P is expressible as

$$\Phi(P) = \int_0^P \mathbf{q}\,.\,d\mathbf{r}, \tag{4.9}$$

where this integral is independent of the path from O to P. We say that all paths from O to P are *reconcilable*. This result depends essentially on the ability of joining any two paths from O to P by a surface lying entirely within the fluid – always possible when the fluid occupies a simply connected region of space – and applying Stokes's Theorem, as proved in Section 1.15.

When the flow region is *not* simply connected, two paths from O to P can be joined by a surface lying entirely within the fluid only if these paths are reconcilable. When the paths are not reconcilable, e.g. if the fluid is exterior to an infinitely long cylinder and the paths together embrace the cylinder, then Stokes's Theorem is not applicable and the velocity potential *may* have more than one value at P, according to the path chosen to link O and P. In this case, the irrotational motion is said to be *cyclic*. When a unique velocity potential exists, the motion is said to be *acyclic*. Clearly only acyclic irrotational motion is possible in a simply connected region.

For a fluid motion to be physically possible, the velocity at each point must be well defined. Thus even if Φ has more than one value at a particular point, the velocity at that point must be single-valued. Hence, although two paths linking O to P may produce two different values of Φ, their values can only differ by a constant. This constant may be identified with the circulation in any one of a family of reconcilable circuits. If $\mathscr{C}$ is any such circuit, (4.9) shows that the circulation about $\mathscr{C}$ is the increase in the value of Φ on describing $\mathscr{C}$ once.

We shall consider in future chapters various examples of cyclic irrotational motion, but in this chapter will restrict ourselves to acyclic motion for which a number of general theorems can be proved, and unless stated otherwise, the flow regions will be assumed to be simply connected.

4.3 KINETIC ENERGY OF IRROTATIONAL FLOW

If the fluid occupies a finite region V, its kinetic energy is given by

$$T = \tfrac{1}{2} \int_V \rho\, \mathbf{q}^2 \, \mathrm{d}V, \tag{4.10}$$

and if the motion is irrotational,

$$\mathbf{q} = \nabla\Phi.$$

But

$$\operatorname{div}\left\{\Phi\, \nabla\Phi\right\} = \Phi\, \nabla^2\Phi + \nabla\Phi\,.\,\nabla\Phi = (\nabla\Phi)^2,$$

by virtue of (4.5). Thus if $S = S_0 + S_1 + \ldots + S_n$ denotes the sum of the outer boundary surface S_0 and the inner boundaries $S_1, S_2, \ldots, S_n$, it follows that the Divergence Theorem transforms (4.10) to

$$T = \tfrac{1}{2}\rho \int_S \Phi\, \frac{\partial\Phi}{\partial n}\, \mathrm{d}S, \tag{4.11}$$

when the unit vector $\mathbf{n}$ normal to S is drawn *out* of the fluid on each boundary and the fluid density is constant.

4.4 KELVIN'S MINIMUM ENERGY THEOREM

The kinetic energy of irrotational motion of a liquid occupying a finite simply connected region is less than that of any other motion of the liquid which is consistent with the same normal velocity of the boundary.

Proof Let T be the kinetic energy of the irrotational motion, and let $\mathbf{q}_1$ be the velocity and T_1 the kinetic energy of any other motion of the liquid consistent with the same kinetic boundary condition. With Φ the velocity potential for the irrotational motion, we may write

$$\mathbf{q}_1 = \nabla\Phi + \mathbf{q}_0,$$

where, to satisfy the equation of continuity,

$$\operatorname{div} \mathbf{q}_0 = 0,$$

and, to satisfy the kinematic boundary condition,

$$\mathbf{n} \cdot \mathbf{q}_0 = 0$$

on the boundary S, which may be a single (exterior) solid boundary or the sum of such boundaries, as in Section 4.3. If the volume of the fluid is V,

$$\begin{aligned} T_1 &= \tfrac{1}{2}\rho \int_V (\nabla\Phi + \mathbf{q}_0)^2 \, \mathrm{d}V \\ &= T + T_0 + \rho \int_V (\mathbf{q}_0 \cdot \nabla\Phi) \, \mathrm{d}V, \end{aligned}$$

where T_0 is the kinetic energy when the fluid velocity is $\mathbf{q}_0$. However,

$$\operatorname{div} (\Phi \, \mathbf{q}_0) = \mathbf{q}_0 \cdot \nabla\Phi + \Phi \operatorname{div} \mathbf{q}_0 = \mathbf{q}_0 \cdot \nabla\Phi.$$

Thus

$$\int_V (\mathbf{q}_0 \cdot \nabla\Phi) \, \mathrm{d}V = \int_S \Phi \, (\mathbf{n} \cdot \mathbf{q}_0) \, \mathrm{d}S = 0.$$

Consequently $T_1 = T + T_0$ with $T_0 > 0$, giving $T < T_1$, which proves the theorem.

4.5 MEAN VALUE OF THE VELOCITY POTENTIAL

Theorem I *The mean value of Φ over any spherical surface S, drawn in the fluid, through whose interior $\nabla^2 \Phi = 0$, is equal to the value of Φ at the centre of the sphere.*

Proof: Let $\Phi(P)$ be the value of Φ at the centre P of a spherical surface S of radius r and let $\bar{\Phi}$ denote the mean value of Φ over S. By definition,

$$\bar{\Phi} = \frac{1}{4\pi r^2} \int_S \Phi \, \mathrm{d}S = \frac{1}{4\pi} \int_S \Phi \, \mathrm{d}\omega, \tag{4.12}$$

where $\mathrm{d}\omega$ is the solid angle subtended at P by $\mathrm{d}S$. With $\mathbf{n}$ the unit vector normal to S directed *out* of the sphere, the Divergence Theorem gives

$$\int_S \frac{\partial\Phi}{\partial r} \, \mathrm{d}S = \int_S \frac{\partial\Phi}{\partial n} \, \mathrm{d}S = \int_V \nabla^2 \Phi \, \mathrm{d}V = 0,$$

with V the volume of the sphere. Hence

$$r^2 \int_S \frac{\partial \Phi}{\partial r}\, \mathrm{d}\omega = 0,$$

giving

$$\frac{\partial}{\partial r} \int_S \Phi\, \mathrm{d}\omega = 0. \tag{4.13}$$

From (4.12), it follows that

$$\frac{\partial}{\partial r}\, \overline{\Phi} = 0,$$

showing that $\overline{\Phi}$ is independent of the choice of r. We may therefore shrink S to the point P and obtain

$$\overline{\Phi} = \Phi\,(P) \tag{4.14}$$

A consequence of this result is that Φ *cannot have a maximum or minimum in any region throughout which* $\nabla^2 \Phi = 0$.

Theorem II *In irrotational motion, the maximum value of the fluid speed occurs at a boundary.*

Proof: Let P be any interior point of the fluid and Q a neighbouring point also in the fluid. Choosing the direction of the x-axis to lie along the direction of $\mathbf{q}$ at P, we have

$$[\mathbf{q}(P)]^2 = \left(\frac{\partial \Phi}{\partial x}\right)_P^2,$$

$$[\mathbf{q}(Q)]^2 = \left(\frac{\partial \Phi}{\partial x}\right)_Q^2 + \left(\frac{\partial \Phi}{\partial y}\right)_Q^2 + \left(\frac{\partial \Phi}{\partial z}\right)_Q^2.$$

Now

$$\nabla^2 \frac{\partial \Phi}{\partial x} = \frac{\partial}{\partial x} \nabla^2 \Phi = 0,$$

so Theorem I shows that $\partial\Phi/\partial x$ cannot be a maximum or minimum at P. Thus Q can be chosen so that

$$\left(\frac{\partial \Phi}{\partial x}\right)_Q^2 > \left(\frac{\partial \Phi}{\partial x}\right)_P^2$$

implying that $[\mathbf{q}(Q)]^2 > [\mathbf{q}(P)]^2$. This means that $|\mathbf{q}(P)|$ cannot be a maximum and the maximum value of $|\mathbf{q}|$ must occur at a boundary.

Note that $\mathbf{q}^2$ can be a minimum at an interior point of a fluid. Such a minimum occurs at a *stagnation point* where $\mathbf{q} = \mathbf{0}$.

The above theorem shows that in steady irrotational flow, the pressure has its *minimum* value on the boundary. This follows from Bernoulli's theorem:

$$\frac{p}{\rho} + \tfrac{1}{2}\mathbf{q}^2 = \text{constant},$$

showing that p is least where $\mathbf{q}^2$ is greatest, i.e. at a boundary. A fluid is presumed to be incapable of sustaining a negative pressure. Thus, for the fluid to remain in contact with a boundary, the maximum possible velocity of the fluid *must* be consistent with at least a zero pressure on the boundary. If, however, fluid becomes detached from the boundary, we say that *cavitation* has occurred. Similarly, the maximum value of p always occurs at a stagnation point.

Theorem III If liquid of infinite extent is in irrotational motion and is bounded internally by one or more closed surfaces S, the mean value of Φ over a large sphere Σ, of radius R, which encloses S, is of the form

$$\bar{\Phi} = \frac{M}{R} + C,$$

where M and C are constants, provided that the liquid is at rest at infinity.

Proof Applying the Divergence Theorem to $\mathbf{q} = \nabla\Phi$ over the region bounded by S and Σ, we obtain

$$\int_\Sigma \frac{\partial\Phi}{\partial R}\,\mathrm{d}\Sigma = \int_S \frac{\partial\Phi}{\partial n}\,\mathrm{d}S = -4\pi M, \tag{4.15}$$

where $-4\pi M$ represents the flux of fluid across Σ or S. But since $\mathrm{d}\Sigma = R^2\,\mathrm{d}\omega$, with $\mathrm{d}\omega$ the solid angle subtended at the centre of Σ by $\mathrm{d}\Sigma$, (4.15) together with (4.12) give

$$\frac{\partial}{\partial R}\int_\Sigma \Phi\,\mathrm{d}\omega = 4\pi\,\frac{\partial\bar{\Phi}}{\partial R} = -\frac{4\pi M}{R^2}.$$

Hence

$$\bar{\Phi} = \frac{M}{R} + C, \tag{4.16}$$

where C is independent of R. To show that C is also independent of the position of the centre of the sphere Σ, consider the rate of change of C when the centre of Σ is displaced by a distance δx in an arbitrary direction while keeping R constant. Clearly from (4.16),

$$\frac{\partial C}{\partial x} = \frac{\partial\bar{\Phi}}{\partial x} = \frac{1}{4\pi}\int_\Sigma \frac{\partial\Phi}{\partial x}\,\mathrm{d}\omega, \tag{4.17}$$

and since the fluid is at rest at infinity, the right-hand side of (4.17) can be made arbitrarily small by choosing R sufficiently large. Hence

$$\frac{\partial C}{\partial x} = 0,$$

showing that C is a constant.

In the important case when S is a solid surface, $M = 0$ since there is then no flow across S. In this case, the mean value of Φ over any sphere enclosing solid boundaries is a constant C. Using equation 1.83 of Section 1.17, the potential at a point of the fluid which is interior to Σ is given by

$$4\pi\,\Phi(P) = \int_S \left\{\Phi\,\frac{\partial}{\partial n}\left(\frac{1}{r}\right) - \frac{1}{r}\,\frac{\partial \Phi}{\partial n}\right\} \mathrm{d}S + \int_\Sigma \left\{\Phi\,\frac{\partial}{\partial n}\left(\frac{1}{r}\right) - \frac{1}{r}\,\frac{\partial \Phi}{\partial n}\right\} \mathrm{d}\Sigma, \tag{4.18}$$

with the direction of **n** outwards from both S and Σ and r is the distance of a variable point from P. Choosing Σ so that its centre is at P, the second integral in (4.18) is

$$\frac{1}{R^2}\int_\Sigma \Phi\,\mathrm{d}\Sigma + \frac{1}{R}\int_\Sigma \frac{\partial \Phi}{\partial R}\,\mathrm{d}\Sigma = 4\pi C,$$

from (4.16) with $M = 0$. Therefore (4.18) gives

$$4\pi\,[\Phi(P) - C] = \int_S \left\{\Phi\,\frac{\partial}{\partial n}\left(\frac{1}{r}\right) - \frac{1}{r}\,\frac{\partial \Phi}{\partial n}\right\} \mathrm{d}S. \tag{4.19}$$

Letting $r \to \infty$, the right-hand side of (4.19) tends to zero, which shows that as $P \to \infty$,

$$\Phi(P) \to C. \tag{4.20}$$

4.6 KINETIC ENERGY OF INFINITE LIQUID

Consider liquid in irrotational motion, bounded internally by solid surface(s) S and at rest at infinity. If the infinite flow region V is simply connected, we consider the kinetic energy T^* of the finite region V^* bounded internally by $S = S_1 + S_2 + \ldots + S_N$ and externally by a large surface Σ enclosing S, as depicted in Fig. 4.1.

Applying the method of Section 4.3, we find that

$$T^* = \frac{1}{2}\rho\int_S \Phi\,\frac{\partial \Phi}{\partial n}\,\mathrm{d}S + \frac{1}{2}\rho\int_\Sigma \Phi\,\frac{\partial \Phi}{\partial n}\,\mathrm{d}S. \tag{4.21}$$

Now div $\mathbf{q} = \nabla^2\Phi = 0$ throughout V^* and the Divergence Theorem accordingly gives

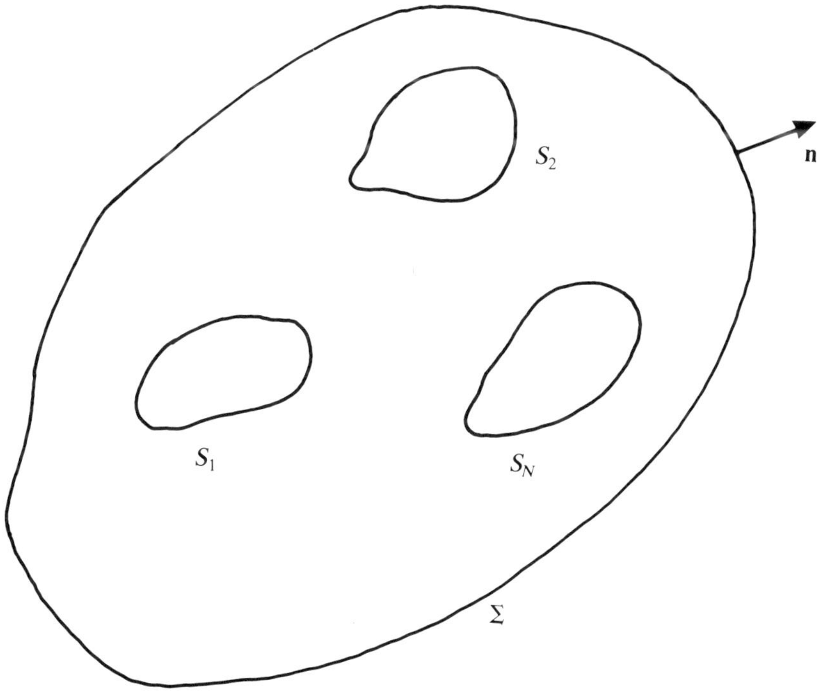

Fig. 4.1

$$\int_S \frac{\partial \Phi}{\partial n}\, dS + \int_\Sigma \frac{\partial \Phi}{\partial n}\, dS = 0\,. \tag{4.22}$$

Combining (4.21) and (4.22), we obtain

$$T^* = \frac{1}{2}\rho \int_S (\Phi - C)\, \frac{\partial \Phi}{\partial n}\, dS + \frac{1}{2}\rho \int_\Sigma (\Phi - C)\, \frac{\partial \Phi}{\partial n}\, d\Sigma, \tag{4.23}$$

for *any* constant C. On choosing C to be the limiting value of Φ at infinity, as established in Section 4.5, and enlarging Σ indefinitely in all directions, $T^* \to T$, the kinetic energy of the entire fluid, and (4.23) gives

$$T = \frac{1}{2}\rho \int_S (\Phi - C)\, \frac{\partial \Phi}{\partial n}\, dS. \tag{4.24}$$

However, the flux of $\mathbf{q}$ across the solid boundaries S is zero. Thus

$$\int_S \frac{\partial \Phi}{\partial n}\, dS = 0.$$

and (4.24) gives

$$T = \frac{1}{2}\rho \int_S \Phi\, \frac{\partial \Phi}{\partial n}\, dS. \tag{4.25}$$

Example [4.1]

Show that acyclic irrotational motion is impossible in a finite volume of fluid bounded by rigid surfaces at rest or in infinite fluid at rest at infinity and bounded internally by rigid bodies at rest.

Suppose that acyclic irrotational motion is possible and let Φ be the velocity potential. From (4.11) or (4.25), the kinetic energy of the fluid is

$$\frac{1}{2}\rho \int_V (\nabla\Phi)^2 \, \mathrm{d}V = \frac{1}{2}\rho \int_S \Phi \frac{\partial \Phi}{\partial n} \, \mathrm{d}S,$$

where S is the sum of all boundaries when V is finite or the sum of internal boundaries when V is infinite. But $\partial\Phi/\partial n = 0$ at each point of S. Thus at each point of V,

$$\nabla\Phi = \mathbf{0},$$

implying that $\Phi =$ constant. There is accordingly no motion of the fluid.

A consequence of this result is that if the solid boundaries in motion are instantaneously brought to rest, the motion of the fluid will instantaneously cease to be irrotational. It does not of course mean that the fluid is brought to *rest* when the boundaries are brought to rest. The ensuing motion is rotational with curl $\mathbf{q} \neq \mathbf{0}$.

Example [4.2]

A rigid surface encloses a finite volume of liquid. Show that it is impossible to move the surface so as to set up a motion of the liquid which will persist if the surface is brought to rest.

Since the liquid is initially at rest, result 3 of Section 3.7 shows that any motion set up in the liquid is irrotational and always remains irrotational. But the previous example shows that irrotational motion is impossible if the boundary is then brought to rest. Hence the fluid must also be brought to rest when the boundary is brought to rest.

4.7 UNIQUENESS THEOREMS

We now prove that the velocity potential for irrotational motion of fluid in simply connected regions of space is unique, subject to certain restrictions.

Theorem I If the region occupied by the fluid is finite, only one irrotational motion of the fluid exists when the boundaries have prescribed velocities.

Proof: Consider $F = \Phi_1 - \Phi_2$, where Φ_1 and Φ_2 are assumed to be two different velocity potentials, corresponding to the two supposedly different irrotational motions. Since the kinematic condition at the boundaries S is satisfied by both flows, at each point of S,

$$\frac{\partial F}{\partial n} = 0.$$

The kinetic energy of the 'difference' motion is

$$\frac{1}{2}\rho \int_V (\nabla F)^2 \, \mathrm{d}V = \frac{1}{2}\rho \int_S F \frac{\partial F}{\partial n} \, \mathrm{d}S = 0.$$

Thus $\nabla F = \mathbf{0}$ at each point of the region V occupied by the fluid, and therefore $\nabla\Phi_1 = \nabla\Phi_2$, showing that the motions are the same. Likewise Φ is unique apart from an additive constant which gives rise to no velocity and without loss of generality can be taken as zero.

Theorem II   If the region occupied by the fluid is infinite and the fluid is at rest at infinity, only one irrotational motion is possible when internal boundaries have prescribed velocities.

Proof: As in the proof of Theorem I, we consider the flow due to the difference potential $F = \Phi_1 - \Phi_2$, where Φ_1 and Φ_2 are the potentials of two supposedly different irrotational motions. Equation (4.25) show that for this 'difference' flow, the kinetic energy is

$$\frac{1}{2}\rho \int_V (\nabla F)^2 \, \mathrm{d}V = \frac{1}{2}\rho \int_S F \frac{\partial F}{\partial n} \, \mathrm{d}S = 0, \tag{4.26}$$

showing that $\nabla F = \mathbf{0}$ at all points of the flow region V. Thus $\nabla\Phi_1 = \nabla\Phi_2$ showing that the irrotational motions are the same.

There is also a unique irrotational motion when infinite fluid is in uniform motion at infinity and the velocities of interior boundaries are prescribed. This can be deduced from Theorem II since the 'difference' flow vanishes at infinity and its kinetic energy is again given by (4.25). Thus (4.26) follows and the uniqueness of the irrotational flow is established.

The above theorems are of great importance in justifying solutions to problems obtained by indirect or even *ad hoc* methods, since the uniqueness theorem allows us to state quite categorically that *the* velocity potential has been found if Laplace's equation and the appropriate boundary conditions are satisfied. For the remainder of this chapter we consider some specific potential flow problems.

4.8 SUBMARINE EXPLOSION

If a spherical cavity of radius R_0 at time $t = 0$, containing gas at pressure p_0, begins to expand rapidly in the surrounding unbounded liquid, the situation closely models the effect of a submarine explosion. During the expansion, the gas is assumed to obey the adiabatic gas law pv^γ = constant, where p is the gas pressure, v the volume of the gas bubble so formed and γ is a constant. The problem is to find the radius R of the gas bubble at time t after the explosion.

Neglecting gravity, this is a problem with *spherical symmetry* so the velocity will be entirely radial from the centre of the bubble and a function only of the radial distance r from the bubble centre and time t. Thus from equation (1.110), the velocity potential $\Phi(r, t)$ satisfies

$$\frac{\partial^2 \Phi}{\partial r^2} + \frac{2}{r}\frac{\partial \Phi}{\partial r} = \frac{1}{r}\frac{\partial^2}{\partial r^2}(r\Phi) = 0,$$

which has the solution

$$\Phi = \frac{F(t)}{r} + G(t),$$

with $F(t)$ and $G(t)$ arbitrary functions of the time t. If it is assumed that the fluid is at rest at $r = \infty$ for $t \geqslant 0$, we may set $G(t) = 0$. The *kinematic* condition at $r = R$ requires that

$$\frac{\partial \Phi}{\partial r} = -\frac{F(t)}{r^2} = \dot{R}, \qquad (r = R),$$

where $\dot{R} = \mathrm{d}R/\mathrm{d}t$. Thus $F(t) = -R^2\dot{R}$, giving

$$\Phi = -\frac{R^2\dot{R}}{r}, \qquad (r \geqslant R,\ t \geqslant 0). \tag{4.27}$$

Bernoulli's equation for unsteady incompressible potential flow with constant density (assumed) under zero body forces is, from (3.39),

$$\frac{p}{\rho} + \frac{1}{2}\mathrm{q}^2 + \frac{\partial \Phi}{\partial t} = H(t), \tag{4.28}$$

where $H(t) = 0$ if we take $p = 0$ where $r = \infty$. Thus (4.28) gives

$$\frac{p}{\rho} + \frac{R^4\dot{R}^2}{2r^4} - \frac{(R^2\ddot{R} + 2R\dot{R}^2)}{r} = 0. \tag{4.29}$$

When $r = R$, (4.29) gives

$$\frac{p}{\rho} = R\ddot{R} + \frac{3}{2}\dot{R}^2. \tag{4.30}$$

At $r = R$, there is also the *dynamic* condition (3.9) requiring continuity of pressure at the surface of the bubble. The adiabatic gas law requires

$$\frac{p}{p_0} = \left(\frac{R_0}{R}\right)^{3\gamma},$$

so (4.30) gives

$$R\ddot{R} + \frac{3}{2}\dot{R}^2 = \left(\frac{R_0}{R}\right)^{3\gamma}\left(\frac{p_0}{\rho}\right). \tag{4.31}$$

Noting that $\ddot{R} = (\mathrm{d}/\mathrm{d}R)\,(\frac{1}{2}\dot{R}^2)$, equation (4.31) is equivalent to

$$\frac{\mathrm{d}}{\mathrm{d}R}(\dot{R}^2) + \frac{3}{R}\,\dot{R}^2 = \frac{2\,R_0^{3\gamma}\,p_0}{\rho\,R^{3\gamma+1}}\,,$$

which gives, after multiplying through by R^3,

$$\frac{\mathrm{d}}{\mathrm{d}R}(R^3\,\dot{R}^2) = \frac{2R_0^{3\gamma}\,p_0}{\rho\,R^{3\gamma-2}}\,. \tag{4.32}$$

On integrating (4.32) from R_0 to R, observing that $R = 0$ when $R = R_0$, we obtain

$$\frac{\mathrm{d}R}{\mathrm{d}t} = \left(\frac{2\,p_0}{3(\gamma-1)\rho}\right)^{1/2}\left[\left(\frac{R_0}{R}\right)^3 - \left(\frac{R_0}{R}\right)^{3\gamma}\right]^{1/2},$$

and the time t taken for the bubble to expand in radius from R_0 to R is accordingly

$$t = \left(\frac{3(\gamma-1)\rho}{2\,p_0}\right)^{1/2}\int_{R_0}^{R}\left[\left(\frac{R_0}{R}\right)^3 - \left(\frac{R_0}{R}\right)^{3\gamma}\right]^{-1/2}\mathrm{d}R. \tag{4.33}$$

For general values of γ, the integral in (4.33) has to be approximated but in the special case when $\gamma = 4/3$, a typical value for many gases, the integral can be evaluated exactly. On making the substitution $R = (1+\lambda)\,R_0$, equation (4.33) reduces to

$$\begin{aligned} t &= \left(\frac{\rho}{2p_0}\right)^{1/2} R_0 \int_0^{\lambda} [\lambda^{-1/2} + 2\lambda^{1/2} + \lambda^{3/2}]\;\mathrm{d}\lambda \\ &= \left(\frac{2\rho}{p_0}\right)^{1/2} R_0\,\lambda^{1/2}\left[1 + \frac{2}{3}\lambda + \frac{1}{5}\lambda^2\right]. \end{aligned} \tag{4.34}$$

Example [4.3]

A bubble of hydrogen sulphide is expanding in sea water of great depth. If initially the bubble radius is 0.5 m and the gas pressure is 1000 atmospheres, find the time taken for the bubble radius to double.

The value of γ for hydrogen sulphide is 1.34, which is close enough to 4/3 for equation (4.34) to apply. We have

$$p_0 = 1000 \text{ atmospheres} = 1.0133 \times 10^3 \text{ Pa} = 1.0133 \times 10^8 \text{ N/m}^2.$$

For sea water, $\rho = 1.026 \times 10^3$ kg/m^3. Thus

$$(2\rho/p_0)^{1/2} \approx 4.5 \times 10^{-3}.$$

On setting $R_0 = 0.5$ and $\lambda = 1$ in (4.34), the bubble radius will have doubled when

$$t = 4.5 \times 10^{-3} \times 0.5 \times (28/15) = 4.2 \times 10^{-3}\ \text{s}.$$

The initial acceleration of the bubble radius may be found from (4.31) with $R = R_0, \dot{R} = 0, \gamma = 4/3$. This is found to be 2×10^5 m/s^2.

4.9 AXIALLY SYMMETRIC FLOWS

Let (r, θ, ϕ) denote spherical polar coordinates with respect to an origin O. A potential flow which is *axisymmetric* about the axis $\theta = 0, \pi$ has the property that the fluid velocity $\mathbf{q}$, and consequently the velocity potential Φ, are the same in any plane ϕ = constant. Thus Φ is independent of the aximuthal angle ϕ and satisfies the axisymmetric form of Laplace's equation:

$$\frac{1}{r^2}\frac{\partial}{\partial r}\left(r^2\frac{\partial\Phi}{\partial r}\right)+\frac{1}{r^2\sin\theta}\frac{\partial}{\partial\theta}\left(\sin\theta\frac{\partial\Phi}{\partial\theta}\right)=0, \tag{4.35}$$

which may be deduced from (1.110). A solution of (4.35) in separable variables r, θ has the form

$$\Phi = R(r)\,.\,\Theta(\theta),$$

which on substituting into (4.35) gives

$$\frac{1}{R}\frac{\mathrm{d}}{\mathrm{d}r}\left(r^2\frac{\mathrm{d}R}{\mathrm{d}r}\right)=-\frac{1}{\Theta\sin\theta}\frac{\mathrm{d}}{\mathrm{d}\theta}\left(\sin\theta\frac{\mathrm{d}\Theta}{\mathrm{d}\theta}\right). \tag{4.36}$$

The left-hand side of (4.36) is a function of r only while the right-hand side is a function of θ only. The equation can therefore be satisfied if and only if either side is a constant, $n(n+1)$ say. In which case,

$$\frac{1}{R}\frac{\mathrm{d}}{\mathrm{d}r}\left(r^2\frac{\mathrm{d}R}{\mathrm{d}r}\right)=n(n+1), \tag{4.37}$$

$$\frac{\mathrm{d}}{\mathrm{d}\theta}\left(\sin\theta\frac{\mathrm{d}\Theta}{\mathrm{d}\theta}\right)+n(n+1)\,\Theta\sin\theta=0. \tag{4.38}$$

Equation (4.37) is easily solved to give

$$R(r)=A_n\,r^n+B_n\,r^{-(n+1)}, \tag{4.39}$$

while (4.38) is reducible to Legendre's Equation:

$$\frac{\mathrm{d}}{\mathrm{d}\mu}\left[(1-\mu^2)\frac{\mathrm{d}\Theta}{\mathrm{d}\mu}\right]+n(n+1)\,\Theta=0, \tag{4.40}$$

on making the substitution $\mu = \cos\theta$. Equation (4.40) possesses a solution, known as the *Legendre Function* of the first kind $P_n(\mu)$, which is regular at $\mu = 0$. For values of n other than an integer or zero, $P_n(\mu)$ is expressible as an infinite series in powers of μ with radius of convergence unity but which does not converge for both $\mu = 1$ and $\mu = -1$, i.e. $\theta = 0$ and $\theta = \pi$. Such a solution is therefore inadmissible in fluid mechanics problems when the flow region contains the axis of symmetry $\theta = 0, \pi$. When n is a positive integer or zero, the solution of (4.40) which is regular at $\mu = 0$ is then a polynomial, the *Legendre Polynomial* of degree in n. These polynomials are normalised so that $P_n(1) = 1$ for all $n \geqslant 0$ and the first three are

$$P_0(\mu) = 1, \quad P_1(\mu) = \mu, \quad P_2(\mu) = \tfrac{1}{2}(3\mu^2 - 1). \tag{4.41}$$

Higher degree polynomials can be derived by use of the recurrence relation:

$$(n + 1)P_{n+1}(\mu) = (2n + 1)\,\mu\, P_n(\mu) - n\, P_{n-1}(\mu). \tag{4.42}$$

If n is a negative integer, $P_n(\mu)$ is also a polynomial by noting from (4.40) that

$$P_n(\mu) = P_{-(n+1)}(\mu).$$

The solution of (4.40) which is independent of $P_n(\mu)$ is the Legendre function of the second kind $Q_n(\mu)$. However, this solution is logarithmically singular at $\mu = 0$ for all values of n, which naturally excludes it as a relevant solution in flow regions which include the plane $\theta = \frac{1}{2}\pi$. The reader may find the foregoing and other properties of Legendre functions in any standard text such as Ref. 21.

It therefore follows that the most general solution of (4.35) in separated variables which is bounded when $\theta = 0, \frac{1}{2}\pi, \pi$ is of the form

$$\Phi(r, \theta) = \sum_{n=0}^{\infty} [A_n r^n + B_n r^{-(n+1)}]\, P_n(\cos\theta). \tag{4.43}$$

4.10 UNIFORM FLOW

Consider the flow which corresponds to a potential given by (4.43) with

$$A_n = U\delta_{1n}, \quad B_n = 0, \qquad (n = 0, 1, 2 \ldots),$$

where U is a constant.
Since $P_1(\cos\theta) = \cos\theta$, (4.43) reduces to

$$\Phi(r, \theta) = Ur\cos\theta \equiv Uz. \tag{4.44}$$

Thus $\mathbf{q} = \nabla\Phi = U\mathbf{k}$ which is a uniform streaming motion of the fluid with speed U along the direction of the positive z-axis or the axis $\theta = 0$. A constant may be added to the potential defined by (4.44) and the resulting potential corresponds to the same uniform streaming flow.

4.11 SPHERE AT REST IN A UNIFORM STREAM

Consider an impermeable solid sphere of radius a at rest with its centre the pole of a system of spherical polar coordinates (r, θ, ϕ). The sphere is immersed in an infinite homogeneous fluid with constant density ρ, which, in the absence of the sphere, would be flowing as a uniform stream with speed U along the direction $\theta = 0$.

The presence of the sphere will produce a 'perturbation' of the uniform streaming motion such that the disturbance diminishes with increasing distance r from the centre of the sphere. We say that the perturbation of the uniform stream *evanesces* or tends to zero, as $r \to \infty$.

We can therefore set up the mathematical problem for the velocity potential as follows: we must find a solution of Laplace's equation $\nabla^2 \Phi = 0$ such that

$$\frac{\partial \Phi}{\partial r} = 0, \qquad (r = a,\ 0 \leqslant \theta \leqslant \pi), \tag{4.45}$$

in order to satisfy the kinematic condition on the sphere's boundary, and

$$\nabla \Phi \to U\mathbf{k}, \qquad (r \to \infty), \tag{4.46}$$

in order to satisfy the uniform stream condition at infinity. The problem is indicated schematically by Fig. 4.2.

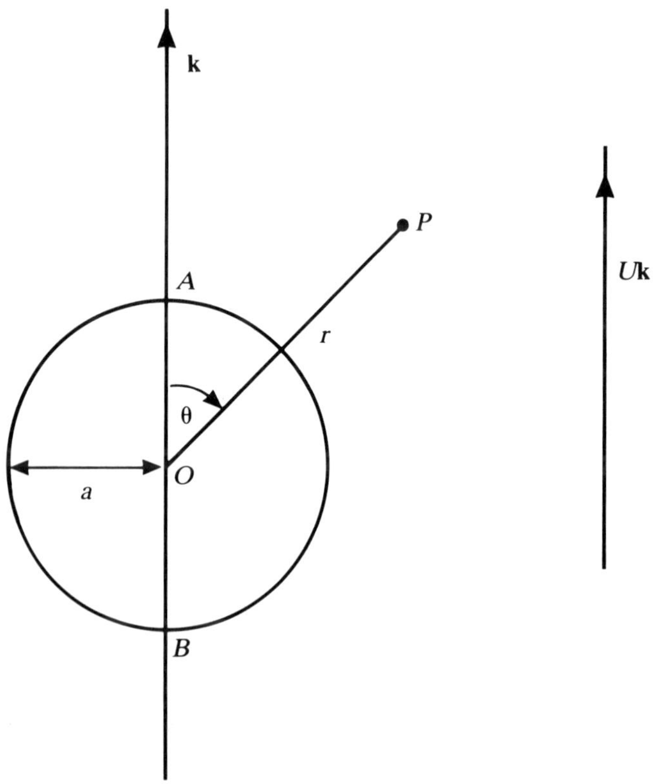

Fig. 4.2

This is an example of axisymmetric flow with the axis of symmetry $\theta = 0, \pi$. Bearing in mind the result of the previous section and the boundary condition (4.46), it is reasonable to write

$$\Phi = Ur\cos\theta + \Phi_1(r, \theta), \qquad (r \geqslant a), \tag{4.47}$$

where Φ_1 satisfies (4.35) together with the boundary conditions

$$\frac{\partial \Phi_1}{\partial r} = -U\cos\theta, \qquad (r = a, 0 \leqslant \theta \leqslant \pi), \tag{4.48}$$

$$|\nabla\Phi_1| \to 0, \qquad (r \to \infty). \tag{4.49}$$

From (4.43), a suitable solution for Φ_1 is

$$\Phi_1 = Br^{-2}\cos\theta,$$

which will satisfy both (4.48) and (4.49) provided that

$$B = U/2a^2.$$

Accordingly, the solution Φ is given by

$$\Phi = U\left(r + \frac{1}{2}\frac{a^3}{r^2}\right)\cos\theta, \qquad (r \geqslant a), \tag{4.50}$$

and this is the *unique* solution to the problem by virtue of the uniqueness Theorem II of Section 4.7.

We could, of course, have chosen the general solution (4.43) to represent Φ rather than (4.47), and in satisfying (4.45) and (4.46) arrived at (4.50). The reader may care to verify this as an exercise. It hinges on the *orthogonal* property of the Legendre polynomials which the reader will find established in standard texts on analysis or differential equations. Briefly, the orthogonal property means that

$$\int_{-1}^{1} P_m(\mu)\, P_n(\mu)\, \mathrm{d}\mu = \frac{2\delta_{mn}}{(2n+1)}, \qquad (n = 0, 1, 2 \ldots), \tag{4.51}$$

and as a consequence, any equation of the type

$$\sum_{n=0}^{\infty} \alpha_n P_n(\cos\theta) = 0$$

for *all* values of θ in the range $0 \leqslant \theta \leqslant \pi$, implies that $\alpha_n = 0, (n = 0, 1, 2, \ldots)$. However, it is *not* necessary, as we have seen, to use the full general solution in a problem of this type and many others which occur in potential theory. In applied mathematics, it pays to be *pragmatic* since there are *clues* to the correct form for Φ in the boundary conditions. In the problem under discussion, the condition at infinity restricts the choice of solutions to one in which positive powers of r greater than unity are not present, and the kinematic condition on the sphere implies that the dependence of the solution on θ can only be of the form $\cos\theta$ once we have 'built in' the far field asymptotic form for a uniform stream, as we did in selecting the form of solution given by (4.47). This in turn leads us to conclude that Φ can only be constructed from the two basic solutions of Laplace's equation which are $r\cos\theta$ and $r^{-2}\cos\theta$. The result (4.50) *inevitably* follows.

The velocity components at $P(r, \theta, \phi)$ for $r \geqslant a$ are

$$\left.\begin{aligned} q_r &= \frac{\partial\Phi}{\partial r} = U\left(1 - \frac{a^3}{r^3}\right)\cos\theta \\ q_\theta &= \frac{1}{r}\frac{\partial\Phi}{\partial\theta} = -U\left(1 + \frac{a^3}{2r^3}\right)\sin\theta \\ q_\phi &= \frac{1}{r\sin\theta}\frac{\partial\Phi}{\partial\phi} = 0 \end{aligned}\right\} \tag{4.52}$$

The *stagnation points* are those points in the flow where the velocity vanishes. These are found by solving the equations

$$U(1 - a^3/r^3)\cos\theta = U(1 + a^3/2r^3)\sin\theta = 0,$$

which are satisfied only by $r = a$, $\sin\theta = 0$, i.e. the points $A(r = a, \theta = 0)$ and $B(r = a, \theta = \pi)$ on the sphere. These are referred to respectively as the *rear* and *forward* stagnation points.

The equations of the streamlines are

$$\frac{\mathrm{d}r}{(1 - a^3/r^3)\cos\theta} = -\frac{r\,\mathrm{d}\theta}{(1 + a^3/2r^3)\sin\theta} = \frac{r\sin\theta\,\mathrm{d}\phi}{0}, \quad (r \geqslant a),$$

giving $\phi = \text{constant}$ and

$$\frac{2r^3 + a^3}{r(r^3 - a^3)}\,\mathrm{d}r = -2\cot\theta\,\mathrm{d}\theta.$$

The latter equation gives

$$\int \frac{2r + (a^3/r^2)}{r^2 - (a^3/r)}\,\mathrm{d}r = -2\ln\sin\theta + C'$$

with C' a constant. On evaluating the integral and simplifying, we obtain

$$r^{-1}(r^3 - a^3)\sin^2\theta = C, \tag{4.53}$$

where $C \geqslant 0$ is a constant. For each choice of C, equation (4.53) gives the trace of a streamline in the plane ϕ = constant. The particular choice of $C = 0$ corresponds to the sphere and the axis of symmetry, which taken together form one of the stream surfaces.

The pressure at any point of the fluid is obtained by applying Bernoulli's equation along the streamline through that point, taking the pressure at infinity to be the constant value p_∞. Thus, in the absence of a body force, the pressure p is given by

$$\frac{p}{\rho} + \frac{1}{2}(\nabla\Phi)^2 = \frac{p_\infty}{\rho} + \frac{1}{2}U^2,$$

which from (4.53) yields

$$p = p_\infty - \frac{1}{2}\rho U^2 \left[\left(1 - \frac{a^3}{r^3}\right)^2 \cos^2\theta + \left(1 + \frac{a^3}{2r^3}\right)^2 \sin^2\theta - 1\right]. \quad (4.54)$$

Of particular interest is the distribution of pressure over the boundary of the sphere. This is given by (4.54) by setting $r = a$, and after simplification is found to be

$$p = p_\infty + \tfrac{1}{8}\rho U^2 \, [9\cos^2\theta - 5]. \quad (4.55)$$

The maximum pressure occurs at the stagnation points where $\theta = 0$ or π. Thus

$$p_{\max} = p_\infty + \tfrac{1}{2}\rho U^2. \quad (4.56)$$

The minimum pressure occurs along the equatorial circle of the sphere where $\theta = \frac{1}{2}\pi$. Thus

$$p_{\min} = p_\infty - \tfrac{5}{8}\rho U^2, \quad (4.57)$$

and from Theorem II of Section (4.5), $p_{\max}$ and $p_{\min}$ given by (4.56) and (4.57) represent the maximum and minimum values taken by the pressure in the fluid *as a whole*. Since $p_{\min} = 0$ when $U = (8p_\infty/5\rho)^{1/2}$, it follows that, for a given value of p_∞, when U is increased from zero, a critical value of U is reached at which $p_{\min} = 0$ along the equatorial circle $r = a$, $\theta = \frac{1}{2}\pi$. At this stage the fluid will tend to break away from the surface of the sphere and *cavitation* is said to occur. A fluid is presumed to be incapable of sustaining a negative pressure implying that p is an intrinsically positive quantity. In the relative motion of a solid boundary and fluid, the fluid can remain in contact with the boundary provided that the pressure at every point of the boundary remains positive. Thus at points where the pressure vanishes, a slight further reduction in pressure would result in the break away of fluid from the boundary and the formation of a vacuum. The phenomenon of cavitation commonly occurs in practice, and is observed, for instance, near the rapidly moving tips of propeller blades and at the rear of fast moving bodies under water, such as torpedoes.

The force $\mathbf{F}$ acting on the sphere due to the flow of fluid past it can be calculated by noting that the force per unit area acting over its surface is simply $-p\mathbf{n}$, where $\mathbf{n}$ is the outward drawn unit normal to the sphere at any point. Thus

$$\mathbf{F} = a^2 \int_{\phi=0}^{2\pi} \int_{\theta=0}^{\pi} \hat{\mathbf{r}}(p)_{r=a} \sin\theta \, d\theta \, d\phi. \quad (4.58)$$

In terms of Cartesian coordinates,

$$\hat{\mathbf{r}} = \mathbf{i} \sin\theta \cos\phi + \mathbf{j} \sin\theta \sin\phi + \mathbf{k} \cos\theta,$$

and since $(p)_{r=a}$ is a function of θ only, the components of $\mathbf{F}$ in the directions of $\mathbf{i}$ and $\mathbf{j}$ are identically zero. On writing $\mu = \cos\theta$ and using (4.55), equation (4.58) reduces to

$$\mathbf{F} = 2\pi a^2 \mathbf{k} \int_{-1}^{1} \left\{ p_\infty + \frac{1}{8}\rho U^2 \,[9\mu^2 - 5] \right\} \mu \, d\mu = 0, \tag{4.59}$$

resulting in the physically unsatisfactory implication that the sphere experiences *no* force due to the motion of the fluid. This result is not simply a consequence of the symmetrical shape of the sphere and we shall show later that a body of arbitrary shape at rest in a uniform stream or moving uniformly through a fluid at rest experiences no force. This is because the neglect of viscosity of the fluid implies the absence of a *shear stress* tangential to the boundary which would contribute to producing a force on the body. The inability of an ideal fluid to produce a force in such circumstances is known as *d'Alembert's Paradox.*

4.12 SPHERE IN MOTION IN FLUID AT REST AT INFINITY

If the sphere translates with constant velocity $-U\mathbf{k}$, we can immediately deduce the velocity potential from the preceding section by adding to the system of sphere and fluid a constant velocity $-U\mathbf{k}$. This has the effect of bringing the fluid to rest at infinity and giving the sphere the translational velocity $-U\mathbf{k}$. Referred to spherical polar coordinates with pole at the centre of the sphere moving with the velocity of the sphere, the boundary value problem for Φ is now to solve

$$\nabla^2 \Phi = 0$$

such that

$$\frac{\partial \Phi}{\partial r} = -U \cos A, \qquad (r = a), \tag{4.60}$$

and

$$|\nabla \Phi| \to 0, \qquad (r \to \infty). \tag{4.61}$$

To solve the boundary value problem for Φ, we first note that since the flow is symmetric about the axis $\theta = 0, \pi$, then $\Phi = \Phi\,(r, \ \theta)$. Also since

$$\cos\theta = P_1(\cos\theta),$$

and the form of the boundary condition (4.6) implies that the dependence of Φ on θ must be like $\cos\theta$, equation (4.43) suggest that Φ has the form

$$\Phi = (Ar + B/r^2)\,P_1(\cos\theta) = (Ar + B/r^2)\cos\theta .$$

However, to satisfy (4.61) it is necessary that $A = 0$. Consequently $B = \frac{1}{2}Ua^3$ for (4.6) to be satisfied. Thus the solution for Φ is accordingly

$$\Phi = \frac{Ua^3}{2r^2}\cos\theta . \tag{4.62}$$

The corresponding velocity components are

$$q_r = -\frac{Ua^3}{r^3}\cos\theta, \quad q_\theta = \frac{Ua^3}{2r^3}\sin\theta, \quad q_\phi = 0. \tag{4.63}$$

The kinetic energy of the fluid can be evaluated using (4.25) of Section 4.6. We have

$$T = \frac{1}{2}\rho\int_S \Phi\,\frac{\partial\Phi}{\partial n}\,\mathrm{d}S = -\frac{1}{2}\rho\int_S \Phi\,\frac{\partial\Phi}{\partial r}\,\mathrm{d}S, \tag{4.64}$$

with S denoting the surface of the sphere. Equation (4.64) becomes, after substituting from (4.62),

$$T = \frac{1}{2}\rho\,a^3U^2\int_{\phi=0}^{2\pi}\int_{\theta=0}^{\pi}\cos^2\theta\,\sin\theta\,\mathrm{d}\theta\,\mathrm{d}\phi$$

$$= \frac{1}{3}\pi\rho a^3U^2 = \frac{1}{4}M'U^2, \tag{4.65}$$

where M' is the mass of fluid displaced by the sphere. The total kinetic energy of sphere and fluid is accordingly $\frac{1}{2}(M + \frac{1}{2}M')U^2$, with M the mass of the sphere. The quantity $M + \frac{1}{2}M'$ is called the *virtual mass* of the sphere.

The solution derived above for Φ is applicable when the sphere translates *unsteadily* along a straight line. Equations (4.63) and (4.65) now give the instantaneous values of the velocity components and kinetic energy at time t with $U = U(t)$. To find the pressure at any point of the fluid, we must remember that whether or not U is a constant, the expression for Φ given by (4.62) is a function of the time t because the frame of reference translates with the fluid. From (3.39) the pressure is given by

$$\frac{p}{\rho} = \frac{p_\infty}{\rho} - \frac{1}{2}(\nabla\Phi)^2 - \frac{\partial\Phi}{\partial t}. \tag{4.66}$$

To find $\partial\Phi/\partial t$, we proceed as follows: first note that (4.62) can be written in the form

$$\Phi = -\frac{1}{2}a^3\,\frac{(\mathbf{U}\,.\,\mathbf{r})}{r^3}, \tag{4.67}$$

where $\mathbf{U} = -U(t)\mathbf{k}$ is the velocity of the sphere. Since $\mathbf{r}$ is the position vector of a *fixed point* of the fluid relative to the moving centre O of the sphere, it follows that

$$\frac{\partial}{\partial t}(-\mathbf{r}) = \mathbf{U}. \tag{4.68}$$

Furthermore, since $r^2 = \mathbf{r} \cdot \mathbf{r}$,

$$r\,\frac{\partial r}{\partial t} = \mathbf{r} \cdot \frac{\partial \mathbf{r}}{\partial t} = -\mathbf{U} \cdot \mathbf{r} = Ur\cos\theta,$$

showing that

$$\frac{\partial r}{\partial t} = U\cos\theta. \tag{4.69}$$

On differentiating (4.67) with respect to t and using (4.68) and (4.69) we obtain

$$\frac{\partial \Phi}{\partial t} = \frac{a^3}{2}\left\{\frac{\dot{U}\cos\theta}{r^2} + \frac{U^2}{r^3} - \frac{3U^2\cos^2\theta}{r^3}\right\}. \tag{4.70}$$

The pressure at any point of the fluid can then be found from (4.66). In particular, at a point on the sphere,

$$\begin{aligned}\frac{p}{\rho} &= \frac{p_\infty}{\rho} - \frac{1}{8}U^2\,[4\cos^2\theta + \sin^2\theta] - \frac{1}{2}[\dot{U}\cos\theta + U^2 - 3U^2\cos^2\theta] \\ &= \frac{p_\infty}{\rho} - \frac{1}{2}\dot{U}\cos\theta + \frac{1}{8}U^2\,[9\cos^2\theta - 5],\end{aligned} \tag{4.71}$$

when the sphere is moving with velocity $-U(t)\mathbf{k}$. The force $\mathbf{F}$ acting on the sphere is given by

$$\mathbf{F} = \int_S \mathbf{n}(p)_{r=a}\,\mathrm{d}S = -\int_S \hat{\mathbf{r}}(p)_{r=a}\,\mathrm{d}S.$$

On comparing (4.71) and (4.55), it immediately follows that

$$\begin{aligned}\mathbf{F} &= \pi a^3 \rho\,\dot{U}\mathbf{k}\int_{-1}^{1}\mu^2\,\mathrm{d}\mu \\ &= \tfrac{2}{3}\pi\rho a^3\,\dot{U}\mathbf{k} = \tfrac{1}{2}M'\dot{U}\mathbf{k},\end{aligned} \tag{4.72}$$

showing that a force acts on an accelerating sphere in the direction opposing the sphere's motion.

4.13 D'ALEMBERT'S PARADOX

The reader might reasonably conclude, from the result proved in the preceding section, that the *zero* resultant force exerted by the fluid on a sphere moving with uniform translational velocity in a quiescent unbounded ideal fluid is a consequence of the particular geometry of the flow configuration and relates only to the sphere. After all, in this particular case the pressure function is symmetric about the diametral plane of the sphere perpendicular to its direction of motion and consequently the force *must* be zero. Regrettably this would be a false conclusion, and we shall now establish the result, often referred to as d'Alembert's Paradox. This states that there is no resultant force exerted by the fluid on a finite body of arbitrary shape which translates steadily through a quiescent unbounded ideal fluid.

Consider a finite body, whose boundary is S, which translates with constant velocity $\mathbf{U}$ in an unbounded fluid at rest at infinity, as indicated in Fig. 4.3.

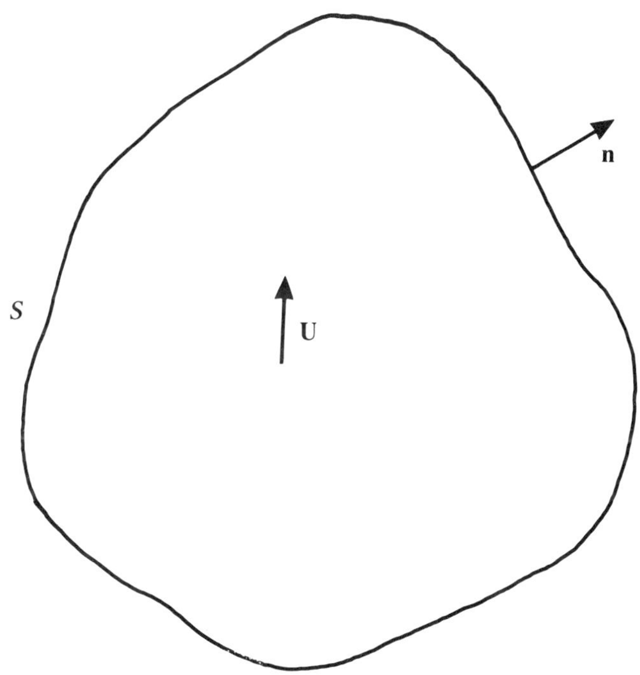

Fig. 4.3

The boundary value problem for the velocity potential Φ is to find a solution of Laplace's equation

$$\nabla^2 \Phi = 0, \tag{4.73}$$

which satisifies the kinematic condition

$$\frac{\partial \Phi}{\partial n} = \mathbf{U} \, . \, \mathbf{n} \tag{4.74}$$

on S, and the condition at infinity

$$|\nabla\Phi| \to 0, \qquad (r \to \infty), \tag{4.75}$$

where r measures distance from an origin O fixed in the body. To prove D'Alembert's Paradox, we do not need to specify the shape of the body and therefore we do not need actually to solve the boundary value problem for Φ which is posed by (4.73), (4.74) and (4.75). We shall need to be more specific about the *rate of decay* of Φ as $r \to \infty$. It is reasonable to say that the disturbance to the flow made by an arbitrarily shaped body as $r \to \infty$ is *qualitatively* similar to that produced by a sphere. If that is accepted, we can say from (4.62) that

$$\left.\begin{aligned} \Phi &= O(r^{-2}) \\ |\nabla\Phi| &= O(r^{-3}) \end{aligned}\right\}. \qquad (r \to \infty). \tag{4.76}$$

The force acting on the body is

$$\mathbf{F} = -\int_S \mathbf{n}p \, \mathrm{d}S, \tag{4.77}$$

and with $\mathbf{n} = l_j \, \mathbf{i}_j$ equation (4.77) is equivalent to

$$F_j = -\int_S l_j \, p \, \mathrm{d}S. \tag{4.78}$$

If p_∞ is the constant pressure at infinity, Bernoulli's equation gives

$$p = p_\infty - \frac{1}{2}\rho\,(\nabla\Phi)^2 - \rho\,\frac{\partial\Phi}{\partial t},$$

and (4.78) consequently yields

$$F_j = \rho \int_S l_j \, \frac{\partial\Phi}{\partial t} \, \mathrm{d}S + \frac{1}{2}\rho \int_S l_j \, (\nabla\Phi)^2 \, \mathrm{d}S - p_\infty \int_S l_j \, \mathrm{d}S, \tag{4.79}$$

However,

$$\int_S l_j \, \mathbf{i}_j \, \mathrm{d}S = \int_\tau \nabla(1) \, \mathrm{d}\tau = \mathbf{0},$$

with τ the volume *enclosed* by S. This establishes that

$$\int_S l_j \, \mathrm{d}S = 0. \tag{4.80}$$

Now $\Phi = \Phi(x_1, x_2, x_3)$ with $x_i = x_i(t)$, $(i = 1, 2, 3)$. Thus

$$\frac{\partial\Phi}{\partial t} = \frac{\partial\Phi}{\partial x_i}\frac{\mathrm{d}x_i}{\mathrm{d}t} = \nabla\Phi \, . \, \frac{\mathrm{d}\mathbf{r}}{\mathrm{d}t} = -\mathbf{U} \, . \, \nabla\Phi, \tag{4.81}$$

from which it follows that

$$\int_S l_j \frac{\partial \Phi}{\partial t} \, dS = -U_i \int_S l_j \frac{\partial \Phi}{\partial x_i} \, dS = -U_i \int_S l_j q_i \, dS \tag{4.82}$$

Since the motion is irrotational,

$$\nabla(\mathbf{q}^2) = 2(\mathbf{q}\,.\nabla)\mathbf{q},$$

and therefore

$$\frac{\partial}{\partial x_i}(\nabla\Phi)^2 = 2q_j \frac{\partial q_i}{\partial x_j} = 2\frac{\partial}{\partial x_j}(q_i q_j) \tag{4.83}$$

since $\partial q_j/\partial x_j = \text{div}\,\mathbf{q} = 0$ for an incompressible fluid.

Let Σ be a large sphere centre O and radius R enclosing S, and V be the volume bounded by S and Σ. It follows that

$$\int_S l_j (\nabla\Phi)^2 \, dS = \int_V \frac{\partial}{\partial x_i}(\nabla\Phi)^2 \, dV - \int_\Sigma l_j (\nabla\Phi)^2 \, d\Sigma,$$

with normals to S and Σ as usual drawn out of the fluid. Under the assumed behaviour (4.76), the integral over Σ vanishes as $R \to \infty$. But from (4.83),

$$\int_V \frac{\partial}{\partial x_i}(\nabla\Phi)^2 \, dV = 2\int_S l_j (q_i q_j) \, dS + 2\int_\Sigma l_j (q_i q_j) \, d\Sigma.$$

Again the integral over Σ vanishes as $R \to \infty$. Therefore

$$\frac{1}{2}\rho \int_S l_j (\nabla\Phi)^2 \, dS = \rho \int_S l_i q_i q_j \, dS$$

$$= \rho U_i \int_S l_i q_j \, dS, \tag{4.84}$$

since (4.74) implies that on S

$$l_i q_i = l_i U_i.$$

combining results (4.80), (4.82), (4.84) with (4.79) yields

$$F_j = \rho U_i \int_S \{l_i q_j - l_j q_i\} \, dS. \tag{4.85}$$

But

$$\int_S \{l_i q_j - l_j q_i\} \, dS$$

$$= \int_V \left\{\frac{\partial q_j}{\partial x_i} - \frac{\partial q_i}{\partial x_j}\right\} dV + \int_\Sigma (l_i q_j - l_j q_i) \, d\Sigma,$$

The integral over Σ vanishes as $R \to \infty$ and the volume integral vanishes identically since curl $\mathbf{q} = \mathbf{0}$. Thus

$$F_j = \rho\, U_i \int_S \{ l_i\, q_j - l_j\, q_i \}\, \mathrm{d}S = 0, \tag{4.86}$$

which establishes the D'Alembert's Paradox for a finite body of arbitrary shape. It also follows that a uniform stream flowing past a finite body of arbitrary shape at rest exerts no force on the body.

4.14 IMPULSIVE MOTION

Impulsive motion occurs in a fluid when there is a very rapid but finite change in the fluid velocity **q** over a short interval of time δt. This may result from a large body force acting over time δt, or a high pressure on a boundary acting over time δt, or the rapid variation in the velocity of a rigid body immersed in the fluid. The situation is effectively modelled mathematically by letting the body force or pressure approach infinity while $\delta t \to 0$ in such a way that the *integral* of body force or pressure over the time interval remains finite in this limit. If the flow is *incompressible*, infinitely rapid propagation of the effect of the impulse takes place, so that an impulsive pressure is produced instantaneously throughout the fluid. The effect of compressibility is that these changes take place in a finite time which depends on the speed of sound in the medium. We shall consider such phenomena in a later volume. Here we shall consider only an incompressible fluid with constant density ρ. The foregoing description implies that the impulsive body force $\mathcal{F}$ and impulsive pressure P are defined respectively by the relations

$$\mathcal{F} = \lim_{\delta t \to 0} \int_t^{t+\delta t} \mathbf{F}\, \mathrm{d}t, \tag{4.87}$$

$$P = \lim_{\delta t \to 0} \int_t^{t+\delta t} p\, \mathrm{d}t. \tag{4.88}$$

Note that finite body forces such as gravity do not contribute to the impulsive body force $\mathcal{F}$. To determine the equation of impulsive motion, first consider Euler's equation:

$$\frac{\partial \mathbf{q}}{\partial t} + (\mathbf{q}\,.\,\nabla)\mathbf{q} = -\frac{1}{\rho}\nabla p + \mathbf{F}. \tag{4.89}$$

This equation is integrated with respect to time from time t to time $t + \delta t$ and the limit as $\delta t \to 0$ is then evaluated. We may assume without loss of generality that the fluid is accelerated impulsively at $t = 0$. Since we expect a *finite* change in **q** as a result of the impulse, the integral of the convective term on the left-hand side of (4.89) vanishes in the limit as $\delta t \to 0$, using (4.87), (4.88), we obtain

$$\mathbf{q}_2 - \mathbf{q}_1 = -\frac{1}{\rho}\nabla P + \mathcal{F}, \tag{4.90}$$

where $\mathbf{q}_1$ and $\mathbf{q}_2$ denote respectively the fluid velocity before and after application of the impulse. In particular, if the fluid is at rest prior to application of the impulse, the velocity $\mathbf{q}$ generated in the fluid by the impulse is given by

$$\mathbf{q} = -\frac{1}{\rho}\nabla P + \mathscr{F}. \tag{4.91}$$

since the motion is started from rest, $\mathbf{q}$ must be irrotational and if $\mathscr{F} = \mathbf{0}$, we have the relation

$$P = -\rho\Phi, \tag{4.92}$$

connecting the impulsive pressure and the velocity potential, when an insignificant constant of integration is ignored. Equation (4.92) may be thought of as providing a physical interpretation of the velocity potential since any acyclic irrotational motion may be started from rest by a suitable distribution of impulsive pressure $-\rho\Phi$. Likewise an irrotational motion can be brought to rest by applying the impulsive pressure $\rho\Phi$ throughout the fluid.

Example [4.4]

Incompressible liquid of constant density ρ is contained within the region bounded by two concentric rigid spherical surfaces of radii a, b ($a < b$). The fluid is initially at rest. If the inner boundary is suddenly given a velocity $U\mathbf{k}$, where $\mathbf{k}$ is a constant vector, show that the outer surface experiences the impulsive force

$$2\pi\,\rho U a^3 b^3\,(b^3 - a^3)^{-1}\,\mathbf{k}.$$

In solving this type of problem it is essential to remember that at the *instant* the impulse is applied, the geometrical configuration is exactly as it was before the application of the impulse. What we seek is the motion set up in the fluid *at that instant.* Thus in this example, the boundaries are concentric spheres at the instant the impulse is applied.

The motion generated in the fluid is irrotational and therefore $\mathbf{q} = \nabla\Phi$, where $\nabla^2\Phi = 0$ to satisfy the equation of continuity. The boundary conditions are

$$\frac{\partial\Phi}{\partial r} = U\cos\theta, \quad (r = a), \qquad \frac{\partial\Phi}{\partial r} = 0, \quad (r = b), \tag{1}$$

with (r, θ, ϕ) spherical polar coordinates and with $\theta = 0$ along the direction of $\mathbf{k}$. The form of the boundary conditions suggest a solution of the form

$$\Phi = (Ar + Br^{-2})\cos\theta \tag{2}$$

which satisfies (1) if

$$A - \frac{2B}{a^3} = U, \quad A - \frac{2B}{b^3} = 0,$$

giving

$$A = -\frac{Ua^3}{(b^3 - a^3)}, \quad B = -\frac{Ua^3 b^3}{2(b^3 - a^3)}.$$

The impulsive force acting on the outer boundary S_b is

$$\int_{S_b} \hat{\mathbf{r}}(p)_{r=b}\, dS = -2\pi\rho b^2 \mathbf{k} \int_0^\pi (\Phi)_{r=b} \cos\theta \sin\theta\, d\theta$$

$$= \frac{3\pi\, \rho\, Ua^3 b^3}{(b^3 - a^3)}\, \mathbf{k} \int_0^\pi \cos^2\theta \sin\theta\, d\theta$$

$$= 2\pi\, \rho\, Ua^3 b^3\, (b^3 - a^3)^{-1}\, \mathbf{k}.$$

4.15 KINETIC ENERGY GENERATED BY IMPULSIVE MOTION

Consider incompressible fluid, initially at rest, which is set in motion by the application of impulsives $\mathbf{I}_1, \mathbf{I}_2, \ldots, \mathbf{I}_n$ to rigid boundaries $S_1, S_2, \ldots, S_n$ respectively. The fluid may be finite or infinite in extent. The kinetic energy of the irrotational motion generated in the fluid is

$$T = \frac{1}{2}\rho \int_S \Phi \frac{\partial \Phi}{\partial n}\, dS, \tag{4.93}$$

where $S = S_1 + S_2 + \ldots + S_n$. Let the velocity given to S_i be $\mathbf{U}_i$ ($i = 1, 2, \ldots, n$). It follows that on S_i,

$$\frac{\partial \Phi}{\partial n} = \mathbf{n} \,.\, \mathbf{U}_i,$$

so (4.93) is equivalent to

$$T = \frac{1}{2}\rho \sum_{i=1}^{n} \mathbf{U}_i \,.\, \int_{S_i} \mathbf{n}\, \Phi\, dS. \tag{4.94}$$

But the impulsive force exerted by the fluid on S_i is $\mathbf{R}_i$ where

$$\mathbf{R}_i = \int_{S_i} \mathbf{n} P\, dS = -\rho \int_{S_i} \mathbf{n}\Phi\, dS. \tag{4.95}$$

Combining (4.95) with (4.94) gives

$$T = -\frac{1}{2} \sum_{i=1}^{n} \mathbf{U}_i \,.\, \mathbf{R}_i. \tag{4.96}$$

In Example [4.4], letting $\mathbf{U}_1$, $\mathbf{U}_2$ be the velocities of spheres radii *a, b* respectively and $\mathbf{R}_1$, $\mathbf{R}_2$ be the corresponding impulsive forces exerted by the fluid, we have

$$\mathbf{U}_1 = U\mathbf{k}, \quad \mathbf{U}_2 = 0,$$

$$\mathbf{R}_1 = -\frac{2}{3}\pi\rho\, Ua^3\,(2a^3 + b^3)\,(b^3 - a^3)^{-1}\,\mathbf{k},$$

$$\mathbf{R}_2 = 2\pi\rho\, Ua^3 b^3\,(b^3 - a^3)^{-1}\,\mathbf{k},$$

giving

$$T = \frac{1}{3}\pi\rho\, U^2 a^3\,\frac{(2a^3 + b^3)}{(b^3 - a^3)}.$$

Let $b \to \infty$, we have

$$T = \tfrac{1}{4} M_1' \, U^2,$$

where M_1' is the mass of fluid displaced by the sphere $r = a$. The impulse $\mathbf{J}_1$ required to set the sphere in motion can be deduced from the principle of conservation of linear momentum, since

$$\mathbf{J}_1 + \mathbf{R}_1 = M_1\,\mathbf{U},$$

with M_1 the mass of the sphere.

4.16 DIRICHLET AND NEUMANN PROBLEMS

So far in this chapter we have considered some special problems of three-dimensional irrotational incompressible flow with constant density. The mathematical problem which has to be solved for such flows is the determination of a velocity potential function Φ satisfying $\nabla^2\Phi = 0$ throughout the flow region, in order to satisfy the equation of continuity, together with prescribed boundary conditions. These boundary conditions have arisen as a result of the kinematic condition which must be satisfied at the surface of a rigid body, and take the form of a prescribed value to be taken by the normal derivative $\partial\Phi/\partial n$. Such a boundary value problem for Φ is termed a *Neumann problem*. In contrast, the typical boundary value problem in electrostatics is that of finding a solution of $\nabla^2\Phi = 0$ for the electrostatic potential Φ at all points not on boundaries when the values of Φ on conducting boundaries are prescribed. This type of boundary value problem for Φ is referred to as a

Dirichlet problem. A third type of boundary value problem occurs in steady heat flow. Here the temperature Φ satisfies Laplace's equation $\nabla^2\Phi = 0$ within a heat conducting medium together with the radiative heat condition $\partial\Phi/\partial n + 2\Phi = 0$ at a boundary. This is called a *mixed* boundary value problem.

Another feature of the flows we have considered so far has been that *singularities* in the velocity and pressure fields have been excluded. We shall now consider some fundamental solutions of the Neumann problem when there are a finite number of mathematical singularities present within the flow region. Again we shall restrict ourselves to potential flows of incompressible fluid of constant density. The study of these solutions is important since not only do the singularities themselves provide a simple means of modelling various physical phenomena in fluid motion, but that combinations of solutions can provide an *indirect* method of solving problems in which singularities in the flow region are not present, e.g. uniform flow past a solid impermeable sphere at rest. This indirect approach to various problems in potential flow is possible because the uniqueness theorems discussed earlier in the chapter *guarantee* that the Neumann problems of the type to be considered have a unique solution Φ apart from an insignificant constant of integration.

4.17 SOURCES, SINKS AND DOUBLETS

Consider a point O of a region occupied by fluid into which more fluid flows radially from O in a symmetrical manner. O is then a point at which fluid enters the flow region and is referred to as a *simple source* or point source or three-dimensional source. If the volume of fluid which is emitted from O is a constant equal to $4\pi m$, we say that m is the *strength* of the simple source. Were the direction of flow reversed so that fluid leaves the flow region at O radially, then O is called a *simple sink*. A simple sink of strength m is a simple source of strength $-m$.

Consider a tank, such as an aquarium, which is fed with fresh water through a fine tube and out of which water flows through another fine tube. If the tubes project slightly into the tank and have multiple perforations, then the inlet and exit flows could be modelled approximately by the combination of a simple source and sink.

Fig. 4.4 shows a simple source of strength m at O in a fluid which is devoid of other sources and sinks and which in the case $m = 0$ would be at rest. P is a field point in the fluid, other than O, and $\overrightarrow{OP} \equiv \mathbf{r}$ while S is the surface of the sphere centred on O and having radius $r = |\mathbf{r}|$. The local fluid velocity $\mathbf{q}$ at P is directed along $\overrightarrow{OP}$ and the speed $q = |\mathbf{q}|$ is constant everywhere on S so that if the flow is steady, $q = q(r)$ and $\mathbf{q} = q(r)\hat{\mathbf{r}}$. The volumetric rate of emission of fluid from O is $4\pi m$ per unit time. The volumetric flux per unit time across the surface of S is $4\pi r^2 q(r)$ and since the fluid is incompressible, these two are equal by virtue of the equation of continuity. Hence

$$q(r) = \frac{m}{r^2}, \tag{4.97}$$

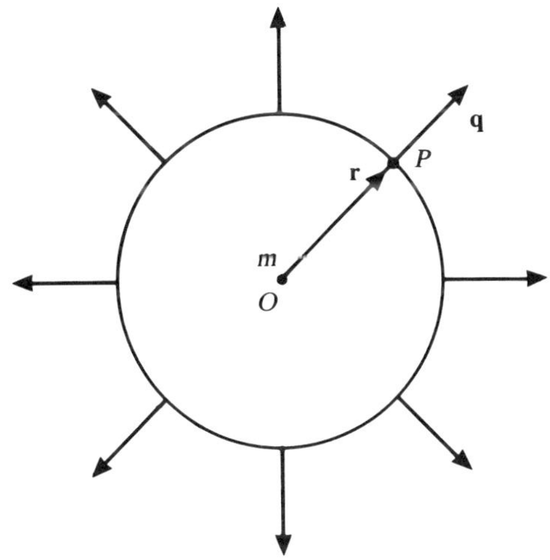

Fig. 4.4

and so

$$\mathbf{q} = \frac{m}{r^2}\,\hat{\mathbf{r}} = \frac{m}{r^3}\,\mathbf{r}. \tag{4.98}$$

It is easily shown that at all points P of the fluid other than O, the form (4.98) satisfies curl $\mathbf{q} = \mathbf{0}$ so that the flow is irrotational. Hence we can find a uniform differentiable scalar Φ such that

$$\mathbf{q} = \nabla\Phi.$$

As the flow pattern is radially symmetric and independent of time, $\Phi = \Phi(r)$. The level surfaces of this scalar function are the concentric spheres $r = \text{constant}$ and so

$$\nabla\Phi = \Phi'(r)\hat{\mathbf{r}}.$$

Hence

$$\Phi'(r) = \frac{m}{r^2},$$

with solution for the velocity potential, in simplest form,

$$\Phi(r) = -\frac{m}{r}. \tag{4.99}$$

For a simple sink of strength m, the velocity potential is m/r.

Fig. 4.5 shows a simple source of strength m at O in a uniform stream having undisturbed velocity $U\mathbf{k}$, $\mathbf{k}$ being the unit vector in $\overrightarrow{OZ}$, the axis of symmetry of

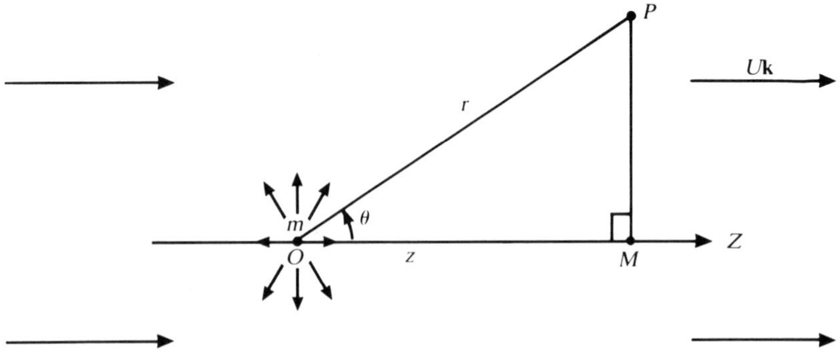

Fig. 4.5

the flow. $OP = r$, $P\hat{O}Z = \theta$ and $OM = z$ is the projection of OP on OZ. The velocity potential of the uniform stream with the source absent is

$$Uz = Ur \cos\theta,$$

on converting from Cartesian to spherical polar coordinates (r, θ, ϕ). The velocity potential of the simple source is $-m/r$ and so the total velocity potential of the combination is

$$\Phi(r, \theta) = -\frac{m}{r} + Ur\cos\theta. \tag{4.100}$$

At $P(r, \theta, \phi)$ the velocity of the fluid is $\mathbf{q} = \nabla\Phi$ giving the spherical polar co-ordinate forms

$$\begin{aligned} q_r &= \frac{\partial\Phi}{\partial r} = \frac{m}{r^2} + U\cos\theta, \\ q_\theta &= \frac{\partial\Phi}{r\partial\theta} = -U\sin\theta, \\ q_\phi &= \frac{\partial\Phi}{r\sin\theta\,\partial\phi} = 0, \end{aligned} \tag{4.101}$$

for the velocity components. Here, $r \geqslant 0$, $0 \leqslant \theta \leqslant \pi$, $0 \leqslant \phi \leqslant 2\pi$ and the only stagnation point, i.e. the point at which $\mathbf{q} = \mathbf{0}$, is given by $r = \sqrt{(m/U)}$, $\theta = \pi$.

We now discuss the field due to a simple source and a simple sink, each of strength m. In Fig. 4.6 we have a simple source of strength m at O_1 and one of strength $-m$ at O_2, where $\overrightarrow{OO_1} \equiv \mathbf{h}$, $\overrightarrow{OO_2} \equiv -\mathbf{h}$, for a chosen origin O. There are no other sources and sinks present so that the pair of singularities alone determine the fluid motion. P is a field point within the moving fluid and $\overrightarrow{OP} \equiv \mathbf{r}$ and $P\hat{O}O_1 = \theta$. We let $\overrightarrow{O_nP} \equiv \mathbf{r}_n$ $(n = 1, 2)$. Then if $h = |\mathbf{h}|$, $r_n = |\mathbf{r}_n|$ $(n = 1, 2)$, the velocity potential at P due to the two point singularities at O_1 and O_2 is

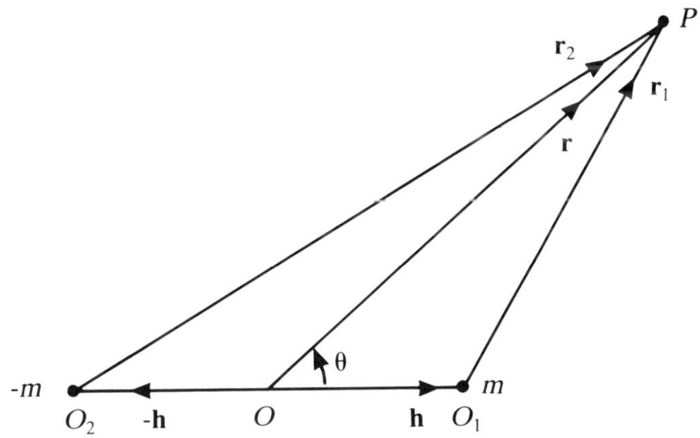

Fig. 4.6

$$\Phi = -\frac{m}{r_1} + \frac{m}{r_2}$$

$$= \frac{m(r_1^2 - r_2^2)}{r_1 r_2 (r_1 + r_2)}$$

$$= \frac{m(\mathbf{r}_1 - \mathbf{r}_2) \cdot (\mathbf{r}_1 + \mathbf{r}_2)}{r_1 r_2 (r_1 + r_2)},$$

since $(\mathbf{r}_1 - \mathbf{r}_2) \cdot (\mathbf{r}_1 + \mathbf{r}_2) = \mathbf{r}_1^2 - \mathbf{r}_2^2 = r_1^2 - r_2^2$. Fig. 4.6 shows $\mathbf{r}_1 - \mathbf{r}_2 = -2\mathbf{h}$ and $\mathbf{r}_1 + \mathbf{r}_2 = 2\mathbf{r}$, since O bisects $O_1 O_2$. Thus

$$\Phi = -\frac{4m\mathbf{h} \cdot \mathbf{r}}{r_1 r_2 (r_1 + r_2)} = -\frac{2\boldsymbol{\mu} \cdot \mathbf{r}}{r_1 r_2 (r_1 + r_2)}. \tag{4.102}$$

where $\boldsymbol{\mu} = 2m\mathbf{h}$.

In (4.102), let us first keep $\boldsymbol{\mu}$ a finite constant and non-zero vector, so that $\mu = |\boldsymbol{\mu}|$ is a finite constant and non-zero scalar. Let $\mathbf{h} \to \mathbf{0}$ along $\overrightarrow{O_1 O}$. Then $m \to \infty$ in such a way that $\boldsymbol{\mu}$ remains the same finite non-zero constant vector. Also r_1 and r_2 both $\to r$ so that, when this special double limiting process takes place, (4.102) gives

$$\Phi = -\frac{\boldsymbol{\mu} \cdot \mathbf{r}}{r^3} = -\frac{\mu \cos\theta}{r^2}. \tag{4.103}$$

The limiting source-sink combination obtained at O when we keep the direction of $\mathbf{h}$ fixed but let $h \to 0$ and $m \to \infty$ with $\mu = 2mh$ remaining a finite non-zero constant is called a three-dimensional *doublet* (or *dipole*). The scalar quantity μ is called the *moment* or *strength* of the doublet. The vector quantity $\boldsymbol{\mu} = \mu\hat{\boldsymbol{\mu}}$ is called the *vector moment* of the doublet and $\hat{\boldsymbol{\mu}}$ (unit vector from O_2 to O_1) determines the direction of the *axis* of the doublet (from sink to source). The second form for Φ in (4.103) shows that $\Phi = \Phi(r, \theta)$ and this states the physically obvious fact that the axis of the doublet is an axis of symmetry. The mathematical doublet obtained as above would be a model for a physical system such as a tank containing an entry and an exit pipe, both ends being very closely juxtaposed.

To determine the velocity field $\mathbf{q}$ of the distribution (4.103), suppose P has the spherical polar coordinates (r, θ, ϕ), where r, θ have the same meanings as before and ϕ is an azimuthal angle. From $\mathbf{q} = \quad \Phi$ and $\mathrm{d}\mathbf{s} = \mathrm{d}r\,\mathbf{r} + \mathbf{r}\,\mathrm{d}\theta\,\hat{\boldsymbol{\theta}} + \mathbf{r}\sin\phi\,\mathrm{d}\phi\,\hat{\boldsymbol{\phi}}$, the components $[q_r, q_\theta, q_\phi]$ in the spherical polar system are

$$
\begin{aligned}
q_r &= \frac{\partial\Phi}{\partial r} = \frac{2\mu\cos\theta}{r^3},\\
q_\theta &= \frac{\partial\Phi}{r\partial\theta} = \frac{\mu\sin\theta}{r^3},\\
q_\phi &= \frac{\partial\Phi}{r\sin\theta\,\partial\phi} = 0.
\end{aligned}
\tag{4.104}
$$

It has previously been shown (Example [2.1]) that for the velocity field (4.104) the equations of the streamlines are

$$
\left.\begin{aligned}
r &= A\sin^2\theta,\\
\phi &= \text{const.}
\end{aligned}\right\}.
\tag{4.105}
$$

Thus the streamlines are due to the doublet lie in planes through the axis of the doublet. They are shown in Fig. 2.3.

Let us now revert to Fig. 4.6 and carry out a different kind of limiting process: one in which r, θ are kept constant, but m and h both made to tend to infinity in such a way that the fluid speed in the neighbourhood of O stays constant. From (4.102)

$$
\Phi = -\frac{4mhr\cos\theta}{r_1 r_2\,(r_1 + r_2)},
$$

where $r_1, r_2 = \sqrt{(h^2 \mp 2hr\cos\theta + r^2)}$. Then

$$
\Phi = \frac{-4mhr\cos\theta}{2h^3\,[1 + O(r/h)]}.
$$

We now let both m and $h \to \infty$ in such a way that $2m/h^2$ remains equal to a finite non-zero constant U having the dimensions of speed, on comparing with (4.98). Ultimately the double limit gives the distribution for which

$$\Phi = -Ur \cos\theta = -Uz, \tag{4.106}$$

where $z = r\cos\theta$. This is the velocity potential due to uniform parallel flow, the fluid velocity everywhere being

$$\mathbf{q} = \nabla\Phi = -U\mathbf{k}. \tag{4.107}$$

Thus uniform parallel flow may be regarded as a kind of limiting source–sink distribution with the source and sink at $\pm\infty$.

Example [4.5]
Doublet in a uniform stream.

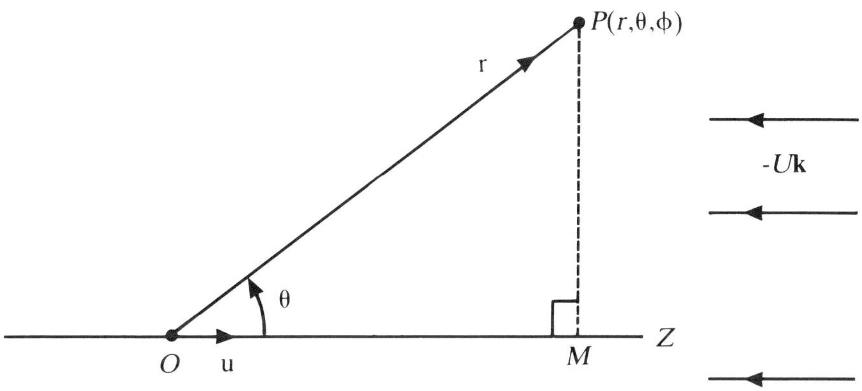

Fig. 4.7

Fig. 4.7 shows a doublet of vector moment $\boldsymbol{\mu} = \mu\mathbf{k}$ at O in a uniform stream whose velocity in the absence of the doublet would be $-U\mathbf{k}$ everywhere ($U =$ const.). P is a point in the field of flow having spherical polar coordinates (r, θ, ϕ), the direction $\overrightarrow{OZ}$ of the doublet's axis being the line $\theta = 0$. PM is the perpendicular on OZ and $OM = x = r\cos\theta$. We obtain the resultant velocity potential as the combination of those of the uniform stream and the doublet. These are respectively:

(i) $-Uz = -Ur\cos\theta$ (stream);
(ii) $-\mu\cos\theta/r^2$ (doublet).

Hence the total velocity potential at P due to the combination is

$$\Phi(r, \theta) = -(Ur + \mu r^{-2})\cos\theta; \qquad r > 0,\ 0 \leqslant \theta \leqslant \pi,\ 0 \leqslant \phi \leqslant 2\pi.$$

The absence of ϕ in the form of Φ is due to the axial symmetry of the flow about OZ.

Since the fluid velocity $\mathbf{q}$ at $P(r, \theta, \phi)$ is $\mathbf{q} = \overline{\nabla}\Phi$ and since the vector arc element at P in the spherical polar coordinate system is

$$\mathrm{d}\mathbf{s} = \mathrm{d}r\,\hat{\mathbf{r}} + r\,\mathrm{d}\theta\,\hat{\boldsymbol{\theta}} + r\sin\theta\,\mathrm{d}\phi\,\hat{\boldsymbol{\phi}},$$

the spherical polar velocity components at P are respectively

$$\begin{aligned} q_r &= \frac{\partial\Phi}{\partial r} = -\left(U - \frac{2\mu}{r^3}\right)\cos\theta, \\ q_\theta &= \frac{\partial\Phi}{r\partial\theta} = \left(U + \frac{\mu}{r^3}\right)\sin\theta, \\ q_\phi &= \frac{\partial\Phi}{r\sin\theta\partial\phi} = 0, \end{aligned} \tag{1}$$

for $r>0, 0\leqslant\theta\leqslant\pi, 0\leqslant\phi<2\pi$.

From (1), we can determine the stagnation points of the flow where the fluid is at rest so that $\mathbf{q} = \mathbf{0}$. Then we require those values of r, θ satisfying $q_r = 0 = q_\theta$ and $r \geqslant 0$, $0 \leqslant \theta \leqslant \pi$. If we write $a = (2\mu/U)^{1/3}$, the only two admissible solutions are:

(i) $r = a, \theta = 0$;
(ii) $r = a, \theta = \pi$.

In each case the value of ϕ is immaterial. There are two stagnation points both on the axis of the doublet.

We observe that, for all admissible θ, $q_r = 0$ on the spherical sphere $r = a$, where $a = (2\mu/U)^{1/3}$ = const. On other surfaces, $q_r = 0$ only when $\theta = \pi/2$. It follows, then, that there is no flow over the surface of the sphere $r = a$. Its poles $r = a$, $\theta = 0$; $r = a$, $\theta = \pi$ are the only stagnation points throughout the field of flow: elsewhere on its surface fluid flows tangentially with speed $q_\theta = \frac{3}{2}U\sin\theta$.

For the region $r \geqslant a$, with $\mu = \frac{1}{2}Ua^3$, we obtain precisely the same velocity potential as that found in Section 4.11 for uniform flow past a stationary impermeable sphere, centre O and radius a. Thus, for $r \geqslant a$, the effect of the sphere is that of a doublet of strength $\mu = \frac{1}{2}Ua^3$ situated at its centre, its axis pointing upstream. So the sphere can be represented by a suitably chosen singularity at its centre.

Example [4.6]

Illustration of line distributions

Fig. 4.8 shows a uniform line souce AB of strength m per unit length. This means that the elemental section of AB at x from A and of length δx is a point source of strength $m\delta x$. If P is a point in the fluid distant r from this element, then the velocity potential at P due to the point source is $-m\delta x/r$. Thus the total velocity potential at P due to the entire line distribution AB is

$$\Phi = -m\int_0^{2l}\frac{\mathrm{d}x}{r}.$$

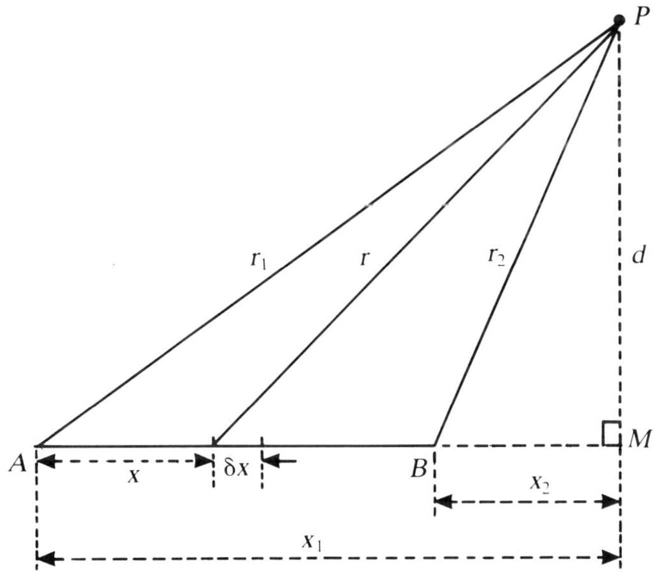

Fig. 4.8

Let $AM = x_1$, $BM = x_2$, where AM is the orthogonal projection of AP on AB. Since $r^2 = (x_1 - x)^2 + d^2 = (x_1 - x)^2 + r_1^2 - x_1^2$,

$$\begin{aligned}\Phi &= -m \int_0^{2l} \frac{dx}{\sqrt{\{(x_1 - x)^2 + r_1^2 - x_1^2\}}} \\ &= -m \ln\{x_1 - x + \sqrt{[(x_1 - x)^2 + r_1^2 - x_1^2]}\} \Big|_{2l}^{0} \\ &= -m\{\ln(x_1 + r_1) - \ln[x_2 + \sqrt{(x_2^2 + r_1^2 - x_1^2)}]\} \\ &= -m \ln\left[\frac{x_1 + r_1}{x_2 + r_2}\right],\end{aligned}$$

using $r_1^2 - x_1^2 = d^2 = r_2^2 - x_2^2$. The same relation also gives

$$\frac{r_1 + x_1}{r_2 + x_2} = \frac{r_2 - x_2}{r_1 - x_1},$$

and each of these fractions is equal to $(r_1 + r_2 + x_1 - x_2)/(r_1 + r_2 + x_2 - x_1)$ by componendo dividendo rules. Hence

$$\begin{aligned}\Phi &= -m \ln\left[\frac{r_1 + r_2 + 2l}{r_1 + r_2 - 2l}\right] \\ &= -m \ln\left[\frac{a + l}{a - l}\right],\end{aligned}$$

where $2a$ is the length of the major axis of the ellipsoid of revolution through P having A and B as foci since for such an ellipsoid $r_1 + r_2$ = const. It follows from this that the equipotentials Φ = const. are precisely the family of confocal ellipsoids $r_1 + r_2 = 2a$ obtained when a is allowed to vary.

The fluid velocity at P is

$$\mathbf{q} = \nabla\Phi = (\partial\Phi/\partial n)\mathbf{n},$$

in the usual notation (Section 1.10). Let P be any point on the ellipsoid specified by parameter a and P' the neighbouring point on the ellipsoid specified by parameter $(a + \delta a)$, where $\overline{PP'} \equiv \delta n\mathbf{n}$ as in Fig. 4.9.

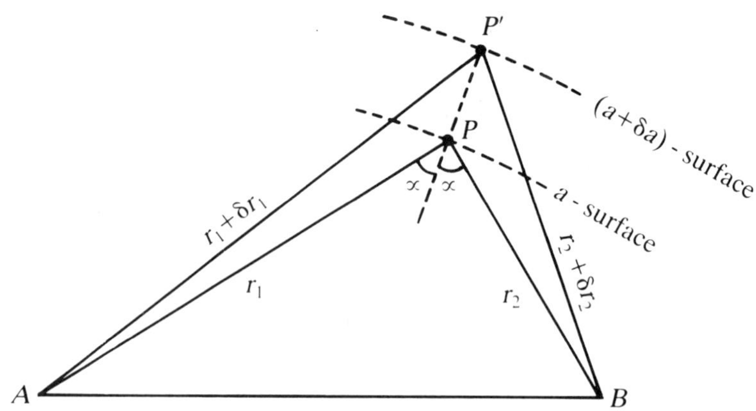

Fig. 4.9

We have

$$\begin{aligned} \mathbf{q} &= -m\,\partial/\partial n\,[\ln(a + l) - \ln(a - l)]\,\mathbf{n} \\ &= +\frac{2lm}{a^2 - l^2}\frac{\partial a}{\partial n}\mathbf{n}. \end{aligned}$$

The normal at P to the a-surface bisects the angle 2α between the focal radii AP, BP. Now

$$(r_1 + \delta r_1)^2 = r_1^2 + 2r_1\delta n\cos\alpha + (\delta n)^2$$

or

$$2r_1\delta r_1 = 2r_1\delta n\cos\alpha + (\delta n)^2 - (\delta r_1)^2.$$

Thus

$$\partial r_1/\partial n = \cos\alpha,$$

and similarly $\partial r_2/\partial n = \cos\alpha$. Thus, as $2a = r_1 + r_2$,

$$\partial a/\partial n = \cos\alpha$$

and so the fluid velocity at P is

$$\mathbf{q} = \left[\frac{2ml\cos\alpha}{a^2 - l^2}\right]\mathbf{n}.$$

4.18 HYDRODYNAMICAL IMAGES FOR THREE-DIMENSIONAL FLOWS

Consider a fluid containing a distribution of sources, sinks and doublets. If a surface S can be drawn in the fluid so as not to pass through any of the singularities distributed on either side of it, then the two sets of singularities on opposites sides of S are said to be *image systems* in the surface S.

4.19 IMAGES IN A RIGID IMPERMEABLE INFINITE PLANE

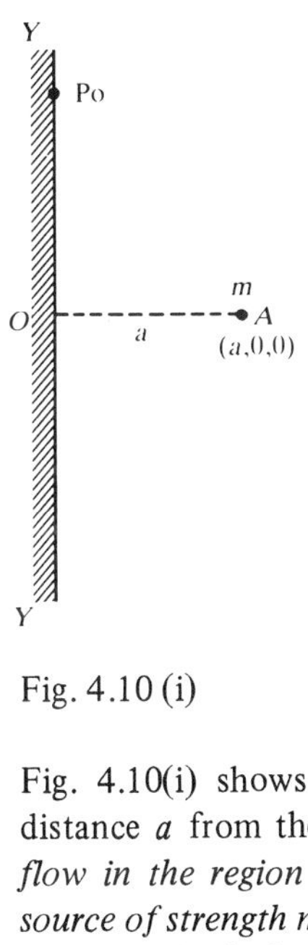

Fig. 4.10 (i)

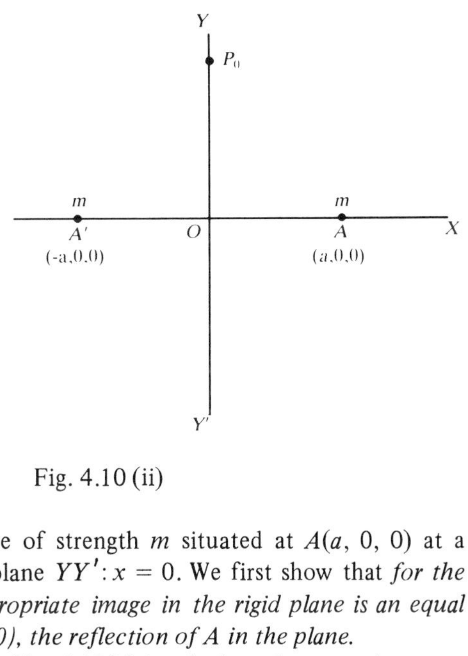

Fig. 4.10 (ii)

Fig. 4.10(i) shows a simple source of strength m situated at $A(a, 0, 0)$ at a distance a from the infinite rigid plane YY': $x = 0$. We first show that *for the flow in the region $x \geqslant 0$ the appropriate image in the rigid plane is an equal source of strength m at $A'(-a, 0, 0)$, the reflection of A in the plane.*

To establish this result, consider Fig. 4.10(ii) in which we have equal sources of strength m at $A(a, 0, 0)$ and $A'(-a, 0, 0)$ with no rigid boundary. Let P_0 be any point on the plane $x = 0$ in Fig. 4.10(ii). Then the fluid velocity at P_0 due to the two sources is

$$\left(\frac{m}{AP_0^3}\right)\overrightarrow{AP}_0 + \left(\frac{m}{A'P_0^3}\right)\overrightarrow{A'P_0}$$

$$= \frac{m}{AP_0^3}\left(\overrightarrow{AP}_0 + \overrightarrow{A'P}_0\right)$$

$$= \left(\frac{2m}{AP_0^3}\right) \overrightarrow{OP}_0, \text{ since } \overrightarrow{AP}_0 + \overrightarrow{A'P}_0 = 2\,\overrightarrow{OP}_0.$$

This shows that at any P_0 of YY', the fluid flows tangentially to the plane $x = 0$ and so there is no transport of fluid across this plane. Thus in Figs 4.10(i), (ii), at all corresponding points P_0 on $x = 0$ we have $\partial\Phi/\partial n = 0$ for the region of flow $x \geqslant 0$. We infer, then, that the image of m at A in the rigid plane YY' in Fig. 4.10(i) is m at A', as required.

Now consider a pair of sources $-m$ at A, $+m$ at B, taken close together and on one side of the rigid impermeable plane YY' (Fig. 4.11(i)). The image system is $-m$ at A', $+m$ at B', where A' and B' are respectively the reflections of A and B in the plane, as shown in Fig. 4.11(ii). In the limit when $B \to A$ along $\overrightarrow{BA}$ in such a way as to form a doublet at A in either diagram, we see that *the image of a doublet in an infinite impermeable rigid plane is a doublet of equal strength and symmetrically disposed to the other with respect to the plane.*

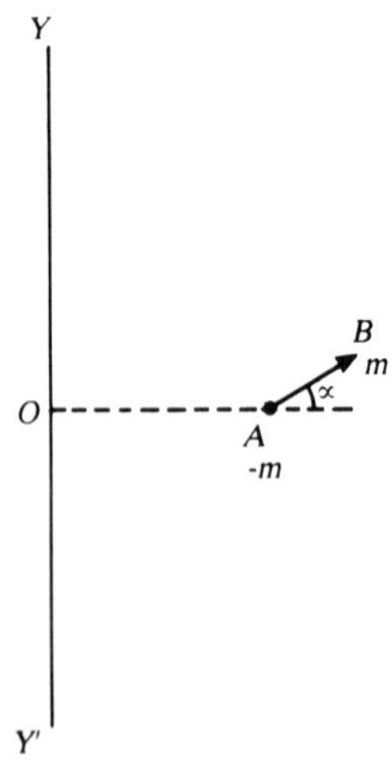

Fig. 4.11(i)

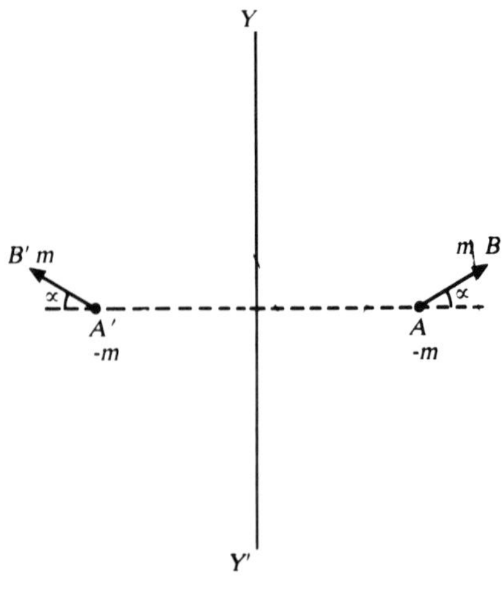

(Fig. 4.11(ii)

Example [4.7]

A three-dimensional doublet of strength μ whose axis is in the direction $\overrightarrow{OZ}$ is distant a from the rigid plane $z = 0$ which is the sole boundary of liquid of constant density ρ, infinite in extent. If p_∞ is the pressure at infinity, show that the pressure on the plane is least at a distance $a\sqrt{5}/2$ from the doublet.

Fig. 4.12(i) shows the actual physical problem with the doublet at A. The image system for the region $z \geqslant 0$ is shown in Fig. 4.12(ii) and it comprises an equal doublet of strength μ at A', the reflection of A in the plane $z = 0$, and with axis along $\overrightarrow{ZO}$. We specify any point P of the fluid by spherical polar coordinates (r, θ, ϕ), where OZ is the line $\theta = 0$ and the plane $x = 0$ is $\theta = \frac{1}{2}\pi$ so that P is confined to the region $0 \leqslant \theta \leqslant \frac{1}{2}\pi$. We let $AP = r_1$, $A'P = r_2$ and

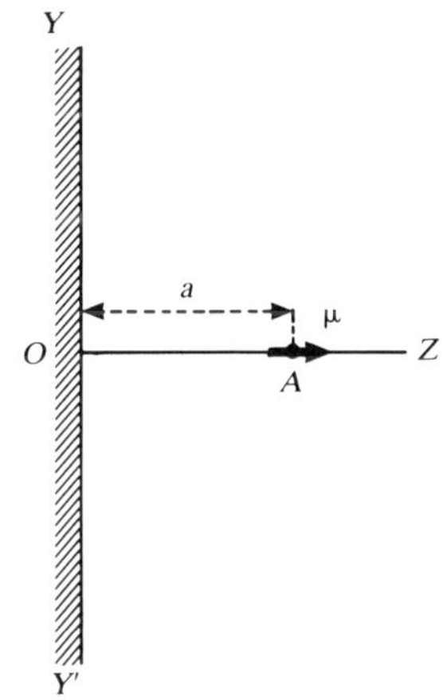

Fig. 4.12(i)

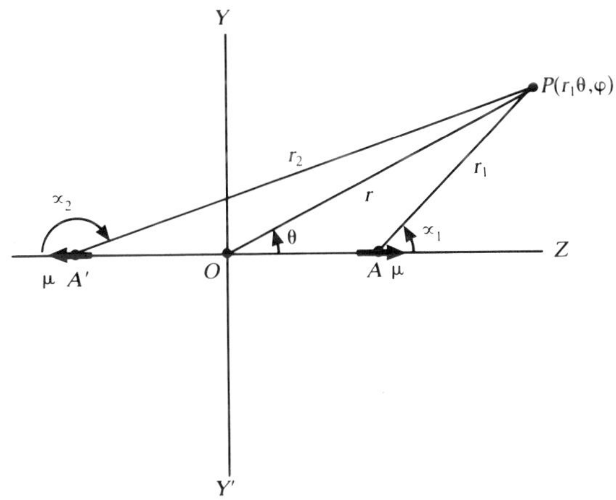

Fig. 4.12(ii)

α_1, α_2 be the angles these lines make with the axes of the doublets as shown. Then the total velocity potential at P is

$$\Phi = -\frac{\mu \cos \alpha_1}{r_1^2} - \frac{\mu \cos \alpha_2}{r_2^2},$$

where

$$r_1^2 = r^2 + a^2 - 2ra \cos \theta; \qquad r_2^2 = r^2 + a^2 + 2ra \cos \theta,$$

and

$$\cos \alpha_1 = (r \cos \theta - a)/r_1; \qquad \cos \alpha_2 = -(r \cos \theta + a)/r_2.$$

Thus

$$\Phi(r, \theta) = \mu\{(r \cos \theta + a)/r_2^3 - (r \cos \theta - a)/r_1^3\}.$$

Since the fluid velocity of P is $\mathbf{q} = \nabla \Phi$, the spherical polar coordinate components of $\mathbf{q}$ are $[q_r, q_\theta, q_\phi]$, where

$$q_r = \frac{\partial \Phi}{\partial r}; \quad q_\theta = \frac{\partial \Phi}{r \partial \theta}; \quad q_\phi = \frac{\partial \Phi}{r \sin \theta \, \partial \phi}.$$

Thus
$$\begin{aligned} q_r = \mu\{&\cos \theta / r_2^3 - 3(\partial r_2/\partial r)(r \cos \theta + a)/r_2^4 \\ &- \cos \theta / r_1^3 + 3(\partial r_1/\partial r)(r \cos \theta - a)/r_1^4\}. \end{aligned}$$

Now $2r_2 \partial r_2/\partial r = 2(r + a \cos \theta)$ and so $\partial r_2/\partial r = (r + a \cos \theta)/r_2$. Similarly, $\partial r_1/\partial r = (r - a \cos \theta)/r_1$. Thus

$$\begin{aligned} q_r = \mu\{&\cos \theta / r_2^3 - 3(r + a \cos \theta)(r \cos \theta + a)/r_2^5 \\ &- \cos \theta / r_1^2 + 3(r - a \cos \theta)(r \cos \theta - a)/r_1^5\}. \end{aligned}$$

Further

$$q_\theta = \frac{\partial\Phi}{r\partial\theta}$$

$$= \frac{\mu}{r}\{-r\sin\theta/r_2^3 - 3(r\cos\theta + a)(\partial r_2/\partial\theta)/r_2^4$$

$$+ r\sin\theta/r_1^3 + 3(r\cos\theta - a)(\partial r_1/\partial\theta)/r_1^4\}.$$

Now $\quad 2r_2\,\partial r_2/\partial\theta = -2ra\sin\theta$ and so $\partial r_2/\partial\theta = -(ra/r_2)\sin\theta$ and

$\quad 2r_1\,\partial r_1/\partial\theta = \quad 2ra\sin\theta$ so that $\partial r_1/\partial\theta = (ra/r_1)\sin\theta$.

Thus

$$q_\theta = \frac{\mu}{r}\{-r\sin\theta/r_2^3 + 3ra\sin\theta\,(r\cos\theta + a)/r_2$$

$$+ r\sin\theta/r_1^3 + 3ra\sin\theta\,(r\cos\theta - a)/r_1\}.$$

Since Φ is independent of ϕ, $q_\phi = 0$. On the plane $\theta = \frac{1}{2}\pi$, we have $r_1^2 = r_2^2 = r^2 + a^2$ and so at $(r, \pi/2, \phi)$,

$$q_r = -6\mu ra\,(r^2 + a^2)^{-5/2}, \quad q_\theta = 0, \quad q_\phi = 0.$$

Along the streamline through this point, Bernoulli's equation is

$$\frac{p}{\rho} + \frac{1}{2}q^2 = \text{const.} = \frac{p_\infty}{\rho},$$

the value of q at infinity being zero. Hence

$$p(r) = p_\infty - 18\rho\mu^2 a^2 r^2\,(r^2 + a^2)^{-5},$$

and so

$$p'(r) = 36\rho\mu^2 a^2 r(4r^2 - a^2)(r^2 + a^2)^{-6}.$$

Then

$$p'(\tfrac{1}{2}a) = 0; \quad p'(\tfrac{1}{2}a+) > 0; \quad p'(\tfrac{1}{2}a-) < 0,$$

which indicates that p is a minimum at $r = \frac{1}{2}a$, $\theta = \frac{1}{2}\pi$ on the plane.

4.20 IMAGES IN IMPERMEABLE SPHERICAL SURFACES

In Section 4.11 we studied the effect of placing a solid impermeable sphere in a uniform stream of incompressible flow. Here we discuss the disturbance produced when such a sphere is placed in a more general flow. We obtain results which do not presuppose axial symmetry of flow.

We first establish some properties of harmonic functions specified by spherical polar coordinates (r, θ, ϕ). If $\Phi(r, \theta, \phi)$ is harmonic then it satisfies Laplace's equation $\nabla^2\Phi = 0$ which reduces on simplifcation to the form

$$\sin\theta\ \frac{\partial}{\partial r}\left(r^2\ \frac{\partial\Phi}{\partial r}\right) + \frac{\partial}{\partial\theta}\left(\sin\theta\ \frac{\partial\Phi}{\partial\theta}\right) + \frac{1}{\sin\theta}\ \frac{\partial^2\Phi}{\partial\phi^2} = 0. \qquad (4.108)$$

Theorem I *If $r^n S_n(\theta, \phi)$ is a harmonic function, so also is $r^{-(n+1)} S_n(\theta, \phi)$.*

Proof Since $\Phi = r^n S_n(\theta, \phi)$ satisfies (4.108),

$$\sin\theta\ S_n\ \mathrm{d}/\mathrm{d}r(nr^{n+1}) + r^n\ \partial/\partial\theta\ (\sin\theta\ \partial S_n/\partial\theta) + (r^n/\sin\theta)\partial^2 S_n/\partial\phi^2 = 0$$

or

$$n(n+1)\sin\theta S_n + \partial/\partial\theta\ (\sin\theta\ \partial S_n/\partial\theta) + \operatorname{cosec}\theta\ \partial^2 S_n/\partial\phi^2 = 0. \qquad (4.109)$$

On substituting $\Phi = r^{-n-1} S_n(\theta, \phi)$ into the left-hand side of (4.108) we obtain the expression

$$\begin{aligned}
&\sin\theta\ S_n\ \mathrm{d}/\mathrm{d}r\,[-(n+1)r^{-n}] + r^{-n-1}\ \partial/\partial\theta(\sin\theta\ \partial S_n/\partial\theta) \\
&\quad + r^{-n-1}\ \operatorname{cosec}\theta\ \partial^2 S_n/\partial\phi^2 \\
&= r^{-n-1}\left\{\sin\theta\ (n+1)n\ S_n + \partial/\partial\theta(\sin\theta\ \partial S_n/\partial\theta)\right. \\
&\quad \left. + \operatorname{cosec}\theta\ \partial^2 S_n/\partial\phi^2\right\} \\
&= 0, \text{ using (4.109).}
\end{aligned}$$

Thus $r^{-n-1} S_n(\theta, \phi)$ is harmonic if $r^n S_n(\theta, \phi)$ is.

(Note also the truth of the converse, i.e. if $r^{-n-1} S_n$ is harmonic, so also is $r^n S_n$.)

Theorem II *If $\Phi(r, \theta, \phi)$ has an expansion of the form*

$$\Phi(r, \theta, \phi) = \sum_{n=0}^{\infty} \alpha_n\, r^n S_n\,(\theta, \phi),$$

the series on the right being uniformly convergent with respect to r, then for constant λ,

$$\frac{1}{r^\lambda}\int_0^r R^{\lambda-1}\ \Phi(R, \theta, \phi)\ \mathrm{d}R = \sum_{n=0}^{\infty} \frac{\alpha_n\, r^n S_n(\theta, \phi)}{n+\lambda}.$$

Proof

$$R^{\lambda-1}\ \Phi(R, \theta, \phi) = \sum_{n=0}^{\infty} \alpha_n\ R^{n+\lambda-1} S_n(\theta, \phi).$$

Hence

$$\int_0^r R^{\lambda-1}\,\Phi(R,\theta,\phi)\,\mathrm{d}R = \sum_{n=0}^{\infty} \frac{\alpha_n r^{n+\lambda} S_n(\theta,\phi)}{n+\lambda}.$$

whence the result follows.

Theorem III Weiss's sphere theorem.

Let $\Phi(r, \theta, \phi)$ be the velocity potential at a point P having spherical polar coordinates (r, θ, ϕ) in an incompressible fluid having irrotational motion and no rigid boundaries. Also suppose Φ has no singularities within the region $r \leqslant a$. Then if a solid impermeable sphere of radius a is introduced into the flow with its centre at the origin of coordinates, the new velocity potential at P in the fluid is

$$\Phi(r,\theta,\phi) + \frac{a}{r}\,\Phi\left(\frac{a^2}{r},\theta,\phi\right) - \frac{1}{a}\int_0^{a^2/r} \Phi(R,\theta,\phi)\,\mathrm{d}R \quad (r > a).$$

Proof: Suppose Φ is regular near the origin: we assume it has an expression of the form

$$\Phi(r,\theta,\phi) = \sum_{n=0}^{\infty} \alpha_n r^n S_n(\theta,\phi),$$

where each $r^n S_n(\theta,\phi)$ is harmonic and each α_n is known.

When the sphere is introduced into the stream, it will produce a perturbation potential which will be zero at infinity and which must be harmonic. A suitable form to choose for the perturbation potential in virtue of Theorem I is $\sum_{n=0}^{\infty} \beta_n r^{-(n+1)} S_n(\theta,\phi)$, where the coefficients β_n are undetermined. Thus the new velocity potential at P is $\Phi^*(r,\theta,\phi)$, where

$$\Phi^*(r,\theta,\phi) = \sum_{n=1}^{\infty} \alpha_n r^n S_n(\theta,\phi) + \sum_{n=0}^{\infty} \beta_n r^{-(n+1)} S_n(\theta,\phi).$$

On $r = a$, we require $\partial\Phi^*/\partial r = 0$. Equating to zero the coefficient of each $S_n(\theta,\phi)$ this boundary condition gives

$$\alpha_n n a^{n-1} - (n+1)\,\beta_n\, a^{-(n+2)} = 0 \qquad (n = 0, 1, 2, \ldots)$$

or

$$\beta_n = \left(1 - \frac{1}{n+1}\right) a^{2n+1} \alpha_n \qquad (n = 0, 1, \ldots).$$

Thus the perturbation potential is

$$\sum_{n=0}^{\infty} \beta_n r^{-(n+1)} S_n = \frac{a}{r} \sum_{n=0}^{\infty} \alpha_n \left(\frac{a^2}{r}\right)^n S_n - \frac{1}{a} \sum_{n=0}^{\infty} \frac{\alpha_n (a^2/r)^{n+1} S_n}{n+1}$$

$$= \frac{a}{r} \Phi\left(\frac{a^2}{r}, \theta, \phi\right) - \frac{1}{a} \int_0^{a^2/r} \Phi(R, \theta, \phi) \, \mathrm{d}R,$$

the integral following by taking $\lambda = 1$ in Theorem II. This establishes the required result.

Example [4.8]
Image of a point source in a solid sphere.

The velocity potential at (r, θ, ϕ) in a fluid due to a simple source of strength m at $(f, 0, 0)$ is

$$\Phi(r, \theta) = -m\,(r^2 - 2rf\cos\theta + f^2)^{-1/2}.$$

Introducing a solid sphere in the region $r \leqslant a$, where $a < f$, we obtain on using Weiss's sphere theorem, a perturbation potential of

$$\frac{-ma/f}{\sqrt{[r^2 - 2r(a^2/f)\cos\theta + (a^2/f)^2]}} + \frac{m}{a} \int_0^{a^2/r} \frac{\mathrm{d}R}{\sqrt{(R^2 - 2Rf\cos\theta + f^2)}}.$$

This shows that the image system of a point source of strength m placed at $f(> a)$ from the centre of the solid sphere consists of a source of strength ma/f at the inverse point in the sphere, together with a continous line distribution of sinks of uniform strength m/a per unit length and extending from the centre to the inverse point.

Example [4.9]
Image of a doublet in a sphere when the axis of the doublet passes through the centre of the sphere.

Fig. 4.13 shows a doublet at A with its axis pointing towards the centre O of a sphere of radius a. Let $OA = f$, $OB = f + \delta f$, where A', B' are the points in the sphere inverse to A, B respectively, so that $OA' = a^2/f$, $OB' = a^2/(f + \delta f)$. At A, B we associate simple sources of strengths m, $-m$ so that the strength of the doublet is $\mu = m\delta f$, where μ is to remain a finite non-zero constant as $m \to \infty$ and $\delta f \to 0$ simultaneously. From the previous example, the image of m at A consists of ma/f at A' together with a continuous line distribution from O to A' of sinks of density m/a per unit length, and the image of $-m$ at B consists of $-ma/(f + \delta f)$ at B' together with a continuous line distribution from O to B' of sources of density m/a per unit length. The total point-source strength within the sphere is, then,

$$\frac{ma}{f} - \frac{ma}{f + \delta f} - \frac{m}{a}\,\delta f \left(\frac{a^2}{f} - \frac{a^2}{f + \delta f}\right) = 0.$$

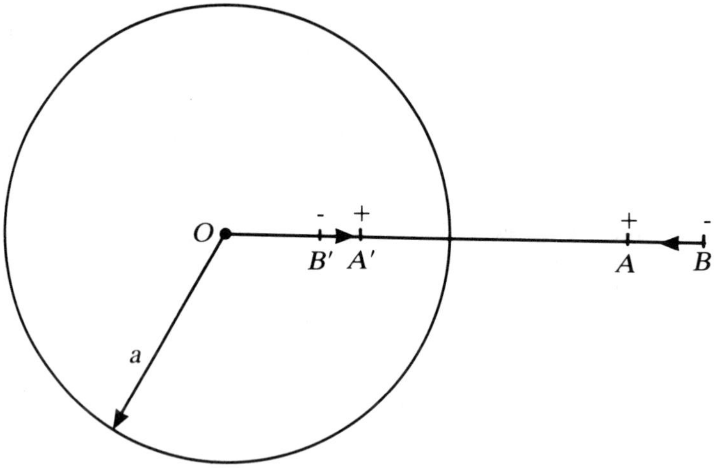

Fig. 4.13

The moment of the distribution within the sphere is

$$\frac{ma}{f} \times B'A' = \frac{ma}{f}\left(\frac{a^2}{f} - \frac{a^2}{f+\delta f}\right)$$

$$= \frac{\mu a^3}{f^2\,(f+\delta f)}$$

$$\to \mu a^3/f^3 \text{ at } A' \text{ as } \delta f \to 0 \text{ and } m \to \infty.$$

Thus, when the double limit is taken, we obtain a doublet at A of strength μ, whose axis points towards O, together with a doublet at the inverse point A' of strength $\mu a^3/f^3$, whose axis is directed away from O.

On letting $f = d + a$ and proceeding to the limit as $a \to \infty$, the sphere becomes an infinite plane and we recover the result of Section 4.19 when the axis of the doublet at A points away from and is perpendicular to the plane.

The reader can obtain the result of Example [4.9] directly using the theorem of Weiss though the integration is fairly heavy.

4.21 TWO-DIMENSIONAL MOTION

Suppose that a fluid moves in such a way that at any given instant the flow pattern in a certain plane within the fluid is the same as that in all other parallel planes within the fluid. Then at the considered instant, the flow is said to be *two-dimensional* or *plane*. Any one of the parallel planes is then termed a *plane of flow*. If at the instant a particular plane of flow is labelled $z = 0$, which is achieved by a suitable choice of rectangular Cartesian axes for specifying coordinates (x, y, z), then at that instant, all physical properties associated with the fluid, i.e. velocity, pressure, density, etc., will be independent of the coordinate z but may have dependence on x and/or y.

If the situation described above persists at all times t, then the flow is plane for all t. In this case, congruence of flow pattern will prevail in each plane of flow. For unsteady plane flow, the pattern will vary from instant to instant, but for steady flow, congruence of flow pattern will result in all planes of flow and will not vary with time.

Plane flow as described cannot be achieved in reality, but in certain important cases close approximation to planarity of flow may occur. For example, if we consider the flow of a stream of water through a rectangular gutter whose sectional dimensions are large, then at locations of the flow field remote from the boundaries the flow may well be regarded as approximately planar. Smooth flow of a deep river whose banks are wide apart affords another case of plane flow at points remote from the bed and the banks, provided the river is free from weirs and obstacles.

We shall consider two-dimensional flows in detail in Chapter 6 but it is useful to list here some results relating to potential flows in two dimensions. The type of rigid bodies which occur in such flows are infinitely long cylinders of constant cross-section and generators perpendicular to the plane of motion. Thus the flow regions when such bodies are present are multiply connected. However, as pointed out in Chapter 1, the two-dimensional form of the Divergence Theorem:

$$\int_C \mathbf{n} \cdot \mathbf{A} \, \mathrm{d}s = \int_S \operatorname{div} \mathbf{A} \, \mathrm{d}S \tag{4.110}$$

holds, provided the unit normal $\mathbf{n}$ is always directed out of the fluid over each finite boundary C_i of the fluid and $C = C_1 + C_2 + \ldots + C_n$. If the fluid is unbounded, equation (4.110) still holds provided that $|\mathbf{A}| \to 0$ faster than R^{-1} as $R \to \infty$, where R denotes distance from a fixed origin in any of the planes of motion.

4.22 KINETIC ENERGY OF ACYCLIC IRROTATIONAL MOTION

Consider two-dimensional acyclic irrotational motion of incompressible liquid bounded internally by a cylinder C_1 and externally by a cylinder C_2, as illustrated in Fig. 4.14.

With S denoting the area of any plane section of the fluid between the cylinders, the kinetic energy of the fluid is

$$T = \frac{1}{2}\rho \int_S (\nabla\Phi)^2 \, \mathrm{d}S = \frac{1}{2}\rho \int_C \Phi \frac{\partial \Phi}{\partial n} \, \mathrm{d}s, \tag{4.111}$$

where $C = C_1 + C_2$ on applying (4.110) with $\mathbf{A} = \Phi\nabla\Phi$. If the cylinder C_2 is at rest, (4.111) gives

$$T = \frac{1}{2}\rho \int_{C_1} \Phi \frac{\partial \Phi}{\partial n} \, \mathrm{d}s \tag{4.112}$$

Equation (4.112) may also be shown to be the expression for the kinetic energy of unbounded fluid at rest at infinity. The proof follows that of Section 4.6 with slight modifications.

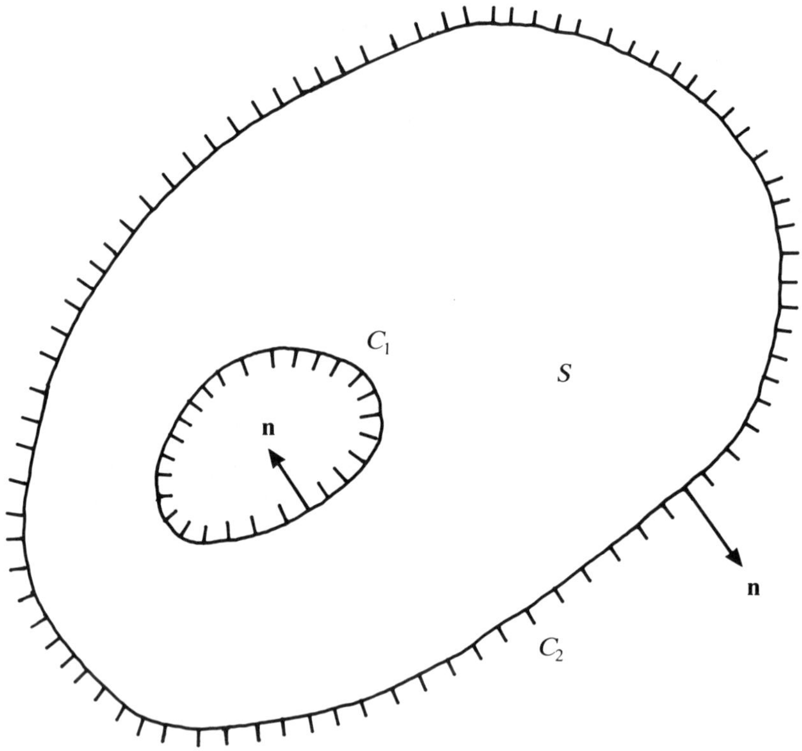

Fig. 4.14

4.23 KINETIC ENERGY OF CYCLIC IRROTATIONAL MOTION

Consider again the flow between cylinders C_1 and C_2. Since the flow region is doubly connected, it is possible for Φ to be multi-valued, which will occur when the circulation about C_1 is non-zero. If Φ denotes a value of the velocity potential and the circulation about C_1 is $2\pi K$ when the cylinder is described in the counter-clockwise sense, the velocity potential accordingly increases to $\Phi + 2\pi K$ after describing C_1 once in the counter-clockwise sense.

For simplicity, consider C_1 and C_2 to be at rest, and let a barrier AB be drawn across the fluid as indicated in Fig. 4.15. The barrier AB now makes the region occupied by the fluid simply connected and if it consists of the same fluid particles, it has no effect on the motion of the fluid. The kinetic energy of the flow is

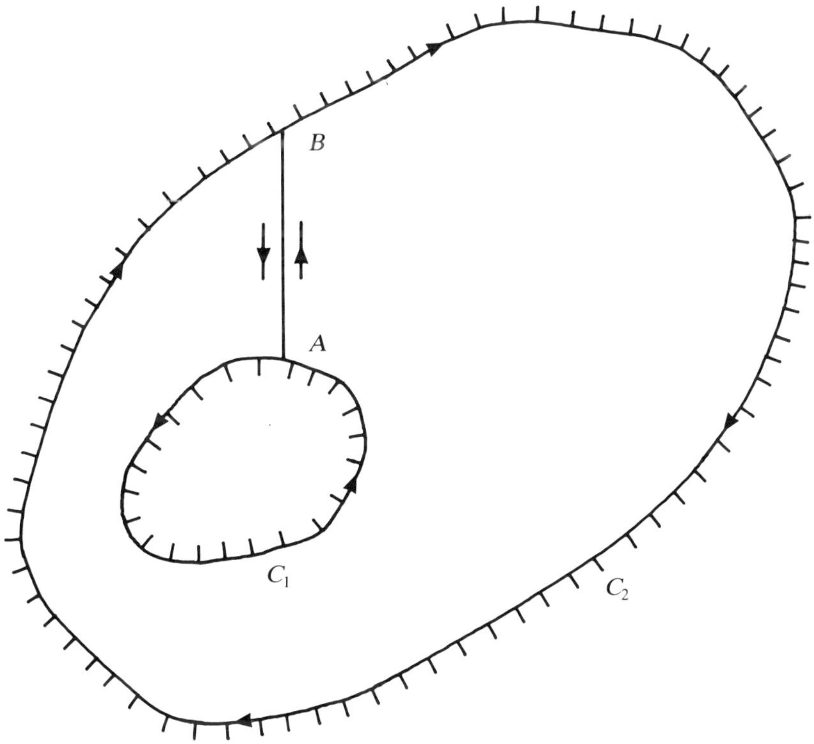

Fig. 4.15

$$T = \frac{1}{2}\rho \oint_C \Phi \frac{\partial \Phi}{\partial n} \, ds,$$

where C is C_1 described counter-clockwise, the straight path $\overrightarrow{AB}$, C_2 described clockwise and the straight path $\overrightarrow{BA}$. Thus

$$T = \frac{1}{2}\rho \int_{AB} \Phi \frac{\partial \Phi}{\partial n} \, ds + \frac{1}{2}\rho \int_{BA} \Phi \frac{\partial \Phi}{\partial n} \, ds, \tag{4.113}$$

with the direction of **n** on AB and BA drawn *out* of the fluid in the simply connected region defined. Equation (4.113) can be written as

$$T = \frac{1}{2}\rho \int_{AB} [\Phi] \frac{\partial \Phi}{\partial n} \, ds, \tag{4.114}$$

where $[\Phi]$ = change in Φ in describing C_1 once counter-clockwise. Thus $[\Phi] = 2\pi K$. Furthermore,

$$\int_{AB} \frac{\partial \Phi}{\partial n} \, ds = m,$$

where m is the flux of the flow between the cylinders. Thus (4.114) reduces to

$$T = \pi\rho K m. \tag{4.115}$$

If C_1 and C_2 are in motion, the kinetic energy is obtained by adding to the right-hand side of (4.115) a term of the form (4.112).

4.24 UNIQUENESS OF ACYCLIC IRROTATIONAL MOTION

The uniqueness theorems for three-dimensional irrotational flow need only slight modifications in their proofs to show that:

1. Acyclic irrotational flow of fluid between cylinders of finite radius has a unique velocity potential.
2. Acyclic irrotational flow of unbounded fluid exterior to cylinders of finite cross-section is unique when the fluid is at rest at infinity or moves as a uniform stream.

4.25 USE OF CYLINDRICAL POLAR COORDINATES

For an incompressible irrotational flow of uniform density the equation of continuity $\nabla^2\Phi = 0$ for the velocity potential $\Phi(R, \phi, z)$ in cylindrical polar coordinates (R, ϕ, z) is

$$\frac{1}{R}\frac{\partial}{\partial R}\left(R\frac{\partial\Phi}{\partial R}\right) + \frac{1}{R^2}\frac{\partial^2\Phi}{\partial\phi^2} + \frac{\partial^2\Phi}{\partial z^2} = 0. \tag{4.116}$$

If the flow is two-dimensional and the coordinate axes are so chosen that all physical quantities associated with the fluid are independent of z then $\Phi = \Phi(R, \phi)$ and (4.116) simplifies to

$$\frac{1}{R}\frac{\partial}{\partial R}\left(R\frac{\partial\Phi}{\partial R}\right) + \frac{1}{R^2}\frac{\partial^2\Phi}{\partial\phi^2} = 0. \tag{4.117}$$

Let us seek solutions of (4.117) by putting

$$\Phi(R, \phi) = f(R)\,g(\phi) \tag{4.118}$$

into it for separation of variables. Then

$$g(\phi)\frac{1}{R}\frac{\mathrm{d}}{\mathrm{d}R}[R\,f'(R)] + \frac{1}{R^2}f(R)g''(\phi) = 0,$$

i.e.

$$\frac{R\,\mathrm{d}/\mathrm{d}R[R\,f'(R)]}{f(R)} = -\frac{g''(\phi)}{g(\phi)}. \tag{4.119}$$

The left-hand side of (4.119) is a function of R only, the right-hand one of ϕ only. As R and ϕ are independent variables, each side of (4.119) is a constant λ, say. Thus the decomposition process has produced the two ordinary differential equations

$$R^2 f''(R) + R f'(R) - \lambda f(R) = 0, \tag{4.120}$$

$$g''(\phi) + \lambda g(\phi) = 0. \tag{4.121}$$

Equation (4.121), the simpler of the two, has periodic solutions when $\lambda > 0$. Normally the physical problem requires that $g(\phi + 2\pi) = g(\phi)$ and this is fulfilled when $\lambda = n^2$ for $n = 1, 2, 3, \ldots$ Then basic solutions of (4.121) are

$$g(\phi) = \begin{matrix}\cos\\ \sin\end{matrix}\, n\phi \qquad (n = 1, 2, 3, \ldots). \tag{4.122}$$

Equation (4.120) is of the Euler homogeneous type and it is reduced to a linear differential equation of constant coefficients by putting $R = e^t$ to give

$$d^2f/dt^2 - n^2 f = 0$$

with basic solutions

$$f = \exp(\pm nt) = R^{\pm n} \qquad (n = 1, 2, \ldots). \tag{4.123}$$

A special solution to (4.117) is obtained by linear superposition of the forms (4.122), (4.123) to give

$$\Phi(R, \phi) = f(R)\, g(\phi) = (A_n R^n + B_n R^{-n})\,(C_n \cos n\phi + D_n \sin n\phi). \tag{4.124}$$

The most general solution of this form is

$$\Phi(R, \phi) = \sum_{n=1}^{\infty} (A_n R^n + B_n R^{-n})\,(C_n \cos n\phi + D_n \sin n\phi) \tag{4.125}$$

since Laplace's equation is linear.

In the special case $n = 0$, rejected in the above development,

$$f = k_1 + k_2 t = k_1 + k_2 \ln R,$$

$$g = k_3 + k_4 \phi$$

so that another solution of (4.115) is

$$\Phi(R, \phi) = (k_1 + k_2 \ln R)\,(k_3 + k_4 \phi). \tag{4.126}$$

If in a physical problem leading to the solution of (4.116) the axis $R = 0$ is excluded from the region of flow, it may be necessary to combine by addition the forms (4.125) and (4.126) but one must choose the constants so that $\Phi(R, \phi) = \Phi(R, \phi + 2\pi)$.

The particular case $n = 1$ yields the important special solutions

$$\Phi = R\cos\phi, \quad \Phi = R\sin\phi, \quad \Phi = R^{-1}\cos\phi, \quad \Phi = R^{-1}\sin\phi. \tag{4.127}$$

We now illustrate the use of these results.

Example [4.10]

Uniform flow past an infinitely long circular cylinder.

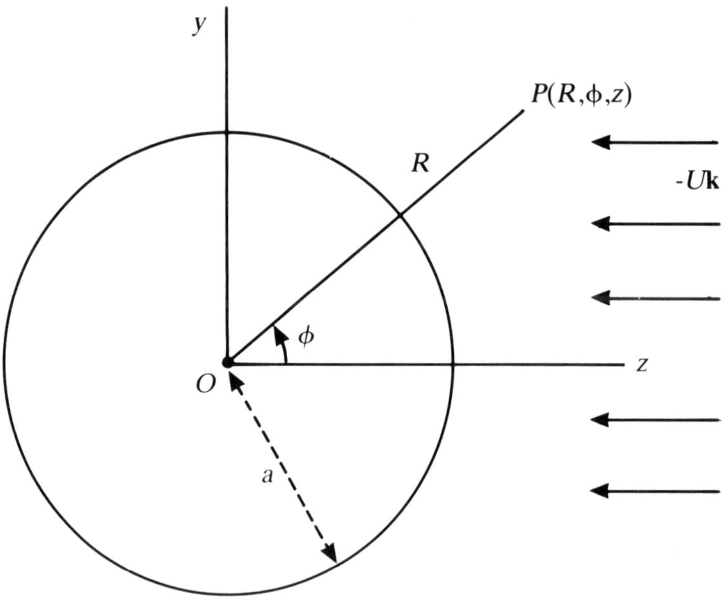

Fig. 4.16

Let P be a point with cylindrical polar coordinates (R, ϕ, z) in the flow region of an unbounded incompressible fluid of uniform density moving irrotationally with uniform velocity $-U\mathbf{i}$ at infinity past the fixed solid cylinder $R \leqslant a$ (Fig. 4.16). The presence of the cylinder perturbs the uniform stream but the perturbation decays to zero with increasing distance R from the axis of the cylinder. This suggests taking the velocity potential for $R > a, 0 \leqslant \phi \leqslant 2\pi$ in the form

$$\Phi(R, \phi) = -UR \cos \phi + AR^{-1} \cos \phi.$$

This form is adopted because the velocity potential of the uniform stream is $-Ux = -UR \cos \phi$ and the perturbation potential due to the presence of the cylinder is taken as $AR^{-1} \cos \phi$ because this function (i) satisfies Laplace's equation, (ii) decays to zero as $R \to \infty$ and (iii) gives rise to a velocity pattern which is symmetric about $\phi = 0, \pi$. Now $R^{-1} \sin \phi$ satifies (i) and (ii) but not (iii).

The kinematic condition on the surface of the cylinder requires that $\partial\Phi/\partial R = 0$, when $R = a$. We obtain at once $A = -Ua^2$, for all ϕ satisfying $0 \leqslant \phi \leqslant 2\pi$. Hence the solution obtained of the Neumann boundary value problem is

$$\Phi(R, \phi) = -U \cos \phi \, (R + a^2 R^{-1}), \qquad R > a, 0 \leqslant \phi \leqslant 2\pi.$$

This solution is unique to within an arbitrary immaterial additive constant and so $\mathbf{q} = \nabla\Phi$ at all $P(R, \phi, z)$ of the fluid. Since in the cylindrical polar coordinate system $\mathrm{d}\mathbf{s} = \mathrm{d}R\ \hat{\mathbf{R}} + R\ \mathrm{d}\phi\,\hat{\boldsymbol{\phi}} + \mathrm{d}z\,\mathbf{k}$, the appropriate scalar velocity components in cylindrical polars are:

$$\begin{aligned} q_R &= \partial\Phi/\partial R &&= -U\cos\phi\,\{1-(a/R)^2\},\\ q_\phi &= \partial\Phi/R\partial\phi &&= U\sin\phi\,\{1+(a/R)^2\},\\ q_z &= \partial\Phi/\partial z &&= 0. \end{aligned}$$

These hold for $R \geqslant a$, $0 \leqslant \phi \leqslant 2\pi$. We note that as $R \to \infty$, $q_R \to -U\cos\phi$, $q_\phi \to U\sin\phi$ which results are consistent with a velocity at infinity of $-U\mathbf{i}$, as is easily verified.

It may have occured to the reader that in place of the perturbation velocity potential $AR^{-1}\cos\phi$, we could have taken the more general form

$$\sum_{n=1}^{N} A_n R^{-n}\cos n\phi,$$

where N could be either a finite positive integer or infinity. This more general form certainly satisfies the requisites (i), (ii), (iii) above. However, the uniqueness theorem shows that there is only one solution and nothing is gained by adding the additional terms. It is easy to show that the more general form simply gives $A_2 = A_3 = \ldots = 0$.

Example [4.11]

A cylinder of infinite length and nearly circular section moves through an infinite volume of liquid with velocity U at right-angles to its axis and in the direction of the positive x-axis. If its section is specified by the equation

$$R = a(1 + \epsilon\cos n\phi),$$

where n is a positive integer and ϵ is small, show that the approximate value of the velocity potential of the fluid is

$$-Ua\left\{\frac{a}{R}\cos\phi + \epsilon\left(\frac{a}{R}\right)^{n+1}\cos(n+1)\phi - \epsilon\left(\frac{a}{R}\right)^{n-1}\cos(n-1)\phi\right\}.$$

Fig. 4.17 illustrates the plane of flow of the cylinder. The tangent to a point P on it makes angles α, $(\pi - \alpha)$ with the radial line OP drawn from O. At large radial distances R from OZ, the fluid velocity becomes vanishingly small. Thus suitable harmonic functions for constructing the velocity potential $\Phi(R, \phi)$ of the fluid have the forms $R^{-k}\ {\cos \atop \sin}\ k\phi$ $(k = 1, 2, \ldots)$. We thus seek a solution of the form

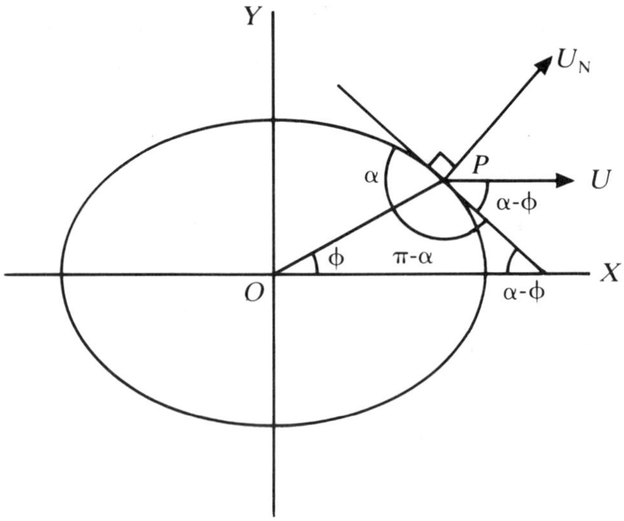

Fig. 4.17

$$\Phi(R, \phi) = \sum_{k=1}^{\infty} R^{-k} (A_k \cos k\phi + B_k \sin k\phi).$$

Were the summation started at $k = 0$ instead of $k = 1$ this would only add on to Φ an arbitrary constant A_0. At $\phi = 0$ and $\phi = \pi$ on the boundary, $q_\phi = 0$ which is satisfied by taking $B_k = 0$ $(k = 1, 2, \ldots)$. The velocity potential now simplifies to the form

$$\Phi(R, \phi) = \sum_{k=1}^{\infty} A_k R^{-k} \cos k\phi,$$

which appropriately remains unaltered on replacing ϕ by $(2\pi - \phi)$.

At any point (R, ϕ, z) of the fluid, the cylindrical polar velocity components are

$$q_R = \frac{\partial \Phi}{\partial R} = -\sum_{k=1}^{\infty} k A_k R^{-(k+1)} \cos k\phi,$$

$$q_\phi = \frac{\partial \Phi}{R \partial \phi} = -\sum_{k=1}^{\infty} k A_k R^{-(k+1)} \sin k\phi,$$

$$q_z = 0.$$

At P on the boundary, since $(\pi - \alpha)$ is the angle between the tangent and OP,

$$\cot(\pi - \alpha) = \frac{1}{R} \frac{dR}{d\phi} = \frac{d}{d\phi} \ln(1 + \epsilon \cos n\phi).$$

Thus

$$\cot \alpha = \frac{\epsilon\, n \sin n\phi}{1 + \epsilon \cos n\phi}.$$

The normal component of velocity U_N of the boundary at P is

$$\begin{aligned} U_N &= U \sin(\alpha - \phi) = U(\sin \alpha \cos \phi - \cos \alpha \sin \phi) \\ &= \frac{U\{\cos \phi\,(1 + \epsilon \cos n\phi) - \sin \phi \,.\, \epsilon n \sin n\phi\}}{\sqrt{\{(1 + \epsilon \cos n\phi)^2 + \epsilon^2 n^2 \sin^2 n\phi\}}}. \end{aligned}$$

As there is no transport of fluid across the surface and no breakaway from it U_N is also the normal velocity component of the fluid. Thus

$$U_N = q_R \sin \alpha + q_\phi \cos \alpha$$

$$= \frac{\sum_{k=1}^{\infty} kA_k a^{-(k+1)}(1 + \epsilon \cos n\phi)^{-(k+1)}\{\cos k\phi\,(1 + \epsilon \cos n\phi) + \sin k\phi \,.\, \epsilon n \sin n\phi\}}{\sqrt{\{(1 + \epsilon \cos n\phi)^2 + \epsilon^2 n^2 \sin^2 n\phi\}}}.$$

Equating the two forms for U_N gives, on simplification,

$$\sum_{k=1}^{\infty} kA_k a^{-(k+1)}\,(1 + \epsilon \cos n\phi)^{-(k+1)}\{\cos k\phi\,(1 + \epsilon \cos n\phi) + \epsilon n \sin k\phi \sin n\phi\}$$

$$= U\{\cos \phi\,(1 + \epsilon \cos n\phi) - \epsilon n \sin \phi \sin n\phi\}.$$

Correct to the first order in ϵ this gives the approximation

$$\sum_{k=1}^{\infty} kA_k a^{-(k+1)}\Big[\cos k\phi + \tfrac{1}{2}\epsilon\{(n-k)\cos(n-k)\phi + (n+k)\cos(n+k)\phi\}\Big]$$

$$\begin{aligned} &\simeq U\Big[\cos \phi + \tfrac{1}{2}\epsilon\{\cos(n+1)\phi + \cos(n-1)\phi\} \\ &\quad - \tfrac{1}{2}\epsilon n\{\cos(n-1)\phi - \cos(n+1)\phi\}\Big]. \end{aligned}$$

Equating coefficients of $\cos \phi$,

$$U = A_1/a^2.$$

Equating coefficients of $\cos(n-1)\phi$ and of $\cos(n+1)\phi$,

$$\begin{aligned} (n-1)A_{n-1}\, a^{-n} + \tfrac{1}{2}A_1 a^{-2}\, \epsilon(n-1) &= -\tfrac{1}{2}U\, \epsilon(n-1), \\ (n+1)A_{n+1}\, a^{-(n+2)} - \tfrac{1}{2}A_1 a^{-2}\, \epsilon(n+1) &= \tfrac{1}{2}U\, \epsilon(n+1). \end{aligned}$$

Thus

$$A_{n-1} = -\epsilon U a^n, \quad A_{n+1} = \epsilon U a^{n+2}.$$

All A_k other than A_1, A_{n-1}, A_{n+2} are zero. Substitution of the determined values of the three non-zero coefficients gives the stated result.

PROBLEMS 4

1. Define irrotational motion and prove that under certain conditions the motion of a frictionless liquid, if once irrotational, is always so. Prove that this theorem remains true, if the motion of each particle be resisted by a force varying as its absolute velocity.

2. If Φ is constant over the boundary of any simply connected region occupied by liquid in irrotational motion, prove that Φ has the same constant value throughout the interior.

3. Prove that, if the normal velocity is zero at every point of the boundary of liquid occupying a simply connected region, and moving irrotationally, Φ is constant throughout the interior of that region.

4. Liquid moving irrotationally occupies a simply connected region bounded partly by surfaces over which Φ is constant, and partly by surfaces over which the normal velocity is zero. Prove that Φ has the same constant value throughout the region.

5. A body moves in a given manner, without change of volume, in an inviscid liquid. T_0 denotes the kinetic energy of the fluid when it has no external boundary and is at rest at infinite distances; T_0' denotes the kinetic energy of that part of the fluid which is outside a closed surface S_0 which is external to the body; T denotes the kinetic energy of the fluid when S_0 is its external boundary and is fixed. Prove that, if the regions occupied by the fluid are simply connected,

$$T > T_0 + T_0'.$$

6. Incompressible liquid is in irrotational motion with velocity potential φ. If a portion of the liquid completely occupying the space $r_1 \leqslant r \leqslant r_2$ between concentric spherical surfaces contains no sources or sinks, show that the kinetic energy of the liquid in this region is

$$\tfrac{1}{2}\rho \int_{r=r_2} \varphi \frac{\partial \varphi}{\partial r}\, \mathrm{d}S - \tfrac{1}{2}\rho \int_{r=r_1} \varphi \frac{\partial \varphi}{\partial r}\, \mathrm{d}S.$$

An infinite liquid contains a solid sphere of mass M of the same density as the liquid, and an impulse I is imparted to the sphere. Show that it begins to move with speed $(2I/3M)$. Show that the kinetic energy of the liquid inside the spherical surface of radius $2^{1/3}a$ concentric with the sphere is equal to one half the total kinetic energy of the liquid.

7. The space between a uniform rigid sphere, of radius a and density σ, and a fixed concentric spherical envelope of radius b is filled with inviscid liquid of constant density ρ. An impulse is applied to the rigid sphere and the system is set in motion from rest. Show that, just after the impulse, the kinetic energy of the liquid and sphere is

$$\tfrac{1}{3}\pi a^3 \rho U^2 \left(\frac{2\sigma}{\rho} + \frac{b^3 + 2a^3}{b^3 - a^3}\right),$$

where U is the velocity of the sphere. Hence or otherwise, determine the impulse.

8. The space between two concentric spherical shells of radii a, b $(a < b)$ is filled with liquid of density ρ. If the shells are set in motion, the inner one with velocity U in the x-direction and the outer with velocity V in the y-direction, show that the initial motion of the liquid is given by the velocity potential

$$\varphi = \frac{\{a^3 U(1 + \tfrac{1}{2}b^3 r^{-3})x - b^3 V(1 + \tfrac{1}{2}a^3 r^{-3})y\}}{b^3 - a^3}$$

where $r^2 = x^2 + y^2 + z^2$, the coordinates being rectangular.

Evaluate the velocity at any point of the liquid and hence, or otherwise, prove that the total momentum communicated has components

$$[\tfrac{4}{3}\pi \rho a^3 U, \tfrac{4}{3}\pi \rho b^3 V, 0].$$

9. The motion of a source-free incompressible fluid of uniform density ρ is irrotational. Prove that the kinetic energy in the singly-connected volume bounded by the surface or surfaces S is $-\tfrac{1}{2}\rho \iint_S \varphi \dfrac{\partial \varphi}{\partial n}\, \mathrm{d}S$, where φ is the velocity potential and n is the normal at the surface drawn into the volume considered.

A sphere of radius a moves through a fluid whose boundaries are fixed at a large distance from the sphere. Show that the kinetic energy of the fluid is $\tfrac{1}{3}\pi a^3 \rho u^3$ at the instant when the sphere has speed u.

Hence or otherwise, show that if the sphere is uniform and of density $\tfrac{1}{2}\rho$, it will rise with an acceleration $\tfrac{1}{2}g$ when released from rest in such a fluid, the system being subject to gravity and the acceleration due to gravity being g.

10. The velocity potential for the motion due to a sphere of radius a moving with variable velocity U in the direction $\theta = 0$ in an infinite uniform inviscid incompressible fluid of density ρ is $\varphi = -(Ua^3/2r^2)\cos\theta$. Assuming a suitable form of Bernoulli's equation, deduce that the pressure at a point on the sphere is given by

$$\frac{p}{\rho} = \frac{p_0}{\rho} + gZ - \tfrac{1}{8}U^2(5 - 9\cos^2\theta) + \tfrac{1}{2}\dot{U}a\cos\theta,$$

when the only body force is gravity and Z is the vertical coordinate measured *downwards* from a fixed origin.

A uniform sphere of mass M, immersed in such a fluid at rest at infinity, is released from rest. Show that the pressure on its horizontal equator will fall below its initial value if $M > 7M'$ where M', which is given to be less than M, is the mass of fluid displaced.

11. A stream of water at great depth flows with constant velocity V over a plane level bottom. A solid hemisphere of radius a is fixed with its base covering a circular area of the bottom. Show that the downward force exerted by the water on the rest of the bottom is thereby increased by $\frac{11}{32}\pi\rho V^2a^2$.

12. The space between two spheres is filled with incompressible fluid. The spheres have radii a, b, $(a < b)$ and move with constant speeds U, V respectively along the line of centres. Show that the instant when the spheres are concentric, the velocity potential is given by

$$\varphi = \frac{\left\{(a^3U - b^3V)r + \tfrac{1}{2}(U - V)a^3b^3r^{-2}\right\}\cos\theta}{b^3 - a^3}$$

and that the equation of the streamlines takes the form

$$\left\{(b^3V - a^3U)r^2 - (V - U)a^3b^3r^{-1}\right\}\sin^2\theta = \text{const.},\quad \psi = \text{const.},$$

where (r, θ, ψ) are spherical polar coordinates with origin at the common centre of the spheres and axis $\theta = 0$ along their line of motion.

13. A solid sphere moves through quiescent frictionless liquid whose boundaries are at a distance from it great compared with its radius. Prove that at each instant the motion in the liquid depends only on the position and velocity of the sphere at that instant. Prove that the liquid streams past the sides of the sphere with half the velocity of the sphere.

14. An infinite ocean of an incompressible perfect liquid of density ρ is streaming past a fixed spherical obstacle of radius a. The velocity is uniform and equal to U except in so far as it is disturbed by the sphere, and the pressure in the liquid at a great distance from the obstacle is Π. Show that the thrust on that half of the sphere on which the liquid impinges is

$$\pi a^2\left\{\Pi - \rho U^2/16\right\}.$$

15. A rigid sphere of radius a is moving in a straight line with velocity u and acceleration f through an infinite incompressible liquid, prove that the resultant fluid pressures over the two hemispheres into which the sphere is divided by a

diametral plane perpendicular to its direction of motion are

$$\Pi\pi a^2 \pm \tfrac{1}{4}Mf - \tfrac{3}{64}Mu^2/a,$$

where Π is the pressure at a great distance, and M is the mass of the fluid displaced by the sphere.

16. A solid sphere is moving through frictionless liquid: compare the velocities of slip of the liquid past it at different parts of its surface.

Prove that when the sphere is in motion with uniform velocity U, the pressure at the part of its surface where the radius makes an angle θ with the direction of motion is increased on account of the motion by the amount

$$\tfrac{1}{16}\rho U^2\,(9\cos 2\theta - 1),$$

where ρ is the density of the liquid.

17. Find the pressure at any point of a liquid, of infinite extent and at rest at a great distance, through which a sphere is moving under no external forces with constant velocity U, and show that the mean pressure over the sphere is in defect of the pressure Π at a great distance of $\tfrac{1}{4}\rho U^2$, it being supposed that Π is sufficiently large for the pressure everywhere to be positive, that is, that $\Pi > \tfrac{5}{8}\rho U^2$.

18. An infinite homogeneous liquid is flowing steadily past a rigid boundary consisting partly of the horizontal plane $y = 0$, and partly of a hemispherical boss $x^2 + y^2 + z^2 = a^2$, with irrotational motion which tends, at a great distance from the origin, to uniform velocity V parallel to the axis of z. Find the velocity potential and the surfaces of equal pressure.

19. A stream of water of great depth is flowing with uniform velocity V over a plane level bottom. A hemisphere of weight w in water and of radius a, rest with its base on the bottom. Prove that the average pressure between the base of the hemisphere and the bottom is less than the fluid pressure at any point of the bottom at a great distance from the hemisphere, if

$$V^2 > 32w/11\pi a^2\rho.$$

20. Prove that at a point on a sphere moving through an infinite liquid the pressure is given by the formula

$$(p - p_0)/\rho = \tfrac{1}{2}af\cos\theta_1 + \tfrac{1}{8}v^2\,(9\cos^2\theta - 5),$$

where v is the velocity, f the acceleration of the sphere, and θ, θ_1 are the angles between the radius and the directions of v, f respectively, and p_0 is the hydrostatic pressure.

21. When a sphere of radius a moves in an infinite liquid show that the pressure at any point exceeds what would be the pressure if the sphere were at rest by

$$\frac{a^3}{2r^2}f - \frac{a^3}{8r^6}(4r^3 + a^3)q^2 + \frac{3}{8}\frac{a^3}{r^6}(4r^3 - a^3)q'^2,$$

where q is the velocity of the sphere and q' and f are the resolved parts of its velocity and acceleration in the direction of r and the density of the liquid is unity.

22. Prove that for liquid contained between two instantaneously concentric spheres, when the outer (radius a) is moving parallel to the axis of x with velocity u and the inner (radius b) is moving parallel to the axis of y with velocity v, the velocity potential is

$$-\frac{1}{a^3 - b^3}\left\{a^3 ux\left(1 + \frac{b^3}{2r^3}\right) - b^3 vy\left(1 + \frac{a^3}{2r^3}\right)\right\},$$

and find the kinetic energy.

23. Liquid density ρ fills the space between a solid sphere of radius a and density ρ' and a fixed concentric spherical envelope of radius b; prove that the work done by an impulse which starts the solid sphere with velocity V is

$$\tfrac{1}{3}\pi a^3 V^2\left(2\rho' + \frac{2a^3 + b^3}{b^3 - a^3}\rho\right).$$

24. The space between two concentric spherical shells of radii a and b $(a > b)$ is filled with an incompressible fluid of density ρ and the shells suddenly begin to move with velocities U, V in the same direction: prove that the resultant impulsive pressure on the inner shell is

$$\frac{2\pi\rho b^3}{3(a^3 - b^3)}\{3a^3 U - (a^3 + 2b^3)V\}.$$

25. Incompressible fluid, of density ρ, is contained between two rigid concentric spherical surfaces, the outer one of mass M_1 and radius a, the inner one of mass M_2 and radius b. A normal blow P is given to the outer surface. Prove that the initial velocities of the two containing surfaces (U for the outer and V for the inner) are given by the equations

$$\left\{M_1 + \frac{2\pi\rho a^3(2a^3 + b^3)}{3(a^3 - b^3)}\right\}U - \frac{2\pi\rho a^3 b^3}{a^3 - b^3}V = P,$$

$$\left\{M_2 + \frac{2\pi\rho b^3 (2b^3 + a^3)}{3(a^3 - b^3)}\right\} V = \frac{2\pi\rho a^3 b^3}{a^3 - b^3} U.$$

26. The motion of an incompressible fluid, being symmetrical with respect to an axis, and the parts of the velocity resolved along and perpendicularly to a radius vector drawn from a point fixed or moving on the axis in any direction making with the axis an angle θ being U and W, prove that if

$$U = \frac{2C}{r^3} \cos\theta + \frac{C'}{4r^4} (1 + 3\cos 2\theta), \ W = \frac{C}{r^3} \sin\theta + \frac{C'}{2r^4} \sin 2\theta,$$

the equation of constancy of mass is satisfied, and $U dr + W r d\theta$ is an exact differential, C and C' being either constants or functions of the time.

Show also that if the fluid be unlimited in extent, and $C' = 0$, the assumed motion would be produced by a sphere moving in any manner with its centre on a fixed straight line.

27. Prove, or verify, that the velocity function

$$\phi = U\left(r + \frac{a^2}{r}\right) \cos\theta$$

represents a streaming motion past a fixed circular cylinder.

The pressure at infinity being given, calculate the resultant fluid action per unit length on half the cylinder lying on one side of a plane through the axis and parallel to the stream.

28. Liquid flows steadily and irrotationally in two dimensions in a space with fixed boundaries, the cross-section of which consists of the two lines $\theta = \pm\,\pi/10$ and the curve $r^5 \cos 5\theta = k^5$.

Prove that, if V is the speed of the liquid in contact with one of the plane boundaries at unit distance from their intersection, the volume of liquid which passes in unit time through a circular ring in the plane $\theta = 0$ is

$$\tfrac{1}{8}\pi V a^2 (a^4 + 12a^2c^2 + 8c^4),$$

where a is the radius of the ring and c the distance of its centre from the intersection of the planes of the boundaries.

5

The stream functions

5.1 TWO-DIMENSIONAL FLOW

In the preceding chapter, the velocity potential Φ was defined in such a way that the condition for irrotationality of the fluid velocity $\mathbf{q}$ was automatically satisfied. In satisfying the equation of continuity, the differential equation which Φ must satisfy was then obtained. In the case of an incompressible fluid, this equation is Laplace's equation $\nabla^2 \Phi = 0$.

When the motion is two-dimensional, as defined in Section 4.21, a second scalar function may be defined by a complementary procedure and this is possible for both rotational and irrotational flows. This function is definable for (a) both steady and unsteady motion of incompressible fluid, and (b) steady motion of compressible fluid. The function is called the *stream function*.

Consider first a flow of incompressible fluid with velocity components $(u, v, 0)$ in a Cartesian system with $u = u(x, y, t)$, $v = v(x, y, t)$. The equation of continuity is

$$\operatorname{div} \mathbf{q} = \frac{\partial u}{\partial x} + \frac{\partial v}{\partial y} = 0. \tag{5.1}$$

Equation (5.1) is satisfied automatically by defining a differentiable function $\psi(x, y)$ such that

$$u = \frac{\partial \psi}{\partial y}, \quad v = -\frac{\partial \psi}{\partial x}. \tag{5.2}$$

The function $\psi(x, y, t)$ is the *stream function* for the flow. The function is given this name because the *streamlines* for the flow are in fact the set of curves $\psi(x, y, t) = f(t)$ with $f(t)$ arbitrary. To prove this, we note that the differential equation of the streamlines is

$$\frac{dx}{u} = \frac{dy}{v}, \tag{5.3}$$

and when (5.2) is substituted into (5.3) we obtain

$$\frac{\partial \psi}{\partial x}\,dx + \frac{\partial \psi}{\partial y}\,dy = 0. \tag{5.4}$$

But the left-hand side of (5.4) is the exact differential $d\psi$. Thus along a streamline $d\psi = 0$, showing that $\psi = f(t)$ along a streamline, and each choice of $f(t)$ defines a different streamline. For steady motion, the function $f(t)$ is a constant.

Another property of the stream function ψ is that the difference in its values taken on two streamlines measures the volume of fluid which flows between these two streamlines per unit depth in unit time. This can be easily established by considering two streamlines corresponding to $\psi = \psi_1$ and $\psi = \psi_2$, as shown in Fig. 5.1, where A, B are any two points on $\psi = \psi_1$ and $\psi = \psi_2$ respectively while $\mathscr{C}$ is any curve drawn in the plane of flow joining A to B. The unit vectors **t** and **n** are respectively tangential and normal to $\mathscr{C}$ at any general point, the direction of **t** chosen so that the curve $\mathscr{C}$ is described positively from A to B and the triad **n**, **t**, **k** right-handedly related in that order, with **k** normal to the plane of flow.

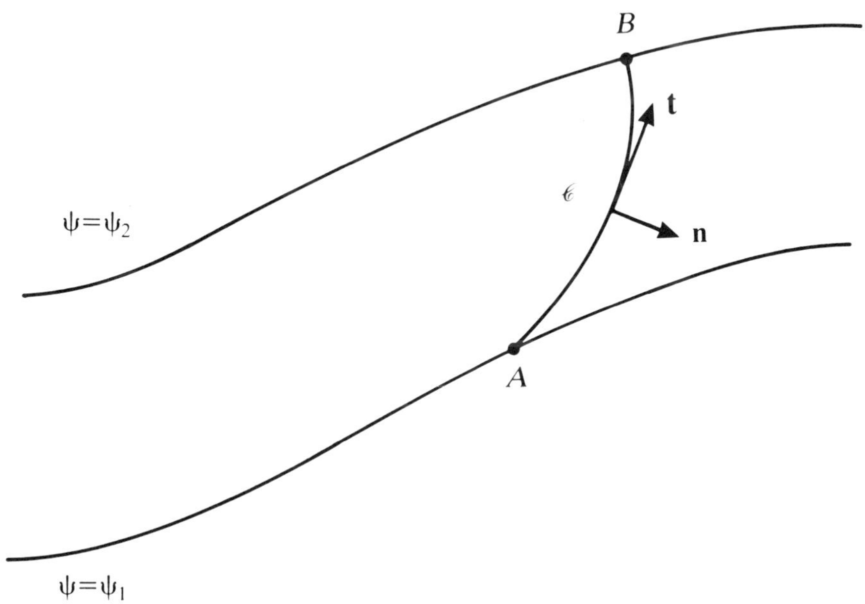

Fig. 5.1

The rate of flow of fluid across $\mathscr{C}$ per unit depth of fluid is

$$Q = \int_A^B (\mathbf{q}\,.\,\mathbf{n})\,\mathrm{d}s = \int_A^B \mathbf{q}\,.\,(\mathrm{d}\mathbf{r}\times\mathbf{k})$$

$$= \int_A^B (u\,\mathrm{d}y - v)\,\mathrm{d}x = \int_A^B \mathrm{d}\psi, \tag{5.5}$$

after substituting from (5.2). Thus

$$Q = \psi(B) - \psi(A). \tag{5.6}$$

Conversely, if A and B are any two points in the flow region and $\mathscr{C}$ is any curve in the plane of flow joining these points, it follows that $\psi(B) - \psi(A)$ measures the amount of fluid per unit depth which flows across $\mathscr{C}$ in unit time at a given instant.

The foregoing results apply to *any* two-dimensional incompressible flow at any time t and at any point where the velocity field is defined. If, however, the flow is irrotational, curl $\mathbf{q} = \mathbf{0}$ and consequently

$$\frac{\partial v}{\partial x} - \frac{\partial u}{\partial y} = -\left(\frac{\partial^2\psi}{\partial x^2} + \frac{\partial^2\psi}{\partial y^2}\right) = 0. \tag{5.7}$$

For two-dimensional irrotational flow we shall denote the velocity potential by ϕ. Equation (5.7) thus shows that for such a flow, both ϕ and ψ satisfy the two-dimensional Laplace equation. Furthermore, since both u and v are expressible in terms of either ϕ or ψ, it follows that

$$\frac{\partial\phi}{\partial x} = \frac{\partial\psi}{\partial y}, \quad \frac{\partial\phi}{\partial y} = -\frac{\partial\psi}{\partial x}, \tag{5.8}$$

showing that

$$\frac{\partial\phi}{\partial x}\frac{\partial\psi}{\partial x} + \frac{\partial\phi}{\partial y}\frac{\partial\psi}{\partial y} = 0, \tag{5.9}$$

or equivalently

$$\nabla\phi\,.\,\nabla\psi = 0. \tag{5.10}$$

This last result expresses the *orthogonality* of the equipotentials ϕ = constant and the streamlines ψ = constant at a general time t.

Although this book is primarily concerned with incompressible flows, it is an easy matter to establish that a stream function can be defined for *steady* two-dimensional flow of a compressible fluid. For such a flow, the equation of continuity (2.31) reduces to

$$\frac{\partial}{\partial x}(\rho u) + \frac{\partial}{\partial y}(\rho v) = 0. \tag{5.11}$$

The stream function ψ is therefore defined so that

$$u = \frac{1}{\rho}\frac{\partial\psi}{\partial y}, \qquad v = -\frac{1}{\rho}\frac{\partial\psi}{\partial x}. \tag{5.12}$$

Let A, B be two points in the flow region and $\mathscr{C}$ be a curve drawn to link A and B. The mass flux across a surface of unit depth and trace $\mathscr{C}$ in a plane of the motion is

$$\begin{aligned} M &= \int_A^B \rho(\mathbf{q}\,.\,\mathbf{n})\,\mathrm{d}s = \int_A^B \rho(u\,\mathrm{d}y - v\,\mathrm{d}x) \\ &= \psi(B) - \psi(A), \end{aligned} \tag{5.13}$$

using (5.12). If $\mathscr{C}$ is part of a streamline, there is no mass flux across $\mathscr{C}$, so (5.13) implies that ψ = constant along a streamline.

5.2 SOME FUNDAMENTAL STREAM FUNCTIONS

In this section we shall again consider some two-dimensional flows which are of a simple form but of considerable importance in fluid mechanics.

Uniform stream

If the stream flows parallel to the x-axis with velocity $U\mathbf{i}$, we have

$$u = \frac{\partial\psi}{\partial y} = U, \qquad v = -\frac{\partial\psi}{\partial x} = 0. \tag{5.14}$$

Integrating the first of these equations gives $\psi = Uy + f(x, t)$, but on substituting this into the second of equations (5.14) we find that f is a function of t only. This arbitrary function of t clearly gives rise to no velocity in the fluid and without loss of generality may be set at zero. The solution of (5.14) is accordingly

$$\psi = Uy = Ur\sin\theta. \tag{5.15}$$

If the uniform stream flows parallel to the y-axis with velocity $U\mathbf{j}$, the stream function is then

$$\psi = -Ux = -Ur\cos\theta. \tag{5.16}$$

By combining (5.15) with (5.16), we may write down the stream function when the stream flows in a direction making an angle α with the positive x-axis. This is given by

$$\psi = Ur\sin(\theta - \alpha).$$

Linear shear flow

In this flow, the fluid moves parallel to an axis, the x-axis for instance, with speed proportional to distance from the axis. Thus

$$u = \lambda y = \frac{\partial \psi}{\partial y}, \quad v = -\frac{\partial \psi}{\partial x} = 0, \tag{5.17}$$

where λ is a constant. Integration of equations (5.17) gives

$$\psi = \tfrac{1}{2}\lambda y^2 = \tfrac{1}{2}\lambda r^2 \sin^2\theta. \tag{5.18}$$

A linear shear flow provides a simple model of the flow of a *real* fluid (when viscosity is not neglected) in the vicinity of a body at rest. The x and y coordinates are then 'local' Cartesian coordinates at a point on the surface of the body. Note that the flow is rotational.

More general shear flows are represented by

$$\mathbf{q} = f(y)\mathbf{i}, \tag{5.19}$$

in which case

$$u = \frac{\partial \psi}{\partial y} = f(y), \quad v = -\frac{\partial \psi}{\partial x} = 0. \tag{5.20}$$

The stream function is then given by

$$\psi = \int f(y)\,\mathrm{d}y. \tag{5.21}$$

A particularly useful example would be

$$f(y) = U_1 + \frac{1}{2}(U_2 - U_1)\,[1 + \tanh(y/a)]$$

where U_1, U_2 and a are constants. The function $f(y)$ is effectively the constant U_1 for $y/a \leqslant -2$ and U_2 for $y/a \geqslant 2$. This shear flow resembles the transition layer between two parallel streams of real fluid.

Linear source and sink

These are the two-dimensional analogues of the three-dimensional simple source and sink described in Section 4.17 of the previous chapter. The flow from an infinite line source is entirely radial along its length and it models approximately a long straight hose, with perforations along its length, of the type commonly used for watering lawns for long periods of time. The velocity is now

$$\mathbf{q} = f(r)\hat{\mathbf{r}}, \tag{5.22}$$

with (r, θ) plane polar coordinates in a plane of the flow.

We have up to now given expressions for the Cartesian resolutes of $\mathbf{q}$ in terms of ψ, but it is easy to show that

$$\mathbf{q} = \nabla\psi \times \mathbf{k}, \tag{5.23}$$

with the unit vector **k** perpendicular to the plane of motion and right-handedly related to $\hat{\mathbf{r}}$ and $\hat{\boldsymbol{\theta}}$. Equation (5.23) allows the resolutes of **q** in *any* planar coordinate system to be expressed in terms of ψ. For polar coordinates,

$$\nabla\psi = \frac{\partial\psi}{\partial r}\hat{\mathbf{r}} + \frac{1}{r}\frac{\partial\psi}{\partial\theta}\hat{\boldsymbol{\theta}}, \tag{5.24}$$

and (5.23) gives

$$q_r = \frac{1}{r}\frac{\partial\psi}{\partial\theta}, \qquad q_\theta = -\frac{\partial\psi}{\partial r}. \tag{5.25}$$

Equation (5.22) accordingly gives

$$\psi = \psi(\theta), \quad \psi'(\theta)/r = f(r).$$

Thus $\psi'(\theta) = \text{constant} = A$ and

$$\psi = A\theta, \quad f(r) = A/r, \qquad (0 \leqslant \theta \leqslant 2\pi).$$

The constant A is determined by the rate of flow out of the line source. This is the flux of **q** across any closed plane curve encircling the source. If the amount of fluid emitted by the source in unit time is $2\pi m$ say, then choosing a circle $\mathscr{C}$ of radius r for evaluating the flux integral, we have

$$2\pi m = \oint_{\mathscr{C}} q_r r \,\mathrm{d}\theta = 2\pi A.$$

Thus, a line source of *strength* m per unit length has stream function

$$\psi = m\theta. \tag{5.26}$$

A line sink of strength m is simply a line source of strength $-m$.

Flow in circles

The fluid moves in concentric circles with speed dependent only on distance from the centre. Thus in polar coordinates,

$$q_r = 0, \quad q_\theta = f(r). \tag{5.27}$$

From (5.25),

$$\frac{\partial\psi}{\partial\theta} = 0, \quad \frac{\partial\psi}{\partial r} = -f(r),$$

giving

$$\psi = -\int f(r)\,\mathrm{d}r. \tag{5.28}$$

Consider the case when $f(r) = Ar$, where A is a constant. The reader familiar with rigid body mechanics will realise that in this flow, the fluid rotates about the axis $r = 0$ *as if it were rigid* with angular velocity A. We say that this flow is a *rigid body motion*, as of course the uniform streaming motion is also. A real fluid contained within a circular cylinder which rotates about its axis can only move steadily in this manner. From (5.28) the stream function for the rigid body rotation

$$q_r = 0, \quad q_\theta = Ar$$

is

$$\psi = -\frac{1}{2}Ar^2. \tag{5.29}$$

Another important flow in circles occurs when $f(r) = A/r$, with A a constant. Now the velocity is singular when $r = 0$, as for a line source, but the flow models that of a real fluid exterior to a rotating circular cylinder in fluid which is otherwise at rest. The singularity along the axis $r = 0$ is called *a line vortex*. A quantity analogous to the strength of a line source is the circulation about the line vortex, which is given by

$$\oint_{\mathscr{C}} \mathbf{q} \cdot \mathrm{d}\mathbf{r} = 2\pi K \tag{5.30}$$

with $\mathscr{C}$ *any* closed curve in the plane of the motion described positively. The fact that K is independent of $\mathscr{C}$ can be appreciated by constructing the closed curve Γ shown in Fig. 5.2. A is a point on a circle γ, radius R, which encloses all points of $\mathscr{C}$, and B is a point on $\mathscr{C}$ lying on the radius to γ through A. The closed curve Γ consists of γ described *clockwise* (i.e. the negative sense), the directed line segment $\overrightarrow{AB}$, the curve $\mathscr{C}$ described counter-clockwise and the directed line segment $\overrightarrow{BA}$.

Now $\mathbf{q} \cdot \mathrm{d}\mathbf{r} = 0$ along $\overrightarrow{AB}$ and $\overrightarrow{BA}$. In addition, on γ and $\mathscr{C}$,

$$\mathbf{q} \cdot \mathrm{d}\mathbf{r} = (\nabla\psi \times \mathbf{k}) \cdot \mathrm{d}\mathbf{r} = (\mathbf{n} \cdot \nabla\psi)\, \mathrm{d}s,$$

with $\mathbf{n}$ directed out of the region of fluid S which is bounded by both curves. On applying the two-dimensional form of the Divergence Theorem, we obtain

$$\int_\Gamma \mathbf{q} \cdot \mathrm{d}\mathbf{r} = \left\{ \oint_{\mathscr{C}} - \oint_\gamma \right\} (\mathbf{n} \cdot \nabla\psi)\, \mathrm{d}s$$

$$= \int_S \nabla^2\psi \;\mathrm{d}S, \tag{5.31}$$

where the line integral notation implies that each curve is described *positively*. But from (5.28),

$$\psi = -A \ln r, \tag{5.32}$$

and therefore

$$\nabla^2\psi = \frac{\mathrm{d}^2\psi}{\mathrm{d}r^2} + \frac{1}{r}\frac{\mathrm{d}\psi}{\mathrm{d}r} = 0, \qquad (r \neq 0).$$

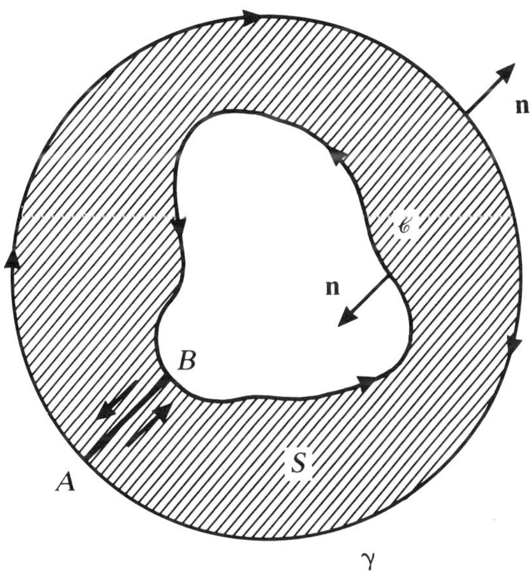

Fig. 5.2

It follows from (5.31) that

$$\oint_{\mathscr{C}} \mathbf{q} \cdot d\mathbf{r} = \oint_{\gamma} \mathbf{q} \cdot d\mathbf{r}, \tag{5.33}$$

with *both* $\mathscr{C}$ and γ described in the positive (counter-clockwise) sense. This establishes that K is independent of the choice of curve $\mathscr{C}$. Evaluation of the line integral around γ leads to

$$2\pi K = A \int_0^{2\pi} d\theta = 2\pi A, \tag{5.34}$$

showing that $A = K$.

When the velocity field is defined by equations (5.27), the vorticity $\boldsymbol{\zeta} = \text{curl}\, \mathbf{q}$ is found to be

$$\boldsymbol{\zeta} = \frac{1}{r} \frac{d}{dr} [r\, f(r)]\, \mathbf{k}.$$

Thus $\boldsymbol{\zeta} = \mathbf{0}$, and the flow irrotational, if and only if $f(r) = A/r$. In other words, the flow in concentric circles is generally rotational for arbitrary choice of $f(r)$, and the one exception is the flow due to a line vortex. The line vortex may be thought of as a mathematical idealisation derived when a vortex tube in which $\boldsymbol{\zeta} \neq \mathbf{0}$ is contracted on to an infinite straight line with the strength of the vortex tube remaining constant and equal to $A = K$. In this limit, there is a line singularity in the vorticity distribution which is everywhere zero when $r \neq 0$.

Other properties and examples of the two-dimensional stream function will be introduced and developed in Chapter 6. In three-dimensional flow, a stream function can be defined when the fluid motion is axisymmetric.

5.3 AXISYMMETRIC FLOW

The introduction of a stream function provides a unified approach to the solving of all, two-dimensional flow problems, both rotational and irrotational, since the determination of the fluid velocity is reduced to determining a *single* scalar function. Unfortunately in the case of general three-dimensional motion this unified method of approach is not possible. There exist, however, a number of classes of three-dimensional flows which can be uniquely characterised by means of a single scalar function. Each of these classes of flow involves a certain mode of symmetry. The most important of these is axisymmetric flow. If (R, ϕ, z) denote cylindrical polar coordinates, the flow is said to be axisymmetric, or axially symmetric, about the z-axis when the velocity components are independent of the azimuthal angle ϕ. In some axisymmetric flows, the azimuthal or 'swirl' component of velocity is zero everywhere. For such flows

$$\partial \mathbf{q}/\partial \phi = 0, \quad \hat{\boldsymbol{\phi}} \cdot \mathbf{q} = 0, \tag{5.35}$$

and the fluid motion is the same in every meridian plane ϕ = constant. It follows that the streamlines lie in meridian planes and the stream surfaces are coaxial surfaces of revolution.

For an incompressible fluid, the equation of continuity for a flow which is symmetric about $R = 0$ is

$$\operatorname{div} \mathbf{q} = \frac{1}{R}\frac{\partial}{\partial R}(R\, q_R) + \frac{\partial q_z}{\partial z} = 0, \tag{5.36}$$

when expressed in cylindrical polar coordinates. This equation is satisfied identically with q_R and q_z given by

$$q_R = \frac{1}{R}\frac{\partial \Psi}{\partial z}, \quad q_z = -\frac{1}{R}\frac{\partial \Psi}{\partial R}, \tag{5.37}$$

where $\Psi(R, z, t)$ is differentiable. This function is called the *Stokes's stream function* and it has properties analogous to ψ for two-dimensional flow.

The equation of the streamlines in a meridian plane ϕ = constant at a fixed time t is

$$\frac{\mathrm{d}R}{q_R} = \frac{\mathrm{d}z}{q_z}, \tag{5.38}$$

which, on substitution from (5.37), gives

$$\frac{\partial \Psi}{\partial R}\,\mathrm{d}R + \frac{\partial \Psi}{\partial z}\,\mathrm{d}z = 0. \tag{5.39}$$

The left-hand side of (5.39) is the exact differential $d\Psi$. Thus at a fixed time the streamlines in a meridian plane ϕ = constant are given by Ψ = constant = C, with each streamline defined by a different value of the constant C.

From (5.37)

$$\mathbf{q} = \frac{1}{R}\left[\hat{\mathbf{R}}\frac{\partial \Psi}{\partial z} - \mathbf{k}\frac{\partial \Psi}{\partial R}\right] = \frac{1}{R}\hat{\boldsymbol{\phi}} \times \nabla \Psi. \tag{5.40}$$

Equation (5.40) is a convenient form for expressing the resolutes of $\mathbf{q}$ in other coordinate systems in which one of the coordinates is the azimuthal angle ϕ. For instance, in spherical polar coordinates (r, θ, ϕ), we have $R = r \sin\theta$, and thus

$$\mathbf{q} = \frac{1}{r\sin\theta}\hat{\boldsymbol{\phi}} \times \left[\hat{\mathbf{r}}\frac{\partial \Psi}{\partial r} + \frac{\hat{\boldsymbol{\theta}}}{r}\frac{\partial \Psi}{\partial \theta}\right]$$

which yields

$$q_r = -\frac{1}{r^2 \sin\theta}\frac{\partial \Psi}{\partial \theta}, \quad q_\theta = \frac{1}{r\sin\theta}\frac{\partial \Psi}{\partial r}. \tag{5.41}$$

More generally, if (ξ, η, ϕ) in that order, constitute a right-handed system of orthogonal curvilinear coordinates of revolution, the line element $d\mathbf{r}$ is given by (1.89) as

$$d\mathbf{r} = h_1\, d\xi\, \hat{\boldsymbol{\xi}} + h_2\, d\eta\, \hat{\boldsymbol{\eta}} + R\, d\phi\, \hat{\boldsymbol{\phi}}$$

and

$$\nabla \psi = \frac{\hat{\boldsymbol{\xi}}}{h_1}\frac{\partial \Psi}{\partial \xi} + \frac{\hat{\boldsymbol{\eta}}}{h_2}\frac{\partial \Psi}{\partial \eta}.$$

Thus (5.40) gives

$$q_\xi = -\frac{1}{Rh_2}\frac{\partial \Psi}{\partial \eta}, \quad q_\eta = \frac{1}{Rh_1}\frac{\partial \Psi}{\partial \xi}. \tag{5.42}$$

The streamlines which pass through the points of a given circle in a plane z = constant and whose centre is on the axis $R = 0$ form a *stream tube*. Each stream tube is a surface of revolution with equation of the form $\Psi = C$, with C a constant. Let A and B be any points in the same azimuthal plane ϕ = constant. The stream tubes passing through A and B have the equations $\Psi = \Psi(A)$, $\Psi = \Psi(B)$ respectively. Let $\mathscr{C}$ be a curve joining the points A and B and let S be the annular surface generated when $\mathscr{C}$ is rotated about the axis of symmetry through 2π radians. The geometrical configuration is sketched in Fig. 5.3.

The volumetric rate of flow of fluid across the surface S is

$$Q = \int_S \mathbf{q} \cdot \mathbf{n}\, dS = \int_{\phi=0}^{2\pi}\left\{\int_A^B (\mathbf{q} \cdot \mathbf{n}) R\, ds\right\} d\phi \tag{5.43}$$

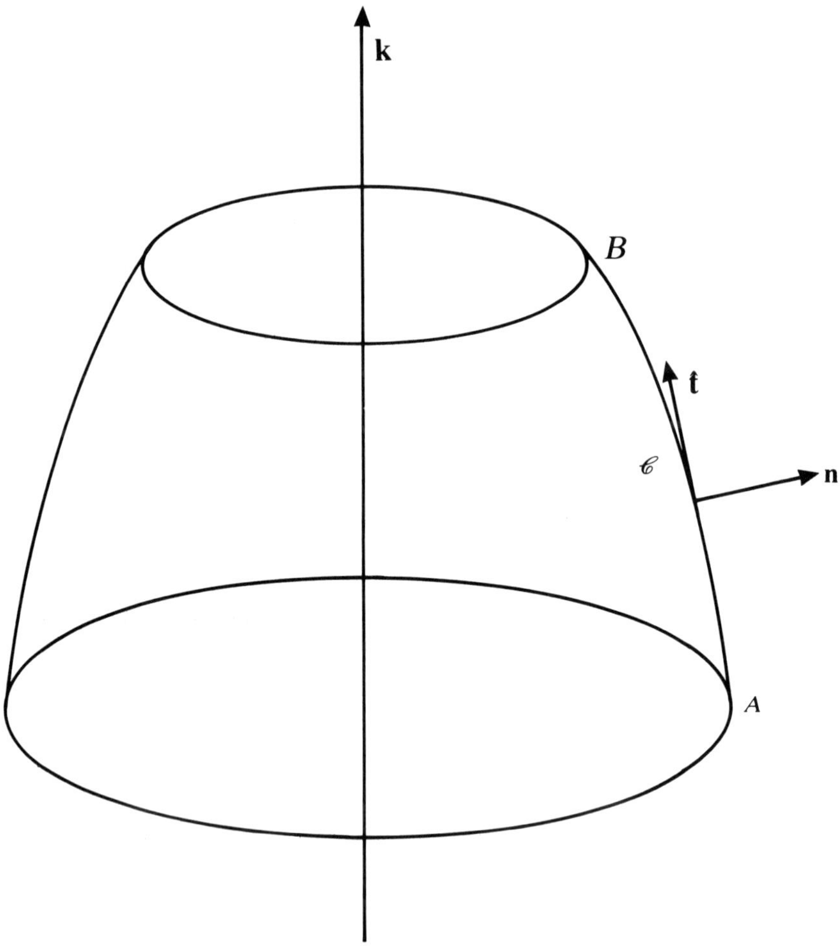

Fig. 5.3

with ds the element of arc length on $\mathscr{C}$ measured from A. The unit vector $\mathbf{n}$ normal to S, and the unit vector $\mathbf{t}$ tangential to $\mathscr{C}$ form a right-handed triad with $\hat{\boldsymbol{\phi}}$ in the order $(\mathbf{t}, \mathbf{n}, \hat{\boldsymbol{\phi}})$. Thus (5.43) can be written as

$$Q = 2\pi \int_A^B R\mathbf{q} \,.\, (\hat{\boldsymbol{\phi}} \times \mathbf{t})\, \mathrm{d}s$$

$$= 2\pi \int_A^B (\hat{\boldsymbol{\phi}} \times \nabla\Psi) \,.\, (\hat{\boldsymbol{\phi}} \times \mathrm{d}\mathbf{r}), \tag{5.44}$$

on using (5.40). Thus

$$Q = 2\pi \int_A^B [(\hat{\phi} \times \nabla\Psi) \times \hat{\phi}] \,.\, d\mathbf{r} = 2\pi \int_A^B \nabla\Psi \,.\, d\mathbf{r}$$

$$= 2\pi \int_A^B d\Psi = 2\pi\,[\Psi(B) - \Psi(A)]\,. \tag{5.45}$$

Equation (5.45) is analogous to (5.6) for two-dimensional flow. Since a constant value of Ψ gives rise to zero velocity, one can choose any particular stream surface to be defined by $\Psi = 0$. Thus $2\pi\,\Psi(P)$ measures the volumetric flow rate between the stream surface through a general point P and the stream surface $\Psi = 0$. The dimensions of Ψ are therefore those of $[\text{length}]^2 \times [\text{velocity}] = L^3T^{-1}$. For the two-dimensional stream function, the dimensions are L^2T^{-1}.

5.4 EQUATION SATISFIED BY STOKES'S STREAM FUNCTION IN IRROTATIONAL FLOW

The results obtained in the preceding section apply to both rotational and irrotational flows of incompressible fluid. When the fluid is irrotational, there is the further condition,

$$\text{curl}\,\mathbf{q} = \hat{\phi}\left[\frac{\partial q_R}{\partial z} - \frac{\partial q_z}{\partial R}\right] = \mathbf{0}. \tag{5.46}$$

After substituting for q_R and q_Z from (5.37) it is found that (5.46) reduces to

$$E^2\Psi \equiv \frac{\partial^2\Psi}{\partial R^2} - \frac{1}{R}\frac{\partial\Psi}{\partial R} + \frac{\partial^2\Psi}{\partial z^2} = 0. \tag{5.47}$$

When the motion is irrotational, a velocity potential Φ exists so that $\mathbf{q} = \nabla\Phi$. Thus

$$\left.\begin{aligned} \frac{\partial\Phi}{\partial R} &= \frac{1}{R}\frac{\partial\Psi}{\partial z}, \\ \frac{\partial\Phi}{\partial z} &= -\frac{1}{R}\frac{\partial\Psi}{\partial R}, \end{aligned}\right\} \tag{5.48}$$

The form of equation (5.47) in general orthogonal coordinates of revolution (ξ, η, ϕ) is found by noting that

$$\zeta = \text{curl}\,\mathbf{q} = \frac{\hat{\phi}}{h_1 h_2}\left[\frac{\partial}{\partial\xi}(h_2\,q_\eta) - \frac{\partial}{\partial\eta}(h_1\,q_\xi)\right],$$

and on substituting for q_ξ and q_η from (5.42), we find that

$$\zeta = R^{-1}\,\hat{\phi}\,E^2\,\Psi, \tag{5.49}$$

where

$$E^2 \equiv \frac{R}{h_1 h_2}\left[\frac{\partial}{\partial\xi}\left(\frac{1}{R}\frac{h_2}{h_1}\frac{\partial}{\partial\xi}\right)+\frac{\partial}{\partial\eta}\left(\frac{1}{R}\frac{h_1}{h_2}\frac{\partial}{\partial\eta}\right)\right]. \tag{5.50}$$

In terms of spherical polar coordinates (r, θ, ϕ), this equation gives

$$E^2 \equiv \frac{\partial^2}{\partial r^2}+\frac{\sin\theta}{r^2}\frac{\partial}{\partial\theta}\left(\frac{1}{\sin\theta}\frac{\partial}{\partial\theta}\right). \tag{5.51}$$

A further result that follows immediately from equation (5.40) is that when the flow is irrotational,

$$\nabla\Phi\,.\,\nabla\Psi = 0,$$

showing that the equipotentials and the stream surfaces intersect orthogonally, which is again analogous to the corresponding property for two-dimensional irrotational flows.

5.5 SOME BASIC STOKES'S STREAM FUNCTIONS

Uniform stream

With $\mathbf{q} = U\mathbf{k}$, the stream function satisfies the equations

$$\frac{\partial\Psi}{\partial R} = UR, \quad \frac{\partial\Psi}{\partial z} = 0,$$

which integrate to give

$$\Psi = \tfrac{1}{2}UR^2. \tag{5.52}$$

The stream surfaces are coaxial cylinders with axes coincident with the z-axis which is the degenerate stream surface $\Psi = 0$. In spherical polar coordinates, (5.52) becomes

$$\Psi = \tfrac{1}{2}Ur^2\sin^2\theta. \tag{5.53}$$

Simple source at origin

We refer to Fig. 4.4 when the flow field is due solely to a simple source at O of strength m. From (4.98) the velocity is given by

$$\mathbf{q} = (m/r^2)\,\hat{\mathbf{r}}, \qquad (r > 0).$$

in spherical polar coordinates (r, θ, ϕ).

The equations satisfied by Ψ are, from (5.41),

$$\frac{\partial \Psi}{\partial \theta} = -m \sin \theta, \qquad \frac{\partial \Psi}{\partial r} = 0,$$

which integrate to give $\Psi = m \cos \theta$. A constant may be added to this solution and this is usually done to make $\Psi = 0$ along the axis of symmetry $\theta = 0$. In which case,

$$\Psi = m (\cos \theta - 1). \tag{5.54}$$

Doublet at origin

Again we assume the flow is due solely to a doublet at O of strength μ. We may take the axis $\theta = 0$ of a system of spherical polar coordinates to coincide with the axis of the doublet. The velocity potential at (r, θ, ϕ) is, from (4.101),

$$\Phi = \frac{\mu \cos \theta}{r^2}, \qquad (r > 0), \tag{5.55}$$

from which we may deduce that $q_\phi = 0$ and

$$q_r = \frac{\partial \Phi}{\partial r} = \frac{-2\mu \cos \theta}{r^3}, \quad q_\theta = \frac{1}{r}\frac{\partial \Phi}{\partial \theta} = \frac{-\mu \sin \theta}{r^3}.$$

The equations satisfied by Ψ are therefore

$$\frac{\partial \Psi}{\partial r} = -\frac{\mu}{r^2} \sin^2 \theta, \qquad \frac{\partial \Psi}{\partial \theta} = 2\mu \frac{\sin \theta \cos \theta}{r}. \tag{5.56}$$

Equations (5.56) give, after integration

$$\Psi = \frac{\mu \sin^2 \theta}{r}. \tag{5.57}$$

Uniform line source

Fig. 5.4 depicts a uniform line source of fluid extending along the straight line segment AB. An element QQ' of length δz at distance $z(= AQ)$ from A is effectively a simple source and its strength may be taken as $m\delta z$ with m the constant source strength per unit length of the distribution along AB. With $QP = r$ and $P\hat{Q}B = \theta$, the stream function $\delta\Psi$ at P for the simple source of strength $m\delta z$ at Q is $m\delta z\ (\cos \theta - 1)$, if we adopt the convention that $\Psi = 0$ on the axis of symmetry through AB. Let l denote the length AB. The value of the stream function Ψ at P due to the *entire* line source AB is accordingly

$$\Psi = m \int_0^l \cos \theta \, \mathrm{d}z - ml$$

$$= m \int_0^l \frac{l + b - z}{\sqrt{\{(l + b - z)^2 + R^2\}}} \, \mathrm{d}z - ml$$

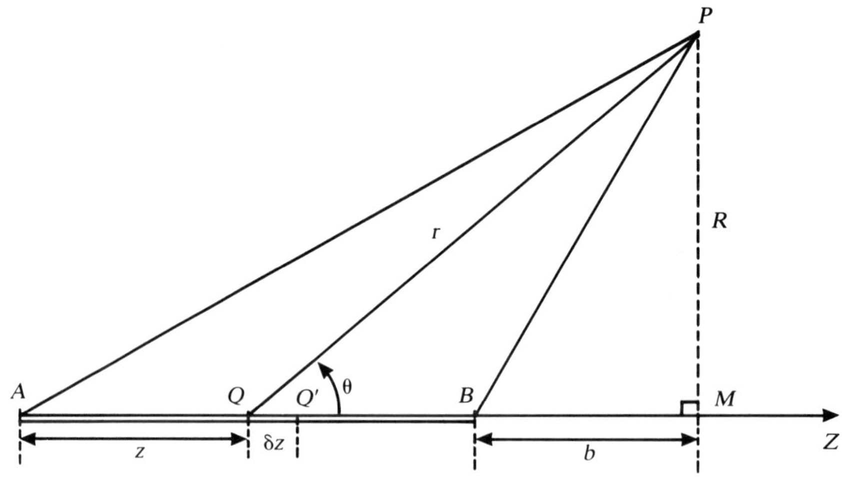

Fig. 5.4

$$\begin{aligned}\psi &= m\left[\sqrt{\{(l+b)^2+R^2\}}-\sqrt{\{l^2+R^2\}}\right]-ml\\ &= m\,(AP-BP-AB).\end{aligned} \tag{5.58}$$

As P is the only variable point in the above, the simpler form $m(AP-BP)$ could be adopted for evaluating velocity components at P. The stream surfaces are $\Psi=$ const. or

$$AP-BP=\text{const.}$$

These are confocal hyperboloids of revolution about AB, with A and B as foci. It was shown earlier (worked example [4.6], Section 4.19) that the equipotentials were confocal ellipsoids of revolution about AB with the same foci. It is well known that the two families of confocals intersect orthogonally.

Example [5.1]

A doublet of vector moment $\mu\mathbf{k}$ is situated at O in a uniform stream whose undisturbed velocity is $-U\mathbf{k}$.

In spherical polar coordinates (r, θ, ϕ), the stream functions for each separate distribution are respectively

$$\Psi_1=\tfrac{1}{2}Ur^2\sin^2\theta,\quad \Psi_2=-(\mu/r)\sin^2\theta.$$

Hence the stream function for the combination is

$$\Psi(r,\theta)=(\tfrac{1}{2}Ur^2-\mu/r)\sin^2\theta.$$

The equations of the stream surfaces are $\Psi(r,\theta)=$ const. In particular the stream surfaces for which $\Psi=0$ are given by

$$\sin\theta=0,$$

or

$$\tfrac{1}{2}Ur^2 - \mu/r = 0,$$

giving $\theta = 0, \pi$, i.e. the z-axis, or $r = (2\mu/U)^{1/3}$, the surface of a sphere centre O and radius $(2\mu/U)^{1/3}$.

Example [5.2]

A point source of strength Ua^2 is introduced at O in a uniform stream whose undisturbed velocity is $U\mathbf{i}$. Show that over the surface of revolution $r \sin\theta = 2a\cos\tfrac{1}{2}\theta$ there is no flow, the system of spherical polar coordinates (r, θ, ϕ) being used with $\theta = 0$ taken as the x-axis. If a rigid blunt-nosed surface of revolution having the above equation is introduced into the flow of velocity $U\mathbf{i}$ after removal of the point source, explain why the two models are hydrodynamically equivalent for corresponding points in the field of flow and determine the pressure distribution as a function of θ on the surface of revolution.

The Ψ-function for the uniform stream is $-\tfrac{1}{2}Ur^2\sin^2\theta$. That for the point source at O is $Ua^2(\cos\theta - 1)$. Thus the total Ψ-function for the combination is

$$\Psi(r, \theta) = -U\{\tfrac{1}{2}r^2\sin^2\theta + a^2(1-\cos\theta)\}$$

for $r > 0$ and $0 \leqslant \theta \leqslant \pi$. The stream surfaces are given by Ψ = constant, or

$$2a^2\cos^2\tfrac{1}{2}\theta - \tfrac{1}{2}r^2\sin^2\theta = \text{constant}.$$

The stream surface for which the constant on the right is zero is

$$r\sin\theta = 2a\cos\tfrac{1}{2}\theta.$$

This last equation is satisfied by:

(i) $\cos\tfrac{1}{2}\theta = 0$, giving $\theta = \pi$,
(ii) $r = a\,\text{cosec}\,\tfrac{1}{2}\theta$.

In a plane through OX, the intersection with the required surface of revolution has the form (ii) in plane polar coordinates (r, θ). To sketch the curve it is helpful to introduce plane Cartesian coordinates (x, y), where $x = r\cos\theta$, $y = r\sin\theta$. Then the parametric equations of the curve (ii) are

$$x(\theta) = a\cos\theta\,\text{cosec}\,\tfrac{1}{2}\theta, \qquad y(\theta) = 2a\cos\tfrac{1}{2}\theta.$$

As $\theta \to 0$, both $r, x \to \infty$ and $y \to 2a$. As $\theta \to 2\pi$, both $r, x \to \infty$, and $y \to -2a$. When $\theta \to \pi$, then $r \to a$, $x \to -a$ and $y \to 0$. Fig. 5.5 shows the profile of the surface of revolution.

As no flow takes place over the surface of revolution $r = a\,\text{cosec}\,\tfrac{1}{2}\theta$, one may introduce a rigid boundary over the surface, excluding the fluid and source within its interior. In effect the hydrodynamical image of the external flow $U\mathbf{i}$ in the surface is the point source Ua^2 at O.

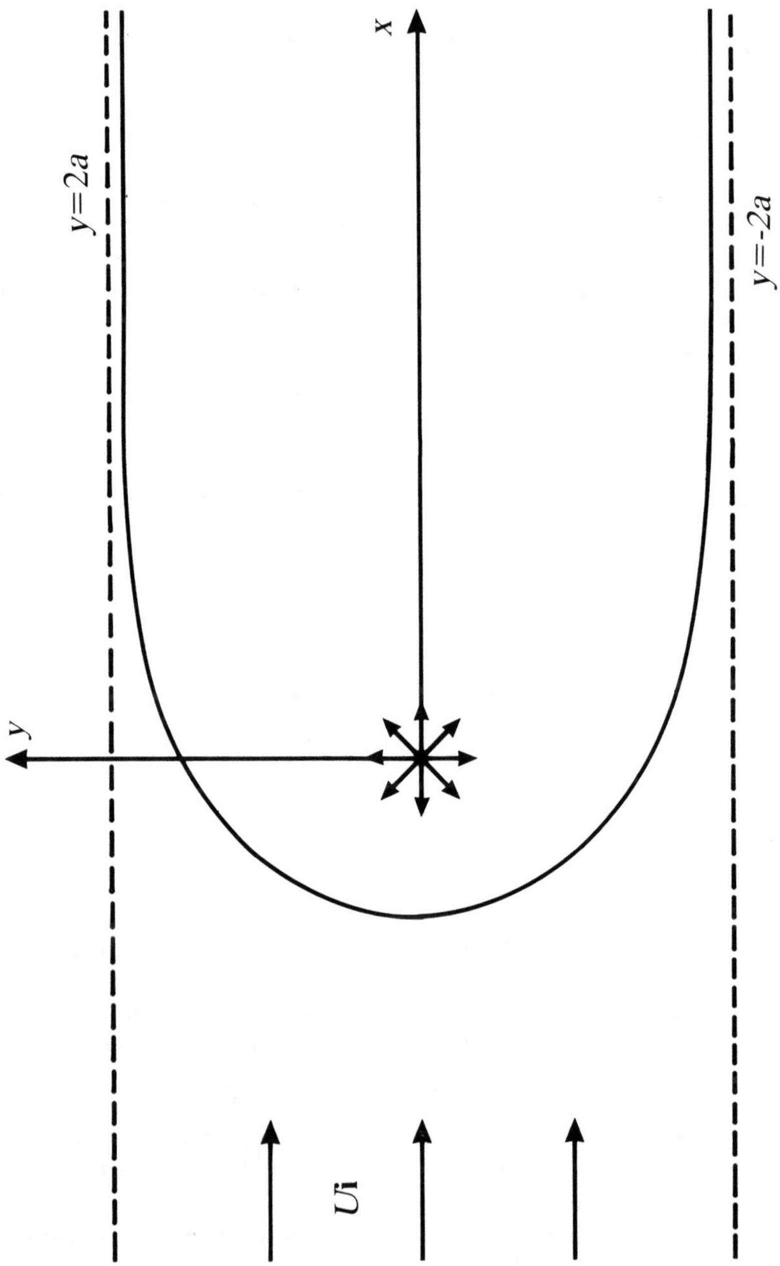

Fig. 5.5

The spherical polar velocity components $[q_r, q_\theta, q_\phi]$ at $P(r, \theta, \phi)$ in the fluid are given by

$$q_r = -\frac{1}{r^2 \sin\theta}\frac{\partial\Psi}{\partial\theta} = U\left\{\frac{a^2}{r^2} + \cos\theta\right\},$$

$$q_\theta = \frac{1}{r\sin\theta}\frac{\partial\Psi}{\partial r} = -U\sin\theta,$$

$$q_\phi = 0.$$

The fluid speed q at P is

$$q = \sqrt{\{q_r^2 + q_\theta^2 + q_\phi^2\}} = U\sqrt{\{1 + 2(a/r)^2\cos\theta + (a/r)^4\}}.$$

On the surface of revolution $r = a \operatorname{cosec} \frac{1}{2}\theta$, the fluid velocity is tangential to the surface and

$$q = \tfrac{1}{2}U\sqrt{\{5 + 2\cos\theta - 3\cos^2\theta\}}.$$

The only stagnation point on the surface is given by $\cos\theta = -1$, corresponding to $\theta = \pi$.

If Bernoulli's equation is applied along a streamline on the entire surface in the appropriate form, we find that

$$\frac{p}{\rho} + \frac{1}{2}q^2 = \text{constant}.$$

Then, if p_0 is the stagnation pressure at the nose $r = a$, $\theta = \frac{1}{2}\pi$,

$$\begin{aligned} p &= p_0 - \tfrac{1}{2}\rho q^2 \\ &= p_0 - \tfrac{1}{8}\rho U^2 (5 + 2\cos\theta - 3\cos^2\theta). \end{aligned}$$

This result may be used to calculate the pressure distribution near the nose of a projectile at rest in a uniform stream whose ambient speed is U: this is also the same as for the case of a projectile moving in the opposite direction at speed U in a fluid at rest.

If the pressure at infinity p_∞ is known, then Bernoulli's equation applied to the central streamline through the nose ($r = a$, $\theta = \pi$) gives

$$p_\infty + \tfrac{1}{2}\rho U^2 = p_0$$

so that p is known in terms of p_∞, ρ, U and θ.

Examples [5.1] and [5.2] illustrate how flows *exterior* to solid bodies may be found *indirectly* by considering a distribution of singularities within the *interior* of the region occupied by the body. In [5.1] the solid body is a sphere while in [5.2] the solid body is a blunt-ended semi-infinite solid of revolution. We may find exterior flows to other semi-infinite bodies of revolution by taking other combinations of sources and sinks along the axis within the body rather than just a single source. If the total strength of the singularities within the body shape is zero, then the body is closed at the downstream end also. This follows since any streamline originating at a singularity must also end at a singularity and there can be no net flux due to the total system of singularities. The method of finding flows about an axisymmetric body using a distribution of singularities along the axis is known as Rankine's method. A solid whose boundary is found in this way is known as a *Rankine solid*. The sphere is an example of a Rankine solid obtained in Example [5.1]. Note that a doublet has total source strength which is zero since it is the limiting case of a source-sink pair with equal and opposite strengths. The following example is relevant in designing airship forms.

Example [5.3]
Now let us combine the three distributions: (i) a uniform stream of velocity $U\mathbf{i}$, (ii) a uniform line sink of total strength m and stretching along the line $0 \leqslant r \leqslant a$, $\theta = 0$, (iii) a point source of strength m at O. The reader will easily obtain the stream function at $P(r, \theta, \phi)$ of the fluid in the form

$$\Psi(r, \theta) = -\frac{1}{2} Ur^2 \sin^2 \theta - \frac{m}{a} (OP - AP - OA) + m(\cos \theta - 1),$$

where O is the origin and A is $(a, 0, 0)$ in spherical polar coordinates.

It is easy to show that $\Psi = 0$ along the whole of the x-axis (including both positive and negative halves). Also $\Psi = 0$ along a closed curve in the plane through OX sketched in Fig. 5.6. This closed curve separates streamlines linking the point source at O with the line source OA from those which extend continuously from $x = -\infty$ to $x = \infty$. The closed curve $\Psi = 0$ gives a surface along which a rigid boundary may be introduced and its shape is like that of an airship. The diagram shows streamlines outside the closed surface $\Psi = 0$ and the distribution of singularities inside the surface. This model, then, may suffice to determine the pressure distribution over the surface of such an airship form which is placed in a uniform stream with velocity $U\mathbf{i}$. Variants on the line sink distribution will produce other airship forms.

5.6 BOUNDARY CONDITIONS SATISFIED BY THE STREAM FUNCTION

Consider axisymmetric flow of ideal fluid past a body of revolution at rest. The boundary condition at the surface of the body is the kinematic condition which describes the impenetrability of the body by the fluid. This implies no normal component of fluid velocity at the body surface. Thus over the body surface the fluid velocity is tangential, which therefore defines the body surface to be a stream surface. Thus on the body, the stream function Ψ satisfies

$$\Psi = \text{constant}. \tag{5.59}$$

In such a flow, the axis of symmetry is always a streamline and it is usual to let $\Psi = 0$ along this axis. Now (5.59) must be satisfied at every point of the body

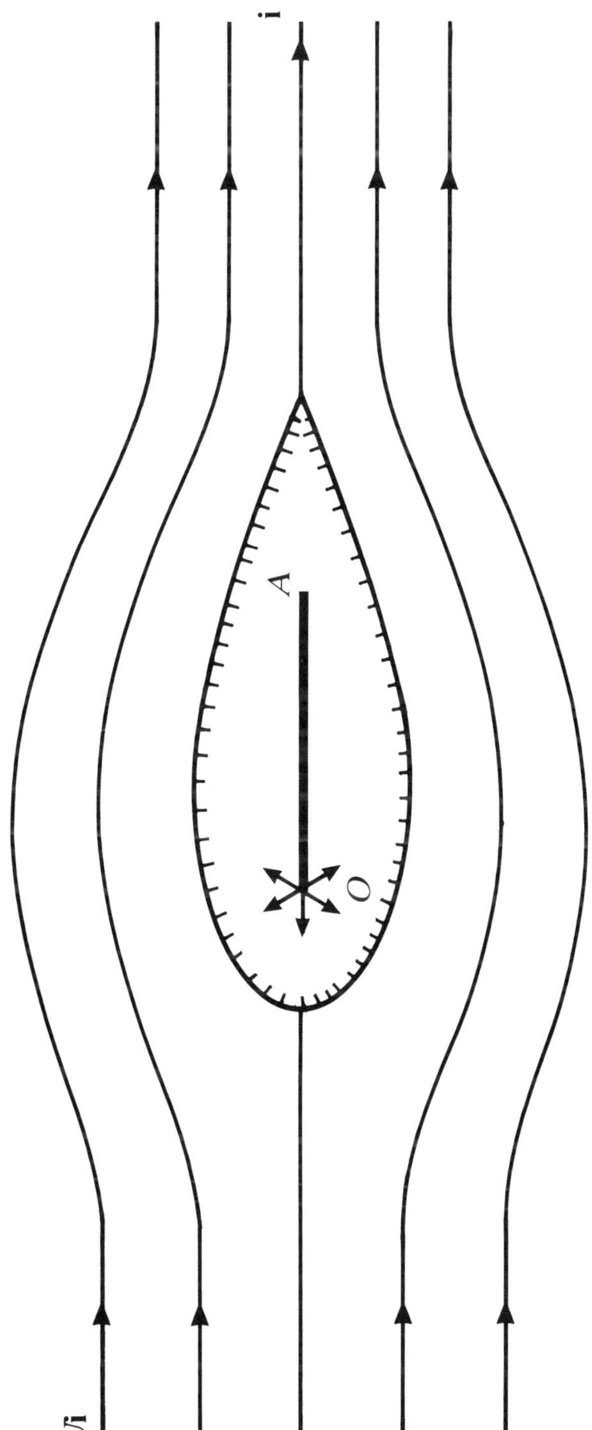

Fig. 5.6

surface. Thus if the body intersects the axis of symmetry, then $\Psi = 0$ on the surface of the body as well as the axis.

It is sometimes useful to describe the velocity at the surface of a body of revolution in terms of a system of coordinates which are intrinsic to the surface itself. Let $\mathbf{n}$ denote the unit vector drawn out of the body into the fluid at a general point on the body surface and let $\mathbf{t}$ be a unit vector tangential to a meridional curve on the body surface, as in Fig. 5.7. The direction of $\mathbf{t}$ is chosen so that $(\mathbf{n}, \mathbf{t}, \hat{\boldsymbol{\phi}})$ are right-handed in that order. Let dn and ds be elements of arc lengths along the direction of $\mathbf{n}$ and $\mathbf{t}$ respectively.

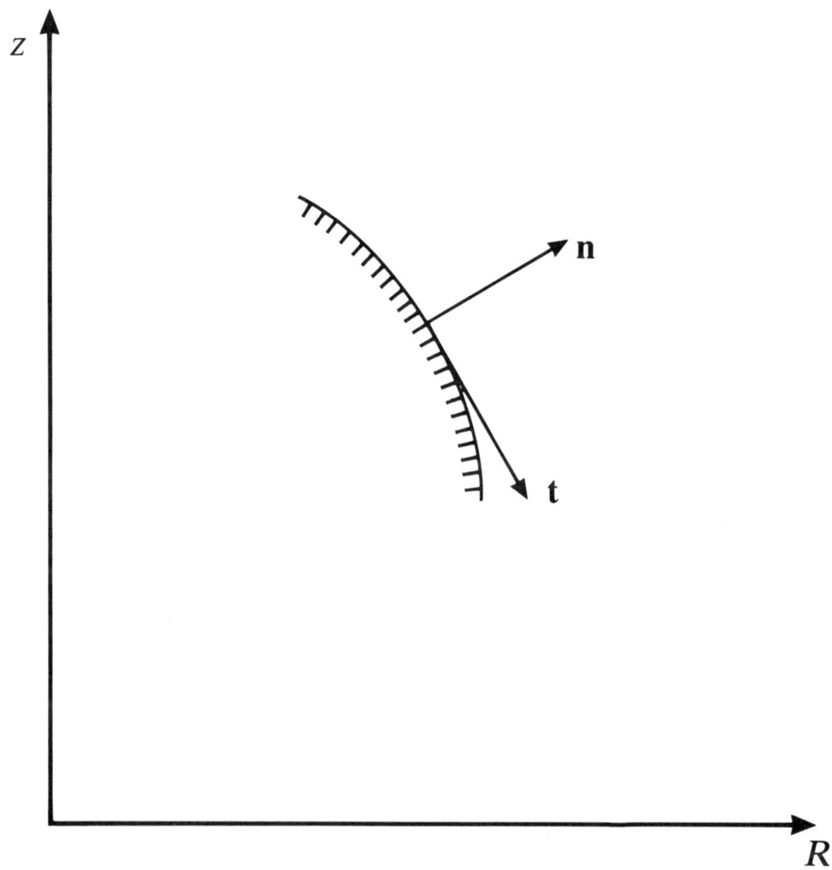

Fig. 5.7

We may regard the foregoing as a system of local orthogonal curvilinear coordinates of revolution of the form (ξ, η, ϕ) defined earlier by assigning the following values

$$\hat{\xi} = \mathbf{n}, \quad \hat{\eta} = \mathbf{t}, \quad h_1 = h_2 = 1, \quad h_3 = R.$$

The expression for the vector gradient operator is then

$$\nabla = \mathbf{n}\,\frac{\partial}{\partial n} + \mathbf{t}\,\frac{\partial}{\partial s} + \frac{\hat{\boldsymbol{\phi}}}{R}\,\frac{\partial}{\partial \phi}. \tag{5.60}$$

The components of the fluid velocity along **n** and **t** can then be found on combining (5.60) and (5.40). This gives

$$q_n = -\frac{1}{R}\frac{\partial \Psi}{\partial s}, \quad q_t = \frac{1}{R}\frac{\partial \Psi}{\partial n}. \tag{5.61}$$

These give respectively the normal and tangential components of **q** on the surface of the body. For the case considered earlier of a body at rest in the fluid, $q_n = 0$ implying that $\partial\Psi/\partial s = 0$ on the body. This means that Ψ does not vary with arc length along the surface of the body, or in other words $\Psi =$ constant on the body surface. The following useful results involving the intrinsic coordinates may be established:

$$\frac{\partial R}{\partial s} = \frac{\partial z}{\partial n}, \quad \frac{\partial R}{\partial n} = -\frac{\partial z}{\partial s}. \tag{5.62}$$

To prove the first, observe that

$$\begin{aligned}\frac{\partial R}{\partial s} &= \mathbf{t}\,.\,\nabla R = \mathbf{t}\,.\,\hat{\mathbf{R}} = \mathbf{t}\,.\,(\hat{\boldsymbol{\phi}} \times \mathbf{k}) \\ &= (\mathbf{t} \times \hat{\boldsymbol{\phi}})\,.\,\mathbf{k} = \mathbf{n}\,.\,\mathbf{k} = \frac{\partial z}{\partial n}.\end{aligned}$$

The second relation in (5.62) can similarly be proved.

Another common type of boundary condition arises when a body of revolution moves with constant velocity along its axis of symmetry. The body will no longer be a stream surface, but the kinematic condition must be satisfied on the surface of the body. If the velocity of the body is $U\mathbf{k}$, this kinematic condition is

$$(\mathbf{q} - U\mathbf{k})\,.\,\mathbf{n} = 0 \tag{5.63}$$

at any point of the body, giving

$$q_n = U\mathbf{k}\,.\,\mathbf{n}. \tag{5.64}$$

Using (5.61) and (5.62), this equation gives

$$\frac{\partial \Psi}{\partial s} = -\,UR\,\frac{\partial R}{\partial s} \tag{5.65}$$

at each point of the body. On integrating (5.65) along a meridional curve of the body, we find that at each point of the body

$$\Psi + \tfrac{1}{2}UR^2 = 0. \tag{5.66}$$

Example [5.4]

Determine the stream function for a sphere of radius a translating with constant velocity $U\mathbf{k}$ in infinite fluid at rest at infinity.

Using spherical polars, we must find Ψ such that

$$E^2 \Psi \equiv \frac{\partial^2 \Psi}{\partial r^2} + \frac{\sin\theta}{r^2} \frac{\partial}{\partial \theta}\left(\frac{1}{\sin\theta} \frac{\partial \Psi}{\partial \theta}\right) = 0, \qquad (r \geqslant a), \tag{1}$$

and such that

$$\Psi = -\tfrac{1}{2} Ua^2 \sin^2\theta, \qquad (r = a), \tag{2}$$

$$(r \sin\theta)^{-1} \nabla \Psi \to \mathbf{0}, \qquad (r \to \infty). \tag{3}$$

The form of the boundary condition (2) suggests we seek a solution of (1) of the form $f(r)\sin^2\theta$. This is a solution for all θ in the range $0 \leqslant \theta \leqslant \pi$ if

$$f''(r) - 2 f(r)/r^2 = 0. \tag{4}$$

The solutions of this equation are $f(r) = r^2$ or r^{-1}. Thus to satisfy (3) we choose

$$\Psi = \frac{A}{r} \sin^2\theta,$$

which will also satisfy (2) if $A = -\frac{1}{2}Ua^3$. Hence the stream function for a translating sphere is

$$\Psi = -\frac{Ua^3}{2r} \sin^2\theta.$$

PROBLEMS 5

1. Define a streamline and say when streamlines are paths of fluid particles.

A point source of strength m is caused to move with uniform speed U through an inviscid incompressible fluid at rest at infinity. If r denotes the distance from the source and θ is the angle that the radius vector from the source makes with the direction of motion, prove that the streamlines are $\theta =$ constant, but that the paths of the particles are given by

$$Ur^2 \sin^2\theta = -2m\cos\theta + A, \text{ where } A \text{ is constant.}$$

2. A simple source, strength m, is fixed at the origin O in a uniform stream of incompressible fluid moving with velocity $U\mathbf{i}$. Show that the velocity potential φ at any point P of the stream is $(m/r) - Ur\cos\theta$, where $OP = r$ and θ is the angle $\overrightarrow{OP}$ makes with the direction $\mathbf{i}$. Find the differential equation of the streamlines and show that they lie on the surfaces

$$Ur^2 \sin^2\theta - 2m\cos\theta = \text{const.}$$

3. An infinite mass of liquid of constant density ρ lies to one side of a rigid impermeable infinite plane and a three-dimensional doublet of strength μ is situated in the liquid, at a distance d from the plane and having its axis directed straight towards the wall. Assuming the pressure in the liquid at infinite distance from the doublet to be zero, show that the doublet is attracted towards the wall with a force $3\pi\mu^2\rho/(2d^4)$.

4. Three-dimensional doublets of strengths μ and $(a/c)^3\mu$ are situated in an infinite expanse of liquid of constant density at the points having Cartesian coordinates $(c, 0, 0)$ and $(a^2/c, 0, 0)$ respectively. Their axes are directed towards and away from the origin respectively. Find the velocity potential and hence show that no fluid crosses the sphere centre $(0, 0, 0)$ and radius a.

5. Derive the stream functions for (i) a uniform flow of magnitude U, (ii) a uniform line source of strength m. Show that a combination of these two stream functions can be used to give the two-dimensional flow past a certain bluff semi-infinite body and express the thickness of the body in terms of m, U.

6. In an irrotational motion of an inviscid liquid of constant density ρ there is a source with a rate of flow $4\pi Ua^2$ at the origin O together with a uniform stream with velocity U; the direction of U is such that the point with coordinates $(a, 0, 0)$ with respect to polar coordinates (r, θ, φ) is a point of stagnation. Show that the velocity at any point of the surface $r = a \sec(\theta/2)$ is tangential to the surface, and if there is no body force, find an expression for the pressure at a general point on this surface.

7. Show that stream functions for a source and a uniform stream can be expressed in the forms $m \cos\theta$ and $-\frac{1}{2}Ur^2 \sin^2\theta$ respectively. A sink and a source, having strengths of equal magnitude $4\pi m$, are situated at $x = \pm a$ respectively in an otherwise uniform stream U parallel to the positive direction of the x-axis. Show that particles originating at distance b from the axis cross the plane $x = 0$ at distance $a\sqrt{3}$ from the x-axis if $b = \sqrt{(3a^2 - 2m/U)}$. Sketch the streamlines and explain the absence of a real value of b in the case $2m > 3Ua^2$.

8. A three-dimensional doublet points in the direction of flow of a uniform stream in which it is situated. Show that one of the equipotential surfaces is a sphere and show that the stagnation points lie in this sphere. Taking the centre of the doublet as the origin and its axis as the line $\theta = 0$, show that the equations of the streamlines in spherical polar coordinates are of the form $(2a^3 + r^3)\sin^2\theta = Cr$, where a is the radius of the sphere.

9. The three-dimensional motion of a liquid is symmetrical with respect to the axis $\theta = 0$ of spherical polar coordinates, and the velocity components are $[u, v]$ in directions associated with the coordinates (r, θ). Obtain the equation of

continuity and deduce that u and v can be expressed in terms of the stream function, ψ.

Show that, if the motion is irrotational, then ψ satisfies the equation

$$r^2 \frac{\partial^2 \psi}{\partial r^2} = \cot\theta \frac{\partial \psi}{\partial \theta} - \frac{\partial^2 \psi}{\partial \theta^2}.$$

Show that if

$$\psi = \tfrac{1}{2} U(a^4/r^2) \sin^2\theta \cos\theta - \tfrac{1}{2} Ur^2 \sin^2\theta,$$

the motion is irrotational.

10. A doublet of strength M is placed at the point $(0, a, 0)$ with its axis parallel to the axis of z. Prove that at points close to the origin the velocity potential of the doublet is approximately

$$\frac{Mz}{a^3} + \frac{3Myz}{a^4},$$

neglecting terms of the order r^3/a^5 and higher powers.

Deduce that if a small sphere of radius c is placed with its centre at the origin, the velocity potential is then increased by the terms

$$\frac{Mc^3 z}{2a^3 r^3} + \frac{2Mc^5 yz}{a^4 r^5}.$$

11. Show that the image of a radial doublet in a sphere is another radial doublet, and compare their magnitudes; show also that the velocity at any point of the sphere is proportional to Rr^{-5}, where r is the distance from the doublet, and R the perpendicular on the diameter on which it lies.

12. Discuss the motion for which the Stokes stream function is given by

$$\psi = \tfrac{1}{2} V\{a^4 r^{-2} \cos\theta - r^2\} \sin^2\theta,$$

where r is the distance from a fixed point and θ is the angle this distance makes with a fixed direction.

13. Find the Stokes stream function ψ where fluid motion is due to a source of strength m (flux $4\pi m$) at a fixed point A, a sink $-m$ at another fixed point B, a translation of the fluid of velocity U in the direction AB being superposed; explain how this solution can be used to deduce the motion of fluid past a certain solid of revolution. If $U = 8m/9a^2$, where $AB = 2a$, prove that the solid is of axial length $4a$, of equatorial radius approximately $1.6a$, and has the same effect on the fluid motion at a great distance as a sphere of radius $a(9/2)^{1/3}$.

14. If AB be a uniform line source, and A and B equal sinks of such strength that there is no total gain or loss of fluid, show that the stream function at any point is

$$C\{(r_1 - r_2)^2 - c^2\}\left(\frac{1}{r_1} - \frac{1}{r_2}\right),$$

where c is the length of AB, and r_1, r_2 are the distances of the point considered from A and B and C is a constant.

15. A solid of revolution is moving along its axis in an infinite liquid; show that the kinetic energy of the liquid is

$$-\tfrac{1}{2}\pi\rho\int\frac{\psi}{R}\frac{\partial\psi}{\partial n}\,ds,$$

where ψ is the Stokes stream function of the motion, R the distance of a point from the axis and the integral is taken once round a meridian curve of the solid. Hence obtain the kinetic energy of infinite liquid due to the motion of a sphere through it with velocity V.

16. There is a source at A, an equal sink at B and AB is the direction of a uniform stream. Determine the form of the streamlines. If A is $(a, 0)$, B is $(-a, 0)$ and the ratio of the flow issuing from A in unit time to the speed of the stream is $2\pi b$, show that the stream function is

$$\psi = Vy - Vb\tan^{-1}\frac{2ay}{x^2 + y^2 - a^2},$$

and that the length, $2l$, and the breadth, $2d$, of the closed wall that forms part of the dividing streamline is given by

$$l = \sqrt{a^2 + 2ab}, \quad \tan\frac{d}{b} = \frac{2ad}{d^2 - a^2},$$

and the locus of the point at which the speed is equal to that of the stream is

$$x^2 - y^2 = a^2 + ab.$$

17. A two-dimensional source of strength m is situated at the point $(a, 0)$, the axis of y being a fixed boundary. Find the points on the boundary at which the fluid velocity is a maximum.

Show that the resultant thrust on that part of the axis of y which lies between $y = \pm\, b$ is

$$2p_0 b - 2m^2\rho\left\{\frac{1}{a}\tan^{-1}\frac{b}{a} - \frac{b}{a^2+b^2}\right\},$$

where p_0 is the pressure at infinity.

18. Two sources, each of strength m, are placed at the points $(-a, 0)$ and $(a, 0)$, and a sink of strength $2m$ is placed at the origin. Show that the streamlines are the curves

$$(x^2+y^2)^2 = a^2(x^2-y^2+\lambda xy),$$

where λ is a variable parameter.

Show also that the fluid speed at any point is

$$\frac{2ma^2}{r_1 r_2 r_3},$$

where r_1, r_2, r_3 are respectively the distances of the point from the sources and the sink.

19. If there is a source at $(a, 0)$ and $(-a, 0)$ and sinks at $(0, a)$, $(0, -a)$, all of equal strength, show that the circle through these four points is a streamline.

20. OX, OY are fixed rigid boundaries and there is a source at (a, b). Find the form of the streamlines and show that the dividing line is

$$xy(x^2-y^2-a^2+b^2) = 0.$$

21. Between the fixed boundaries $\theta = \pi/4$, $\theta = -\pi/4$, the two-dimensional motion is due to a source of strength m at $r = a$, $\theta = 0$, and an equal sink at $r = b$, $\theta = 0$. Show that the stream function is

$$-m\tan^{-1}\frac{r^4(a^4-b^4)\sin 4\theta}{r^8 - r^4(a^4+b^4)\cos 4\theta + a^4b^4}.$$

22. Show that the velocity components given by

$$u = U\left(1 - \frac{ay}{x^2+y^2} + \frac{b^2(x^2-y^2)}{(x^2+y^2)^2}\right),$$

$$v = U\left(\frac{ax}{x^2+y^2} + \frac{2b^2xy}{(x^2+y^2)^2}\right),$$

represent a possible fluid motion in two dimensions.

23. A source is placed midway between two planes whose distance from one another is $2a$. Find the equation of the streamlines when the motion is in two dimensions, and show that those particles, which at an infinite distance are distant $\frac{1}{2}a$ from one of the boundaries, issued from the source in a direction making an angle $\pi/4$ with it.

24. If a circle be cut in half by the y-axis, forming a rigid boundary, and a source, of strength m, be on the x-axis at a distance a, equal to half the radius, from the centre, prove that the streamlines are given by

$$(16a^4 + r^4)\cos 2\theta - 17a^2r^2 = (16a^4 - r^4)\sin 2\theta \cot(\psi/m).$$

ψ being a suitably adjusted value of the stream function.

Show that the streamline $\psi = m\pi/2$ leaves the source in a direction perpendicular to Ox and enters the sink at an angle $\pi/4$ with Ox, and sketch the streamlines.

25. In the two-dimensional motion of an infinite liquid there is a rigid boundary consisting of that part of the circle $x^2 + y^2 = a^2$ which lies in the first and fourth quadrants and the parts of the axis of y which lie outside the circle. A simple source of strength m is placed at the point $(f, 0)$ where $f > a$. Prove that the speed of the fluid at the point $(a\cos\theta, a\sin\theta)$ of the semicircular boundary is

$$\frac{4maf^2 \sin 2\theta}{a^4 + f^4 - 2a^2f^2 \cos 2\theta}.$$

Find at what points of the boundary the pressure is least.

26. Water enters a circular enclosure of radius a at the centre O and escapes by a small hole at the point A of the boundary into the region outside which is also occupied by water and is unbounded. The motion being considered two-dimensional, prove (i) that the asymptotes of the streamlines pass through a fixed point; (ii) that the tangent at O to a streamline and the corresponding asymptote are equally inclined at OA; (iii) that the streamline has a double point at A, the tangents at which are perpendicular. Sketch one of the streamlines.

6

Two-dimensional flow

6.1 INTRODUCTION

In this chapter we return to two-dimensional flows and examine the use of complex variable techniques. Comprehensive details on the complex analysis will be found in the references cited. Here we shall give a broad outline of the subject which will be sufficient to develop the applications of it to flows that are two-dimensional, incompressible and irrotational.

6.2 RÉSUMÉ OF THE MAIN FEATURES OF COMPLEX ANALYSIS

Suppose that $z = x + iy$ and that

$$w = F(z) = f(x, y) + i\,g(x, y) \tag{6.1}$$

where $i = \sqrt{(-1)}$ and x, y, $f(x, y)$, $g(x, y)$ are all real. Then z and w are functions of the complex variable z. Following standard practice, we write

$$f(x, y) = \text{Re}(w); \quad g(x, y) = \text{Im}(w).$$

We further suppose that within a given region R of the complex z-plane the functions f and g and their first derivatives $\partial f/\partial x$, $\partial f/\partial y$, $\partial g/\partial x$, $\partial g/\partial y$ exist and are everywhere continuous. If at any point P of R specified by z, the derivative $\mathrm{d}w/\mathrm{d}z = f'(z)$ is unique, then w is said to be ***analytic*** or ***regular*** at that point. If the derivative is unique at all P of R, then w is said to be analytic or regular throughout the region.

Let $w = F(z)$ be an analytic function defined as in (6.1). Then $w = w(x, y)$ and

$$\frac{\partial w}{\partial x} = \frac{\mathrm{d}F}{\mathrm{d}z}\frac{\partial z}{\partial x} = F'(z)$$

or

$$\frac{\partial f}{\partial x} + i\,\frac{\partial g}{\partial x} = F'(z). \tag{6.2}$$

Further

$$\frac{\partial w}{\partial y} = \frac{\mathrm{d}F}{\mathrm{d}z}\frac{\partial z}{\partial y} = i\,F'(z)$$

or

$$\frac{\partial f}{\partial y} + i\,\frac{\partial g}{\partial y} = i\,F'(z)$$

which simplifies to

$$\frac{\partial g}{\partial y} - i\,\frac{\partial f}{\partial y} = F'(z). \tag{6.3}$$

Since $F'(z)$ is unique for an analytic $F(z)$, equations (6.2) and (6.3) give

$$\frac{\partial f}{\partial x} + i\,\frac{\partial g}{\partial x} = \frac{\partial g}{\partial y} - i\,\frac{\partial f}{\partial y}.$$

Equating the real and imaginary parts in this equation gives the *Cauchy-Riemann equations* in the form

$$\frac{\partial f}{\partial x} = \frac{\partial g}{\partial y}; \quad \frac{\partial f}{\partial y} = -\,\frac{\partial g}{\partial x}. \tag{6.4}$$

The functions f and g are called *conjugate functions*. From (6.4) and the results

$$\frac{\partial^2 f}{\partial x \partial y} = \frac{\partial^2 f}{\partial y \partial x}; \quad \frac{\partial^2 g}{\partial x \partial y} = \frac{\partial^2 g}{\partial y \partial x},$$

it is easily established that

$$\nabla^2 f(x, y) = 0. \tag{6.5}$$

$$\nabla^2 g(x, y) = 0 \tag{6.6}$$

where $\nabla^2 \equiv \partial^2/\partial x^2 + \partial^2/\partial y^2$. Also from (6.4),

$$\frac{\partial f}{\partial x}\frac{\partial g}{\partial x} + \frac{\partial f}{\partial y}\frac{\partial g}{\partial y} = 0$$

or

$$\nabla f \, . \, \nabla g = 0 \tag{6.7}$$

which shows that at each point of intersection (x, y) the curves

$$f(x, y) = \text{const.}; \quad g(x, y) = \text{const.}$$

cut orthogonally. The harmonic and orthogonal properties expressed by (6.5), (6.6), (6.7) underlie the applications of complex analysis to two-dimensional incompressible irrotational flows.

The Cauchy-Riemann equations (6.4) furnish a necessary condition for $w = f + ig$ to be analytic. When augmented with the further conditions of continuity of the first-order derivatives $\partial f/\partial x$, $\partial f/\partial y$, $\partial g/\partial x$, $\partial g/\partial y$, they provide a set of sufficient conditions for w to be analytic. Thus if $\phi(x, y)$, $\psi(x, y)$ have continuous first-order partial derivatives and if the functions satisfy the Cauchy-Riemann equations, then $\phi + i\psi$ is an analytic function of z.

Example [6.1]
Show that z^2 is an analytic function of z.

$$F(z) = z^2 = (x + iy)^2 = x^2 - y^2 + 2ixy\,.$$

Then

$$f(x, y) = \text{Re}\{F(z)\} = x^2 - y^2\,;$$
$$g(x, y) = \text{Im}\{F(z)\} = 2xy\,.$$

Hence

$$\partial f/\partial x = 2x,\ \partial f/\partial y = -2y,\ \partial g/\partial x = 2y,\ \partial g/\partial y = 2x.$$

The Cauchy-Riemann equations (6.4) are thus satisfied and so z^2 is analytic.

Example [6.2]
Show that $\bar{z} = x - iy$ is not analytic.

With $f(x, y) = x$, $g(x, y) = -y$,

$$\partial f/\partial x = 1,\ \partial f/\partial y = 0,\ \partial g/\partial x = 0,\ \partial g/\partial y = -1.$$

Thus $\partial f/\partial x \neq \partial g/\partial y$ and so the function is not an analytic function of z.

Example [6.3]
Show that $F(z, \bar{z})$ cannot be analytic function of z.

Let us assume that it is. As before we take

$$F(z, \bar{z}) = f(x, y) + ig(x, y),$$

where f, g are both real when x, y are. Now $z = x + iy$, $\bar{z} = x - iy$, and so $x = \frac{1}{2}(z + \bar{z})$, $y = \frac{1}{2}i(\bar{z} - z)$. Hence

$$\frac{\partial F}{\partial \bar{z}} = \left(\frac{\partial f}{\partial x} + i\,\frac{\partial g}{\partial x}\right)\frac{\partial x}{\partial \bar{z}} + \left(\frac{\partial f}{\partial y} + i\,\frac{\partial g}{\partial y}\right)\frac{\partial y}{\partial \bar{z}}$$

$$= \tfrac{1}{2}\left(\frac{\partial f}{\partial x} + i\,\frac{\partial g}{\partial x}\right) + \tfrac{1}{2}i\left(\frac{\partial f}{\partial y} + i\,\frac{\partial g}{\partial y}\right)$$

$$= \tfrac{1}{2}\left(\frac{\partial f}{\partial x} - \frac{\partial g}{\partial y}\right) + \tfrac{1}{2}i\left(\frac{\partial f}{\partial y} + \frac{\partial g}{\partial x}\right).$$

Since F is assumed to be an analytic function of z, the Cauchy-Riemann equations (6.4) hold and so

$$\partial F/\partial \bar{z} = 0.$$

This means that F is independent of $\bar{z}$, contrary to the given information. The contradiction thus violates the assumption of analyticity of $F(z, \bar{z})$.

Cauchy's Integral Theorem

Let $\mathscr{C}$ be a closed curve in the complex z-plane and suppose it encloses an area S. Then if $F(z)$ is an analytic function of z on S and on $\mathscr{C}$,

$$\oint_{\mathscr{C}} F(z)\, \mathrm{d}z = 0.$$

The integral on the left-hand side is termed a 'contour integral'.

Proof: With $z = x + iy$, $F(z) = f(x, y) + ig(x, y)$, as in the above notation, we have

$$\oint_{\mathscr{C}} F(z)\, \mathrm{d}z = \oint_{\mathscr{C}} (f + ig)(\mathrm{d}x + i\mathrm{d}y)$$

$$= \oint_{\mathscr{C}} (f\,\mathrm{d}x - g\,\mathrm{d}y) + i \oint_{\mathscr{C}} (g\,\mathrm{d}x + f\,\mathrm{d}y)$$

$$= -\int_S\!\!\int\left(\frac{\partial g}{\partial x} + \frac{\partial f}{\partial y}\right)\mathrm{d}x\,\mathrm{d}y + i\int_S\!\!\int\left(\frac{\partial f}{\partial x} - \frac{\partial g}{\partial y}\right)\mathrm{d}x\,\mathrm{d}y.$$

Here Stokes's theorem for two dimensions in the form of equation (1.72) has been applied to each of the contour integrals around $\mathscr{C}$ to produce the two integrals over S. But the Cauchy-Riemann equations (6.4) reduce the integrands of each of these double integrals to zero and so, since S is a finite domain each double integral vanishes. Thus the contour integral around $\mathscr{C}$ vanishes.

Continuing with basic complex analysis we assert that if $F(z)$ is analytic at and within a neighbourhood of a point $z = z_0$, then for any point z in the neighbourhood, when $|z - z_0|$ is sufficiently small,

$$F(z) = a_0 + \sum_{n=1}^{\infty} a_n(z - z_0)^n, \tag{6.8}$$

where $a_0 = F(0)$ and $n\,!\,a_n = F^{(n)}(z_0)$ with $F^{(n)}(z) = \mathrm{d}^n F/\mathrm{d}z^n$ $(n = 1, 2, \ldots)$. Equation (6.8) expresses the *Taylor series* for the function. Full discussion of this is found in the references.

However, many of the functions which we shall meet in this chapter will have expansions of the form

$$F(z) = a_0 + a_1(z - z_0) + a_2(z - z_0)^2 + \ldots + a_n(z - z_0)^n + \ldots$$
$$+ \frac{b_1}{z - z_0} + \frac{b_2}{(z - z_0)^2} + \ldots + \frac{b_n}{(z - z_0)^n} + \ldots \tag{6.9}$$

A function $F(z)$ of the form (6.9) is said to be *meromorphic* and the point $z = z_0$ is called a *pole*. If $b_n \neq 0$ but $b_m = 0$ for all $m > n$, then $F(z)$ is said to have a *pole of order n* at $z = z_0$. The coefficient b_1 in (6.9) is called the *residue* of $F(z)$ at $z = z_0$. Knowledge of the residues of a meromorphic function $F(z)$ within a closed contour $\mathscr{C}$ is crucial to the evaluation of the contour integral of $F(z)$ round $\mathscr{C}$. To this end we give a brief résumé of *Cauchy's residue theorem*. Fuller particulars will be found in the appropriate references.

Cauchy's Residue Theorem

Suppose that $\mathscr{C}$ is a closed curve in the complex z-plane and that $F(z)$ is pole-free on $\mathscr{C}$ and meromorphic within its interior S. Let the function have poles at the points $z = z_1, z_2, \ldots, z_n$ within $\mathscr{C}$ and suppose that the residues of $F(z)$ at these points are $k_1, k_2, \ldots, k_n$ respectively. Then

$$\oint_{\mathscr{C}} F(z)\,\mathrm{d}z = 2\pi i\,(k_1 + k_2 + \ldots + k_n). \tag{6.10}$$

Proof: (i) If $z = z_r$ is taken as the centre of the circle γ_r, of radius ϵ_r we first evaluate the contour integral around γ_r. Here $z = z_r$ is a particular one of the n poles of $F(z)$. Scrutiny of (6.9) shows that within the neighbourhood of $z = z_r$, $F(z)$ may be expressed in the form

$$F(z) = \phi(z) + \sum_{s=1}^{m} b_s(z - z_r)^{-s},$$

where m may be either a finite positive integer or infinity, and $\phi(z)$ is an analytic function of the form $a_0 + \sum_{n=1}^{\infty} a_n(z - z_r)^n$. Then the countour integral of $F(z)$ around γ_r is

$$\oint_{\gamma_r} F(z)\,\mathrm{d}z = \oint_{\gamma_r} \phi(z)\,\mathrm{d}z + \sum_{s=1}^{m} b_s \oint_{\gamma_r} (z - z_r)^{-s}\,\mathrm{d}z. \tag{6.11}$$

The contour integral of $\phi(z)$ around γ_r which features in the first term on the right-hand side of (6.11) is zero as $\phi(z)$ is analytic within and on γ_r.

We next consider the series of contour integrals around γ_r expressed in (6.11). On γ_r, $z - z_r = \epsilon_r \, e^{i\theta}$, where $0 \leqslant \theta \leqslant 2\pi$ and so $dz = i\epsilon_r \, e^{i\theta} \, d\theta$. Thus, for $s = 1, 2, \ldots, m$,

$$\oint_\gamma \frac{dz}{(z - z_r)^s} = i \int_0^{2\pi} \frac{\epsilon_r \, e^{i\theta} \, d\theta}{\epsilon_r^s \, e^{is\theta}}$$

$$= \frac{i}{\epsilon_r^{s-1}} \int_0^{2\pi} \exp(1 - s)\, i\theta \; d\theta .$$

When $s = 1$, the last integral is 2π, so that

$$\oint_\gamma \frac{dz}{z - z_r} = 2\pi i.$$

When $s = 2, 3, \ldots,$

$$\int_0^{2\pi} \exp(1 - s) i\theta \; d\theta = \frac{i}{s - 1} \Big[\cos (s - 1)\theta - i \sin (s - 1)\theta \Big]_0^{2\pi} = 0.$$

It follows at once that

$$\sum_{s=1}^{m} b_s \oint_{\gamma_r} (z - z_r)^{-s} \, dz = 2\pi i b_1 .$$

This implies that the value of the contour integral on the left-hand side of (6.11) is simply $2\pi i$ times the residue of $F(z)$ at $z = z_r$, the centre of the small circle γ_r.

(ii) To complete the proof of the residue theorem, we consider Fig. 6.1 in which small circles γ_r are drawn centred at each z_r and connected to $\mathscr{C}$ by parallel lines l_r, l_r' which are very close together and which are ultimately made to coincide ($r = 1, 2, \ldots, n$). Let Γ be the closed contour shown in Fig. 6.1 comprising $\mathscr{C}$, the parallel connections l_r, l_r' and the small circles γ_r ($r = 1, \ldots, n$). This composite contour Γ excludes all residues of $F(z)$. Hence by Cauchy's integral theorem,

$$\oint_\Gamma F(z) \, dz = 0. \tag{6.12}$$

With Γ described in the sense shown in Fig. 6.1, $\mathscr{C}$ is described in the positive sense and each circle γ_r in the negative sense. Then, as each $l_r' \to l_r$ ($r = 1, \ldots, n$),

$$\oint_\Gamma F(z) \, dz = \oint_\mathscr{C} F(z) \, dz + \sum_{r=1}^{n} \left\{ \int_{l_r} F(z) \, dz + \int_{\gamma_r} F(z) \, dz + \int_{l_r'} F(z) \, dz \right\} \tag{6.13}$$

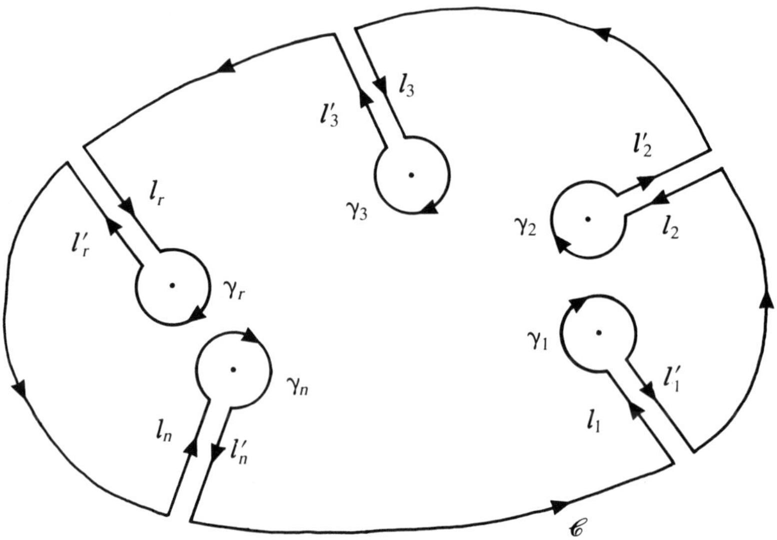

Fig. 6.1

In (6.13), the left-hand side is zero by virtue of equation (6.12). Further, we note that the paths l_r and l'_r are parallel but in opposite senses. Hence, as each $l'_r \to l_r$ $(r = 1, \ldots, n)$,

$$\left(\int_{l_r} + \int_{l'_r}\right) F(z)\,\mathrm{d}z \to 0 \qquad (r = 1, \ldots, n). \tag{6.14}$$

For the contour integrals around each circle γ_r, we use (i) and recall that the sense of description of each γ_r is negative. Since k_r is the residue of $F(z)$ at the centre $z = z_r$ of γ_r,

$$\oint_{\gamma_r} F(z)\,\mathrm{d}z = -2\pi i\, k_r \qquad (r = 1, \ldots, n). \tag{6.15}$$

Thus using (6.12), (6.14), (6.15), equation (6.13) simplifies to

$$0 = \oint_{\mathscr{C}} F(z)\,\mathrm{d}z - \sum_{r=1}^{n} 2\pi i k_r,$$

as required.

It thus follows that if we know the residues of a meromorphic function $F(z)$ within a pole-free closed contour, then the contour integral of $F(z)$ in the positive sense round that contour is simply $2\pi i$ times the sum of the residues at the enclosed poles. This is a simple rule and the main difficulty in applying it is the actual evaluation of the residues. Some examples are now given to indicate how this is achieved. Many more cases will be encountered when the method is applied to various two-dimensional flows.

It is important to stress that the contour integrals are the same about *any* closed curve $\mathscr{C}$ which is pole-free and which encloses the same poles as the circles in Fig. 6.1.

Example [6.4]

Find $\displaystyle\int_{\mathscr{C}} \frac{z^2\,dz}{(z+1)(3z-2)(z+8)}$, when $\mathscr{C}$ is the circle $|z| = 4$.

Denoting the integrand by $F(z)$, its poles are given by the zeros of $1/F(z)$. Thus the poles occur at $z = -1$, $z = \frac{2}{3}$, $z = -8$. Of these only the first two occur within $\mathscr{C}$ and so we need to find the residues of $F(z)$ at $z = -1$ and $z = \frac{2}{3}$.

In this case $F(z)$ may be expressed in terms of partial fractions of the form

$$F(z) \equiv \frac{A}{z+1} + \frac{B}{z-\frac{2}{3}} + \frac{C}{z+8}.$$

A, B, C are the residues of $F(z)$ at $z = -1, \frac{2}{3}, -8$. Then

$$A = \lim_{z\to-1} \{(z+1)F(z)\} = \lim_{z\to-1} \left\{\frac{z^2}{(3z-2)(z+8)}\right\} = -\frac{1}{35};$$

$$B = \lim_{z\to\frac{2}{3}} \{(z-\tfrac{2}{3})F(z)\} = \lim_{z\to\frac{2}{3}} \left\{\frac{z^2}{(z+1)\,3(z+8)}\right\} = -\frac{2}{195}.$$

Hence the value of the contour integral is

$$2\pi i(A+B) = -10\pi i/273.$$

Example [6.5]

Find $\displaystyle\int_{\mathscr{C}} \frac{dz}{z^2(z^2+10z+16)}$, $\mathscr{C}$ is $|z| = 3$.

Denote by $F(z)$ the integrand. The poles of $F(z)$ are the zeros of $1/F(z)$ and are given by $z^2(z+2)(z+8) = 0$, i.e.

$$z = 0 \text{ (double)}, \quad z = -2 \text{ (simple)}, \quad z = -8 \text{ (simple)}.$$

The double pole at $z = 0$ and the simple one at $z = -2$ are within $\mathscr{C}$, but that at $z = -8$ is outside it.

The partial fractions of $F(z)$ assume the form

$$F(z) \equiv \frac{A}{z} + \frac{B}{z^2} + \frac{C}{z+2} + \frac{D}{z+8}$$

from which A, C, D are the residues at $z = 0, -2, -8$.

Clearly

$$C = \lim_{z\to -2} \{(z+2)F(z)\} = \lim_{z\to -2} \left\{\frac{1}{z^2(z+8)}\right\} = \frac{1}{24}.$$

Now

$$(16+10z+z^2)^{-1} = \frac{1}{16}\left[1+\frac{5z}{8}+0(z^2)\right]^{-1} = \frac{1}{16}\left[1-\frac{5z}{8}+0(z^2)\right],$$

and so

$$\frac{1}{z^2(16+10z+z^2)} = \frac{1}{16}\left[\frac{1}{z^2}-\frac{5}{8z}+\ldots\right]$$

which shows that

$$A = -\frac{5}{128}.$$

Hence the value of the contour integral is $2\pi i(A + C)$, etc.

6.3 THE COMPLEX POTENTIAL

Here and in the remainder of the chapter we confine attention to irrotational plane flows of incompressible fluid of uniform density for which the velocity potential $\phi(x, y)$ and the stream function $\psi(x, y)$ exist. Here (x, y) specify two-dimensional Cartesian coordinates in a plane of flow. Let us write

$$w = \phi(x, y) + i\psi(x, y). \tag{6.16}$$

We suppose all four first-order partial derivatives of ϕ and ψ with respect to x, y exist and are continuous throughout the plane of flow. Now the velocity $\mathbf{q} = [u, v]$ has components satisfying

$$u = \partial\phi/\partial x = \partial\psi/\partial y; \qquad v = \partial\phi/\partial y = -\partial\psi/\partial x. \tag{6.17}$$

Thus ϕ and ψ satisfy the Cauchy-Riemann equations and so w must be an analytic function of z which we write as $w(z)$. Equation (6.16) may now be written

$$w(z) = \phi(x, y) + i\psi(x, y). \tag{6.18}$$

This function $w(z)$ is called the *complex potential* of the plane flow.

From the form (6.18),

$$\frac{\partial w}{\partial x} = \frac{dw(z)}{dz}\frac{\partial z}{\partial x} = w'(z)$$

and so

$$w'(z) = \partial\phi/\partial x + i\,\partial\psi/\partial x,$$

or, using (6.17),

$$w'(z) = u - iv. \tag{6.19}$$

If at $P(x, y)$ in the plane of flow the local fluid velocity $\mathbf{q}$ is inclined at angle θ to the positive x-axis and if $q = |\mathbf{q}|$, then $u = q \cos\theta$, $v = q \sin\theta$ and (6.19) gives

$$w'(z) = q\,e^{-i\theta}. \tag{6.20}$$

Thus

$$\left.\begin{aligned} q &= |w'(z)|, \\ \theta &= -\arg w'(z). \end{aligned}\right\} \tag{6.21}$$

The forms (6.19), (6.20), (6.21) show the usefulness of the complex potential. For a given plane flow for which ϕ, ψ are known, $w(z)$ is immediately constructed using (6.18). The flow field is then found in Cartesian components using (6.19) and in polar components using (6.21). It follows, then, that a detailed analysis of an incompressible and irrotational plane flow requires finding its complex potential $w(z)$. Very often the particular two-dimensional flow under consideration is a combination of two or more basic flows. The basic flows which will first be considered include a uniform stream and the flows produced by such hydrodynamical singularities as inifinite line sources and sinks, infinite line doublets and infinite line vortices. When the given flow is a combination of two or more of these basic distributions, $w(z)$ is obtained merely by superposing the separate complex potentials for the basic flows.

Example [6.6]

Discuss the flow for which the complex potential is $w = z^2$.

We have

$$\phi + i\psi = (x + iy)^2 = (x^2 - y^2) + 2ixy.$$

Thus

$$\phi(x, y) = x^2 - y^2; \quad \psi(x, y) = 2xy.$$

The equipotentials, $\phi = \text{const.}$, are the rectangular hyperbolae

$$x^2 - y^2 = \text{const.}$$

having asymptotes $y = \pm x$. The streamlines, $\psi = \text{const.}$, are the rectangular hyperbolae

$$xy = \text{const.}$$

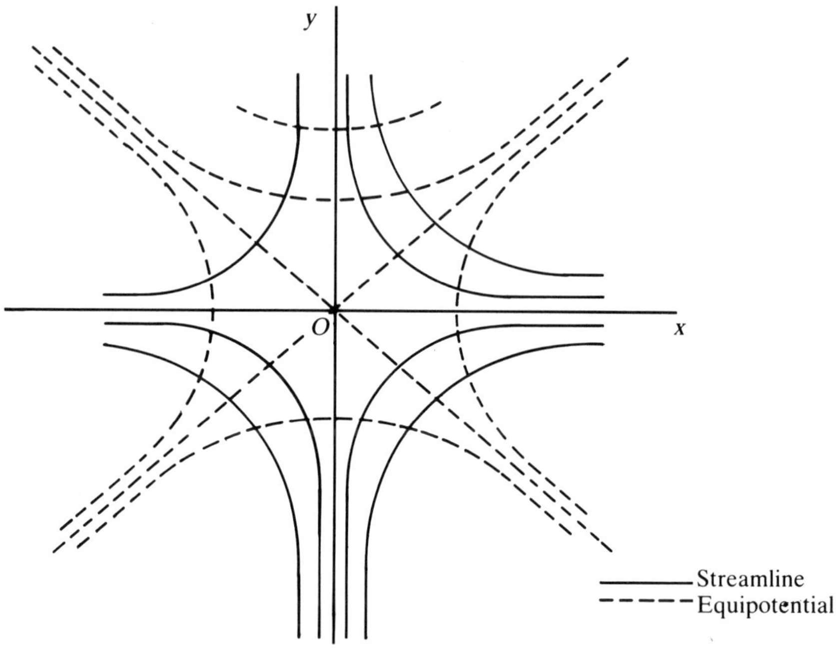

Fig. 6.2

having the axes $x = 0 = y$ as asymptotes. Fig. 6.2 shows both families of curves in the plane of flow.

Since there is no flow over a streamline, any one of them may be replaced by a rigid barrier. In particular, the axes $x = 0$, $y = 0$ may be taken as rigid boundaries.

As $\mathrm{d}w/\mathrm{d}z = u - iv = 2z$ is zero only at the origin, this is the only stagnation point in the region of flow.

It is easy to confirm that the two families of rectangular hyperbolae cut orthogonally in accordance with general theory.

It is also easy to show that $\nabla^2\phi = 0 = \nabla^2\psi$, where $\nabla^2 \equiv \partial^2/\partial x^2 + \partial^2/\partial y^2$. Again this accords with general theory which shows that ϕ and ψ are both two-dimensional harmonic functions.

6.4 EVALUATION OF STANDARD COMPLEX POTENTIALS

(i) Uniform stream

We first consider the case of a uniform stream whose velocity has constant magnitude U, its direction being inclined at angle α to the positive direction of the x-axis. Then, in standard notation,

$$u = U\cos\alpha, \qquad v = U\sin\alpha$$

and so

$$w'(z) = u - iv = U\,e^{-i\alpha}.$$

The simplest form for $w(z)$, ignoring integration constants, is therefore

$$w(z) = Uz\,e^{-i\alpha}. \tag{6.22}$$

From the form (6.22),

$$\begin{aligned}\phi(x, y) &= \text{Re}\,\{w(z)\}\\ &= U\,\text{Re}\,\{(x + iy)(\cos\alpha - i\sin\alpha)\}\\ &= U(x\cos\alpha + y\sin\alpha);\\ \psi(x, y) &= \text{Im}\,\{w(z)\}\\ &= U(y\cos\alpha - x\sin\alpha).\end{aligned}$$

Thus the equations of the equipotentials are

$$x\cos\alpha + y\sin\alpha = \text{const.}$$

These equations represent a family of parallel straight lines. The equations of the streamlines are

$$y\cos\alpha - x\sin\alpha = \text{const.}$$

These equations represent another family of parallel straight lines and they are inclined at α to the positive x-direction. The two families of straight lines intersect orthogonally in accordance with general theory.

(ii) Uniform line sources and links

Line sources and sinks were first introduced in Section 5.2. For convenience a somewhat different approach is taken here though the two are quite consistent.

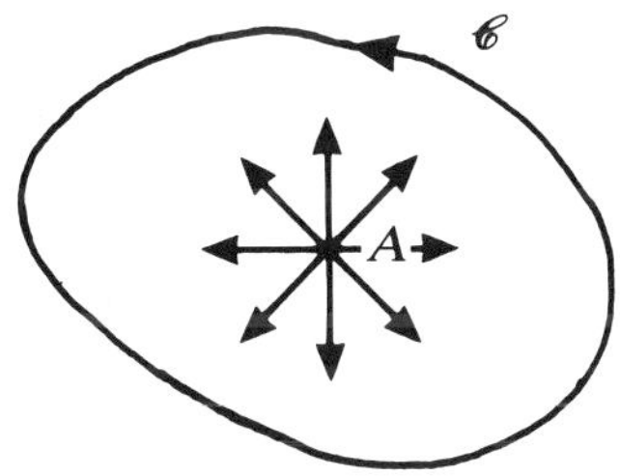

Fig. 6.3

In Fig. 6.3 let A be any point in the considered plane of flow and $\mathscr{C}$ **be** any closed curve embracing it. The surface of the infinite cylinder having its generators through the points of $\mathscr{C}$ and normal to the plane of flow is constructed. Suppose that in each plane of flow, fluid is emitted radially and symmetrically from all points on the infinite line through A normal to the plane of flow and such that the rate of emission from all such points as A is the same. The the line through A is called a *line source*. If fluid drains away through such a line and under the same conditions of symmetry, then the line is called a *line sink.*

Suppose the infinite line through A is a source and that it emits fluid at the rate of $2\pi m\rho$ units of mass per unit time, ρ being the fluid density (constant). Then this is the mass flux per unit time over unit length of the surface of the infinite cylinder through $\mathscr{C}$. We define the *strength of the line source to be m* which is clearly independent of $\mathscr{C}$. If for we now take the circle centre A and radius r, the cylinder has circular section and the speed of flow q is everywhere the same on it and the mass flux per unit time over unit length of the cylinder is $2\pi r\rho q$. By conservation of mass, this is equal to $2\pi m\rho$ and so

$$q = m/r. \tag{6.23}$$

On the surface of the cylinder the velocity potential $\phi = \phi(r)$ and

$$\mathrm{d}\phi/\mathrm{d}r = m/r$$

giving, in its simplest form,

$$\phi(r) = m \ln r. \tag{6.24}$$

If the considered field-point P on the circle $\mathscr{C}$ has plane polar coordinates (r, θ) with respect to A (origin), and if $\overline{AP} \equiv z = r\,\mathrm{e}^{i\theta}$, then since $\ln r = \ln|z|$,

$$\phi = \mathrm{Re}\,\{w(z)\} = m \ln|z|$$

and so the appropriate form for the complex potential at P due to the line source would seem to be

$$w(z) = m \ln z. \tag{6.25}$$

This is appropriate since $ln\ z$ is analytic. Since $ln\ z = ln\ r + i\theta$, this gives

$$\phi = m \ln r; \qquad \psi = m\theta$$

so that the equipotentials and streamlines have the respective forms

$$r = \text{const.}; \qquad \theta = \text{const.}$$

These accord with the physical reality of the situation. Reference back to equation (5.27) shows the value of ψ obtained there by rather different methods agrees with that obtained here using the complex potential of the uniform line source.

For a line source of uniform strength m per unit length through $z = z_0$ instead of $z = 0$, since $r = |z - z_0|$ now, the appropriate complex potential is

$$w(z) = m \ln(z - z_0). \tag{6.26}$$

The reader can verify that this form satisfies the Cauchy–Riemann equations and is analytic.

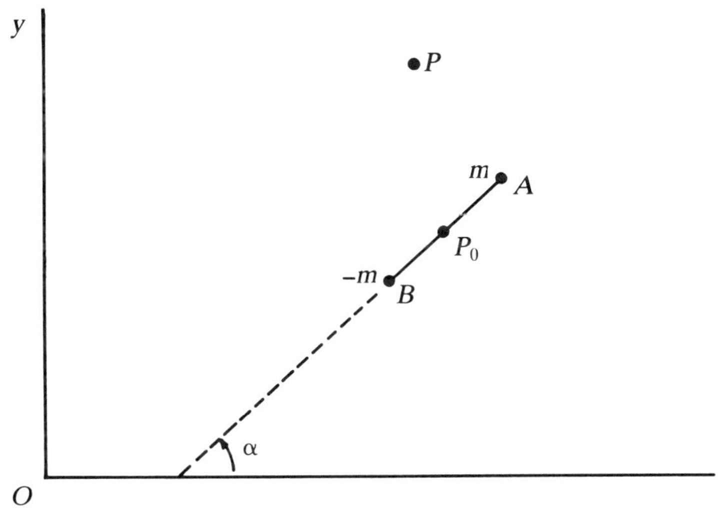

Fig. 6.4

(iii) Line source–sink pair and line doublet

A uniform line source of strength m per unit length runs through A perpendicularly to the plane of flow and an equal line sink, which is a line source of strength $-m$ per unit length, goes through B. The two constitute a *line source–sink pair* when the separation distance AB and m are both finite and non-zero. P_0 is the mid-point at AB and we take

$$\overrightarrow{OP}_0 \equiv z_0, \ \overrightarrow{OA} \equiv z_0 + h\,\mathrm{e}^{i\alpha}, \qquad \overrightarrow{OB} \equiv z_0 - h\,\mathrm{e}^{i\alpha},$$

where $BA = 2h$ and α is the angle between $\overrightarrow{Ox}$ and $\overrightarrow{BA}$. If P is a general field point in the plane of flow and $\overrightarrow{OP} \equiv z$, then

$$\overrightarrow{AP} \equiv z - z_0 - h\,\mathrm{e}^{i\alpha},$$

$$\overrightarrow{BP} \equiv z - z_0 + h\,\mathrm{e}^{i\alpha}$$

and so the complex potential at P due to the line pair is

$$w(z) = m\ \ln(z - z_0 - h\,\mathrm{e}^{i\alpha}) - m\ \ln(z - z_0 + h\,\mathrm{e}^{i\alpha}). \tag{6.27}$$

The *moment of the line source–sink pair* is μ defined to be

$$\mu = 2mh. \tag{6.28}$$

In the above, m, h, α are all real. We discuss two cases: follows.

(a) Let us first keep constant μ, z_0, α and let $m \to \infty$ and $h \to 0$ in such a way that μ remains constant. We have

$$w(z) = \mu\{\ln(z - z_0 - h\,\mathrm{e}^{i\alpha}) - \ln(z - z_0 + h\,\mathrm{e}^{i\alpha})\}/(2h).$$

We evaluate the limit of this as $h \to 0$, using l'Hôpital's rule to give

$$\lim_{h\to 0} w(z) = \frac{\mu}{2} \lim_{h\to 0} \left\{ \frac{-e^{i\alpha}}{z - z_0 - h\,e^{i\alpha}} - \frac{e^{i\alpha}}{z - z_0 + h\,e^{i\alpha}} \right\}$$

$$= -\frac{\mu\,e^{i\alpha}}{z - z_0}. \tag{6.29}$$

This is a finite limit and such a limiting distribution through P_0 perpendicular to the plane of flow is called a *line doublet of strength μ per unit length.* The *axis* of the line doublet is along $\overline{BA}$.

(b) Next, let us see what happens to the source–sink pair distribution when we let both m and $h \to \infty$, keeping z_0 and α fixed and also maintaining the constancy of the ratio m/h. We have, from (6.19) in standard notation,

$$w'(z) = u - iv = m \left\{ \frac{1}{z - z_0 - h\,e^{i\alpha}} - \frac{1}{z - z_0 + h\,e^{i\alpha}} \right\}$$

$$= \frac{-m\,e^{-i\alpha}}{h} \left\{ \left[1 - \frac{(z - z_0)\,e^{-i\alpha}}{h}\right]^{-1} + \left[1 + \frac{(z - z_0)\,e^{-i\alpha}}{h}\right]^{-1} \right\}$$

$$= -U\,e^{-i\alpha} \left[1 + O\left(\frac{1}{h^2}\right)\right],$$

where $U = m/h$, a finite constant. In the limit as $h \to \infty$,

$$w'(z) \to -U\,e^{-i\alpha}$$

and so

$$w(z) \to -Uz\,e^{-i\alpha}. \tag{6.30}$$

Comparison of this with (6.22) shows that this type of flow is uniform parallel flow making angle α with $\overrightarrow{xO}$.

Example [6.7]

Discuss the flow due to a uniform line doublet at O of strength μ per unit length its axis being along $\overrightarrow{OX}$.

The complex potential at $P(x, y)$ is

$$w = -\frac{\mu}{z} = -\frac{\mu(x - iy)}{x^2 + y^2}$$

so that

$$\phi = \frac{-\mu x}{x^2 + y^2},$$

$$\psi = +\frac{\mu y}{x^2 + y^2}.$$

Thus the equipotentials, $\phi = \text{const.}$, are the coaxial circles

$$x^2 + y^2 = 2k_1 x$$

and the streamlines $\psi = \text{const.}$, are the coaxial circles

$$x^2 + y^2 = 2k_2 y.$$

The first family have centres $(k_1, 0)$ and radii k_1: the second centres $(0, k_2)$ and radii k_2. Figure 6.5 shows some of each diagrammatically. The two families are mutually orthogonal.

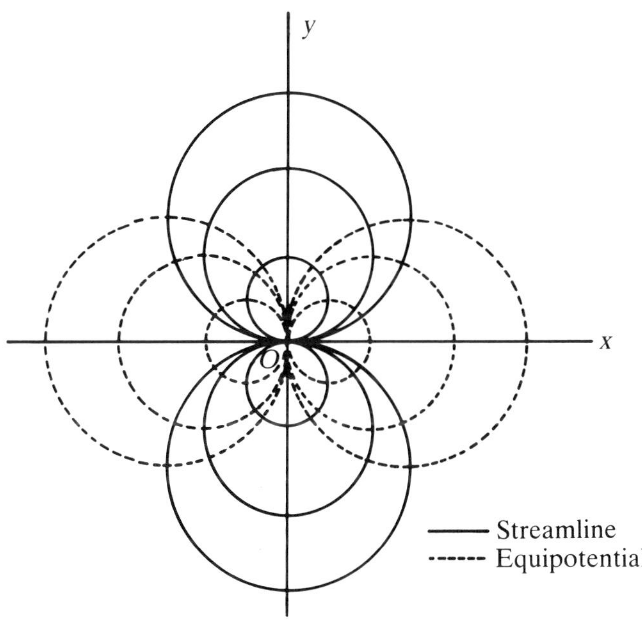

Fig. 6.5

Example [6.8]

Find the equations of the streamlines due to uniform line sources of strength m per unit length through the points $A(-c, 0)$, $B(c, 0)$ and a uniform line sink of strength $2m$ per unit length through the origin.

Let $P(x, y)$ be chosen in the plane of flow so that

$$\overrightarrow{OP} \equiv z = x + iy, \qquad \overrightarrow{AP} \equiv z + c, \qquad \overrightarrow{BP} \equiv z - c.$$

Then the complex potential at P due to the distribution is

$$
\begin{aligned}
w &= m\ln(z+c) + m\ln(z-c) - 2m\ln z \\
&= m\ln\left[\frac{z^2-c^2}{z^2}\right] \\
&= m\ln\left[\frac{x^2-y^2-c^2+2ixy}{x^2-y^2+2ixy}\right] \\
&= m\ln\left[\frac{(x^2-y^2-c^2+2ixy)\,(x^2-y^2-2ixy)}{(x^2-y^2)^2+4x^2y^2}\right] \\
&= m\ln\left[\frac{(x^2+y^2)^2-c^2\,(x^2-y^2)+2ic^2xy}{(x^2-y^2)^2+4x^2y^2}\right].
\end{aligned}
$$

Thus

$$
\psi = \mathrm{Im}(w) = m\tan^{-1}\left[\frac{2c^2xy}{(x^2+y^2)^2-c^2\,(x^2-y^2)}\right].
$$

The equations of the streamlines are given by ψ = const., i.e. $\tan(\psi/m)$ = const. On simplification this leads to

$$
(x^2+y^2)^2 = c^2(x^2-y^2+kxy),
$$

where k = const.

(iv) **Line vortices**

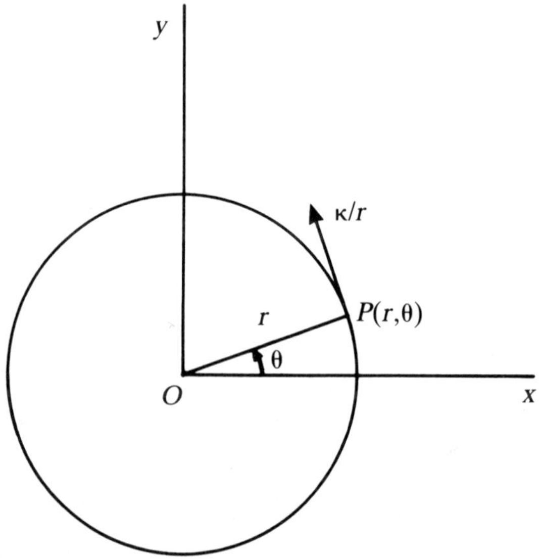

Fig. 6.6

In Section 5.2 the notion of a line vortex was introduced. From the equations (5.27–5.34) inclusive we frame the following definition of a *uniform line vortex*. Suppose that an infinite line runs through the point O perpendicularly to the plane of flow in Fig. 6.6 and such that the fluid velocity $\mathbf{q}$ at $P(r, \theta)$ has the radial and transverse components

$$q_r = 0, \qquad q_\theta = \kappa/r, \qquad (\kappa = \text{const.}) \tag{6.31}$$

The streamlines in the plane of flow are the concentric circles $r = \text{const.}$ and we note that as $r \to \infty$, $q_\theta \to 0$. Also as $r \to 0$, $q_\theta \to \infty$. The particles of fluid describe concentric circles centred on O in the plane of flow. The speeds of the particles increase to infinity as we pass from one concentric circle to another inwards towards the origin. Such a flow is said to be due to a uniform line vortex through O with κ denoting the *strength of the line per unit length*. The circulation around a closed curve embracing the line vortex is found to be $2\pi\kappa$ in accordance with equation (5.30). The circulation around a closed curve not embracing the line vortex is zero.

It is easy to show that if $\mathbf{q}$ is the fluid velocity at all points $P(r, \theta)$, other than $r = 0$, for the above distribution, then curl $\mathbf{q} = \mathbf{0}$. This means that we can find a velocity potential $\phi(r, \theta)$ such that $\mathbf{q} = \nabla\phi$, giving

$$\left.\begin{aligned} q_r &= \frac{\partial\phi}{\partial r} = 0 \\ q_\theta &= \frac{1}{r}\frac{\partial\phi}{\partial\theta} = \frac{\kappa}{r} \end{aligned}\right\}. \tag{6.32}$$

The first of the two equations (6.32) shows that $\phi = \phi(\theta)$ and so the second gives $\phi'(\theta) = \kappa$. Thus

$$\phi = \phi(\theta) = \kappa\theta. \tag{6.33}$$

Let us now seek an analytic function $w(z)$ whose real part is $\kappa\theta$. Since $z = r\,e^{i\theta}$, $\ln z = \ln r + i\theta$ and so $-i \ln z = \theta - i \ln r$. The suggested form for the complex potential $w(z) = \phi + i\psi$ is then

$$w(z) = -\,i\kappa \ln z. \tag{6.34}$$

On equating the real and imaginary parts in (6.34), we recover

$$\phi = \kappa\,\theta, \qquad \psi = -\kappa \ln r. \tag{6.35}$$

The form obtained for ψ in (6.35) agrees with that in (5.32). Thus (6.34) is the appropriate complex potential for the uniform line vortex through O. The equations (6.35) show that the equipotentials for the line vortex are the straight rays $\theta = \text{const.}$ through O and the streamlines are the concentric circles $r = \text{const.}$, centred on O and cutting the rays orthogonally. The form (6.34) is analytic at all parts of the z-plane save at the origin $z = 0$ through which the line vortex passes and which gives rise to a logarithmic type of singularity.

Summarising the cardinal features of the line vortex and the field of flow it produces, as obtained in Section 5.2 and here, we may say that *a two-dimensional hydrodynamical distribution having the complex potential* $w(z) = -i\kappa \ln z$, *where* κ *is a real constant, gives a circulation round any closed curve* $\mathscr{C}$ *embracing O of amount* $2\pi\kappa$. *Also round any other closed curve* $\mathscr{C}$ *which does not contain O the circulation is zero. Further, the streamlines are the concentric circles* $r =$ *const. on which the fluid speed increases to infinity as* $r \to 0$. *The equipotentials are the radial lines through O intersecting the circles orthogonally.*

Example [6.9]

Two parallel line vortices of strengths κ_1, κ_2 per unit length ($\kappa_1 + \kappa_2 \neq 0$) in unlimited liquid cross a plane of flow at A, B respectively. The centroid of point masses κ_1 at A and κ_2 at B is G. If the motion of the liquid is due solely to these vortices, show that G is a fixed point around which A and B move in circular orbits with constant angular velocity.

Show also that the fluid speed at any point P in the plane is

$$(\kappa_1 + \kappa_2)PC/(AP\,.\,PB),$$

where C is the centroid of point masses κ_2 at A and κ_1 at B.

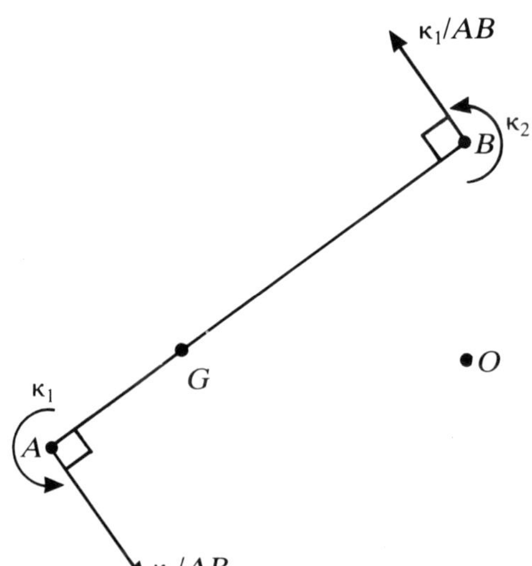

Fig. 6.7

From the equations (6.32), we see that

κ_1 at A gives to B a velocity κ_1/AB perpendicular to AB;

κ_2 at B gives to A a velocity κ_2/AB perpendicular to AB.

The senses of these velocities are shown in Fig. 6.7. Associating point masses κ_1 at A, κ_2 at B and allowing them to have the respective fluid velocities at A and B the total linear momentum of the two particles measured in the direction of the velocity of B is

$$\kappa_2(\kappa_1/AB) + \kappa_1(-\kappa_2/AB) = 0.$$

Thus the centroid of the masses does not move.

Fig. 6.7 shows that the velocity of B relative to A is $(\kappa_1 + \kappa_2)/AB$. Thus the line AB rotates about G with angular velocity $(\kappa_1 + \kappa_2)/AB^2$.

Let O be the fixed origin of coordinates in the plane of flow and suppose $\overrightarrow{OA} \equiv a$, $\overrightarrow{OB} \equiv b$, $\overrightarrow{OP} \equiv z$, where a, b, z are complex and P is any point in the plane. Then $\overrightarrow{AP} \equiv z - a$, $\overrightarrow{BP} \equiv z - b$ and so the complex potential at P is

$$w = -i[\kappa_1 \ln(z-a) + \kappa_2 \ln(z-b)].$$

Thus

$$\frac{dw}{dz} = -i\left[\frac{\kappa_1}{z-a} + \frac{\kappa_2}{z-b}\right] = -i\,\frac{[\kappa_1(z-b) + \kappa_2(z-a)]}{(z-a)(z-b)}.$$

Since C is the centroid of point masses κ_2 at A and κ_1 at B,

$$\kappa_2\,\overrightarrow{AP} + \kappa_1\,\overrightarrow{BP} = (\kappa_1 + \kappa_2)\,\overrightarrow{CP}$$

or

$$\kappa_2(z-a) + \kappa_1(z-b) = (\kappa_1 + \kappa_2)(z - \bar{z}),$$

where $\overrightarrow{OC} \equiv \bar{z}$. Thus

$$\frac{dw}{dz} = -i\,\frac{(\kappa_1 + \kappa_2)(z - \bar{z})}{(z-a)(z-b)},$$

and so the speed of P is

$$\left|\frac{dw}{dz}\right| = \frac{(\kappa_1 + \kappa_2)\,|z - \bar{z}|}{|z-a| \times |z-b|} = \frac{(\kappa_1 + \kappa_2)PC}{AP.\,PB}.$$

6.5 IMAGE SYSTEMS IN PLANE FLOWS

Image systems for two-dimensional flows may be determined. They are analogues of such systems in three dimensions but whereas such image systems comprise point or finite line singularities, the two-dimensional images necessarily involved infinite line singularities, perpendicular to the plane of flow, and include line sources, line sinks and line doublets. We first find the line images for a number of such cases. The methods used will afford good illustrations of many of the hydrodynamical principles that have been presented. Thus, for example, if we recall that no fluid crosses a streamline, then the flow is unaltered by introducing a rigid impermeable cylindrical barrier through the streamline with its generators perpendicular to the plane of flow and extending to infinity in either direction. As will become evident, this leads to establishing the appropriate system of images in the barrier for the flow on either side. Mathematical justification of the process is covered by appropriate uniqueness theorems.

In the following the plane of flow is specified by rectangular coordinates (x, y) and by polar coordinates (r, θ).

(i) Image of a line source in a plane

Without loss of generality we take the rigid impermeable plane to be $x = 0$ and perpendicular to the plane of flow. Suppose a uniform line source of strength m per unit length runs through the point $z = z_1$ parallel to the rigid plane, where $\mathrm{Re}(z_1) > 0$. Let us now remove the barrier and assume that the fluid occupies the whole of the z-plane and introduce an equal line source through $z = -\bar{z}_1$, the reflection of $z = z_1$ in the plane $x = 0$. We assert that so far as the fluid on the same side of $x = 0$ as the line source through $z = z_1$ is concerned, the flow patterns are identical. We establish this as follows.

The complex potential due to the two line sources is

$$w(z) = m \ln (z - z_1) + m \ln (z + \bar{z}_1),$$

a function analytic in all finite parts of the plane except at the singularities $z = z_1$, $z = -\bar{z}_1$. Then

$$dw/dz = u - iv = m\{(z - z_1)^{-1} + (z + \bar{z}_1)^{-1}\}.$$

At the point $z = iy$ on the plane $x = 0$,;

$$\begin{aligned} u - iv &= m\{(iy - z_1)^{-1} + (iy + \bar{z}_1)^{-1}\} \\ &= m\{(\bar{z}_1 + iy)^{-1} - (z_1 - iy)^{-1}\}. \end{aligned}$$

The latter form is the difference between two complex conjugates, which is necessarily pure imaginary. Hence at any point on the plane $x = 0$, $u = 0$ and $v \neq 0$ (in general). Thus no fluid crosses any point of the plane $x = 0$ and so the flow pattern on either side of $x = 0$ is unaltered by introducing a rigid impermeable plane through $x = 0$ perpendicular to the plane of flow.

We have established, then, that the *image of a uniform line source in a rigid impermeable infinite plane parallel to it is another line source of equal strength through the reflection of the original line in the plane.*

(ii) Image of a line-doublet in a plane

Fig. 6.8 shows the plane of flow for uniform line sources of strengths $-m$ through A, A' and $+m$ through B, B', where A and B are in the region $x > 0$ and the primed points are the reflections of the corresponding unprimed ones in the plane $x = 0$ through Oy perpendicular to the plane of flow. Thus A' and B' lie in the region $x < 0$. The line $\overrightarrow{AB}$ makes angle α with $\overrightarrow{Ox}$, so that $\overrightarrow{A'B'}$ makes

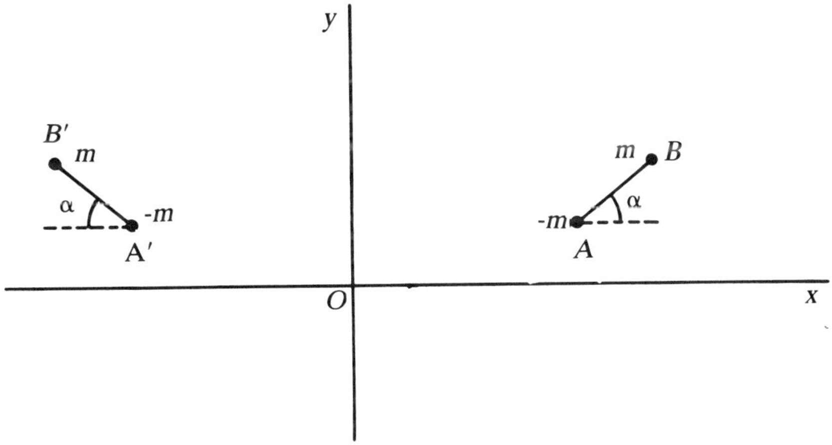

Fig. 6.8

angle $(\pi - \alpha)$ with $\overrightarrow{Ox}$. The hydrodynamical distribution $-m$ through A' is the image of $-m$ through A in the plane $x = 0$ and m through B' is the image of m through B in the plane $x = 0$. There is no flow over the plane $x = 0$ and so a rigid impermeable barrier of infinite extent may be introduced over this plane without disturbing the flow. If we keep α fixed and carry out the double limiting process described in Section 6.4 (iii) (letting $m \to \infty$ and B to approach A along $\overrightarrow{BA}$ and B' to approach A' along $\overrightarrow{B'A'}$), then ultimately we obtain at A and A' a pair of equal line doublets with their axes making angles α, $\pi - \alpha$ to the direction $\overrightarrow{Ox}$. Either line doublet is the hydrodynamical image of the other in the infinite rigid impermeable plane $x = 0$.

(iii) Image of a line vortex in a plane

Let us find the image in a plane of a uniform line vortex of strength κ per unit length through $z = z_1$ in the plane of flow, where $\text{Re}(z_1) > 0$. Introduce a line vortex of strength $-\kappa$ through $z = -\bar{z}_1$, the reflection of z_1 in the plane $x = 0$. The complex potential due to the two line vortices with the fluid now occupying the whole of the z-plane, is

$$w(z) = -i\kappa\left\{\ln(z - z_1) - \ln(z + \bar{z}_1)\right\}.$$

Thus

(z $$w'(z) = -i\kappa\left\{z - z_1)^{-1} - (z + \bar{z}_1)^{-1}\right\}.$$

At the point $z = iy$ on the plane $x = 0$,

$$\begin{aligned} w'(z) &= -i\kappa\left\{(iy - z_1)^{-1} - (iy + \bar{z}_1)^{-1}\right\} \\ &= i\kappa\left\{(z_1 - iy)^{-1} + (\bar{z}_1 + iy)^{-1}\right\}. \end{aligned}$$

Since $(z_1 - iy)^{-1} + (\bar{z}_1 + iy)^{-1}$ is the sum of the two complex conjugates this sum is real and so $w'(z)$ is pure imaginary. Thus on $x = 0$,

$$u = \text{Re}\{w'(z)\} = 0,$$

i.e. no fluid crosses the plane $x = 0$ and so a rigid impermeable infinite boundary may be introduced at this location. *Thus the hydrodynamical image of a line vortex in a rigid impermeable infinite plane is an equal and opposite line vortex at the reflection of the original line in the plane.*

An alternative procedure is the following. From the form $w(z)$ cited above for the two line vortices,

$$\begin{aligned}\psi &= \text{Im}\{w(z)\} \\ &= -\kappa \text{ Re}\{\ln(z - z_1) - \ln(z - \bar{z}_1)\} \\ &= -\kappa \{\ln|z - z_1| - \ln|z - \bar{z}_1|\} \\ &= 0 \text{ when } z = iy,\end{aligned}$$

since any point $z = iy$ on $x = 0$ is equi-distant from the complex conjugates z_1, $\bar{z}_1$. Thus $x = 0$ is a streamline across which there is no flow and through which a rigid impermeable infinite plane barrier may be introduced. For if two systems of singularities, one on either side of a curve $\mathscr{C}$, are image systems in $\mathscr{C}$, then necessarily $\psi =$ const. along $\mathscr{C}$.

Example [6.10]

A two-dimensional doublet of strength $\mu\mathbf{i}$ per unit length is at the point $z = ia$ in a uniform stream of velocity $-V\mathbf{i}$ in a semi-infinite liquid of constant density occupying the half-plane $y > 0$ and having $y = 0$ as a rigid impermeable boundary, $\mathbf{i}$ being the unit vector in the positive x-axis. Find the complex potential of the motion. If $0 < \mu < 4a^2 V$, show that there are no stagnation points on the boundary. Show also that the pressure on it is a minimum at the origin and a maximum at the points $(\pm a\sqrt{3}, 0)$.

Fig. 6.9(i) displays the actual flow. Fig. 6.9(ii) the equivalent flow due to an image system consisting of equal line doublets of strength $\mu\mathbf{i}$ per unit length at $A(0, a)$, $A'(0, -a)$. The velocity potential due to the stream is $-Vx = -V\,\text{Re}(z)$, and so its complex potential is the analytic function $-Vz$. The complex potentials due to the line doublets through A, A' are respectively $-\mu(z - ia)^{-1}$, $-\mu(z + ia)^{-1}$, since $\overline{AP} \equiv z - ia$, $A'P \equiv z + ia$. Thus the total complex potential due to the distribution in Fig. 6.9(ii) is

$$\begin{aligned}w(z) &= -Vz - \mu(z - ia)^{-1} - \mu(z + ia)^{-1} \\ &= -Vz - 2\mu z(z^2 + a^2)^{-1}.\end{aligned}$$

Thus at $(x, 0)$ on the boundary,

$$w'(x) = -V - 2\mu(a^2 - x^2)(a^2 + x^2)^{-2},$$

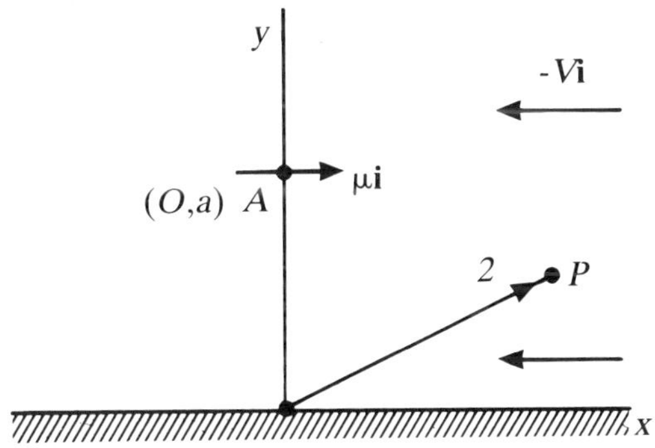

Fig. 6.9 (i) Actual flow.

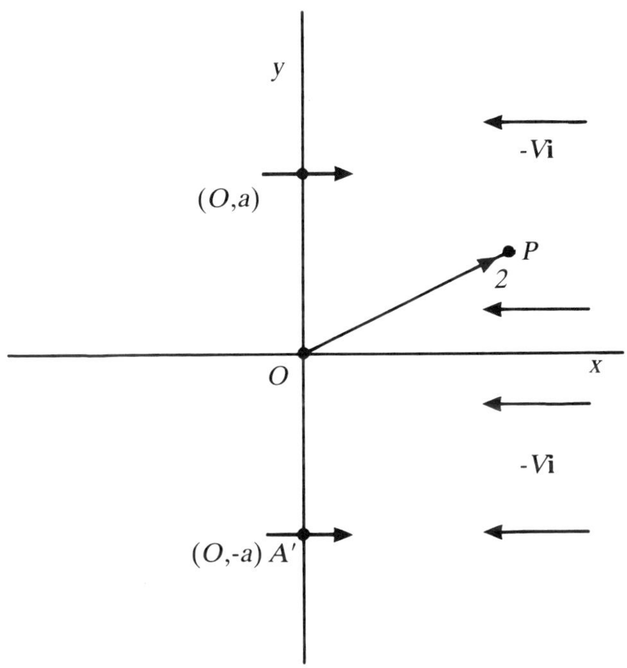

Fig. 6.9(ii) Equivalent flow

which gives the speed of the fluid at $(x, 0)$. Stagnation points on the boundary, if they exist, are given by $w'(x) = 0$, i.e. by

$$Vx^4 + 2(Va^2 - \mu)x^2 + (Va^2 + 2\mu)a^2 = 0.$$

This is a quadratic equation in x^2 and its discriminant is

$$\Delta = 4\{(Va^2 - \mu)^2 - V(Va^2 + 2\mu)a^2\} = 4\mu(\mu - 4a^2 V).$$

Clearly $\Delta < 0$ when $0 < \mu < 4a^2 V$, showing that the quadratic equation has no real roots when this latter inequality holds. This means there are then no stagnation points on the boundary $y = 0$.

Applying Bernoulli's equation along the streamline $y = 0$ we have, at $(x, 0)$ on the boundary,

$$\frac{p}{\rho} + \frac{1}{2}\left\{V + \frac{2\mu(a^2 - x^2)}{(a^2 + x^2)^2}\right\}^2 = \text{const.}$$

Thus p is a maximum when $\{V + 2\mu(a^2 - x^2)(a^2 + x^2)^{-2}\}^2$ is a minimum and conversely. Write $y(x)$ for this last expression: then

$$y^{1/2} = V + 2\mu(a^2 - x^2)(a^2 + x^2)^{-2},$$

and so

$$\tfrac{1}{2}y^{-1/2}y' = -4\mu x(3a^2 - x^2)(a^2 + x^2)^{-3}.$$

Thus $y' = 0$ when $x = 0$ or $\pm a\sqrt{3}$. The last expression shows that $y' < 0$ when $x = 0+$ and $y' > 0$ when $x = 0-$. Thus y is a maximum and hence p is a minimum at the origin. The same expression shows that when $x = a\sqrt{3}+$, $y' > 0$ and when $x = a\sqrt{3}-, y' < 0$ so that at $x = a\sqrt{3}, y$ is a minimum and so p is a maximum. At $x = -a\sqrt{3}+, y' > 0$ and at $x = -a\sqrt{3}-, y' > 0$ showing p is also a maximum at $x = -a\sqrt{3}$.

6.6 THE MILNE–THOMSON CIRCLE THEOREM

In the previous section images in a plane of specified flow distribution were found. It is of some importance to obtain the images of such distributions in an impermeable circular cylindrical boundary the generators of which extend to infinity in either direction and are perpendicular to the considered plane of flow. For this purpose we establish a theorem due to L. M. Milne–Thomson. Prior to this, however, we explain some terminologies and develop some properties of conjugate complex functions.

In the following, z is complex and x is real. Let us first suppose that $f(x)$ is a complex function of the real variable x where

$$f(x) = U(x) + iV(x), \tag{6.36}$$

$U(x)$ and $V(x)$ both being real functions of the real variable x. Then

$$U(x) = \text{Re}\{f(x)\}, \quad V(x) = \text{Im}\{f(x)\}$$

and the conjugate function $\bar{f}(x)$ is defined to be

$$\bar{f}(x) = U(x) - iV(x). \tag{6.37}$$

If we now replace x by z then, following the pattern of (6.36) and (6.37), we define $f(z)$ and $\bar{f}(z)$ to be

$$f(z) = U(z) + iV(z), \tag{6.38}$$

$$\bar{f}(z) = U(z) - iV(z). \tag{6.39}$$

Further, if $f(\bar{z}), \bar{f}(\bar{z})$ are defined to be

$$f(\bar{z}) = U(\bar{z}) + iV(\bar{z}), \tag{6.40}$$

$$\bar{f}(\bar{z}) = U(\bar{z}) - iV(\bar{z}). \tag{6.41}$$

For if $z = x + iy$, then $\bar{f}(\bar{z})$ is the complex conjugate of $f(z)$ and has the value $\bar{f}(x - iy)$, since $\bar{z} = x - iy$. We may write

$$\bar{f}(\bar{z}) = \overline{f(z)}. \tag{6.42}$$

By way of example, if

$$f(z) = Uz\, e^{i\alpha} \qquad (u,\ \alpha \text{ real}),$$

then

$$\bar{f}(z) = Uz\, e^{-i\alpha}.$$

This 'partial complex conjugate' is formed by taking the complex conjugate of everything save z. Then the '[illegible]al complex conjugate' is

$$\bar{f}(\bar{z}) = U\bar{z}\, e^{-i\alpha} = \overline{f(z)}. \tag{6.43}$$

Let us next recall that each of the following functions has a singularity at $z = z_1$:

(i) $m \ln(z - z_1)$, (ii) $-\mu\, e^{i\alpha}(z - z_1)^{-1}$, (iii) $-i\,\kappa \ln(z - z_1)$.

The first and last are logarithmic singularities while the second is a simple pole with residue $-\mu e^{i\alpha}$. Considered as complex potentials for two-dimensional distributions with m, μ, α, κ all real, they represent the respective distributions

(i) line source of strength m per unit length;
(ii) line doublet of strength μ per unit length with axis at α to $\overrightarrow{Ox}$;
(iii) line vortex of strength κ per unit length.

In each case the line distribution is through $z = z_1$ and is normal to the plane of flow.

Next, we note that the analytical singularities of $f(z)$ are the aggregate of those of both $U(z)$ and $V(z)$ and hence they are also those of $\bar{f}(z)$. Moreover, these analytical singularities are, as we have seen, easily identifiable as corresponding hydrodynamical distributions.

Statement of the Milne–Thomson circle theorem

Let $f(z)$ be the complex potential for a flow having no rigid boundaries and such that there are no singularities within the circle $|z| = a$. Then on introducing the *solid cylinder $|z| = a$, with impermeable boundary, into the flow, the new complex potential for the fluid outside the cylinder is given by $w = f(z) + \bar{f}(a^2/z)$, for $|z| \geqslant a$.*

Proof: All singularities of $f(z)$ (through which line sources, line sinks, line doublets or line vortices may be present) occur in the region $|z| > a$. Hence the singularities of $f(a^2/z)$ occur in the region $a^2/|z| > a$, or $|z| < a$. Thus the singularities of $\bar{f}(a^2/z)$ also lie in the region $|z| < a$. It follows that within the region $|z| > a$, the functions $f(z)$ and $f(z) + \bar{f}(a^2/z)$ both have the same analytical singularities. Thus both functions considered as complex potentials represent the same hydrodynamical distributions in the region $|z| > a$.

The proof of the theorem is now completed by considering what happens on the circular boundary $|z| = a$. To this end we write $z = a\,e^{i\theta}$, on the boundary where θ is real. Then $a^2/z = a\,e^{-i\theta} = \bar{z}$ on the circular boundary. Thus on the boundary $|z| = a$,

$$w = f(z) + \bar{f}(a^2/z) = f(z) + \bar{f}(\bar{z}).$$

The latter expression is entirely real, being the sum of two complex conjugates. Hence on the boundary,

$$\psi = \operatorname{Im}\{w\} = 0.$$

This shows that the circular boundary is a streamline across which no fluid flows. Hence $|z| = a$ is a possible boundary for the flow specified by the complex potential $w(z) = f(z) + \bar{f}(a^2/z)$.

We have now shown that $w(z)$ is appropriate as the complex potential not only for the region of fluid flow $|z| > a$ but also for the boundary $|z| = a$, as was required.

We note that $\bar{f}(a^2/z)$ represents the perturbation to the complex potential $f(z)$ caused by the insertion of the cylinder $|z| = a$. It also gives the distribution of the image system within the cylinder's interior $|z| < a$ for the actual distribution in the external fluid region $|z| > a$.

Example [6.11]

Uniform flow past a stationary cylinder.

The velocity potential due to an undisturbed uniform stream having velocity $-U\mathbf{i}$ (U real) is $-Ux = -U\,\mathrm{Re}(z)$. Since z is an analytic function, the corresponding complex potential is

$$f(z) = -Uz.$$

Hence

$$\bar{f}(z) = -\bar{U}z = -Uz,$$

and so

$$\bar{f}(a^2/z) = -Ua^2/z.$$

With the cylinder $|z| = a$ present, the complex potential for the fluid region $|z| \geqslant a$ is

$$w(z) = f(z) + \bar{f}(a^2/z) = -U(z + a^2/z).$$

Taking $z = r\,e^{i\theta}$, where $r \geqslant a$, equating real and imaginary parts of w gives

$$\phi = \mathrm{Re}\{w(z)\} = -U\cos\theta\,(r + a^2/r),$$

$$\psi = \mathrm{Im}\{w(z)\} = -U\sin\theta\,(r - a^2/r).$$

The problem was treated by another method in Example [4.10].

The perturbation term $\bar{f}(a^2/z) = -Ua^2/z$ gives the image of the flow in the cylinder. This image represents a uniform line doublet of strength Ua^2 per unit length and axis in the direction **i**.

Example [6.12]
Cylinder $|z| = a$ in ambient uniform stream at angle α to $\overrightarrow{xO}$.

The undisturbed velocity potential of the flow is

$$f(z) = U\,e^{-i\alpha}z.$$

Then

$$\bar{f}(z) = U\,e^{i\alpha}z,$$

on forming the conjugate of everything except z. Hence

$$\bar{f}(a^2/z) = U\,e^{i\alpha}a^2/z.$$

With the cylinder $|z| = a$ present, the complex potential for the flow in $|z| \geqslant a$ is $|z| \geqslant a$ is

$$\begin{aligned} w(z) &= f(z) + \bar{f}(a^2/z) \\ &= U(e^{-i\alpha}z + e^{i\alpha}a^2/z). \end{aligned}$$

Example [6.13]
Image of a line source in a circular cylinder.

Suppose a uniform line source of strength m per unit length passes through $z = d$ in the plane of flow, where $d > a$. The complex potential due solely to this distribution is

$$f(z) = m\ln(z - d).$$

For this function, since m and d are real,

$$\bar{f}(z) = m\ln(z - d),$$

and so

$$\bar{f}(a^2/z) = m \ln(a^2/z - d)$$
$$= m \ln(z - a^2/d) - m \ln z + \text{const.}$$

The constant $\ln(-d)$ is immaterial since velocities are found from the derivative of the complex potential. The last form gives the image system of the line source at $z = d\ (> a)$ in the circular cylinder $|z| = a$: this image system consists of:

(i) an equal line source of strength m per unit length through the inverse point $z = a^2/d$ in the circular section;
(ii) a line sink of strength m per unit length through the centre $z = 0$ of the circular section.

With the cylinder $|z| = a$ present, the total complex potential for the fluid region $|z| \geqslant a$ is

$$w = f(z) + \bar{f}(a^2/z)$$
$$= m \ln(z - d) + m \ln(z - a^2/d) - m \ln z + \text{const.}$$

Example [6.14]
Line doublet parallel to the axis of a right circular cylinder.

Suppose a uniform doublet of strength μ per unit length runs through the point $z = d(> a)$ perpendicularly to the plane of flow. Then if the axis of the line doublet makes angle α with $\overrightarrow{Ox}$, the complex velocity potential due solely to this distribution is

$$f(z) = -\frac{\mu\,e^{i\alpha}}{z - d}.$$

Then

$$\bar{f}(z) = -\frac{\mu\,e^{-i\alpha}}{z - d} \qquad (\mu, \alpha, d \text{ real}),$$

and so

$$\bar{f}\left(\frac{a^2}{z}\right) = -\frac{\mu\,e^{-i\alpha}}{(a^2/z) - d} = \frac{\mu\,e^{-i\alpha}}{d}\left[1 + \frac{a^2/d}{z - (a^2/d)}\right].$$

Thus

$$\bar{f}\left(\frac{a^2}{z}\right) = \left(\frac{a}{d}\right)^2 \mu\,e^{-i\alpha}\,\frac{1}{z - (a^2/d)} + \text{const.}$$
$$= -(a/d)^2\,\mu \exp\{i(\pi - \alpha)\}/\{z - (a^2/d)\} + \text{const.}$$

The constant is immaterial and the above form is recognised as the complex potential due to a uniform line doublet of strength $\mu(a/d)^2$ per unit length passing through $z = a^2/d$, the point inverse in the circle to the original distribution through $z = d$, and the axis of this doublet is at $\pi - \alpha$ to $\overline{Ox}$. This is the image of the given line doublet in the cylinder.

Example [6.15]

A line vortex of circulation $2\pi\kappa$ is at rest at the point $z = na (n > 1)$, in the presence of a plane circular impenetrable boundary $|z| = a$, around which there is circulation $2\pi\lambda\kappa$. Show that $\lambda = 1/(n^2 - 1)$.

Show that there are two stagnation points $z = a\,e^{i\theta}$ on the circular boundary, symmetrically placed about the real axis in the quadrants nearest to the vortex, the appropriate values of θ, $2\pi - \theta$ satisfying $\cos\theta = (3n^2 - 1)/(2n^3)$. Prove that θ is real.

The strength per unit length of the line vortex through $z = na$ is $2\pi\kappa$ and so the complex potential due to it alone is

$$f(z) = -i\kappa\, ln\,(z - na).$$

Since κ, n, a and the function form ln are all real,

$$\bar{f}(z) = i\kappa \ln\,(z - na),$$

and so

$$\bar{f}(a^2/z) = i\kappa \ln\{(a^2/z) - na\}.$$

When the solid cylinder $|z| = a$ is introduced, N.B. the velocity potential in the region $|z| \geqslant a$ becomes $f(z) + \bar{f}(a^2/z)$, in the absence of other distributions. However, there is a circulation $2\pi\lambda\kappa$ round the cylinder and as this is equivalent to a line vortex through $z = 0$ of strength $2\pi\lambda\kappa$ per unit length, we must add on to this last expression an additional term $-i\lambda\kappa \ln z$ giving for the fluid region $|z| \geqslant a$ the total complex potential

$$\begin{aligned} w(z) &= -i\kappa \ln\,(z - na) + i\kappa \ln\{(a^2/z) - na\} - i\lambda\kappa \ln z \\ &= -i\kappa \ln\,(z - na) - i\kappa\,(1 + \lambda) \ln z + i\kappa \ln\,(z - a/n) + \text{const.} \end{aligned}$$

This form for $w(z)$ for $|z| \geqslant a$ is equivalent to the following hydrodynamical distributions:

(i) a line vortex through $z = na$ (point A) of strength $2\pi\kappa$ per unit length;
(ii) a line vortex through $z = 0$ (point O) of strength $2\pi(1 + \lambda)\kappa$ per unit length;
(iii) a line vortex through $z = a/n$ (point A') of strength $-2\pi\kappa$ per unit length.

This distribution is shown in Fig. 6.10.

The velocity of A due to the other two distributions is

$$2\pi(1 + \lambda)\kappa/OA - 2\pi\kappa/A'A$$

perpendicular to the vortex line. As the vortex at A is at rest,

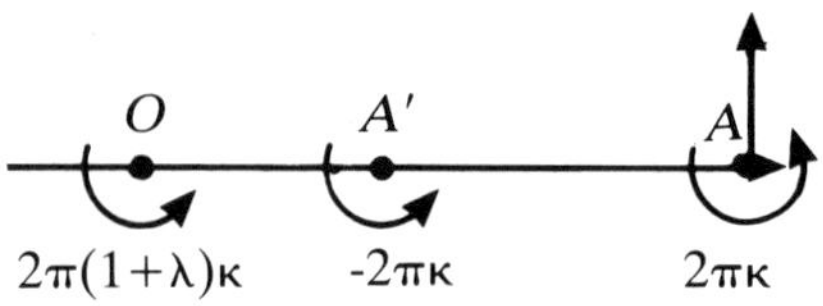

Fig. 6.10

$$(1+\lambda)AA' = OA \qquad (AA' = na - a/n,\ OA = na).$$

Thus

$$\lambda = 1/(n^2-1).$$

From $w(z)$,

$$\frac{\mathrm{d}w}{\mathrm{d}z} = -i\kappa\left\{\frac{1}{z-a} + \frac{\lambda+1}{z} - \frac{1}{z-(a/n)}\right\}.$$

Putting $z = a\,\mathrm{e}^{i\theta}$ and simplifying gives

$$\frac{\mathrm{d}w}{\mathrm{d}z} = (i\kappa\,\mathrm{e}^{i\theta})\,\frac{2n^3\cos\theta - 3n^2 + 1}{a(n^2-2n\cos\theta+1)(n^2-1)}.$$

The stagnation points on the circle, if any, are given by $\mathrm{d}w/\mathrm{d}z = 0$ for $z = a\,\mathrm{e}^{i\theta}$. This requires that

$$\cos\theta = (3n^2-1)/(2n^3).$$

The plausibility of this requires that the right-hand side lies between -1 and $+1$. Let us write

$$f(n) = \frac{3}{2n} - \frac{1}{2n^3}.$$

Then

$$f(1) = 1.$$

Also,

$$f'(n) = -\frac{3}{2n^2} + \frac{3}{2n^4} = \frac{3}{2n^4}(1-n^2).$$

Then

$$f'(n) < 1 \text{ for } n > 1.$$

Thus for $n > 1, f(n)$ decreases monotonically from 1 at $n = 1$ to 0 as $n \to \infty$. For all $n > 1$, then, real values of θ are obtained from $\cos\theta = (3n^2-1)/(2n^3)$. Two distinct values of θ are obtained for any given $n > 1$: one is $\theta = \alpha$, where $0 < \alpha \leqslant \pi/2$, the other is $\theta = 2\pi - \alpha$.

6.7 EXTENSION OF THE CIRCLE THEOREM

The following is an immediate extension of the Milne–Thomson circle theorem. Its proof is similar to that of the main theorem and will not be reproduced. The statement is:

Let $f(z)$ be the complex potential for a flow having no rigid boundaries and such that there are no singularities in the flow outside the circle $|z| = a$. Then on introducing the rigid impermeable cylindrical surface $|z| = a$ into the flow and leaving the distribution of singularities unchanged within its interior, the new complex potential for the fluid motion within the boundary becomes $w = f(z) + \bar{f}(a^2/z)$, for $|z| \leqslant a$.

Example [6.16]

A line source and a line sink of equal strength are placed at the points $(\pm \frac{1}{2}a, 0)$ within the fixed circular boundary $x^2 + y^2 = a^2$. Show that the equations of the streamlines are given by

$$(r^2 - \tfrac{1}{4}a^2)(r^2 - 4a^2) - 4a^2y^2 = Ky(r^2 - a^2), \text{ where } r^2 = x^2 + y^2.$$

where K is a constant.

For a line source of strength m per unit length through $(\frac{1}{2}a, 0)$ and a line sink of the same strength through $(-\frac{1}{2}a, 0)$, the complex potential in the absence of the circular boundary is

$$f(z) \equiv m \ln(z - \tfrac{1}{2}a) - m \ln(z + \tfrac{1}{2}a).$$

As $\bar{f}(z) = f(z)$, with m and a real,

$$\begin{aligned}\bar{f}(a^2/z) &= m \ln\{(a^2/z) - \tfrac{1}{2}a\} - m \ln\{(a^2/z) + \tfrac{1}{2}a\}\\ &= m \ln(2a - z) - m \ln(2a + z).\end{aligned}$$

With the circular boundary inserted, the complex potential within its interior becomes

$$\begin{aligned}w(z) &= f(z) + \bar{f}(a^2/z)\\ &= m\,[\ln(x - \tfrac{1}{2}a + iy) - \ln(x + \tfrac{1}{2}a + iy)\\ &\quad + \ln(2a - x - iy) - \ln(2a + x + iy)].\end{aligned}$$

The stream function $\psi = \operatorname{Im}\{w(z)\}$, and so

$$\begin{aligned}\psi/m &= \left\{\tan^{-1}\left(\frac{y}{x - \frac{1}{2}a}\right) - \tan^{-1}\left(\frac{y}{x + \frac{1}{2}a}\right)\right\}\\ &\quad - \left\{\tan^{-1}\left(\frac{y}{2a - x}\right) + \tan^{-1}\left(\frac{y}{2a + x}\right)\right\}.\end{aligned}$$

We now use the identities

$$\tan^{-1} A \pm \tan^{-1} B = \tan^{-1} \{(A \pm B)/(1 \mp AB)\}.$$

Then

$$\psi/m = \tan^{-1}\left[\frac{\dfrac{y}{x-\frac{1}{2}a} - \dfrac{y}{x+\frac{1}{2}a}}{1+\dfrac{y^2}{x^2-\frac{1}{4}a^2}}\right] - \tan^{-1}\left[\frac{\dfrac{y}{2a-x} + \dfrac{y}{2a+x}}{1-\dfrac{y^2}{4a^2-x^2}}\right]$$

$$= \tan^{-1}\left(\frac{ay}{r^2-\frac{1}{4}a^2}\right) + \tan^{-1}\left(\frac{4ay}{r^2-4a^2}\right)$$

$$= \tan^{-1}\left[\frac{ay(r^2-4a^2)+4ay(r^2-\frac{1}{4}a^2)}{(r^2-\frac{1}{4}a^2)(r^2-4a^2)-4a^2y^2}\right]$$

$$= \tan^{-1}\left[\frac{5ay(r^2-a^2)}{(r^2-\frac{1}{4}a^2)(r^2-4a^2)-4a^2y^2}\right].$$

The streamlines are given by $\psi/m =$ const. which leads at once to the required result.

6.8 CIRCULAR CYLINDER IN UNIFORM STREAM WITH CIRCULATION

Suppose that an infinitely long circular cylinder of radius a is placed in a uniform stream having velocity $-U\mathbf{i}$. The complex potential of the undisturbed stream is $-Uz$ and with the cylinder $|z| = a$ inserted the circle theorem shows that the complex potential is $-U(z + a^2/z)$. Now suppose the fluid surrounding the cylinder is given a circulation of $2\pi\kappa$. Then in accordance with Example [6.11], a term $-i\kappa \ln z$ has to be added to the complex potential of the stationary cylinder in the uniform stream. Thus the total complex potential is now

$$w(z) = -U(z + a^2/z) - i\kappa \ln z. \tag{6.44}$$

At a stagnation point in the flow, $u = 0$ and $v = 0$ and so $dw/dz = 0$, or

$$U[1-(a/z)^2] + (i\kappa/z) = 0. \tag{6.45}$$

This equation has two roots, given by

$$z = a\left[-i\left(\frac{\kappa}{2Ua}\right) \pm \sqrt{\left\{1-\left(\frac{\kappa}{2Ua}\right)^2\right\}}\right]. \tag{6.46}$$

The flow pattern depends very much on the magnitude of κ. In this connection we discuss separately the three distinct cases:

(i) $\kappa < 2Ua$; (ii) $\kappa = 2Ua$; (iii) $\kappa > 2Ua$.

(i) Let us first suppose that $\kappa < 2Ua$. Then there exists a real acute angle β such that $\sin\beta = (\kappa/2Ua)$. Thus two stagnation points occur on the circle. They are given by

$$z = a(-i\sin\beta \pm \cos\beta),$$

i.e.

$$z_1 = a\,e^{-i\beta}, \qquad z_2 = -a\,e^{i\beta}.$$

These points are distinct. The pattern of streamline flow in this case is shown in Fig. 6.11. A_1 and A_2 are the stagnation points.

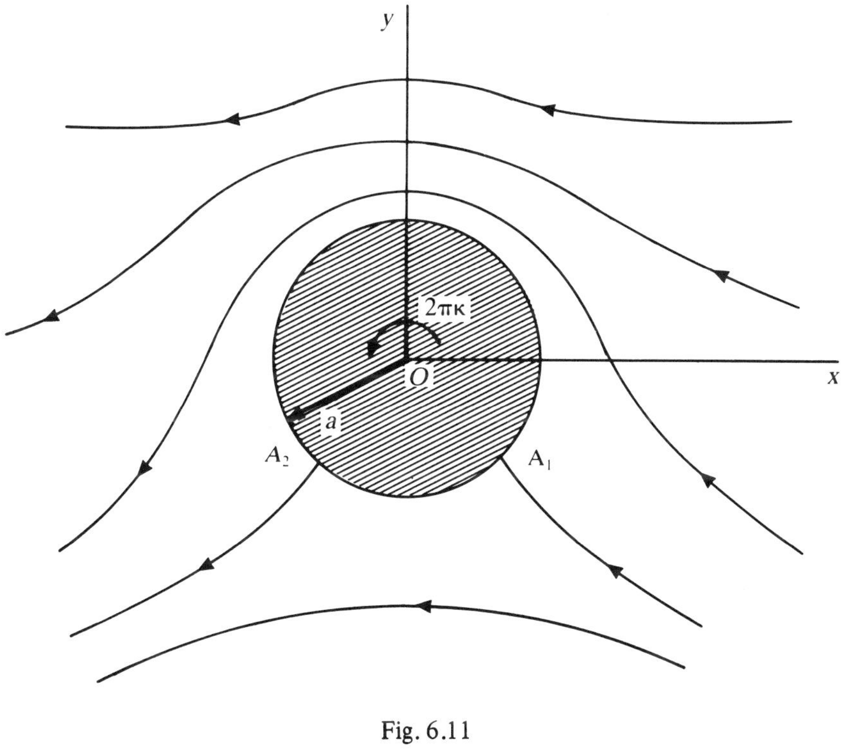

Fig. 6.11

At A_1, A_2 the pressure is a maximum by virtue of Bernoulli's equation $(p/\rho) + \frac{1}{2}q^2 =$ const. along a streamline. The effect on the cylinder is a tendency to lift it in the direction $\overrightarrow{Oy}$. In the case when $\kappa = 0$ the streamline flow is symmetrical about $y = 0$ and there is no such tendency. The lifting tendency when $\kappa \neq 0$ is called the ***Magnus effect***. As κ increases beyond zero, the tendency is for A_1, A_2 to move downwards. We return again to the Magnus effect in Example [6.17].

The stream function is

$$\psi = \mathrm{Im}(w) = -Uy(1 - a^2r^2) - \kappa \ln r,$$

where $z = r\,\mathrm{e}^{i\theta} = x + iy$, $r = \sqrt{(x^2 + y^2)}$. This is unaltered when x is replaced by $-x$ and so the streamline-pattern is symmetrical about $x = 0$.

(ii) When $\kappa = 2Ua$, $\beta = \pi/2$ and the stagnation points A_1, A_2 of the previous case now coincide at the single point C, the lowest point of the circle (Fig. 6.12).

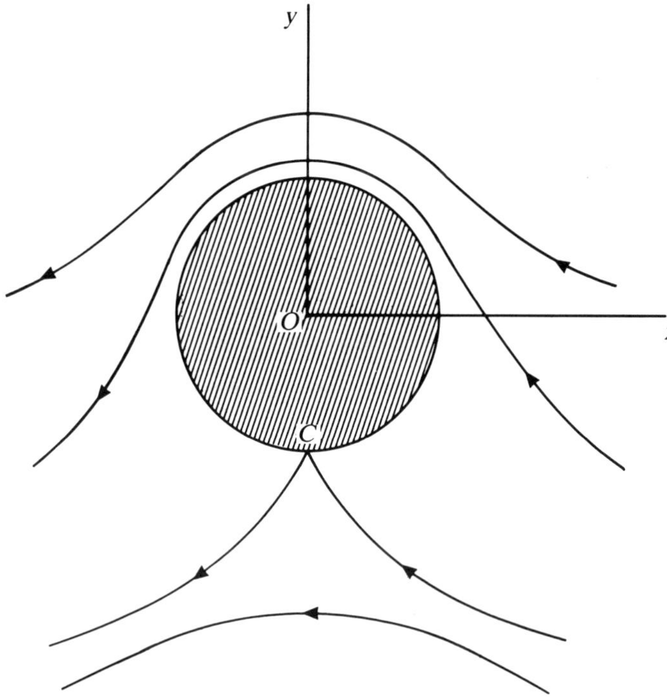

Fig. 6.12

(iii) When $\kappa > 2Ua$ we can find a real constant γ such that $\cosh\gamma = (\kappa/2Ua)$. The quadratic equation (6.45) now simplifies to the form

$$[(z/a) + i\,\cosh\gamma]^2 = 1 - \cosh^2\gamma = -\sinh^2\gamma$$

with roots

$$z_1 = -ai\,\mathrm{e}^{\gamma}, \qquad z_2 = -ai\,\mathrm{e}^{-\gamma}.$$

The point z_1 is outside the circle since $|z_1| = a\,\mathrm{e}^{\gamma} > 1$ and the point z_2 is its inverse in the circle and z_2 does not belong to the flow region. In Fig. 6.13, D represents the point z_1. A critical streamline passes through D separating closed loop streamlines, which enclosed the circle, from streamlines which are not closed and extend to infinity.

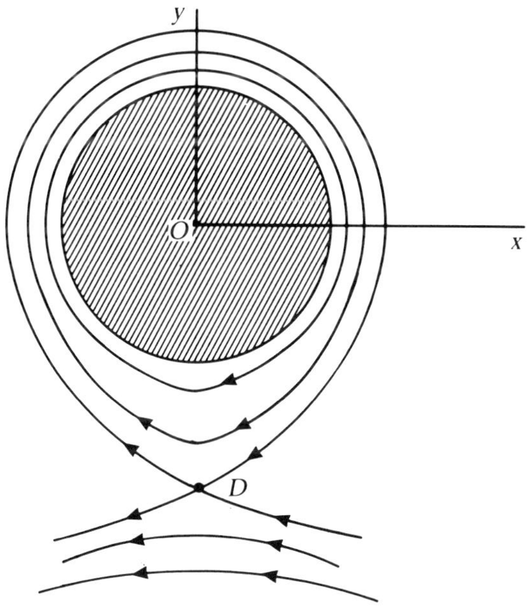

Fig. 6.13

6.9 THE THEOREM OF BLASIUS

When a cylinder of arbitrary closed section and infinite length is placed with its generators perpendicular to the plane of flow of a two-dimensional fluid motion, in general the action of the fluid is to cause translation and rotation of the cylinder. We now establish a theorem due to Blasius which enables one to evaluate the resultant force and couple exerted on unit length of the cylinder by the moving fluid. We confine attention to the case when the fluid is incompressible, of uniform density and its motion irrotational. The theorem is useful in subsonic aerodynamics for evaluating the force and couple exerted on an aerofoil.

The enunciation of the theorem due to Blasius is as follows.

An incompressible fluid moves steadily and irrotationally under no external forces parallel to the complex z-plane past a fixed cylinder whose section in that plane is bounded by a closed curve $\mathscr{C}$. The complex potential for the flow is $w(z)$. Then the action of the fluid pressure on the cylinder is equivalent to a force per unit length having components $[X, Y]$ and a couple per unit length of moment M, where

$$Y + iX = \frac{\rho}{2} \oint_{\mathscr{C}} \left(\frac{dw}{dz}\right)^2 dz; \qquad M = \mathrm{Re}\left\{-\frac{\rho}{2} \oint_{\mathscr{C}} z \left(\frac{dw}{dz}\right)^2 dz\right\}.$$

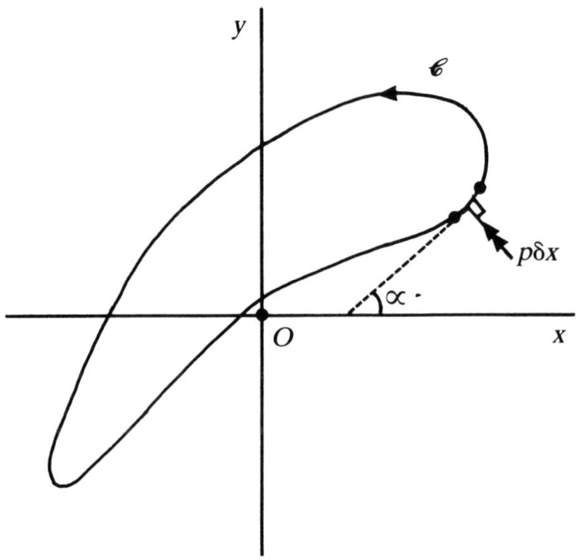

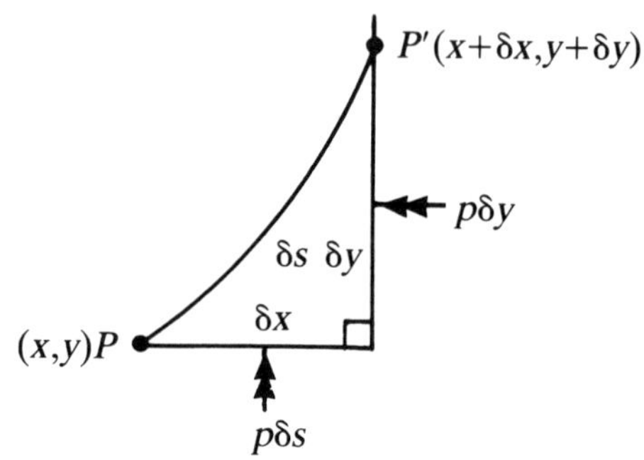

Fig. 6.14(i) Fig. 6.14(ii)

Proof Fig. 6.14(i) shows the section $\mathscr{C}$ of the cylinder in the plane of flow. $P(x, y)$; $P'(x + \delta x, y + \delta y)$ are neighbouring points on $\mathscr{C}$ and the arc PP' is of length δs. Then if p denotes the fluid pressure at P, the force on unit length of the section δs of the cylinder is $p\delta s$ normal to $\mathscr{C}$. If α is the angle which the tangent at P to $\mathscr{C}$ makes with $\overline{Ox}$, then the x- and y-components of this force are $[\delta X, \delta Y]$ where

$$\delta X = -p\delta s \sin\alpha = -p\delta y; \qquad \delta Y = p\delta s \cos\alpha = p\delta x,$$

since $\delta x = \delta s \cos\alpha$, $\delta y = \delta s \sin\alpha$ (Fig. 6.14(ii)) and the moment δM of these forces about the normal to the plane of flow through O is

$$\delta M = p(x\delta x + y\delta y).$$

Since

$$\delta Y + i\delta X = p(\delta x - i\delta y) = p\,\delta\bar{z} \qquad (z = x + iy; \bar{z} = x - iy),$$

we have

$$Y + iX = \oint_{\mathscr{C}} p\,\mathrm{d}\bar{z}.$$

As no fluid crosses $\mathscr{C}$ it is a streamline and along such a streamline Bernoulli's equation holds in the form $(p/\rho) + \frac{1}{2}q^2 = C$. Thus, as $\rho =$ const., $C =$ const.,

$$Y + iX = \rho C \oint_{\mathscr{C}} \mathrm{d}\bar{z} - \frac{\rho}{2} \oint_{\mathscr{C}} q^2 \, \mathrm{d}\bar{z}$$

$$= -\frac{\rho}{2} \oint_{\mathscr{C}} q^2 \, \mathrm{d}\bar{z}.$$

Now $\mathrm{d}w/\mathrm{d}z = u - iv$ and so

$$q^2 = u^2 + v^2 = (u - iv)(u + iv) = (\mathrm{d}w/\mathrm{d}z)(\mathrm{d}\bar{w}/\mathrm{d}\bar{z}).$$

Then

$$Y + iX = -\frac{\rho}{2} \oint_{\mathscr{C}} \frac{\mathrm{d}w}{\mathrm{d}z} \, \mathrm{d}\bar{w},$$

since $(\mathrm{d}\bar{w}/\mathrm{d}\bar{z})\mathrm{d}\bar{z} = \mathrm{d}\bar{w}$. But on the streamline $\mathscr{C}$, $\psi =$ const. and so $\mathrm{d}w = \mathrm{d}\bar{w}$. Thus $\mathrm{d}\bar{w} = (\mathrm{d}w/\mathrm{d}z)\mathrm{d}z$ on $\mathscr{C}$ and so

$$Y + iX = -\frac{\rho}{2} \oint_{\mathscr{C}} \left(\frac{\mathrm{d}w}{\mathrm{d}z}\right)^2 \mathrm{d}z. \tag{6.47}$$

From the form for δM, the total moment per unit length of the cylinder is

$$M = \oint_{\mathscr{C}} p(x \, \mathrm{d}x + y \, \mathrm{d}y).$$

Now

$$x\delta x + y\delta y = \tfrac{1}{2}\delta(x^2 + y^2)$$

and using again $p = -\frac{1}{2}\rho q^2 + \rho C$ along $\mathscr{C}$,

$$M = -\frac{\rho}{2} \oint_{\mathscr{C}} q^2(x \, \mathrm{d}x + y\mathrm{d}y) + \frac{\rho C}{2} \oint_{\mathscr{C}} \mathrm{d}(x^2 + y^2)$$

$$= -\frac{\rho}{2} \oint_{\mathscr{C}} q^2 \, (x \, \mathrm{d}y + y \, \mathrm{d}y).$$

Since $z = x + iy$, $\mathrm{d}\bar{z} = \mathrm{d}x - i \, \mathrm{d}y$, we see from the product $z\mathrm{d}\bar{z}$ that

$$x \, \mathrm{d}x + y \, \mathrm{d}y = \mathrm{Re}(z\mathrm{d}\bar{z}).$$

Thus

$$M = \mathrm{Re}\left\{-\frac{\rho}{2} \oint_{\mathscr{C}} q^2 z \, \mathrm{d}\bar{z}\right\}.$$

But $q^2 = (\mathrm{d}w/\mathrm{d}z)(\mathrm{d}\bar{w}/\mathrm{d}\bar{z})$, giving

$$M = \text{Re}\left\{-\frac{\rho}{2}\oint_{\mathscr{C}} z\,\frac{dw}{dz}\,d\bar{w}\right\}$$

$$= \text{Re}\left\{-\frac{\rho}{2}\oint_{\mathscr{C}} z\,\frac{dw}{dz}\,dw\right\},$$

since $d\bar{w} = dw$ on $\mathscr{C}$. Finally $dw = (dw/dz)\,dz$ gives the required form

$$M = \text{Re}\left\{-\frac{\rho}{2}\oint_{\mathscr{C}} z\left(\frac{dw}{dz}\right)^2 dz\right\}. \tag{6.48}$$

The contour integrals in (6.47), (6.48) are evaluated using the residue calculus. Equating real and imaginary parts in (6.47) gives Y and X respectively.

Example [6.17]

Infinite cylinder of circular section in uniform stream, with circulation. (Magnus effect again.)

In Section 6.8 this problem was formulated and led to equation (6.44), i.e.

$$w(z) = -U(z + a^2/z) - i\kappa \ln z.$$

Then

$$w'(z) = -U(1 - a^2/z^2) - (i\kappa/z),$$

and so

$$Y + iX = -\frac{\rho}{2}\oint_{\mathscr{C}} [w'(z)]^2\, dz; \qquad \mathscr{C}: |z| = a.$$

Thus

$$Y + iX = -\frac{\rho}{2}\oint_{\mathscr{C}}\left[U + \frac{i\kappa}{z} - U\frac{a^2}{z^2}\right]^2 dz.$$

The integrand is a meromorphic function of z whose only singularity is a pole of order 4 at $z = 0$ (within $\mathscr{C}$). The residue of the integrand there is the coefficient of $(1/z)$ in its expansion, i.e. the residue is $2iU\kappa$. Hence

$$Y + iX = -\tfrac{1}{2}\rho \times 2\pi i \times 2iU\kappa = 2\rho U\kappa\pi.$$

Equating real and imaginary parts gives $Y = 2\rho U\kappa\pi$, $X = 0$.

The moment of the couple per unit length of cylinder is

$$M = -\frac{\rho}{2}\,\text{Re}\oint_{\mathscr{C}} z[w'(z)]^2\, dz.$$

The coefficient of $(1/z)$ in the integrand is the coefficient of $(1/z^2)$ in $[U+(i\kappa/z)-U(a^2/z^2)]^2$. This is $-\kappa^2 - 2U^2a^2$, and so

$$M = -\frac{\rho}{2} \times \text{Re}\left\{-2\pi i\,(\kappa^2 + 2U^2a^2)\right\} = 0.$$

Thus the action of the fluid on the circular cylinder produces a lifting force in the y-direction of amount $2\pi\rho U\kappa$ per unit length and zero moment. Common sense dictates that $X = 0$, $M = 0$. For the streamline pattern is, as we have seen, symmetrical about $x = 0$, so that at all pairs of points $(\pm x, y)$ in the stream the hydrodynamical pressures will be the same. At equal surface elements δS at the points $(\pm x, y)$ on the cylindrical surface, the forces will have equal and opposite components in the x-direction, leading to a resultant force in the y-direction, passing through O and hence having no moment about O. In this problem the presence of circulation produces the lift force and this fact is taken into account in the design of aeroplane wings. Essentially the principle of the rotor ship is the Magnus effect. Likewise the same principle in three dimensions explains the behaviour of 'cut' tennis balls and 'sliced' golf balls.

Example [6.18]

Verify that $w(z) = i\kappa \ln\{(z - ia)/(z + ia)\}$, κ and a both real, is the complex potential of a steady flow of liquid about a circular cylinder, the plane $y = 0$ being a rigid boundary. Find the force and the couple exerted on unit length of the cylinder.

Putting $w = \phi(x, y) + i\psi(x, y)$, we have

$$\psi = \kappa \ln \left|\frac{z - ia}{z + ia}\right|.$$

The streamlines, found from $\psi =$ const., are thus given by

$$|z - ia|/|z + ia| = \lambda = \text{const.},$$

or

$$PA/PB = \lambda,$$

where $\overrightarrow{OP} \equiv z = x + iy$, $\overrightarrow{OA} \equiv ia$, $\overrightarrow{OB} \equiv -ia$. When $\lambda \neq 1$, this last equation represents a family of non-intersecting coaxal circles having A and B for limiting points, i.e. circles of zero radius. $\lambda = 1$ gives the straight line which is the perpendicular bisector of the line segment AB and it is the radical axis of the coaxal system. No fluid crosses a streamline and so a rigid boundary may be introduced along any cirlce $\lambda =$ const. of the coaxal system, including the perpendicular bisector $\lambda = 1$. This is the plane $y = 0$ in three dimensions. Hence if we introduce rigid boundaries along:

(i) a particular circle $\lambda =$ const. $(\neq 1)$;
(ii) along a plane $y = 0$ $(\lambda = 1)$

then the first part of the question is answered.

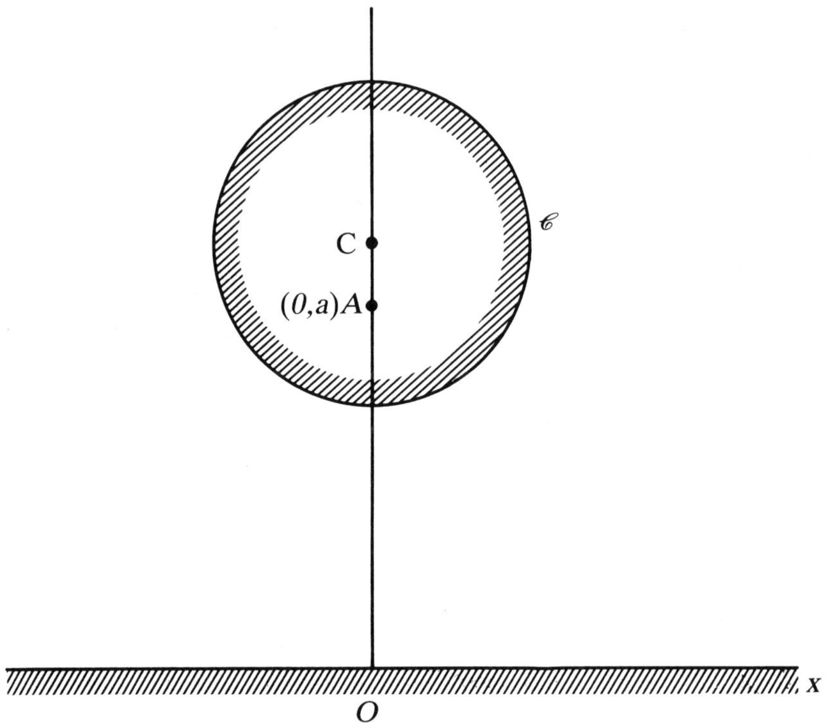

Fig. 6.15

In Fig. 6.15, $\mathscr{C}$ is any circle of the λ-system and it encloses the point $A(0, a)$, but $B(0, -a)$ is external to it. Rigid boundaries are introduced along the surface of $\mathscr{C}$ and along the plane $y = 0$.

Since

$$w(z) = i\kappa\{\ln(z - ia) - \ln(z + ia)\},$$

and

$$w'(z) = i\kappa\{(z - ia)^{-1} - (z + ia)^{-1}\}.$$

Now

$$Y + iX = -\frac{\rho}{2}\oint_{\mathscr{C}} [w'(z)]^2 \, dz,$$

where

$$\oint_{\mathscr{C}} [w'(z]^2 \, dz = -\kappa^2 \oint_{\mathscr{C}} \left[\frac{1}{(z - ia)^2} - \frac{2}{(z - ia)(z + ia)} + \frac{1}{(z + ia)^2}\right] dz.$$

The integrand has double poles at $z = \pm ia$: as $z = ia$ is the only pole within $\mathscr{C}$ this is the sole point at which the residue is found. It is only the middle term of the integrand which furnishes a non-zero contribution to the contour integral and the appropriate residue at $z = ia$ is

$$\lim_{z \to ia} \left\{ -\frac{2}{(z-ia)(z+ia)} \times (z-ia) \right\} = \frac{i}{a}.$$

Hence

$$\oint_{\mathscr{C}} [w'(z)]^2 \, dz = -\kappa^2 \, [2\pi i \times (i/a)] = -2\pi\kappa^2/a,$$

and so

$$Y + iX = \rho\pi\kappa^2/a,$$

giving

$$X = 0, \qquad Y = \rho\pi\kappa^2/a.$$

For the couple exerted on unit length of the cylinder,

$$M = -\frac{\rho}{2} \operatorname{Re} \oint_{\mathscr{C}} z[w'(z)]^2 \, dz.$$

Now

$$\oint_{\mathscr{C}} z[w'(z)]^2 \, dz = \kappa^2 \oint_{\mathscr{C}} z \left[\frac{1}{(z-ia)^2} - \frac{2}{(z-ia)(z+ia)} + \frac{1}{(z+ia)^2} \right] dz$$

$$= -\kappa^2 \oint_{\mathscr{C}} \left[\frac{z}{(z-ia)^2} - \frac{2z}{(z-ia)(z+ia)} \right] dz,$$

the term $z(z+ia)^{-2}$ contributing nothing to the value of the integral.

Now

$$\frac{z}{(z-ia)^2} = \frac{(z-ia)+ia}{(z-ia)^2} = \frac{1}{z-ia} + \frac{ia}{(z-ia)^2},$$

which has residue of 1 at $z = ia$. Also $-2z/\{(z-ia)(z+ia)\}$ has a residue at $z = ia$ given by

$$\lim_{z \to ia} \left(\frac{-2z}{z+ia} \right) = -1.$$

The sum of the residues within $\mathscr{C}$ is zero and so therefore is the contour integral and clearly $M = 0$.

PROBLEMS 6

1. Evaluate the following integrals when $\mathscr{C}$ is the circle $|z| = 2$.

(i) $\displaystyle\int_c \frac{(z+1)\,dz}{z(z-1)}$, (ii) $\displaystyle\int_c \frac{(2z+4)\,dz}{z^2(z+8)}$.

2. Evaluate the following when $\mathscr{C}$ is the square with vertices $\pm 1 \pm i$.

(i) $\displaystyle\int_c \frac{z^2+1}{z(z-4)(2z+1)}\,dz$, (ii) $\displaystyle\int_{\mathscr{C}} \frac{e^z\,dz}{z^2+z+1}$.

3. In a two-dimensional motion, a source of strength m is placed at each of the points $(-1, 0)$ and $(1, 0)$, and a sink of strength $2m$ is placed at the origin. Show that the streamlines are the curves

$$(x^2+y^2)^2 = x^2 - y^2 + kxy,$$

where k is a parameter.

4. A two-dimensional motion of a liquid has the complex potential

$$w = U\left(z + \frac{a^2}{z}\right) + ik \ln\left(\frac{z}{a}\right),$$

where the constants U, k, a are real and positive. Show that

(i) the velocity at infinity is U in the positive sense of the real axis;
(ii) the circle $|z| = a$ is a streamline;
(iii) there are, in general, two stagnation points;
(iv) there is a circulation $2\pi k$ about the circle.

5. Two equal line sources are placed at A, B. Prove that at any point P the speed of the fluid is proportional to $OP/(AP.\ BP)$, where O is the middle point of AB.

6. A line source is placed at A and a similar line sink of half the strength at B. Prove that the direction of flow is parallel to AB at points on a certain circle of radius $AB\sqrt{2}$.

7. A line source is placed at the point $(d, 0)$ in front of an infinite plane $x = 0$. Prove that the equations of the stream lines are $x^2 + kxy - y^2 = d^2$, where k is any constant.

8. A line source of strength m per unit length is at a distance c from the infinite plane which bounds a semi-infinite liquid of density ρ. Find the pressure at any point of the plane and hence show that the total force, on unit length of the plane parallel to the line source, is $\pi\rho m^2/c$. What is its direction?

9. A line doublet whose axis lies along $\overrightarrow{Ox}$ is placed at the point $(a, 0)$ in front of the infinite plane $x = 0$. Prove that the equations of the streamlines are

$$(x^2 - y^2 - z^2)^2 + 4x^2y^2 + \lambda xy = 0,$$

where λ is a constant.

10. The space on one side of an infinite plane wall, $y = 0$, is filled with inviscid incompressible fluid of constant density ρ, moving at infinity with speed U in the direction of the positive x-axis. The motion of the fluid is steady and two-dimensional in the (x, y)-plane. A uniform line doublet of strength μ whose axis points at an angle α to the positive x-axis is situated at the point $(0, a)$.

(i) Show that if $U = (2\mu \cos \alpha)/a^2$ there is a stagnation point at $(0, 0)$.

(ii) If $\alpha = \pi/2$ and $U = 2\mu/a^2$, show that the pressure at $(x, 0)$ on the wall is $p_\infty - 8\rho\mu^2 \left[\dfrac{a^2x^2}{(a^2 + x^2)^4} - \dfrac{x}{a(a^2 + x^2)^2}\right]$ where p_∞ is the pressure at infinity. Is the origin a stagnation point in this case?

11. Two-dimensional incompressible flow is bounded by perpendicular rigid walls OX, OY. There is a line vortex of strength κ at (x, y) in the fluid. Show that the equation of the path of the line vortex has the form

$$x^2 + y^2 = x^2y^2/a^2 \ (a = \text{const.})$$

12. Obtain the complex potential function appropriate to uniform streaming in two dimensions past a circular cylinder of radius a, the speed at great distance being U. If a marked particle of liquid is at distance b directly upstream of the axis of the cylinder O at time $t = 0$, prove that at time t it will be at distance r from O where

$$Ut - b + r = \tfrac{1}{2} a \ln \left\{\frac{(r + a)(b - a)}{(r - a)(b + a)}\right\}.$$

Use this result to show that if the marked particle moves a distance s while the liquid at infinity moves a distance d, then

$$s = \frac{b^2 - a^2}{b + a \coth\{(d - s)/a\}}.$$

13. Describe, with the aid of a diagram, the flow pattern given by

$$w(z) = Uz + m \ln (z - 2a),$$

where U, m, a are all real constants.

If a circular cylinder of radius a is inserted with its centre at the origin, find the complex potential of the disturbed motion, there being no circulation about the cylinder.

14. A line vortex of strength κ lies at a distance $b(> a)$ from the axis of a circular cylinder of radius a. If the vortex remains stationary find the circulation around the cylinder.

15. Describe, with the aid of a diagram, the flow pattern whose complex potential is

$$-Uz + m \ln (z - 2a),$$

where U, m, a are all real constants.

If a solid circular cylinder of radius a is inserted with its centre at the origin, find the complex potential of the disturbed motion, there being no circulation about the cylinder. Show that the points $z = \pm a$ are stagnation points and, further, that if $m = -\frac{5}{2}aU$, then the points $z = \pm ia$ are also stagnation points.

16. Liquid moves two-dimensionally without circulation in the region exterior to both the circles

$$|z - i| = \sqrt{2}, \quad |z + i| = \sqrt{2}.$$

Find the complex potential given that there is a unit source at $z = 1$ and that the liquid is at rest at infinity. Find the velocity at the point $3i$.

17. A two-dimensional flow is due to a line source of strength m at the point $(2a, 0)$ and a vortex filament of strength m at $(-2a, 0)$. The solid circular cylinder $|z| = a$ is inserted into the flow. Find the complex potential at any point and confirm that there are no finite stagnation points anywhere on the line $y = 0$.

18. A source and a sink of equal strength are placed at the points $(a, 0)$, $(-a, 0)$ respectively, within a fixed circular boundary $|z| = 2a$. Show that the streamlines are given by

$$16a^2y^2 + \lambda v(r^2 - 4a^2) = (r^2 - 16a^2)(r^2 - a^2)$$

where λ is a constant, and that the velocity at the point $(2a, \theta)$ is

$$20m \sin\theta / a(17 - 8\cos 2\theta) .$$

19. *Electrostatic form of the circle theorem*

Let $f(z) = \phi(x, y) + i\psi(x, y)$ be the complex potential for a given two-dimensional distribution which is analytic in the region $|z| \geqslant a$. Suppose that the boundary $|z| = a$ is now introduced into the potential field and that $\phi = 0$ on this boundary. Show that the new complex potential for the region $|z| \geqslant a$ is now given by

$$w = f(z) - \bar{f}(a^2/z).$$

The reader conversant with electrostatic theory will recognize that this result gives the complex electrostatic potential when the earthed conducting sphere $|z| = a$ is introduced into a field whose complex potential was originally $f(z)$. Then $\text{Re}(w)$ gives the electrostatic potential of the field, the lines of force being given by $\text{Im}(w) = \text{const.}$

20. For the complex potential

$$w = \pi a V \coth(\pi a/z) \qquad (z = x + iy)$$

where a, V, x, y are all real, show that

(i) The flow in regions for which $|z| \to \infty$ has uniform velocity $V\mathbf{i}$;
(ii) The only stagnation point is $z = 0$.

By considering the streamline $\psi = 0$, or otherwise, verify that the given form for w represents the flow past a cylindrical log of diameter $2a$, lying on the bed of an infinite ocean, the origin being taken at the point of contact of the log and the bed of the ocean.

21. Show that the force exerted on the circular cylinder $|z| = a$ in the irrotational flow produced by a line source of strength m at $z = 3a$ is given by $[X, Y]$ per unit length, where

$$X = \frac{\rho \pi m^2}{12a}, \qquad Y = 0.$$

22. Find the resultant force on a fixed circular boundary $\mathscr{C}$ of radius a with centre at the origin if the fluid motion is caused entirely by a vortex of strength $2\pi\kappa$ at $2a + 0i$ and there is no circulation round $\mathscr{C}$.

23. The complex potential of a steady two-dimensional fluid motion in the z-plane is

$$\sum_{n=1}^{\infty} b_n z^n \quad (b_n \text{ real constants}).$$

If the rigid cylindrical obstacle $|z| = a$ is inserted into the flow, prove that the cylinder will experience a thrust of magnitude

$$2\pi\rho \sum_{n=1}^{\infty} a^{2n}\, n(n+1)\, b_n b_{n+1}$$

per unit length. In which direction does the thrust act?

24. An infinite circular cylinder of radius a and density σ is enclosed by a fixed infinite coaxial circular cylinder of radius b, the space between the

cylinders being filled by a liquid of constant density ρ. Prove that the impulse per unit length required to start the inner cylinder translating with velocity V is

$$\frac{\pi a^2}{b^2 - a^2} \; [(\sigma + \rho) b^2 - (\sigma - \rho) a^2] \, V.$$

25. The region between two fixed coaxial circular cylinders, of radii a and b ($> a$), which lies between parallel planes perpendicular to the cylinders a distance c apart, is occupied by liquid of constant density ρ. Show that the velocity potential of a motion, whose kinetic energy is T, is given by $A\,\theta$ where

$$\Pi \rho A^2 c \ln (b/a) = T.$$

26. Show that when a circular cylinder of radius a moves uniformly along a straight line in an infinite liquid, the path of any fluid particle is given by the equations

$$\frac{\mathrm{d}z}{\mathrm{d}t} = \frac{Va^2}{(\bar{z} - Vt)^2}, \qquad \frac{\mathrm{d}\bar{z}}{\mathrm{d}t} = \frac{Va^2}{(z - Vt)^2}$$

where $z = x + iy$, $\bar{z} = x - iy$, with x, y measuring distance from the initial location of the cylinder axis along the perpendicular to its direction of motion.

27. If a long circular cylinder of radius a moves in a straight line at right angles to its length in liquid at rest at infinity, show that when a fluid particle, initially at distance b ($> a$) ahead of the axis of the cylinder on the line of motion of the cylinder, has moved a distance c, the cylinder will have moved a distance

$$c + \frac{b^2 - a^2}{b + a \coth(c/a)}.$$

7

Conformal transformation and its applications

7.1 RÉSUMÉ OF BASIC FEATURES OF CONFORMAL TRANSFORMATION

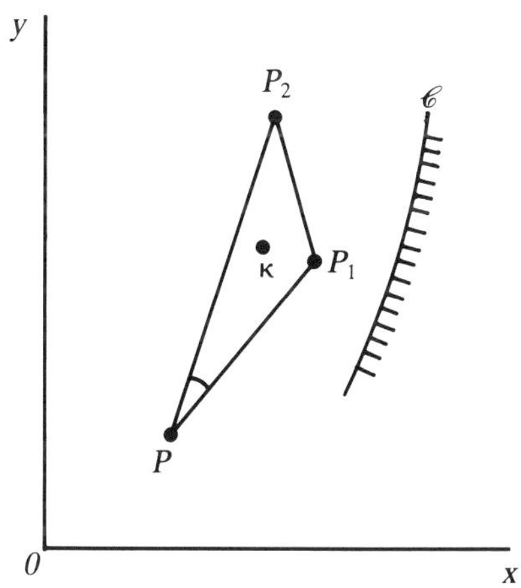

Fig. 7.1(i)

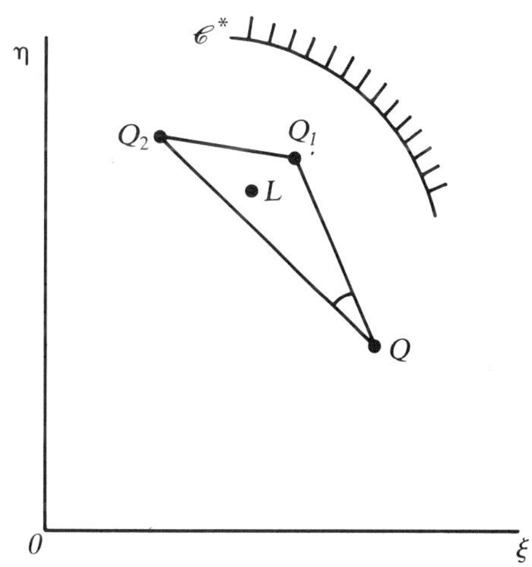

Fig. 7.1(ii)

Suppose z and t are two complex variables defined by

$$z = x + iy, \qquad t = \xi + i\eta,$$

where x, y, ξ, η are real variables. Then we can form two Argand diagrams for the loci of z and t. In the first (Fig. 7.1(i)), x is plotted horizontally and y vertically. In the second (Fig. 7.1(ii)), ξ is plotted horizontally and η vertically.

Now let z describe a certain curve $\mathscr{C}$ in the (x, y)-plane and suppose that t is related to z by a transformation of the form

$$t = g(z). \tag{7.1}$$

If $g(z)$ is a single-valued function of z, then to each point in the z-plane there corresponds a unique point in the t-plane. Thus the transformation (7.1) enables the curve $\mathscr{C}$ in the z-plane to be mapped into a curve $\mathscr{C}^*$ in the t-plane.

Let us now make the supposition that $g(z)$ is analytic: then $g'(z)$ is unique at each point z in the (x, y)-plane. Let P, P_1, P_2 be neighbouring points in the z-plane so that

$$\overrightarrow{OP} \equiv z, \qquad \overrightarrow{OP_1} \equiv z + \delta z_1, \qquad \overrightarrow{OP_2} \equiv z + \delta z_2.$$

Under the transformation (7.1), suppose that P, P_1, P_2 map into the respective points Q, Q_1, Q_2 in the t-plane, where

$$\overrightarrow{OQ} \equiv t, \qquad \overrightarrow{OQ_1} \equiv t + \delta t_1, \qquad \overrightarrow{OQ_2} \equiv t + \delta t_2.$$

It is assumed that $|\delta z_1|$, $|\delta z_2|$, $|\delta t_1|$, $|\delta t_2|$ are all small.

Because $g(z)$ is analytic, $\mathrm{d}t/\mathrm{d}z$ is unique at each P. Thus, to a first order of smallness, at P

$$\frac{\delta t_1}{\delta z_1} = \frac{\delta t_2}{\delta z_2},$$

or

$$\delta t_1/\delta t_2 = \delta z_1/\delta z_2.$$

If two complex quantities are equal, then their moduli and arguments are equal and the last result gives

$$\left|\frac{\delta t_1}{\delta t_2}\right| = \left|\frac{\delta z_1}{\delta z_2}\right|,$$

and
$$\arg \delta t_1 - \arg \delta t_2 = \arg \delta z_1 - \arg \delta z_2.$$

These results may further be expressed:

$$\frac{QQ_1}{QQ_2} = \frac{PP_1}{PP_2} \tag{7.2}$$

and

$$Q_2\hat{O}Q_1 = P_2\hat{P}P_1. \tag{7.3}$$

The pair of equations (7.2), (7.3) express similarity of the elementary triangles Q_2QQ_1, P_2PP_1. Otherwise expressed: *within the neighbourhood of any point P in the z-plane and its corresponding mapping Q in the t-plane, angles remain unaltered (equation (7.3)) and so do ratios of corresponding linear dimensions* *(equation (7.2)) under a transformation of the form (7.1), where $g(z)$ is analytic.* Such a transformation is said to be *conformal*.

Now suppose K is any point within the triangle PP_1P_2 (Fig. 7.1(i)). Suppose K maps into L under the conformal transformation (7.1) (Fig. 7.1(ii)). Then the pairs of triangles KPP_1, LQQ_1; KP_1P_2, LQ_1Q_2; KP_2P, LQ_2Q are equiangular and so we infer that L lies within the triangle QQ_1Q_2. Since K is *any* point in the triangle PP_1P_2, we see that the whole of the interior of this triangle maps into the interior of triangle QQ_1Q_2. Very clearly, then, *the interior of any closed curve $\mathscr{C}$ in the z-plane will map into the interior of the corresponding closed curve $\mathscr{C}^*$ in the t-plane under the conformal transformation (7.1).*

The *magnification* of the conformal transformation is M, defined to be

$$M = QQ_1/PP_1 = |\mathrm{d}t/\mathrm{d}z| = |g'(z)|. \tag{7.4}$$

In this connection, it is important to distinguish locations for which $M = 0$ and those for which $M \neq 0$. Since $g(z)$ is analytic, the Cauchy–Riemann equations give

$$\frac{\partial \xi}{\partial x} = \frac{\partial \eta}{\partial y}, \qquad \frac{\partial \eta}{\partial x} = -\frac{\partial \xi}{\partial y}.$$

The Jacobian J of the transformation is defined to be

$$\begin{aligned} J = \frac{\partial(\xi, \eta)}{\partial(x, y)} &= \frac{\partial \xi}{\partial x}\frac{\partial \eta}{\partial y} - \frac{\partial \eta}{\partial x}\frac{\partial \xi}{\partial y} \\ &= \left(\frac{\partial \xi}{\partial x}\right)^2 + \left(\frac{\partial \eta}{\partial x}\right)^2, \end{aligned}$$

on using the Cauchy–Riemann equations. As

$$g'(z) = (\partial\xi/\partial x) + i(\partial\eta/\partial x),$$

$$M = |g'(z)| = J^{1/2}. \tag{7.5}$$

If $g'(z) = 0$ then $M = 0$, $J = 0$ and this implies the existence of a functional relationship between ξ and η in the form $\eta = \eta(\xi)$. In such circumstances a region of the (x, y)-plane in which $g'(z) = 0$ will be mapped on to the single curve $\eta = \eta(\xi)$ in the (ξ, η)-plane. Thus the conformal property described above will break down. Points of the z-plane for which $g'(z) = 0$ are called *singular points* of the transformation. A point at which $g'(z) \neq 0$ is called an *ordinary point*.

If, however, $g'(z) \neq 0$, then $J \neq 0$ and no region of the z-plane throughout which this holds can be mapped on to a curve in the (ξ, η)-plane. In particular, a rectangular element of area $\delta x \delta y$ will map into another elemental rectangle of area $\delta\xi\delta\eta = J\delta x\delta y$. It follows, then, that for the conformal transformation (7.1) to work we must impose on it the condition $g'(z) \neq 0$ at the considered point or region of the z-plane. (Ref. 7)

7.2 APPLICATIONS OF CONFORMAL MAPPINGS TO POTENTIAL FLOWS

Now suppose that there is a two-dimensional steady irrotational flow of incompressible liquid of uniform density ρ. The plane of flow is the z-plane of Fig. 7.1(i). On applying the conformal transformation $t = g(z)$ the new plane of flow becomes the t-plane of Fig. 7.1(ii). A rigid boundary $\mathscr{C}$ in the z-plane maps into a curve $\mathscr{C}^*$ in the t-plane. Let the complex velocity potential for the z-plane be

$$w = f(z) = \phi(x, y) + i\psi(x, y)$$

where ϕ and ψ are both real functions of x, y and represent the usual velocity potential and stream function respectively. By means of the transformation $t = g(z)$ we can express w as a function $f^*(t)$ of the form

$$w = f^*(t) = \phi^*(\xi, \eta) + i\psi^*(\xi, \eta).$$

At corresponding points t and z, the function w takes the same values, so that

$$\phi(x, y) = \phi^*(\xi, \eta),$$

$$\psi(x, y) = \psi^*(\xi, \eta).$$

Now $\mathscr{C}$ is a rigid boundary and so it is also a streamline for which ψ = const. Thus $\psi^*(\xi, \eta)$ = const. along $\mathscr{C}^*$, showing that $\mathscr{C}^*$ is a streamline and we may introduce a rigid boundary along it.

In theory w can be expressed as a function of t by eliminating z from $w = f(z)$ and $t = g(z)$ but it is often more convenient to use the result

$$\frac{dw}{dt} = \frac{dw}{dz}\,\frac{dz}{dt} = \frac{1}{g'(z)}\,\frac{dw}{dz}.$$

Hence

$$\left|\frac{dw}{dz}\right|^2 = \left|\frac{dw}{dt}\right|^2 \times \left|\frac{dt}{dz}\right|^2.$$

Referring to Fig. 7.1(i), the kinetic energy of unit length of the liquid element within the triangle PP_1P_2 is

$$\begin{aligned}\delta T &= \tfrac{1}{2}\rho \times \text{Area of triangle } PP_1P_2 \times (\text{speed at } P)^2 \\ &= \rho \times \Delta PP_1P_2 \times \left|\frac{dw}{dz}\right|^2 \\ &= \rho \times \Delta PP_1P_2 \times \left|\frac{dw}{dt}\right|^2 \times \left|\frac{dt}{dz}\right|^2 \\ &= \rho \times \Delta QQ_1Q_2 \times \left|\frac{dw}{dt}\right|^2,\end{aligned}$$

where ΔPP_1P_2 means the area of the triangle PP_1P_2, etc. The last expression is the kinetic energy of unit depth of the liquid element within the triangle QQ_1Q_2 of Fig. 7.1(ii). By summation over the entire fluid domains within the z- and t-planes we clearly obtain:

Total K.E. of unit depth of liquid in z-plane

= Total K.E. of unit depth of liquid of same density in t-plane.

In general, $|\mathrm{d}w/\mathrm{d}z| \neq |\mathrm{d}w/\mathrm{d}t|$, so that velocities at corresponding points in the z- and t-planes are not equal. *The total kinetic energies of the fluids over corresponding areas are the same, but their distributions over these areas differ.*

Practical applications of conformal transformation to a given problem involving hydrodynamical singularities such as line sources, line doublets and line vortices, requires knowing how such distributions themselves transform. We therefore make these investigations at this stage.

Let us first consider a uniform line source of strength m per unit length through the point $z = z_0$ in the plane of flow. Under the conformal transformation (7.1), $z = z_0$ in the z-plane maps into $t = t_0$ in the t-plane. Let $\mathscr{C}$ be a closed curve in the z-plane enclosing the point $z = z_0$. Then z_0 maps into a point $t = t_0$ in the t-plane and $t = t_0$ is enclosed by a closed curve $\mathscr{C}^*$. Since $\psi(x, y) = \psi^*(\xi, \eta)$ at corresponding points (x, y) of $\mathscr{C}$ and (ξ, η) of $\mathscr{C}^*$,

$$\oint_{\mathscr{C}} \mathrm{d}\psi = \oint_{\mathscr{C}^*} \mathrm{d}\psi^*. \tag{7.6}$$

These relations show that the volumetric flux per unit time over unit length of the cylinder of section $\mathscr{C}$ in the z-plane is equal to that over unit length of the cylinder of section $\mathscr{C}^*$ in the t-plane. If $z = z_0$ is an ordinary point of the conformal transformation (7.1) and if $\mathscr{C}$, enclosing z_0, is described once, then $\mathscr{C}^*$ encloses t_0 and is also described once. Hence if through z_0 there is a uniform line source of strength m per unit length giving a volumetric flux $2\pi m$ units of volume per unit length of the line soure per unit time, then this is also the volumetric flux across the two cylinders per unit length and with sections $\mathscr{C}$, $\mathscr{C}^*$ per unit time. But this flux across unit length of the cylinder of section $\mathscr{C}^*$ in the t-plane means that there is a uniform line source through $t = t_0$ perpendicular to the t-plane and of strength m per unit length. *Hence under conformal transformation of the form $t = g(z)$, with $g'(z) \neq 0$, a uniform line source of strength m per unit length through $z = z_0$ in the z-plane maps into an equal line source of strength m per unit length through $t = t_0$ in the t-plane.*

But what happens if $z = z_0$ is a singular point at which $g'(z_0) = 0$? In these circumstances, it may so happen that as z describes $\mathscr{C}$ once in the z-plane, t will describe $\mathscr{C}^*$ exactly n times in the t-plane. Then the relation (7.6) gives, in an obvious notation,

$$2\pi m = n \times 2\pi m^*$$

or

$$m^* = m/n.$$

Thus the source of strength m per unit length through $z = z_0$ maps into one of strength m/n per unit length through $t = t_0$. (The reader will easily see that if $g'(z_0) = \infty$ and if n has the same meaning as before, then we now have $m^* = nm$ through $t = t_0$.)

The case of transformation of a line doublet is easily treated by taking line sources $-m$, $+m$ through $z = z_1, z_2$ which map into the line sources $-m$, $+m$ through $t = t_1, t_2$ respectively, where $t_1 = g(z_1)$, $t_2 = g(z_2)$ and all points are ordinary so that $g'(z) \neq 0$. If $|z_2 - z_1| = h$, then $|t_2 - t_1| = Mh$, where $M = \sqrt{J}$ is the magnification. Thus *a line doublet of strength μ per unit length through $z = z_1$ in the z-plane maps into one of strength $M\mu$ per unit length through $t = t_1$ in the t-plane where M is the magnification of the conformal transformation*. Note, however, that the orientations of the doublets generally differ. The orientations are found from the geometries of the configurations. One has to discuss the two limits.

$$\arg(z_2 - z_1) \text{ as } z_2 \to z_1, \qquad \arg(t_2 - t_1) \text{ as } t_2 \to t_1.$$

If $z_2 - z_1 = \delta z_1$ and $t_2 - t_1 = \delta t_1 = g'(z_1)\delta z_1$, then

$$\arg \delta t = \arg g'(z_1) + \arg \delta z_1,$$

which shows that under the conformal transformation (7.1) the axis of the line doublet at $z = z_1$ has undergone a rotation in the positive sense of amount $\arg g'(z_1)$.

For the transformation of line vortices, we observe that, in an obvious notation,

$$\oint_{\mathscr{C}^*} \mathbf{q}^* \cdot d\mathbf{s}^* = \oint_{\mathscr{C}^*} d\phi = \int_{\mathscr{C}} d\phi = \oint_{\mathscr{C}} \mathbf{q} \cdot d\mathbf{s} = 2\pi\kappa.$$

Thus a uniform line vortex of strength κ per unit length through an ordinary point $z = z_0$ transforms into one of equal strength through $t = t_0 = g(z_0)$ under the conformal transformation (7.1).

Example [7.1]

A uniform line doublet of strength μ per unit length is situated at the point $t = i$ in the t-plane its axis pointing in the positive direction of the real axis. There is a rigid boundary along the entire length of the real axis and fluid occupies the region $\text{Im}(t) \geqslant 0$. By applying the conformal transformation $t = e^z$, determine the corresponding fluid motion in the z-plane.

Fig. 7.2(i) shows the physical model for flow in the half-plane $\text{Im}(t) \geqslant 0$ with $t = \xi + i\eta$. Fig. 7.2(ii) shows the entire t-plane; the corresponding mathematical model has equal line doublets of strength μ per unit length through $t = i$ (actual) and $t = -i$ (image), their axes both being along the direction of the positive real axis $O\xi$. The complex potential at t in Fig. 7.2(ii) is thus

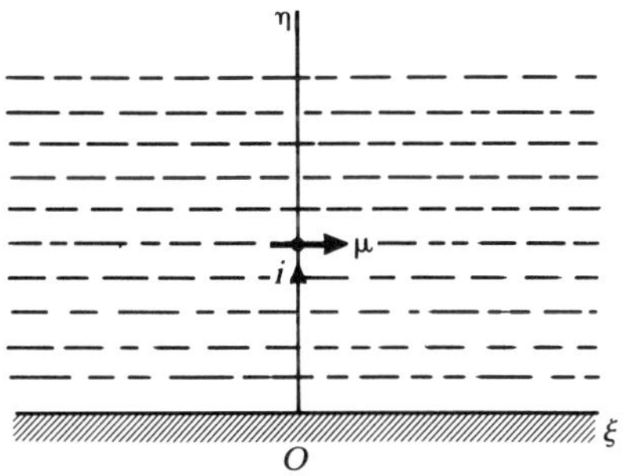

Fig. 7.2(i)

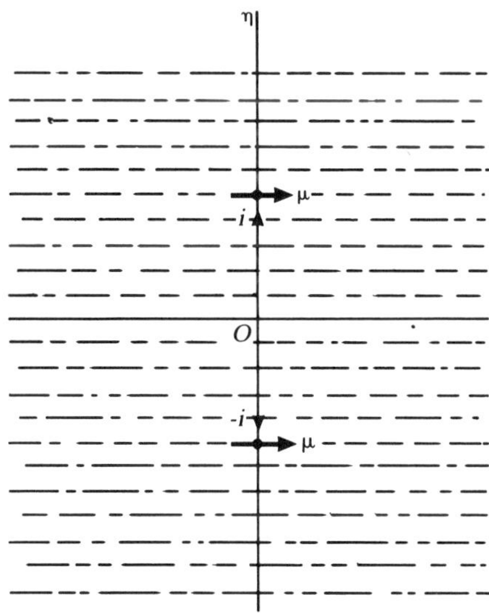

Fig. 7.2(ii)

$$w = -\frac{\mu}{t-i} - \frac{\mu}{t+i} = -\frac{2\mu t}{t^2+1}.$$

This is also the complex potential at t in Fig. 7.2(i), where $\text{Im}(t) \geqslant 0$.

Now let us apply the given transformation to the half-plane $\text{Im}(t) \geqslant 0$ of Fig. 7.2(i). The conformal transformation may be written in the alternative form

$$z = \ln t = \ln(\xi + i\eta)$$

for which $t = 0$ is the only singular point in the finite part of the t-plane. The complex potential in the z-plane is

$$w = -\frac{2\mu e^z}{e^{2z}+1} = -\mu \operatorname{sech} z.$$

The boundary $\eta = 0$ transforms into

$$z = \ln \xi = \ln|\xi| + i \arg \xi,$$

so that

$$\left.\begin{aligned} y &= \text{Im}(z) = \arg \xi, \\ \text{i.e.} \qquad y &= 0 \text{ for } \xi > 0, \\ &= \pi \text{ for } \xi < 0. \end{aligned}\right\}$$

The point $t = i$ maps into

$$z = \ln i = \ln|i| + i \arg i = 0 + \tfrac{1}{2}\pi i.$$

This point is mid-way between the two planes $y = 0$, $y = \pi$ and so the entire fluid region $\eta \geqslant 0$ for Fig. 7.2(ii) maps into the infinite strip $0 \leqslant y \leqslant \pi$ of the z-plane.

Since $dz/dt = 1/t = -i$ at $t = i$ and so the magnification at $t = i$ is $|dz/dt| = 1$. Thus the line doublet of strength μ at $t = i$ maps into one of equal strength at $z = \tfrac{1}{2}\pi i$. Furthermore, at $t = i$, $\arg(dz/dt) = \arg(-i) = -\tfrac{1}{2}\pi$, which shows that under the conformal transformation the axis of the doublet undergoes a negative (clockwise) rotation of $\tfrac{1}{2}\pi$.

We have shown, then, that the complex potential $w = -\mu \operatorname{sech} z$ determines the fluid motion between the rigid impermeable planes $y = 0$, $y = \pi$ when a uniform line doublet of strength μ per unit length and having its axis pointing in the negative y-direction is introduced at the point $(0, \tfrac{1}{2}\pi)$. Fig. 7.3 shows the configuration.

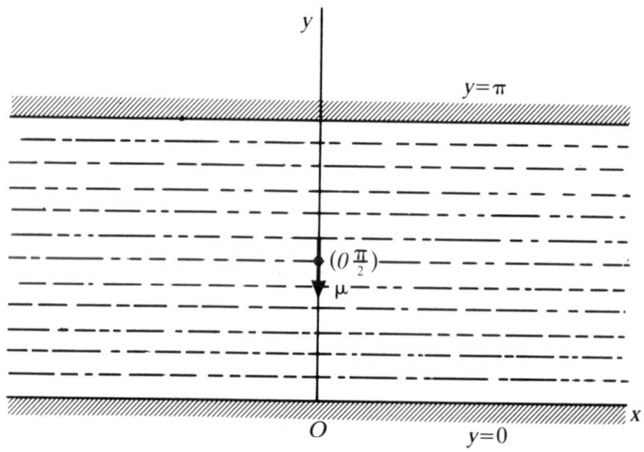

Fig. 7.3

Example [7.2]

A plane flow of liquid in the t-plane ($t = \xi + i\eta$) is a steady stream of speed V parallel to the ξ-axis in a channel defined by $\sqrt{b} \leqslant \eta \leqslant \sqrt{a}$. Find the complex potential of the motion.

Apply the conformal transformation $t = \sqrt{z}$ to this flow and describe the boundaries and streamlines of the flow in the z-plane. Show that, in the absence of body forces, the concentric circles $|z| = \text{const.}$ are curves of equal pressure.

If the pressure at infinity is just sufficient to make the pressure in the channel positive everywhere, show that the pressure at points in the channel on the circle $|z| = a$ is $\tfrac{1}{8}\rho V^2\,(a-b)/(ab)$.

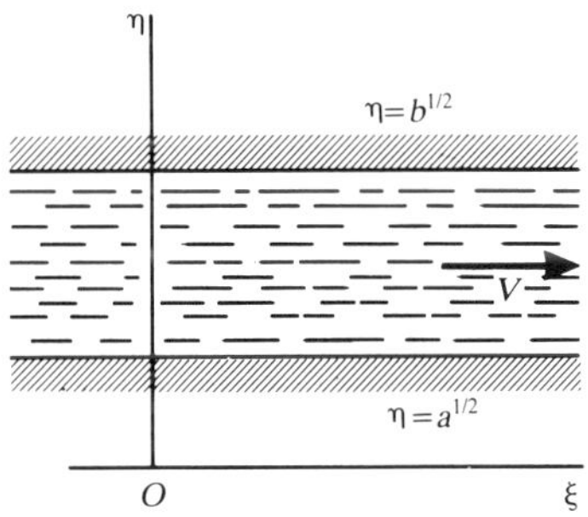

Fig. 7.4(i)

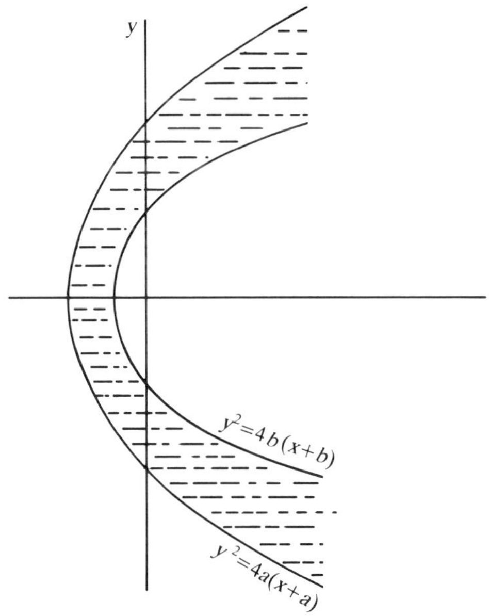

Fig. 7.4(ii)

The velocity potential of the flow in the t-plane (Fig. 7.4(i)) is $V\xi = \text{Re}(Vt)$. Hence the complex potential of this flow is

$$w = Vt = V(\xi + i\eta).$$

Applying the transformation $t = \sqrt{z}$,

$$w = V\sqrt{z} = V\sqrt{(x + iy)}.$$

When $\eta = \sqrt{b}$,

$$z = x + iy = t^2 = (\xi + i\sqrt{b})^2,$$

whence

$$x = \xi^2 - b, \qquad y = 2\xi\sqrt{b}.$$

The eliminant of ξ from these is

$$y^2 = 4b(x + b),$$

which is the equation of a parabola, vertex $(-b, 0)$ and focus $(0, 0)$. Since $-\infty \leqslant \xi \leqslant \infty$, the parametric equations of the curve show that the entire line $\eta = \sqrt{b}$ maps into the entire parabola $y^2 = 4b(x + b)$.

Similarly the line $\eta = \sqrt{a}$ maps into the entire parabola

$$y^2 = 4a(x + a).$$

The region of flow $\sqrt{b} \leqslant \eta \leqslant \sqrt{a}$ maps into the region between the two parabolas (Fig. 7.4(ii)).

With the usual notations, we take $z = r\,e^{i\theta}$, $w = \phi + i\psi$. Then

$$\phi + i\psi = w = V\sqrt{z} = Vr^{1/2}\exp(\tfrac{1}{2}i\theta),$$

and so

$$\phi = Vr^{1/2}\cos(\theta/2), \qquad \psi = Vr^{1/2}\sin(\theta/2).$$

The streamlines are the curves $\psi/V = c$ (= const.) or

$$r\sin^2(\theta/2) = c^2$$

or

$$2c^2 = r(1 - \cos\theta) = \sqrt{(x^2 + y^2)} - x.$$

Thus, in simplified form, the streamlines are

$$y^2 = 4c^2(c^2 + x).$$

These are confocal parabolas with (0, 0) as focus. The boundaries belong to the system with $c = a$ and $c = b$.

Now

$$\mathrm{d}w/\mathrm{d}z = \tfrac{1}{2}Vz^{-1/2} \qquad (z = r\,e^{i\theta}),$$

and so the speed of the fluid is

$$|\mathrm{d}w/\mathrm{d}z| = \tfrac{1}{2}Vr^{-1/2} = q, \text{ say.}$$

Application of Bernoulli's equation along a streamline gives, in the usual notation

$$(p/\rho) + \tfrac{1}{2}q^2 = K \text{ (= const.)}.$$

On the circle $|z| = a$, the speed $q = \frac{1}{2}Va^{-1/2}$ and so

$$p = \rho\{K - \tfrac{1}{8}(V^2/a)\},$$

which is essentially constant, so that the circle $|z| = a$ is an iso-pressure curve.

Along any streamline, $(p/\rho) + \frac{1}{2}q^2 = K$, and so

$$\begin{aligned} p &= \rho\{K - \tfrac{1}{8}(V^2/r)\} \\ &= p_\infty - \tfrac{1}{8}(\rho V^2/r), \end{aligned}$$

since $p \to p_\infty$ as $r \to \infty$, so that $p_\infty = \rho K$. Since $\min r = b$,

$$\min p = p_\infty - \tfrac{1}{8}(\rho V^2/b).$$

This is zero if $p_\infty = \frac{1}{8}(\rho V^2/b)$, so that

$$p = \frac{\rho V^2}{8}\left(\frac{1}{b} - \frac{1}{r}\right).$$

The result now follows on taking $r = a$.

7.3 SINGLE INFINITE ROW OF LINE VORTICES

Let us now consider an infinite row of line vortices, each of strength κ through the points $z = na$ ($n = 0, \pm 1, \pm 2, \ldots$, a real). The conformal transformation

$$t = \sin(\pi z/a) \tag{7.7}$$

will map the z-plane on to a certain region in the t-plane. We note that $dt/dz = 0$, when $z = (2n + 1)a/2$ ($n = 0, \pm 1, \pm 2, \ldots$): these values determine the singular points of the transformation. The region

$$-\tfrac{1}{2}a < x < \tfrac{1}{2}a, \qquad -\infty < y < \infty \tag{7.8}$$

excludes the singularities of the transformation and the whole of the t-plane may be mapped conformally on to this region. If we introduce a line vortex of strength κ through $t = 0$ the complex potential due to it is

$$w = -i\kappa \ln t. \tag{7.9}$$

Under the conformal transformation (7.7), a line vortex of strength κ through $t = 0$ maps into the infinite row of equal line vortices, each of strength κ, through $z = na$ ($n = 0, \pm 1, \pm 2, \ldots$) in the z-plane. The complex potential of this distribution is obtained by substitution from (7.7) into (7.9) to give

$$w = -i\kappa \ln \sin(\pi z/a). \tag{7.10}$$

As an alternative we may obtain (7.10) without using a conformal transformation as follows. The complex potential due to the line vortex of strength κ through $z = na$ is $-i\kappa \ln(z - na)$. Thus the total complex potential for the entire row is obtained by superposition to give

$$\begin{aligned} w &= -i\kappa \sum_{n=-\infty}^{\infty} \ln(z - na) \\ &= -i\kappa \ln\{z(z^2 - a^2)(z^2 - 4a^2)\ldots\} \\ &= -i\kappa \ln\left\{\frac{\pi z}{a}\left(1 - \frac{z^2}{a^2}\right)\left(1 - \frac{z^2}{4a^2}\right)\left(1 - \frac{z^2}{9a^2}\right)\ldots\right\} + \text{const.} \end{aligned}$$

In accordance with Ref. 28, Section 7.5, $\sin\theta$ may be expressed as the infinite product

$$\sin\theta = \theta\left(1 - \frac{\theta^2}{\pi^2}\right)\left(1 - \frac{\theta^2}{4\pi^2}\right)\left(1 - \frac{\theta^2}{9\pi^2}\right)\ldots$$

Hence, omitting the immaterial constant, the last expression for w reduces to the form (7.10).

From Kelvin's Theorem (Section 3.7) we know that any particular vortex will move with the fluid. No single vortex will move of its own accord and so the motion of any particular line vortex in the above row will be due to that of the others. Let us consider the motion of the mth filament in the row. The complex potential at the mth filament due to all the others is

$$w_m = -i\kappa \left[\sum_{n=-\infty}^{\infty} {}^{(m)} \ln (z - na) \right]_{z=ma},$$

where the index (m) means that the term $\ln(z - ma)$ is omitted from the summation. Hence the velocity components $[u_m, v_m]$ of this filament are given by

$$u_m - iv_m = \frac{dw_m}{dz} = -i\kappa \left[\sum_{n=-\infty}^{\infty} {}^{(m)} \frac{d}{dz} \ln (z - na) \right]_{z=ma}$$

$$= i\kappa \sum_{n=-\infty}^{\infty} {}^{(m)} \frac{1}{(m-n)a}.$$

The infinite series converges to zero, so that *all line vortices in the row are at rest.*

The fluid velocity at any point z of the plane other than a line vortex is readily found from (7.10):

$$\frac{dw}{dz} = u - iv = -\frac{i\kappa\pi}{a} \cot\left(\frac{\pi z}{a}\right). \tag{7.11}$$

This shows that the points on the line of vortices midway between adjacent pairs are stagnation points, i.e. the stagnation points are the points $z = (2n + 1)a/2$ $(n = 0, \pm 1, \pm 2, \ldots)$. They are the only stagnation points of the flow.

By writing $z = x + iy$ in (7.11) and equating the real and imaginary parts it is easily shown that

$$[u, v] = \frac{\pi\kappa[-\sinh(2\pi y/a), \sin(2\pi x/a)]}{a\,[\cosh(2\pi y/a) - \cos(2\pi x/a)]}. \tag{7.12}$$

From these relations it is easily seen that as $y \to \pm\infty$,

$$u \to \mp \pi\kappa/a, \qquad v \to 0.$$

Thus at large distances y above the line of vortices the flow approximates closely to that of a uniform stream of velocity $\pi\kappa/a$ in the negative x-direction and at large distances below the line of vortices it approximates to uniform flow of equal and opposite velocity. The motion may be likened to that of two rigid blocks sliding over a layer of rollers separating them.

Example [7.3]

Liquid fills the doubly infinite strip between rigid barriers along the lines

$x = \pm \frac{1}{2}\pi$ and a line vortex of strength κ is placed at the origin. Determine the fluid motion between the barriers.

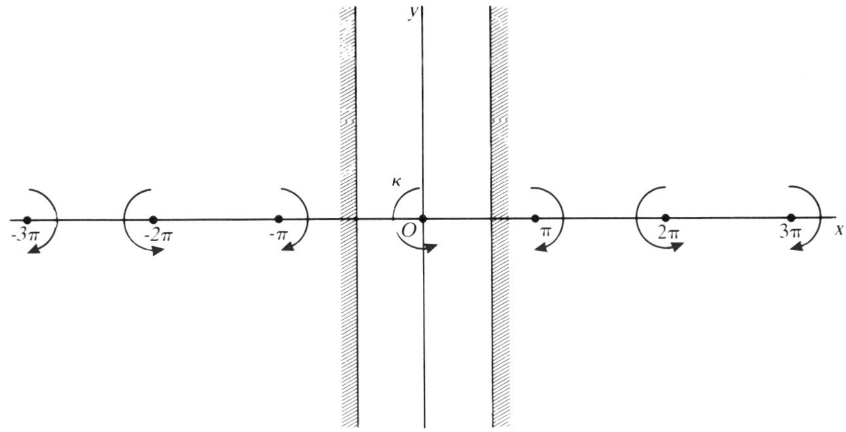

Fig. 7.5

In Fig. 7.5, we have a line vortex of strength κ through $z = 0$. This produces infinitely many images in the two planes. The image system will consist of line vortices of strength $-\kappa$ through the points $x = \pm\,\pi, \pm 3\pi, \pm 5\pi, \ldots$ and of strength $+\kappa$ through the points $x = \pm 2\pi, \pm 4\pi, \pm 6\pi, \ldots$ With the barriers removed and the distribution of line vortices $+\kappa$ at $x = 0, \pm 2\pi, \pm 4\pi, \ldots$ and $-\kappa$ at $x = \pm\pi, \pm 3\pi, \pm 5\pi, \ldots$ the complex potential is

$$\begin{aligned} w &= -i\kappa \ln z - i\kappa \sum_{n=-\infty}^{\infty} \ln(z - 2n\pi) + i\kappa \sum_{n=-\infty}^{\infty} \ln\{z - (2n+1)\pi\} \\ &= -i\kappa \ln z - i\kappa \sum_{n=1}^{\infty} \ln\{z^2 - (2n\pi)^2\} \\ &\qquad + i\kappa \ln(z+\pi) + i\kappa \sum_{n=1}^{\infty} \ln\{(z+\pi)^2 - (2n\pi)^2\} \\ &= -i\kappa \ln\left\{\frac{z}{z+\pi} \prod_{n=1}^{\infty} \frac{1 - (z/2n\pi)^2}{1 - [(z+\pi)/2n\pi]^2}\right\}. \end{aligned}$$

Using again

$$\sin\theta = \theta \prod_{n=1}^{\infty} \left(1 - \frac{\theta^2}{n^2\pi^2}\right),$$

the expression for w simplifies to

$$w = -i\kappa \ln \frac{\sin (z/2)}{\sin [(z+\pi)/2]} = -i\kappa \ln \tan \left(\frac{z}{2}\right).$$

This is also the complex potential for the fluid in the region $|x| < \frac{1}{2}\pi$ with the rigid boundaries $x = \pm \frac{1}{2}\pi$ and the line vortex κ through $z = 0$.

The streamlines of the flow are given by $\psi = \text{const}$. Now

$$\begin{aligned}
\psi = -\operatorname{Im}(w) &= \kappa \ln |\tan (\tfrac{1}{2}z)| \\
&= \frac{\kappa}{2} \ln \left\{ \tan\left(\frac{z}{2}\right) \tan\left(\frac{\bar{z}}{2}\right)\right\} \\
&= \frac{\kappa}{2} \ln \left\{ \frac{\cos \frac{1}{2}(z-\bar{z}) - \cos \frac{1}{2}(z+\bar{z})}{\cos \frac{1}{2}(z-\bar{z}) + \cos \frac{1}{2}(z+\bar{z})} \right\} \\
&= \frac{\kappa}{2} \ln \left\{ \frac{\cosh y - \cos x}{\cosh y + \cos x} \right\}.
\end{aligned}$$

Thus the equations of the streamlines are given by

$$\cosh y = C \cos x \qquad (C = \text{const.}).$$

These curves are ovals symmetrical about both axes.

To determine the motion of the vortex at O, we use the auxiliary complex potential

$$w' = w + i\kappa \ln z$$

in which the logarithmic singularity at $z = 0$ is removed. Thus

$$\frac{dw'}{dz} = i\kappa \left(\frac{1}{z} - \frac{1}{\sin z}\right).$$

It is an easy exercise to show that $(1/z) - \operatorname{cosec} z \to 0$ as $z \to 0$. Hence

$$\lim_{z \to 0} \left(\frac{dw'}{dz}\right) = 0,$$

which shows that the velocity of the line vortex of strength κ placed at 0 is zero. This vortex will remain at rest.

7.4 THE KARMAN VORTEX STREET

Fig. 7.6 shows two parallel rows $y = \pm \frac{1}{2}b$ of line vortices. Those on $y = \frac{1}{2}b$ are each of strength κ and are situated at the points $(na, \frac{1}{2}b)$ $(n = 0, \pm 1, \pm 2, \ldots)$. Those on $y = -\frac{1}{2}b$ are each of strength $-\kappa$ and are situated at $((n + \frac{1}{2})a,$

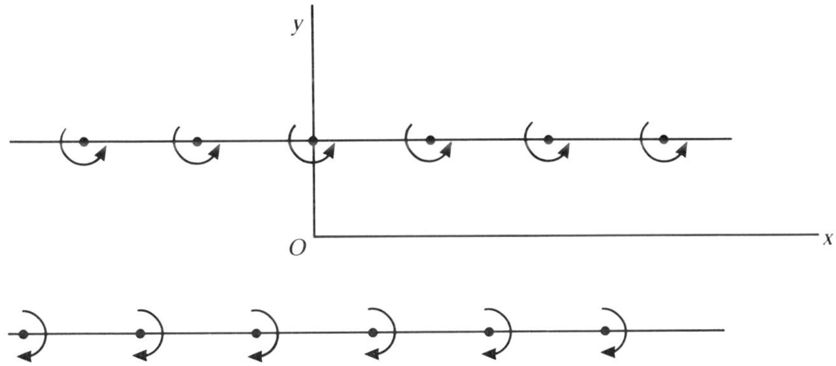

Fig. 7.6

$-\frac{1}{2}b))$ $(n = 0, \pm 1, \pm 2, \ldots)$. Using the theory of the previous section, the complex potential, obtained by superposing the separate contributions from the two parallel lines is

$$w = -i\kappa \ln \sin \frac{\pi}{a}\left(z - \frac{ib}{2}\right) + i\kappa \ln \sin \frac{\pi}{a}\left(z - \frac{a}{2} + \frac{ib}{2}\right). \qquad (7.13)$$

The first term on the right arises from the vortices, each of strength $+\kappa$, in the line $y = \frac{1}{2}b$ and is derived from (7.10) by shifting the origin from $z = 0$ to $z = \frac{1}{2}ib$, so that z is replaced by $z - \frac{1}{2}ib$. The next term follows in the same by replacing κ in (7.10) by $-\kappa$ and moving the origin from $z = 0$ to $z = \frac{1}{2}a - \frac{1}{2}ib$. The velocity of the line vortex at $z = \frac{1}{2}(a - ib)$ is obtained from the auxiliary complex potential

$$w' = w - i\kappa \ln \sin \frac{\pi}{a}\left(z - \frac{a}{2} + \frac{ib}{2}\right),$$

which is regular at $z = \frac{1}{2}(a - ib)$. Then

$$w' = -i\kappa \ln \sin \frac{\pi}{a}\left(z - \frac{ib}{2}\right)$$

and the velocity $[u, v]$ at the vortex $z = \frac{1}{2}(a - ib)$ is found from

$$u - iv = \left(\frac{\mathrm{d}w'}{\mathrm{d}z}\right)_{z=\frac{1}{2}(a-ib)} = \frac{\pi\kappa}{a} \tanh\left(\frac{\pi b}{a}\right). \qquad (7.14)$$

This expression depends only on κ, a, b and not on the position of the line vortex in the row $y = -\frac{1}{2}b$. Consequently, all line vortices in this row move with the same velocity

$$u = (\pi\kappa/a)\tanh(\pi b/a), \quad v = 0.$$

Similarly it can be shown that all the line vortices on $y = +\frac{1}{2}b$ advance with the same velocity. So the entire system, which is known as a *Kármán vortex street*, advances in the positive x-direction with velocity $(\pi\kappa/a)\,\text{th}(\pi b/a)$, the pattern remaining unaltered at all times.

A Kármán vortex street provides a model in ideal flow for what is observed when a uniform stream of real fluid flows past a circular cylinder at rest. When the speed of the stream is small, the streamline pattern is essentially symmetrical both fore and aft of the cylinder, but as the speed of the stream is increased, asymmetry of the flow develops. Eddies are observed to form in regions of separated flow attached to the downstream side of the cylinder. With further increase in the speed of the stream, these eddies detach from the cylinder and are carried downstream. The eddies are shed alternately from the top and bottom of the cylinder on its downstream side and their pattern is closely simulated by the Kármán vortex street. However, the phenomenon observed in a real fluid is strictly a result of the action of viscosity within the fluid. The effect and significance of the role played by viscosity in fluid motion is considered in detail in the companion volume to this book.

7.5 THE JOUKOWSKI TRANSFORMATION

A most important conformal transformation, especially for applications in aerodynamics, is due to Joukowski. The *Joukowski transformation* is defined to be

$$t = z + (c^2/z), \qquad c > 0, \tag{7.15}$$

where $z = x + iy = r\,\mathrm{e}^{i\theta}$, $t = \xi + i\eta$ and $x, y, r, \theta, \xi, \eta$ are all real. Equating real and imaginary parts,

$$\xi = (r + c^2/r)\cos\theta, \qquad \eta = (r - c^2/r)\sin\theta. \tag{7.16}$$

Consequently the confocal ellipses

$$\frac{\xi^2}{(a + c^2/a)^2} + \frac{\eta^2}{(a - c^2/a)^2} = 1, \tag{7.17}$$

are the transforms of the concentric circles $r = a$. These ellipses have semi-axes of lengths $a \pm c^2/a$ and the focal distances from the centre are

$$\sqrt{\{(a + c^2/a)^2 - (a - c^2/a)^2\}} = 2c,$$

so that the focal points in (ξ, η)-coordinates are $(\pm\, 2c, 0)$. The circle $r = c$ is seen from (7.16) to transform into

$$\xi = 2c\cos\theta, \qquad \eta = 0,$$

i.e. it transforms into the finite line segment

$$-2c \leqslant \xi \leqslant 2c, \qquad \eta = 0. \tag{7.18}$$

Essentially this is a degenerate ellipse and part of the confocal system (7.17).

Reverting to (7.15), it is clear that the transformation may be regarded as a quadratic equation in z. Its roots, in terms of t, are given by

$$z = \tfrac{1}{2}\{t \pm \sqrt{(t^2 - 4c^2)}\}. \tag{7.19}$$

Thus, excepting for the points $z = \pm c$; $t = \pm 2c$, to each point in the t-plane there are two points in the z-plane. The point $\xi = 2c \cos\theta$, $\eta = 0$ on the line segment (7.18) corresponds to the two points $z = c \exp(\pm i\theta)$ on the circle $|z| = c$. Let z_1, z_2 be those values of z in (7.19) corresponding to the plus and minus signs respectively so that

$$z_1 = \tfrac{1}{2}\{t + \sqrt{(t^2 - 4c^2)}\}, \tag{7.20}$$

$$z_2 = \tfrac{1}{2}\{t - \sqrt{(t^2 - 4c^2)}\}. \tag{7.21}$$

From these forms,

$$z_2 = c^2/z_1 \tag{7.22}$$

and it follows at once that as $|t| \to \infty$, $|z_1| \to \infty$ and $|z_2| \to 0$. Thus (7.20) maps the entire region $|z| \geqslant a$ on to the entire t-plane, and (7.21) maps the entire region $|z| \leqslant a$ on to the entire t-plane. In fact the t-plane is said to be a Riemann surface that is cut along the finite line segment $-2c \leqslant \xi \leqslant 2c$, $\eta = 0$. In subsequent applications of the Joukowski transformation, we shall employ the form (7.20) and not (7.21), thereby ensuring that the region $|z| \geqslant a$ is mapped on to the entire t-plane and that there is (1–1) correspondence between the two regions.

The complex potential for uniform flow in the t-plane and with velocity U in the positive ξ-direction is

$$w = Ut. \tag{7.23}$$

This remains unaltered for inviscid flow in the t-plane when the flat plate $-2c \leqslant \xi \leqslant 2c$, $\eta = 0$ is introduced into the flow since it in no way disturbs the flow. Applying the Joukowski transformation (7.15) gives

$$w = U(z + c^2/z), \tag{7.24}$$

which represents uniform flow, without circulation, round the circular cylinder $|z| = c$, as we have seen in Example [6.11]. The equations of the streamlines in the z-plane are given by $\mathrm{Im}(w) = \text{const.}$, i.e. by

$$(r - c^2/r)\sin\theta = K,$$

or

$$y(x^2 + y^2 - c^2) = K(x^2 + y^2). \tag{7.25}$$

As we are taking $z = z_1$ from (7.20), these streamlines are outside the circle $|z| = c$. Also the form (7.25) shows that y is an even function of x so that the streamlines are symmetrical about the y-axis. The correspondences of the t- and z-planes are shown in Figs 7.7(i), (ii).

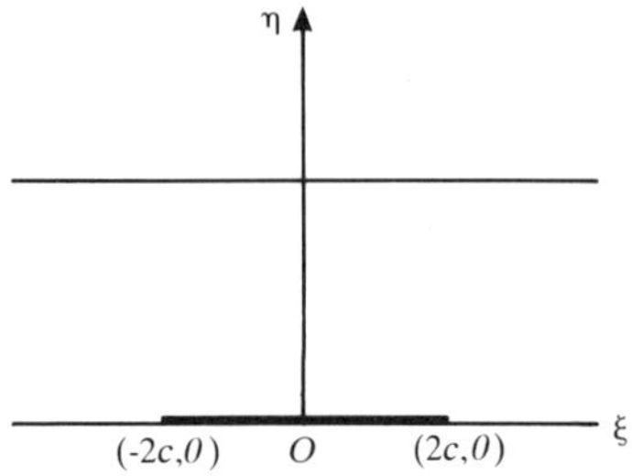

Fig. 7.7(i)

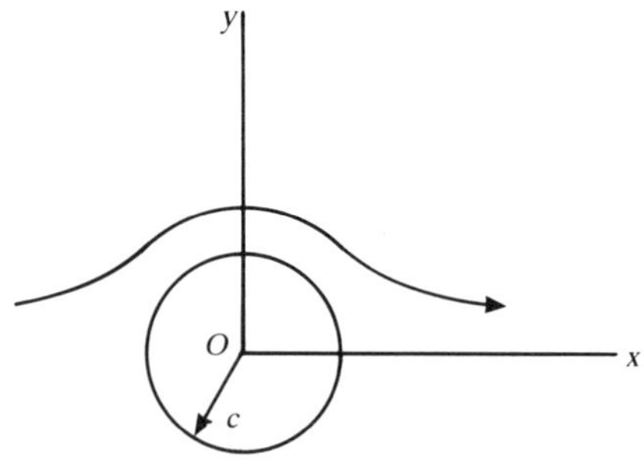

Fig. 7.7(ii)

Next, we recall from Section 6.8 that the complex potential for circular flow with positive circulation κ about the circular cylinder $|z| = a$ is

$$w = (-i\kappa/2\pi) \ln z. \tag{7.26}$$

Taking $a \geqslant c$, and the form (7.20), this formula becomes

$$w = -\frac{i\kappa}{2\pi} \ln\left\{\frac{t + \sqrt{(t^2 - 4c^2)}}{2}\right\}. \tag{7.27}$$

This gives the appropriate complex potential for a flow with circulation κ round the surface of the elliptic cylinder whose cross-section is given by (7.17) if $a > c$, and by the flat plate (7.18) if $a = c$. The streamlines for (7.26) are the concentric circles

$$x^2 + y^2 = b^2 \tag{7.28}$$

and these transform, for $b > a$, into the confocal ellipses

$$\frac{\xi^2}{(b + c^2/b)^2} + \frac{\eta^2}{(b - c^2/b)^2} = 1. \tag{7.29}$$

Superposing the two flows characterised by (7.24), (7.26) we obtain

$$w(z) = U\left(z + \frac{c^2}{z}\right) - \frac{i\kappa}{2\pi} \ln z \tag{7.30}$$

for the complex potential of a uniform stream of velocity U in the positive x-direction past the cylinder $|z| = c$ around which there is circulation κ in the positive or anticlockwise sense. The nature of this flow has been fully discussed in Section 6.8. The methods used there show that if $|4\pi Uc| > |\kappa|$ there are two stagnation points both on $|z| = c$ and they are symmetrically disposed with respect to the y-axis. If $|4\pi Uc| < |\kappa|$, each root of $w'(z) = 0$ lies on the y-axis, one lying within $|z| = c$ and the other outside it. From the point of view of the flow under investigation only the root outside $|z| = c$ is relevant.

Under the conformal transformation (7.15), equation (7.30) becomes

$$w = Ut - (i\kappa/2\pi)\ln\left\{\tfrac{1}{2}[t + \sqrt{(t^2 - 4c^2)}]\right\}, \tag{7.31}$$

which gives the complex potential for uniform flow with velocity U in the direction of the positive ξ-axis past the flat plate $-2c \leqslant \xi \leqslant 2c$, $\eta = 0$ around which there is an anticlockwise circulation of amount κ.

Now let us change the directions of the coordinate axes. If the frame specified by Ox, Oy undergoes a negative or clockwise rotation α to new positions specified by OX, OY, then

$$\overrightarrow{OP} \equiv z = x + iy = r\,e^{i\theta} \text{ in the first frame;}$$

$$\overrightarrow{OP} \equiv Z = X + iY = r\exp[i(\theta + \alpha)] \text{ in the second frame.}$$

Thus the rotation is achieved through the conformal transformation

$$z = e^{-i\alpha} Z. \tag{7.32}$$

Under this transformation, the complex potential (7.24) transforms into

$$w = U(e^{-i\alpha} Z + e^{i\alpha} c^2/Z), \tag{7.33}$$

which represents flow without circulation round a circular cylinder of radius c of a uniform stream of velocity U at an angle α to the positive X-axis. This result is also easily obtained using the Milne–Thomson circle theorem.

Now suppose the conformal transformation

$$t = Z + c^2/Z, \quad \text{or} \quad Z = \tfrac{1}{2}\{t + \sqrt{(t^2 - 4c^2)}\} \tag{7.34}$$

is made from the Z- to the t-plane, t being different from that used earlier in the section. Then the elliptic cylinder is again transformed into the flat plate $-2c \leqslant \xi \leqslant 2c$, $\eta = 0$ of a uniform stream making angle α with the positive ξ-direction ($t = \xi + i\eta$). The value of w, as given by (7.33), becomes on simplification

$$w = U\{t\cos\alpha - i\sqrt{(t^2 - 4c^2)}\sin\alpha\}. \tag{7.35}$$

On this we may impose a positive circulation κ round the plate represented in the usual notation by

$$w = -(i\kappa/2\pi)\ln\left\{\tfrac{1}{2}[t + \sqrt{(t^2 - 4c^2)}]\right\}. \tag{7.36}$$

From the form (7.35) the reader will easily confirm that with or without the added circulation, $dw/dt \to Ue^{-i\alpha}$ as $|t| \to \infty$. Since $dw/dt = \bar{q}$, it follows that $q \to Ue^{i\alpha}$ as $|t| \to \infty$. This accordingly represents the undisturbed uniform flow with velocity U at α to the positive ξ-axis.

7.6 FLOW ROUND AEROFOILS

From the analysis of the previous section, the Joukowski transformation

$$t = z + c^2/z, \qquad c > 0, \tag{7.37}$$

transforms the circle

$$|z| = a > c \tag{7.38}$$

in the z-plane into the ellipse

$$\xi = (a + c^2/a)\cos\theta, \qquad \eta = (a - c^2/a)\sin\theta \tag{7.39}$$

in the t-plane, where $t = \xi + i\eta$. For an ellipse with major and minor semi-axes A, B respectively, the constants a, c are given by

$$a = \tfrac{1}{2}(A + B), \qquad c^2 = \tfrac{1}{4}(A^2 - B^2). \tag{7.40}$$

Conversely, from (7.20), the transformation

$$z = \tfrac{1}{2}\{t + \sqrt{(t^2 - 4c^2)}\} \tag{7.41}$$

maps the region consisting of the exterior and periphery of the ellipse $\xi^2/A^2 + \eta^2/B^2 = 1$ in the t-plane on to the region comprising the exterior and periphery of the circle $x^2 + y^2 = a^2$ in the z-plane. It therefore follows that the complex potential

$$w(z) = Uz\,\mathrm{e}^{-i\alpha} + (Ua^2\,\mathrm{e}^{i\alpha}/z) + (i\kappa/2\pi)\ln z, \tag{7.42}$$

together with the transformation (7.41) will give the complex potential $W(t)$ which describes the flow of a uniform stream of speed U past an ellipse, when the direction of the stream makes an angle α with the major axis of the ellipse and around which there is a circulation $-\kappa$.

The stagnation points of the flow are given by the roots of the equation $\mathrm{d}W/\mathrm{d}t = 0$, or equivalently $\mathrm{d}w/\mathrm{d}z = 0$ since $\mathrm{d}z/\mathrm{d}t$ does not vanish at points either on or exterior to the ellipse. The latter equation is therefore

$$U\,\mathrm{e}^{-i\alpha} - (Ua^2\,\mathrm{e}^{i\alpha}/z^2) + (i\kappa/2\pi z) = 0. \tag{7.43}$$

When B/A is small the ellipse is thin and the downstream end of the major axis given by $t = A = a + c^2/a$ may be thought of as the trailing edge. This thin ellipse is an example of an *aerofoil*, which is a two dimensional shape used in the design of aeroplane wings. The trailing edge of the ellipse corresponds to $z = a$ in the z-plane. Thus from (7.43) if κ is chosen to be

$$\kappa = 4\pi a U \sin\alpha \tag{7.44}$$

we ensure that the trailing edge of the ellipse is a stagnation point. This is an example of the *Kutta–Joukowski condition* which requires suitable choice of the circulation κ to provide a smooth air flow off the trailing edge of an aerofoil. Such a condition on the circulation round a two-dimensional aerofoil models very closely what is observed in practice. As an aircraft begins to move before take-off, the flow about the wing is close to that predicted by potential flow theory without circulation. The stagnation point is on the upper surface of the wing and the flow near the trailing edge is at high speed. With increasing speed, separation of the flow from the wing takes place at the trailing edge and a vortex is shed behind the trailing edge. Since this vortex is left behind by the wing, a circulation of opposite sense about the wing must be induced in order to satisfy Kelvin's theorem on the constancy of circulation about any closed circuit moving with the fluid. As there is no circulation about a large circuit embracing the wing, there can be no total circulation about the same circuit when it embraces both wing and vortex. When the vortex has become detached from the wing the stagnation point moves to the trailing edge in accordance with the Kutta–Joukowski condition.

The force and torque acting on the elliptical aerofoil considered above can be easily found using the theorem of Blasius established in Section 6.9. The force is given in the notation of that section by

$$Y + iX = \frac{\rho}{2} \oint_{\Gamma} \left(\frac{\mathrm{d}W}{\mathrm{d}t}\right)^2 \mathrm{d}t,$$

where Γ is the boundary of the ellipse in the t-plane. Transforming to the corresponding integral in the z-plane we have

$$Y + iX = \frac{\rho}{2} \oint_{\mathscr{C}} \left(\frac{\mathrm{d}w}{\mathrm{d}z}\right)^2 \left(\frac{\mathrm{d}z}{\mathrm{d}t}\right) \mathrm{d}z, \tag{7.45}$$

where $\mathscr{C}$ is the circle $|z| = a$. On substituting from (7.41), (7.42), we obtain

$$Y + iX = -\frac{\rho}{2} \oint_{\mathscr{C}} \left[U\,\mathrm{e}^{-i\alpha} - (Ua^2\,\mathrm{e}^{i\alpha}/z^2) + (i\kappa/2\pi z)\right]^2 \left(1 - \frac{c^2}{z^2}\right)^{-1} \mathrm{d}z. \tag{7.46}$$

The singularities of the integrand of (7.46) are within the circle $\mathscr{C}$, so that the integral has the same value round *any* curve enclosing $\mathscr{C}$ and in particular we may choose a large circle with $z = 0$ as centre. For such a contour,

$$(1 - c^2 z^{-2})^{-1} = 1 + c^2 z^{-2} + \mathrm{O}(c^4 z^{-4})$$

and on letting the radius of the circle tend to infinity, the only term which gives rise to a non-zero contribution to the integral is that involving z^{-1}. Hence

$$Y + iX = \rho\kappa U\,\mathrm{e}^{-i\alpha}$$

from which we obtain

$$X = -\rho\kappa U \sin\alpha, \qquad Y = \rho\kappa U \cos\alpha. \tag{7.47}$$

The two force components $[X, Y]$ are along the ξ- and η-axes. However, it is more elucidating to evaluate the components which are directed along and perpendicular to the direction of flow of the stream. Such components are referred to as the *drag* and *lift* respectively. The lift force L per unit length is accordingly

$$L = Y\cos\alpha - X\sin\alpha = \rho\kappa U = 4\pi a\rho U^2 \sin\alpha \tag{7.48}$$

when the Kutta–Joukowski condition (7.44) is applied. The drag force D per unit length is

$$D = X\cos\alpha + Y\sin\alpha = 0. \tag{7.49}$$

This result (7.49) accords with d'Alembert's Paradox. Although the prediction of the drag force is alas physically unrealistic, the ideal potential flow theory provides a good approximation to the lift force when the angle of incidence α is small. Aeroplane flight is made possible by having a suitably designed wing held at a small angle of incidence α to the direction of motion of the aircraft in still air. The lift on the wing is then the component of force perpendicular to the relative velocity of flow. In a real fluid a small value of α is essential. For when α exceeds a few degrees, a large wake forms downstream of the wing due to the separation of the flow from the wing and this results in little or no lift and a large drag. The wing is then said to be in a condition of *stall*.

Fig. 7.8 shows an elliptic wing at small incidence α in a stream when the Kutta–Joukowski condition applies. Notice the smooth way in which fluid leaves the wing at the trailing edge.

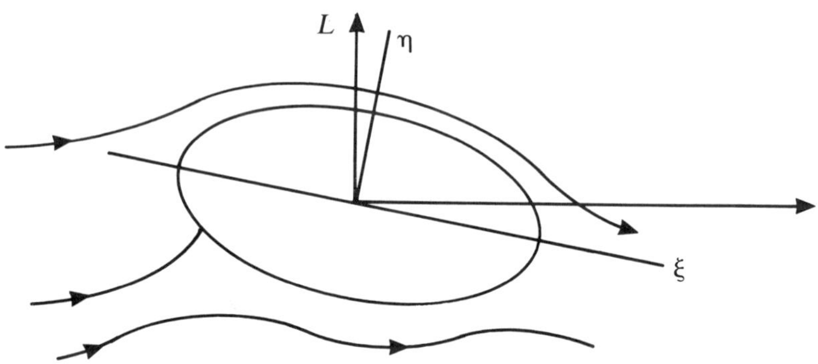

Fig. 7.8

The torque acting on the wing section, when moments of the pressure forces are taken about the origin, is given by the Blasius formula

$$M = -\frac{\rho}{2}\,\mathrm{Re}\oint_\Gamma t\left(\frac{\mathrm{d}W}{\mathrm{d}t}\right)^2 \mathrm{d}t. \tag{7.50}$$

On transforming this into an integral in the z-plane we find

$$M = -\frac{\rho}{2}\,\mathrm{Re}\oint_{\mathscr{C}}\left(z + \frac{c^2}{z}\right)\left(1 - \frac{c^2}{z^2}\right)^{-1}\left(\frac{\mathrm{d}w}{\mathrm{d}z}\right)^2 \mathrm{d}z. \tag{7.51}$$

Proceeding as in the case of the force integral, we find

$$M = -2\pi\rho U \sin 2\alpha \tag{7.52}$$

when the Kutta–Joukowski condition is applied. The sign of M shows that the angle α tends to increase as the aircraft moves, indicating that $\alpha = 0$ is an unstable position.

If the constant c of (7.37) takes the value a, the radius of the circle (7.38) in the z-plane, then the ellipse in the t-plane degenerates into the strip

$$-2a \leqslant \xi \leqslant 2a, \qquad \eta = 0. \tag{7.53}$$

Because the flat plate, as defined by this limiting configuration, has angle of incidence α to the stream, the stagnation points do not coincide with the leading and trailing edges when there is no circulation, as illustrated in Fig. 7.9.

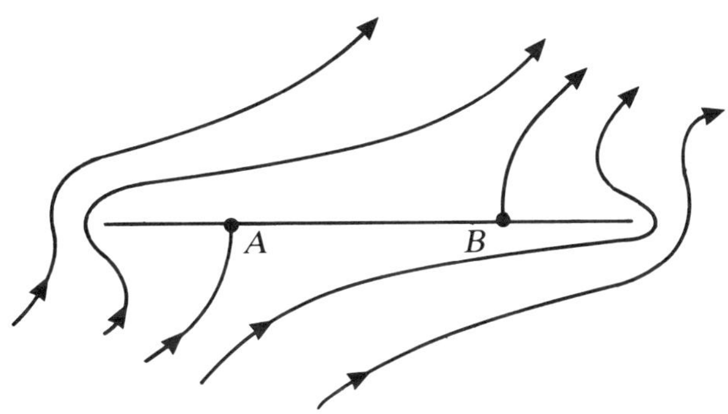

Fig. 7.9

Fig. 7.9 shows that the upstream stagnation point is located on the lower surface of the plate while the downstream one is located on the upper surface. The locations of these stagnation points are given by $\xi = \pm 2a \cos \alpha$. Furthermore, the velocity at the leading and trailing edges is singular since $\mathrm{d}z/\mathrm{d}t$ has singularities at these points. Clearly this is unrealistic physically and is not encountered in practice since real aerofoils have finite thickness and hence finite radius of curvature at the leading edge. However, the trailing edge is usually quite sharp, so the existence of very large, if not infinite, velocity components at the trailing edge is certainly a possibility. This difficulty may be overcome by transferring the stagnation point B in Fig. 7.9 to the trailing edge. This is achieved by adding to the flow a specific circulation, as in the case for the ellipse. This is again an example of the Kutta–Joukowski condition, and the reader will have no difficulty in verifying that the value of the circulation required to ensure that B is transferred to the trailing edge of the flat plate has exactly the same magnitude and sense as were found for the ellipse. Accordingly $\kappa = 4\pi aU \sin \alpha$ and the circulation about the flat plate is clockwise and the fact that κ is independent of c indicates that its value remains unaltered when $c = a$.

The trailing edge of the flat plate, unlike that of the ellipse, is not actually a stagnation point when the Kutta–Joukowski condition is applied, since dz/dt has a singularity at the trailing edge. To determine the structure of the fluid velocity near this region we need only consider the value of dW/dt close to the trailing edge, where $t = 2a$ or $z = a$. Let us write $z = a + \epsilon$, where $|\epsilon| \ll a$. Then from (6.90), (6.92),

$$\begin{aligned} w'(a+\epsilon) &= U\,e^{-i\alpha} - Ua^2\,e^{i\alpha}(a+\epsilon)^{-2} + 2U\,ia\sin\alpha\,(a+\epsilon)^{-1} \\ &= 2U\,(\epsilon/a)\cos\alpha + O(\epsilon/a)^2. \end{aligned} \tag{7.54}$$

Also since $z'(t) = (1 - a^2/z^2)^{-1}$,

$$z'(a+\epsilon) = (a/2\epsilon) + O(1). \tag{7.55}$$

Thus near the trailing edge,

$$\frac{dW}{dt} = \frac{dw}{dz}\cdot\frac{dz}{dt} = U\cos\alpha + O(\epsilon/a), \tag{7.56}$$

showing that the fluid leaves the flat plate in a direction parallel to the plate and with finite velocity of magnitude $U\cos\alpha$. Thus, for flow past a flat plate, the Kutta–Joukowski condition has achieved the requirement of eliminating an infinite velocity at the trailing edge. The flow pattern is sketched in Fig. 7.10.

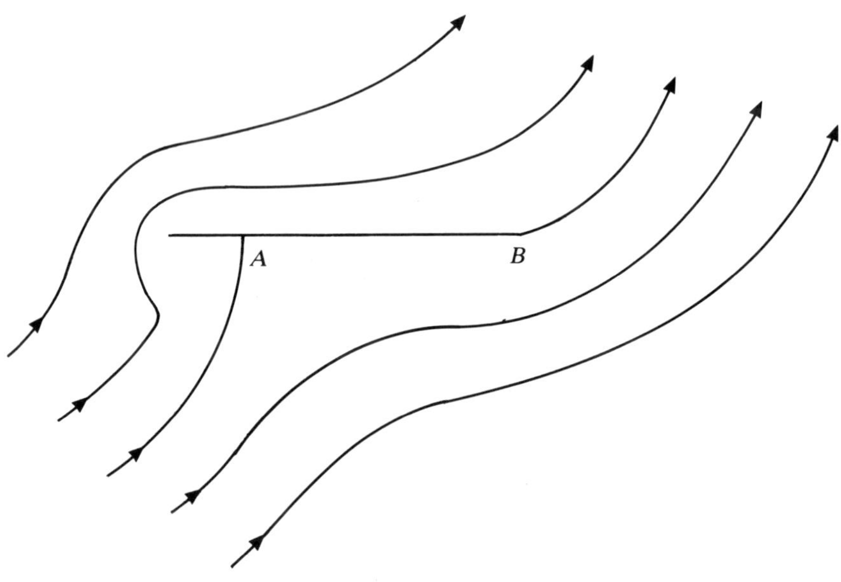

Fig. 7.10

We have seen that with the Kutta–Joukowski condition applied, the trailing edge of a thin elliptic aerofoil is a stagnation point and the velocity is finite at all points of the ellipse. For a flat plate aerofoil, the Kutta–Joukowski condition ensures that the velocity at the trailing edge is finite but the velocity at the leading edge is infinite. In designing an aeroplane wing section it is necessary to round off the leading edge to reduce the velocity there but desirable to sharpen the trailing edge since this reduces the size of the wake and reduces the drag in a real fluid.

We now consider how to design an aerofoil meeting these two requirements. The circle $|z+d|=a+d$ in the z-plane, with a and d positive real numbers, has $z=-d$ for its centre and is of radius $a+d$. All points on the circle are given by

$$z=-d+(a+d)\,\mathrm{e}^{i\theta}, \qquad 0\leqslant\theta\leqslant 2\pi. \tag{7.57}$$

Under the transformation $t=z+a^2/z$, this circle is transformed into a curve in the t-plane specified parametrically by

$$\left.\begin{aligned}\xi &= 2a\cos\theta-d(1-\cos 2\theta)+\mathrm{O}(d/a)^2,\\ \eta &= 2d\sin\theta-d\sin 2\theta+\mathrm{O}(d/a)^2,\end{aligned}\right\} \tag{7.58}$$

when $d\ll a$, corresponding to an aerofoil section whose thickness is small compared with its chord length. The shape of this aerofoil is sketched in Fig. 7.11. It is symmetrical about the axis $\eta=0$. The upper surface is defined by choosing θ to satisfy $0\leqslant\theta\leqslant\pi$ and on the lower surface θ satisfies $\pi\leqslant\theta\leqslant 2\pi$. The maximum thickness is approximately $3\sqrt{3}\,d$ and it occurs when $\theta=2\pi/3$ and hence $\xi=-a$. The leading and trailing edges are given by

$$\xi=\mp 2a+\mathrm{O}(d/a)^2, \qquad \eta=\mathrm{O}(d/a)^2 \tag{7.59}$$

respectively, and the slope of the aerofoil surface is

$$\frac{\mathrm{d}\eta}{\mathrm{d}\xi}=\frac{\mathrm{d}\eta}{\mathrm{d}\theta}\,\frac{\mathrm{d}\theta}{\mathrm{d}\xi}\mathrel{\hat{=}}\frac{\cos\theta-\cos 2\theta}{\sin\theta-(d/a)\sin 2\theta}. \tag{7.60}$$

Thus $\mathrm{d}\eta/\mathrm{d}\xi=0$ at the trailing edge where $\theta=0$ and the aerofoil has a cusp at this edge while the leading edge is blunt since $\mathrm{d}\eta/\mathrm{d}\xi=\infty$, when $\theta=\pi$.

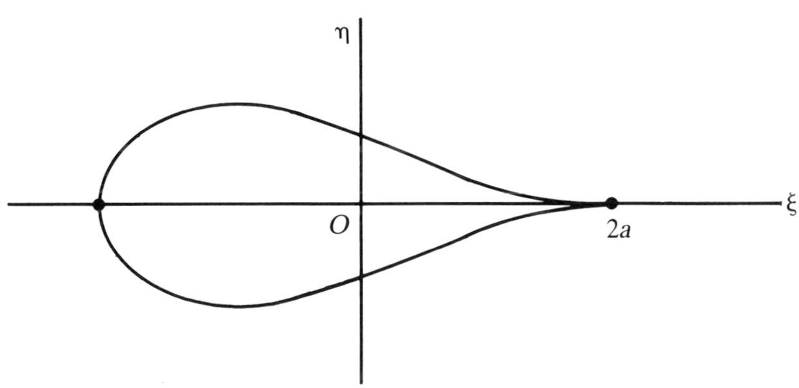

Fig. 7.11

It is now easy to determine the complex potential for flow past an aerofoil of this shape. We first note that the complex potential for a uniform stream of velocity U flowing in a direction inclined at angle α to the positive x-axis past the circle $|z+d| = a+d$ about which there is a circulation $-\kappa$, is given by

$$w(z) = U\,\mathrm{e}^{-i\alpha}(z+d) + U(a+d)^2\,\mathrm{e}^{i\alpha}(z+d)^{-1} + (i\kappa/2\pi)\ln(z+d). \quad (7.61)$$

With $z = \frac{1}{2}\{t+\sqrt{(t^2-4a^2)}\}$, equation (7.56) determines the complex potential $W(t)$ for a uniform stream of velocity U and flowing, in a direction making angle α with the ξ-axis, past the aerofoil whose shape is defined by the equations (7.58) and about which there is a circulation $-\kappa$. The Kutta–Joukowski condition at the trailing edge requires that $w'(a) = 0$, giving

$$U\,\mathrm{e}^{-i\alpha} - U\,\mathrm{e}^{i\alpha} + (i\kappa/2\pi)(a+d)^{-1} = 0,$$

or

$$\kappa = 4\pi U(a+d)\sin\alpha. \quad (7.62)$$

This equation has the same form as (7.44) except that the radius $(a+d)$ of the new circle in the z-plane replaces a in that equation. The velocity components in the flow are found from

$$\frac{\mathrm{d}W}{\mathrm{d}t} = \frac{\mathrm{d}w}{\mathrm{d}z}\,\frac{\mathrm{d}z}{\mathrm{d}t}. \quad (7.63)$$

In particular, it should be noted that both velocity components at the leading and trailing edges of the aerofoil are finite. We may examine the flow near the trailing edge in the same way as for the flat plate, since the trailing edge is defined by $t \triangleq 2a$ or $z = a$. Thus near the trailing edge we write $z = a + \epsilon$, where $|\epsilon| \ll a$. Then (7.61) gives

$$w'(a+\epsilon) = 2\epsilon(a+d)^{-1}\,U\cos\alpha + O(\epsilon/a)^2 \quad (7.64)$$

and

$$z'(t) = (a/2\epsilon) + O(1), \quad (7.65)$$

when $z = a + \epsilon$. Thus near the trailing edge,

$$\frac{\mathrm{d}W}{\mathrm{d}t} = w'(a+\epsilon)z'(t) = \frac{Ua\cos\alpha}{a+d} + O\left(\frac{\epsilon}{a}\right), \quad (7.66)$$

showing that the flow leaves the trailing edge in a direction parallel to the ξ-axis and with finite velocity as indicated in Fig. 7.12

The lift on the aerofoil is obtained in the same way as for the ellipse and is given by (7.48). The lift per unit length is now $4\pi(a+d)\rho U^2 \sin\alpha$.

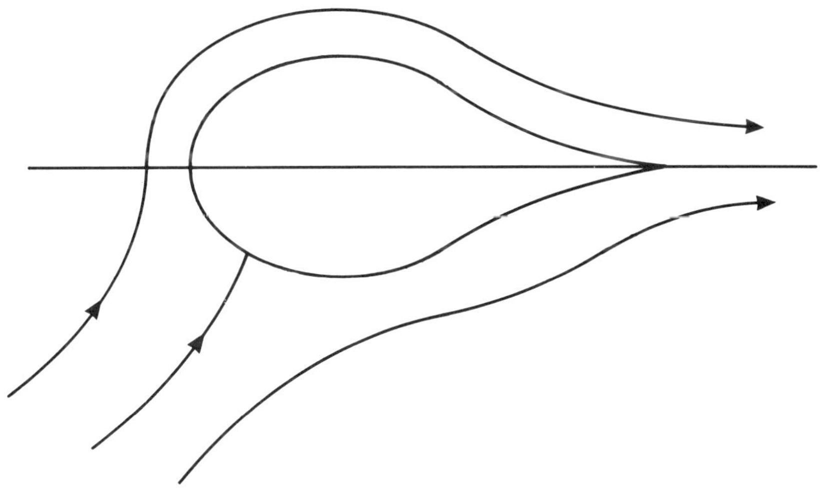

Fig. 7.12

7.7 THE JOUKOWSKI AEROFOIL

The three aerofoil shapes considered in the last section, namely the ellipse, flat plate and the third symmetrical type with cusped trailing edge are all examples of *Joukowski aerofoils*. Their shapes are the transforms of various circles in the z-plane using the Joukowski transformation (7.37) with various values of c. All three of these aerofoils are straight. The purpose of the wing on an aeroplane is to generate lift and the wing shape is so chosen as to assist the change in direction of air flow which will produce circulation and maximise this lift and minimise the drag. We have observed that the minimum drag is not the zero value for the case of a real fluid as predicted by ideal potential flow theory. It is therefore reasonable to suppose that if *camber* is introduced into the wing shape, the necessary circulation about the wing may be induced more effectively than when the wing is straight. A cambered aerofoil shape can also be produced using the Joukowski transformation when the centre of the circle in the z-plane does not lie on the x-axis.

We consider the effect of the Joukowski transformation (7.37) on a circle in the z-plane passing through the point $z = a(> 0)$ and centred on $z = z_1$. This circle Γ_1 is

$$\Gamma_1 : |z - z_1| = |z_1 - a| = c. \tag{7.67}$$

Let us apply the transformation

$$z = a^2/\sigma \tag{7.68}$$

where $z = x + iy = r\,e^{i\theta}$, $\sigma = p + iq$ and where x, y, r, θ, p, q are all real. The point z maps into $\sigma = (a^2/r)\,e^{-i\theta}$. This means that if the complex planes of z and

σ coincide, so that the x- and p-axes are in alignment and also the y- and q-axes, then σ is obtained from z by first inverting in the circle centre O and radius a and then reflecting in the real axis. We recall that circles invert into either circles or straight lines and that these, in turn, reflect into circles and lines respectively. Then clearly Γ_1 will, in general, transform into another circle Γ_2. We first investigate the geometrical relations between Γ_1 and Γ_2.

The circle Γ_1 may be written in the alternative form

$$(z - z_1)(\bar{z} - \bar{z}_1) = c^2 = (z_1 - a)(\bar{z}_1 - a), \tag{7.69}$$

recalling that for any complex variable Z, $|Z|^2 = Z\,\bar{Z}$. On applying (7.68) the form (7.69) becomes

$$(a^2 - z_1\sigma)(a^2 - \bar{z}_1\sigma) = c^2\,\sigma\bar{\sigma}.$$

Thus

$$(z_1\bar{z}_1 - c^2)\,\sigma\bar{\sigma} - a^2(z_1\sigma + \bar{z}_1\bar{\sigma}) + a^4 = 0.$$

Also from (7.69), $c^2 = z_1\bar{z}_1 - a(z_1 + \bar{z}_1) + a^2$, and so the above equation becomes

$$(z_1 + \bar{z}_1 - a)\,\sigma\bar{\sigma} - az_1\sigma - a\bar{z}_1\bar{\sigma} + a^3 = 0.$$

or

$$\sigma\bar{\sigma} = \frac{az_1\sigma}{z_1 + \bar{z}_1 - a} - \frac{a\bar{z}_1\bar{\sigma}}{z_1 + \bar{z}_1 - a} = -\frac{a^3}{z_1 + \bar{z}_1 - a}.$$

Thus

$$\left[\sigma - \frac{a\bar{z}_1}{z_1 + \bar{z}_1 - a}\right]\left[\bar{\sigma} - \frac{az_1}{z_1 + \bar{z}_1 - a}\right] = R^2, \tag{7.70}$$

where

$$R^2 = \frac{a^2 z_1\bar{z}_1 - a^3(z_1 + \bar{z}_1 - a)}{(z_1 + \bar{z}_1 - a)^2} = \frac{a^2(z_1 - a)(\bar{z}_1 - a)}{(z_1 + \bar{z}_1 - a)^2} = \frac{a^2c^2}{(z_1 + \bar{z}_1 - a)^2}.$$

The form of R^2 is that of a real positive quantity. At this stage, we impose the restriction that

$$\operatorname{Re}(z_1) = \operatorname{Re}(\bar{z}_1) < 0. \tag{7.71}$$

Then (7.70) may be written in the form

$$\left|\sigma + \frac{a\bar{z}_1}{a - z_1 - \bar{z}_1}\right| = R = \frac{ac}{a - z - \bar{z}_1}. \tag{7.72}$$

The restriction (7.71) ensures that $a - z - \bar{z}_1 > 0$ always so that $R > 0$. Equation (7.72) is recognised as that of a circle Γ_2 whose centre C_2 is $\sigma = -a\bar{z}_1/(a - z_1 - \bar{z}_1)$ and whose radius is $R = ac/(a - z_1 - \bar{z}_1)$.

When we put $\sigma = a$, the left-hand side of (7.72) becomes

$$\frac{a\,|a - z_1|}{a - z_1 - \bar{z}_1} = \frac{ac}{a - z_1 - \bar{z}_1},$$

using (7.67).

Thus $\sigma = a$ satisfies (7.72) and so Γ_2 goes through this point.

Denoting by C_1 the centre of Γ_1 and by A the point $\sigma = a$, we have

$$\overrightarrow{AC_1} \equiv z_1 - a,$$

and

$$\overrightarrow{AC_2} \equiv -\frac{a\bar{z}_1}{a - z_1 - \bar{z}_1} - a = \frac{a(z - a)}{a - z_1 - \bar{z}_1}.$$

Thus

$$\overrightarrow{AC_2} = \lambda\overrightarrow{AC_1}, \qquad \text{where } \lambda = a/(a - z_1 - \bar{z}_1) > 0.$$

This relationship shows that as λ is real, the points A, C_1, C_2 are collinear and C_1, C_2 are on the same side of A since $\lambda > 0$. Hence the line of centres of Γ_1, Γ_2 passes through A. Further, $AC_1 = |z_1 - a| = c$ is the radius of Γ_1 and $AC_2 = ac/(a - z_1 - \bar{z}_1) + R$, the radius of Γ_2. Hence Γ_1 and Γ_2 touch at A.

For C_2, $\sigma = -\lambda\bar{z}_1$ and the reflection of this point in the y-axis is C_2^* for which $\sigma = \lambda z_1$, since λ is real. Since $\lambda > 0, z_1$ and λz_1 have the same argument and as the y-axis bisects internally the angle $C_2 O\, C_2^*$, it follows that the y-axis also bisects internally the angle $C_1 O\, C_2$.

To summarise the construction shown in Fig. 7.13, we draw the circle Γ_1 with given centre C_1, where $z = z_1$ and radius $C_1A = c = |z_1 - a|$, A being the point $z = a$ on the positive x-axis. We then locate C_2 on AC_1 using the fact that Oy bisects $C_1\hat{O}C_2$ internally. Γ_2 is then drawn having centre C_2 and radius $C_2A = R = ac/(a - z_1 - \bar{z}_1) = \lambda c$.

Suppose R_1 is the point $z = r\,e^{i\theta}$ on Γ_1. R_2 is the corresponding point $\sigma = a^2/z = (a^2/r)\,e^{-i\theta}$ on Γ_2, the axis Ox bisecting $R_1\hat{O}R_2$ which enables R_2 to be located on Γ_2 for a given R_1 on Γ_1.

Now suppose we make the further conformal transformation

$$t = z + \sigma = z + (a^2/z). \tag{7.73}$$

In Fig. 7.13, $\overrightarrow{OR_1} \equiv z_1$, $\overrightarrow{OR_2} \equiv \sigma$ and $\overrightarrow{OR_3} \equiv z + \sigma$, where R_3 is constructed so that $OR_2R_3R_1$ is a parallelogram. As R_1 moves once round the circle Γ_1, R_3 describes a closed loop containing O. In fact it traces out a curve which is a Joukowski aerofoil. Its shape varies considerably with the choices of z_1 and c. In many cases the shape is similar to the section of an aeroplane wing which is why the above theory is of some importance in wing design.

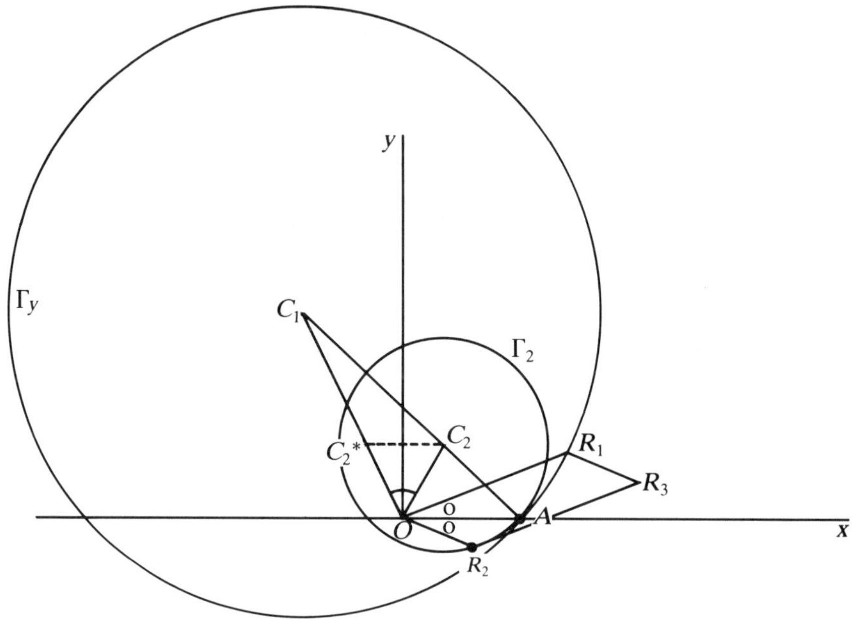

Fig. 7.13

We can easily write down the complex potential for uniform flow in a chosen direction past the circular cylinder Γ_1 specified by equation (7.67), with a known circulation around it. The Joukowski transformation (7.73) enables us to determine the corresponding flow around the aerofoil.

The singular points of the transformation (7.73) are given by $\mathrm{d}t/\mathrm{d}z = 0$ and they are clearly $z = \pm a$ in the z-plane. The corresponding points in the t-plane are $t = \pm 2a$. The point $z = a$ is the contact point of the circles Γ_1 and Γ_2. Let us investigate the neighbourhood of its mapping $t = 2a$. To this end we rewrite the transformation (7.73) in the form

$$\frac{t-2a}{t+2a} = \left(\frac{z-a}{z+a}\right)^2. \tag{7.74}$$

To investigate correspondence within the neighbourhoods of the singular points $z = a$, $t = 2a$, we put

$$z = a + r\,\mathrm{e}^{i\theta}, \qquad t = 2a + s\,\mathrm{e}^{i\phi},$$

where r, s, θ, ϕ are real, r and s being small and positive. Substitution into (7.74) gives

$$s\,\mathrm{e}^{i\phi}/(4a + s\,\mathrm{e}^{i\phi}) = r^2\,\mathrm{e}^{2i\theta}/(2a + r\,\mathrm{e}^{i\theta})^2.$$

To the first order of approximation this gives

$$as/r^2 = \exp i(2\theta - \phi).$$

The left-hand side is positive so that

$$\left.\begin{aligned} 2\theta &= \phi \\ \text{or} \qquad 2\delta\theta &= \delta\phi \end{aligned}\right\}. \tag{7.75}$$

In moving round the point of contact A of the circles Γ_1, Γ_2 the change in θ is π and so the corresponding change in ϕ at $t = 2a$ is 2π. The aerofoil has two branches which touch at $t = 2a$, the trailing edge (Figs 7.14(i), (ii)). The point $t = 2a$ is a cusp.

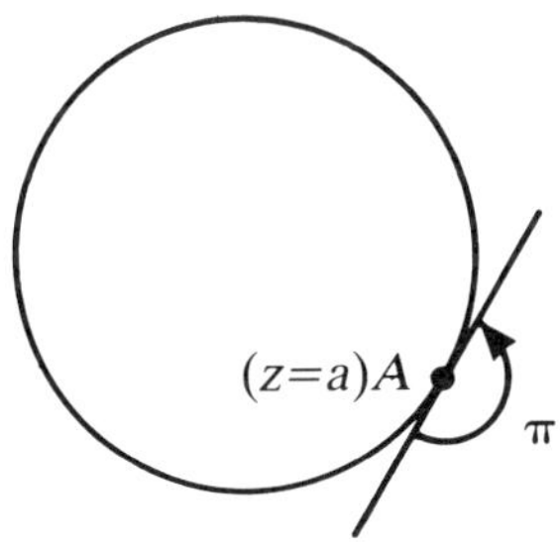

Fig. 7.14(i)

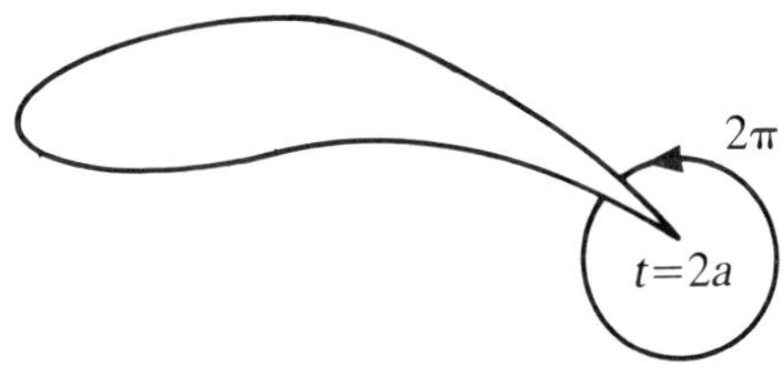

Fig. 7.14(ii)

If q is the speed at A on the circle Γ_1 and q' that at the cusp $t = 2a$, then

$$q' = \left|\frac{\mathrm{d}W}{\mathrm{d}t}\right| = \left|\frac{\mathrm{d}w}{\mathrm{d}z}\right| \times \left|\frac{\mathrm{d}z}{\mathrm{d}t}\right| = q\left|\frac{\mathrm{d}z}{\mathrm{d}t}\right|,$$

where $w(z)$ or $W(t)$ is the complex potential. At the trailing edge, $\mathrm{d}t/\mathrm{d}z = 0$ and so the above analysis implies $q' = \infty$ there. This can be prevented only if $q = 0$ at the point A on the circle Γ_1 and this is accomplished by adding to the flow a circulation of suitable strength.

Let us next find the circulation κ around the aerofoil when the Kutta–Joukowski condition is satisfied at the trailing edge $z = a$. In Fig. 7.13, let $C_1\hat{A}O = \beta$: then $a - z_1 = c\,\mathrm{e}^{-i\beta}$. If α is the angle of incidence of the oncoming stream to the positive x-axis, then since the equation of the circle Γ_1 is $|z - z_1| = c$ we have, for a clockwise (or negative) circulation κ around Γ_1, the complex potential

$$w(z) = U\,\mathrm{e}^{-i\alpha}(z - z_1) + Uc^2\,\mathrm{e}^{i\alpha}(z - z_1)^{-1} + (i\kappa/2\pi)\ln(z - z_1). \tag{7.76}$$

In (7.76), the first term on the right is the complex potential of the undisturbed oncoming stream, the second is the perturbation caused by the circular cylinder of section Γ_1 and is obtained by application of the Milne–Thomson circle theorem, and the last term is the complex potential due to a uniform line vortex $-\kappa$ through $z = z_1$. Following the procedure of Section 7.6, the appropriate Kutta–Joukowski condition is obtained from $w'(a) = 0$. Using $a - z_1 = c\,\mathrm{e}^{-i\beta}$, we find

$$U\left[e^{-i\alpha} - \frac{e^{i\alpha}}{e^{-2i\beta}}\right] + \frac{i\kappa}{2\pi}\frac{1}{c\,e^{-i\beta}} = 0$$

which simplifies to

$$\kappa = 4\pi Uc \sin(\alpha + \beta). \tag{7.77}$$

Finally we obtain the lift force L per unit length on the aerofoil. Let $z = z_1$ map into $t = t_1$. Since $z = z_1$ is within the circle Γ_1, $t = t_1$ is within the aerofoil. In the t-plane the complex potential of a uniform stream at angle of incidence α to the positive real axis is of the form $W = Ut\,e^{-i\alpha}$. The presence of the aerofoil, having a circulation $-\kappa$ at $t = t_1$, will produce a perturbation which must vanish as $|t| \to \infty$. Thus $t = t_1$ is a singular point and $W(t)$ must be so chosen that

$$W'(t) = U\,e^{-i\alpha} + A_1(t - t_1)^{-1} + A_2(t - t_1)^{-2} + \ldots$$

which satisfies the condition that $W'(t) \to U\,e^{-i\alpha}$ as $|t| \to \infty$. Then

$$W(t) = Ut\,e^{-i\alpha} + A_1 \ln(t - t_1) - A_2(t - t_1)^{-1} + \ldots$$

The complex potential at points for which $|t - t_1|$ is small is approximately $(i\kappa/2\pi)\ln(t - t_1)$ and so $A_1 = i\kappa/2\pi$ and hence

$$W'(t) = U\,e^{-i\alpha} + (i\kappa/2\pi)(t - t_1)^{-1} + A_2(t - t_1)^{-2} + \ldots$$

If $\mathscr{C}$ denotes the section of the aerofoil, then by the theorem of Blasius we have in the usual notation

$$Y + iX = -\frac{\rho}{2}\oint_{\mathscr{C}}\left(\frac{dW}{dt}\right)^2 dt$$

$$= -\frac{\rho}{2}\oint_{\mathscr{C}}\left[U\,e^{-i\alpha} + \frac{i\kappa}{2\pi(t - t_1)} + \frac{A_2}{(t - t_1)^2} + \ldots\right]^2 dt.$$

The integrand is meromorphic having $t = t_1$ as a pole of infinite order. The residue at this pole is $iU\kappa\,e^{-i\alpha}/\pi$ and so

$$Y + iX = \rho U\kappa(\cos\alpha - i\sin\alpha),$$

and

$$L = Y\cos\alpha - X\sin\alpha = \rho U\kappa.$$

The drag D per unit length is zero. The above method is quite general and one can show that the lift force on any cylinder in ideal flow with uniform velocity U and circulation κ is always $\rho U\kappa$ per unit length and there is no drag force. These results are independent of the geometry of the cylinder and also of the direction of the uniform incident stream, though, of course, the direction of L is perpendicular to that of the stream.

In the case of the aerofoil considered above when the Kutta–Joukowski condition prevails the lift force per unit length becomes $L = 4\pi c\rho U^2 \sin(\alpha + \beta)$.

7.8 THE SCHWARZ–CHRISTOFFEL TRANSFORMATION

It is convenient to map the vertices of an n-sided plane polygon on to a straight line, usually the real axis of the new complex plane. The interior of the polygon may be mapped on to either half-plane which the straight line determines. The *Schwarz–Christoffel transformation* is a conformal type of transformation which enables the procedure to be effected in general terms. This type of transformation will first be described and then applied to certain problems of plane flow.

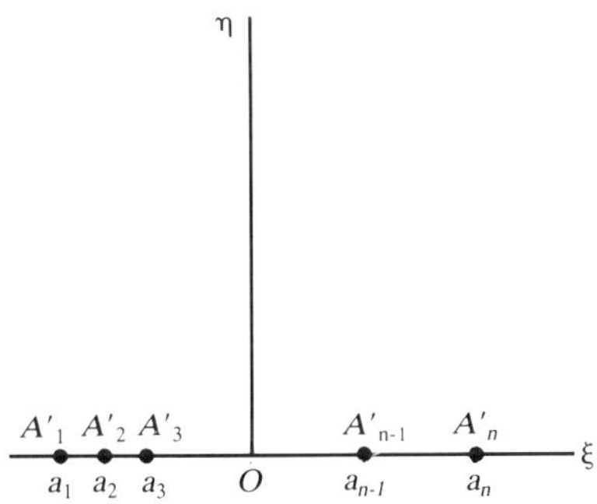

Fig. 7.15(i)

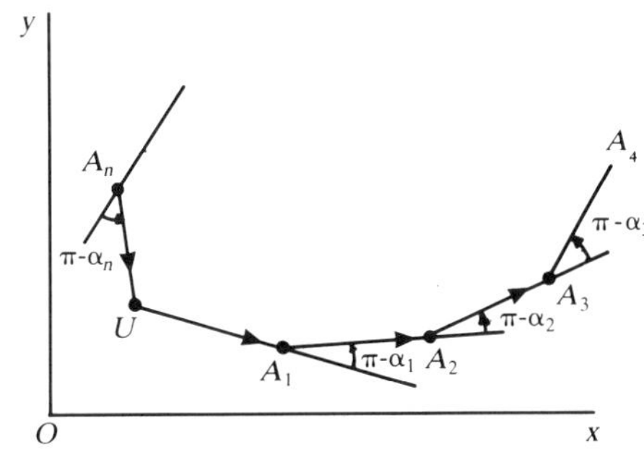

Fig. 7.15(ii)

Fig. 7.15(i) shows the complex t-plane, where $t = \xi + i\eta$ with ξ, η both real, and the n points A'_r for which $t = a_r$ are taken on the real axis ($r = 1, 2, \ldots, n$). The ordering is such that $a_1 < a_2 < \ldots < a_{n-1} < a_n$. Now suppose the t-plane is mapped on to the z-plane ($z = x + iy$) by means of the transformation

$$\mathrm{d}z/\mathrm{d}t = K(t - a_1)^{(\alpha_1/\pi)-1}\,(t - a_2)^{(\alpha_2/\pi)-1} \ldots (t - a_n)^{(\alpha_n/\pi)-1}. \tag{7.78}$$

The transformation is conformal everywhere except at the n singularities $t = a_r$ ($r = 1, \ldots, n$) in the finite part of the t-plane. The numbers $\alpha_r (r = 1, \ldots, n)$ are real constants, but K is a constant which may be complex.

Let δt be a change made in t and δz the attendant change in z so that

$$\delta z = \delta t\, K(t - a_1)^{(\alpha_1/\pi)-1} \ldots (t - a_n)^{(\alpha_n/\pi)-1}.$$

Then

$$\arg \delta z = \arg \delta t + \arg K + [(\alpha_1/\pi) - 1] \arg(t - a_1) + \ldots$$
$$+ [(\alpha_n/\pi) - 1] \arg(t - a_n). \tag{7.79}$$

The result (7.79) will now be employed to study the changes in arg δz produced as t travels along the real axis $\eta = 0$ from $\xi = -\infty$ to $\xi = +\infty$ through the singularities of the transform (7.78). We shall see that on passing through each singularity $t = a_r$ $(r = 1, \ldots, n)$, there is a change produced in arg δz, which means a change in direction in the corresponding z-plane (Fig. 7.15(ii)).

First suppose that t travels from $-\infty$ to a_1- along $\eta = 0$. Then

$$\arg \delta t = 0; \quad \arg (t - a_1) = \arg (t - a_2) = \ldots = \arg (t - a_n) = \pi.$$

Hence from (7.79),

$$(\arg \delta z)_1 = \arg K + (\alpha_1 - \pi) + (\alpha_2 - \pi) + \ldots + (\alpha_n - \pi) = \text{const.} \qquad (7.80)$$

Equation (7.80) shows that z describes a straight-line segment UA_1 as t travels from $-\infty$ to a_1- along $\eta = 0$.

Next, let t travel from a_1+ to a_2- along $\eta = 0$. Then

$$\arg \delta t = 0; \; \arg (t - a_1) = 0; \; \arg (t - a_2) = \ldots = \arg (t - a_n) = \pi.$$

Hence

$$(\arg \delta z)_2 = \arg K + 0 + (\alpha_2 - \pi) + \ldots + (\alpha_n - \pi) = \text{const.} \qquad (7.81)$$

Equation (7.81) indicates that as t moves from a_1+ to a_2- asong $\eta = 0$, z describes a straight line segment A_1A_2 in Fig. 7.15(ii).

From (7.80), (7.81), we obtain

$$(\arg \delta z)_2 - (\arg \delta z)_1 = \pi - \alpha_1, \qquad (7.82)$$

which shows that the angle between the directions $\overrightarrow{UA_1}$ and $\overrightarrow{A_1A_2}$ is $\pi - \alpha_1$, as marked in Fig. 7.15(ii). This means that as t travels along the axis $\eta = 0$ from $-\infty$ past A_1', the direction of motion of z changes by $\pi - \alpha_1$ in the positive sense, as the course changes from $\overrightarrow{UA_1}$ to $\overrightarrow{A_1A_2}$.

Similarly, as t moves from a_2+ to a_3- along $\eta = 0$,

$$\arg \delta t = \arg (t - a_1) = \arg (t - a_2) = 0;$$

$$\arg (t - a_3) = \ldots = \arg (t - a_n) = \pi$$

and so

$$(\arg \delta z)_3 = \arg K + 0 + 0 + (a_3 - \pi) + \ldots + (\alpha_n - \pi) = \text{const.} \qquad (7.83)$$

Equation (7.83) shows that z describes a straight-line segment A_2A_3 as t travels from a_2+ to a_3- along $\eta = 0$. From (7.82), (7.83),

$$(\arg \delta z)_3 - (\arg \delta z)_2 = \pi - \alpha_2. \qquad (7.84)$$

This shows that the direction of z changes by $\pi - \alpha_2$ in the positive sense as z passes through A_2.

Continuing in this way, we see that z *describes the sides*

$$UA_1, A_1A_2, A_2A_3, \ldots, A_{n-1}A_n, A_nU$$

of a closed polygon in the z-plane as t moves along the real axis in the t-plane from $-\infty$ *to* a_1, a_2 *to* a_2, a_2 *to* $a_3, \ldots, a_{n-1}$ *to* a_n, a_n *to* $+\infty$ *respectively*. The angles turned through in the positive sense at the n points A_r are $\pi - \alpha_r$ so that the internal angles at these points are $\alpha_r (r = 1, 2, \ldots, n)$.

Now suppose the internal angles α_r are so chosen that

$$\alpha_1 + \alpha_2 + \ldots + \alpha_n = (n-2)\pi. \tag{7.85}$$

Then $A_1A_2, \ldots, A_n$ is a closed polygon, the portion $A_n U A_1$ being a straight line. Further, since, the half-plane $\mathrm{Im}(t) > 0$ is on the left of an observer travelling along the positive direction of the axis $\eta = 0$, the corresponding area in the z-plane is the interior of the polygon.

In summary, then, we may say: *If $a_1 < a_2 < \ldots < a_n$ are distinct points on the line* $\mathrm{Im}(t) = 0$ *in the t-plane, and if $\alpha_1, \alpha_2, \ldots, \alpha_n$ are n real constants satisfying*

$$\alpha_1 + \alpha_2 + \ldots + \alpha_n = (n-2)\pi,$$

where n is an integer > 2, then the transformation

$$dz/dt = K(t-a_1)^{(\alpha_1/\pi)-1}(t-a_2)^{(\alpha_2/\pi)-1}\ldots(t-a_n)^{(\alpha_n/\pi)-1}$$

or

$$z = L + K\int_{t_0}^{t}(t-a_1)^{(\alpha_1/\pi)-1}\ldots(t-a_n)^{(\alpha_n/\pi)-1}\,dt \tag{7.86}$$

where t_0, K, L are constants which may be complex, maps the entire line $\mathrm{Im}(t) = 0$ *on to the boundary $A_1A_2, \ldots, A_nA_1$ of a closed polygon of n sides in such a way that the n points $t = a_r$ transform into A_r at which the internal angles are α_r $(r = 1, 2, \ldots, n)$ and the half-plane* $\mathrm{Im}(t) > 0$ *transforms into the interior of the polygon.*

The above presupposes that each point $t = a_r (r = 1, 2, \ldots, n)$ lies on the finite part of the line $\eta = 0$ in Fig. 7.15(i). Suppose, however, that $a_n = +\infty$. In the transformation (7.78), put $K = A(-a_n)^{1-(\alpha_n/\pi)}$. Then

$$K(t-a_n)^{(\alpha_n/\pi)-1} = A\left\{1-(t/a_n)\right\}^{(\alpha_n/\pi)-1} \to A \text{ as } a_n \to \infty.$$

The transformation (7.78) now simplifies to the form

$$dz/dt = A(t-a_1)^{(\alpha_1/\pi)-1}\ldots(t-a_{n-1})^{(\alpha_{n-1}/\pi)-1}.$$

This shows that letting $a_n \to +\infty$ has the effect of suppressing the factor $(t - a_n)^{(\alpha_n/\pi)-1}$ in (7.78). Similarly, letting $a_1 \to -\infty$ has the effect of suppressing the factor $(t-a_1)^{(\alpha_1/\pi)-1}$ in (7.78).

Suppose, then, that we wish to transform a given closed polygon of n sides in the z-plane into the real axis $\eta = 0$ of the t-plane $(n = 3, 4, 5, \ldots)$. The internal angles $\alpha_1, \alpha_2, \ldots, \alpha_n$ of the polygon satisfy (7.85) and the type of transform which is appropriate is given by (7.78). First, it is necessary to choose the points $a_1 < a_2 < \ldots < a_n$ on $\eta = 0$ to give the correct shape of figure in the z-plane. Since the α_r are all known, the mapping from the t-plane to the z-plane will certainly give a polygon where n angles are correct, whatever the chosen points

$a_1, a_2, \ldots, a_n$ may be. However, for integral values of $n > 3$ specification of angles alone is inadequate from the point of view of obtaining the correct shape of the figure. For $n = 3$, two equiangular triangles are similar. For $n = 4$, two equiangular quadrilaterals may not be. For if $P_1P_2P_3P_4$, $Q_1Q_2Q_3Q_4$ are two equiangular quadrilaterals such that the angles at the corresponding vertices P_r and Q_r are equal ($r = 1, 2, 3, 4$), then in general the figures are not similar, i.e. correspondingly lettered sides are not in the same proportion. If, however, we augment the equal angle property with a single relation of the form

$$P_1P_2/Q_1Q_2 = P_2P_3/Q_2Q_3,$$

then it is easy to prove that

$$P_1P_2/Q_1Q_2 = P_2P_3/Q_2Q_3 = P_3P_4/Q_3Q_4 = P_4P_1/Q_4Q_1,$$

i.e. that proportionality of all sides has been obtained so that the equiangular figures have now become similar. (The above three relations may easily be obtained from any given one.) In the case of equiangular pentagons, two relations need to be given: thus if $P_1P_2P_3P_4P_5$ and $Q_1Q_2Q_3Q_4Q_5$ are equiangular pentagons they will not be similar unless two relations between the lengths of their corresponding sides are given. If we are told for example

$$P_1P_2/Q_1Q_2 = P_2P_3/Q_2Q_3 = P_3P_4/Q_3Q_4$$

(two distinct relations) together with equality of angles at each P_r, Q_r ($r = 1, 2, \ldots, 5$), then the two figures are similar. It is easy to see that for two equiangular n-gons ($n = 3, 4, 5, \ldots,$), there must be $(n - 3)$ independent relations of the above kind connecting their sides to ensure geometrical similarity.

It follows, then, that for (7.78) to represent the correct mapping of an n-gon in the z-plane on to the line $\eta = 0$ in the t-plane, we have a free choice for fixing any three of the points $a_1, a_2, \ldots, a_n$, but the others have to accord with $n - 3$ relations between the lengths of sides. In (7.86), L may be absorbed into z giving the single term $z - L$ which is achieved by suitable choice of origin. K is but a magnification factor that is at our disposal and we may take $t_0 = 0$ – again by a suitable choice of origin. Let us now look at some worked examples to see how the various points that have been made are called into play.

Example [7.4]

Let us start with a rectangle in the z-plane with angles $\alpha_1 = \alpha_2 = \alpha_3 = \alpha_4 = \pi/2$. If its vertices are mapped on to the points $t = a_r$ ($r = 1, 2, 3, 4$), then the appropriate Schwarz–Christoffel transformation has the form

$$\frac{\mathrm{d}z}{\mathrm{d}t} = \frac{K}{\sqrt{\{(t-a_1)(t-a_2)(t-a_3)(t-a_4)\}}}.$$

We are free to choose any three of the a_r but the fourth is not at our disposal. Its choice is restricted by a single relationship between the lengths of the sides of the rectangle. We discuss different aspects of this case.

(i) Mapping of a semi-infinite strip. Let us take $a_2 = -1, a_3 = +1, a_4 = \infty$. Clearly two sides of the rectangle are infinite and we must take $a_1 = -\infty$. In accordance with observations made above, the factors $t - a_1$, $t - a_2$ are suppressed so that

$$\frac{\mathrm{d}z}{\mathrm{d}t} = \frac{K}{\sqrt{(t^2 - 1)}}.$$

Then

$$z = K \cosh^{-1} t + L.$$

If we move the origin suitably, we can choose $L = 0$ so that

$$t = \cosh(z/K).$$

This gives the correspondence tabulated:

t	1	-1	∞	$\infty + i\pi K$
z	0	$i\pi K$	∞	$-\infty$

The shaded portions of the diagrams Figs 7.16(i), (ii) correspond.

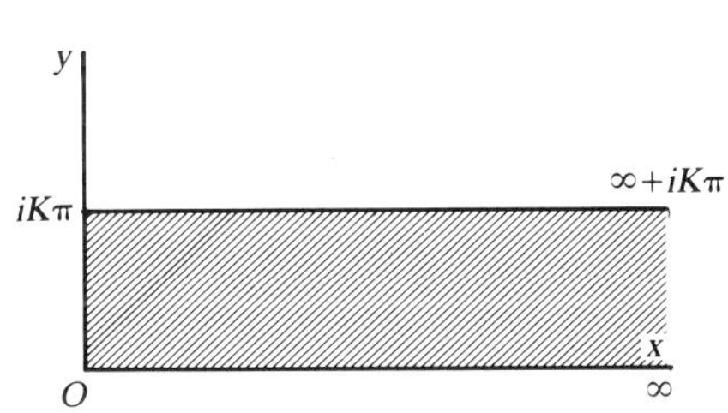

Fig. 7.16(i)

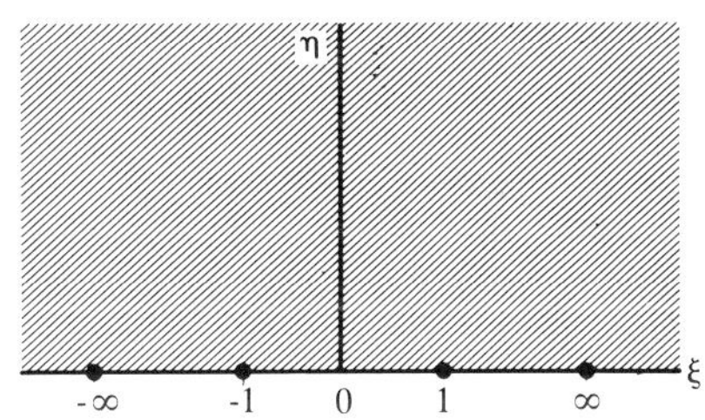

Fig. 7.16(ii)

(ii) An alternative way of mapping Fig. 7.16(i) on to the real axis $\eta = 0$ of the t-plane, is to regard the semi-infinite strip as a triangle whose angles are $\pi/2$, $\pi/2, 0$. Then the appropriate Schwarz–Christoffel transform is

$$\frac{\mathrm{d}z}{\mathrm{d}t} = \frac{C}{t\sqrt{(t^2 - 1)}},$$

if we map $z = 0$ on $t = 1$, $z = iK\pi$ on $t = -1$, the zero angle at $x = \infty$ on $t = 0$. We have

$$\mathrm{d}z = \frac{-C\,\mathrm{d}(1/t)}{\sqrt{(1 - 1/t^2)}}$$

and so

$$z = B + C \cos^{-1}(1/t).$$

When $t = 1, z = 0$ and so $B = 0$. When $t = -1, z = iK\pi$ and so $C = iK$ from which we find the appropriate transformation is

$$t = \operatorname{sech}(z/K).$$

When we take $z = x + iy$ and let $x \to \infty$, we obtain $t = 0$ appropriately.

Note that in this case we are free to choose any three values for t_1, t_2, t_3 (in the notation of the general theory) since the polygon is a triangle.

Example [7.5]
Mapping of a doubly infinite strip.

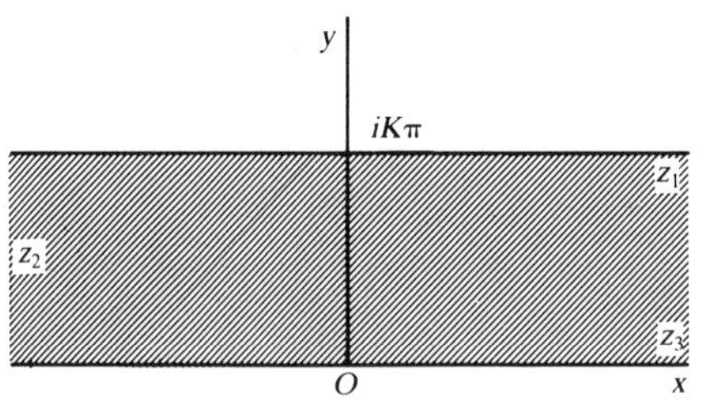

Fig. 7.17(i)

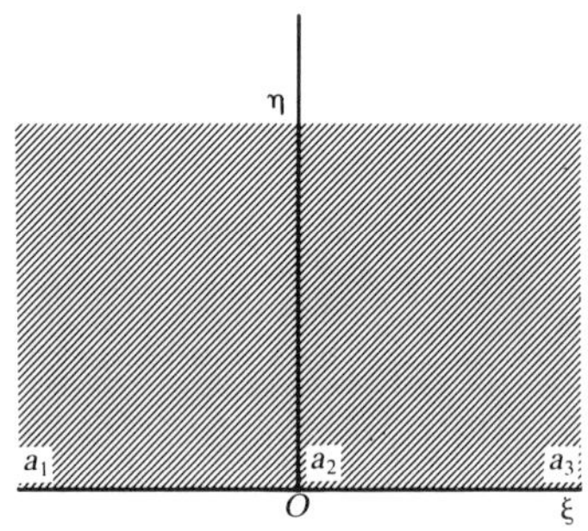

Fig. 7.17(ii)

We map the doubly infinite strip in the z-plane

$$0 \leqslant y \leqslant K\pi, \qquad -\infty \leqslant x \leqslant \infty$$

on to the real axis $\eta = 0$ of the t-plane. Regard the infinite strip (Fig. 7.17(i)) as a triangle with angles $\pi/2$ at $z_1 = \infty + iK\pi$, zero at $z_2 = -\infty$, $\pi/2$ at $z_3 = \infty$. The corresponding transforms in the t-plane (Fig. 7.17(ii)) are $a_1 = -\infty$, $a_2 = 0$, $a_3 = +\infty$. The factors $(t - a_1)$ and $(t - a_3)$ will be suppressed, and so the Schwarz–Christoffel transform is

$$\frac{dz}{dt} = \frac{C}{t}$$

or

$$z = B + C \ln t.$$

$t_1 = -\infty$ gives $z_1 = \infty + C\pi i$, $t_2 = 0$ gives $z_2 = -\infty$; $t_3 = \infty$ gives $z_3 = \infty$. Thus $C = K$ and B is real. The simplest form of the transform is clearly

$$z = K \ln t.$$

7.9 APPLICATIONS OF THE SCHWARZ–CHRISTOFFELL TRANSFORMATION TO POTENTIAL FLOWS

Many of the problems of two-dimensional irrotational flow involve a plane of flow in which the moving liquid is confined to an area that is polygonal. The Schwarz–Christoffel transformation is eminently suited to this type of problem. The boundaries are mapped on to a straight line and the fluid on to a half-plane. Behaviour of the fluid under such a conformal transformation – which in this case is non-singular everywhere except at points corresponding to the vertices of the polygon – has been treated and so the solution proceeds according to the rules. Hydrodynamical applications usually involve polygons which extend to infinity.

Example [7.6]

An incompressible liquid moves irrotationally in two dimensions within a region that has two boundaries C_1 and C_2 defined as follows:

$$C_1: \; y = a, \qquad -\infty < x < \infty;$$

$$C_2: \begin{cases} y = 0, & -\infty < x \leqslant 0; \\ \;\; = -x \tan \gamma\pi, & 0 \leqslant x \leqslant x_1; \\ \;\; = -(b-a), & x_1 \leqslant x < \infty, \end{cases}$$

where $b - a = x_1 \tan \gamma\pi$ for $x_1 > 0$ and $0 < \gamma < \frac{1}{2}$.

At $-\infty$ the liquid is flowing in the positive x-direction with speed U and it always remains in contact with any side walls. Show that, with the usual notation, the flow may be determined from the equations

$$t = \exp\left(-\frac{\pi w}{aU}\right), \qquad z = \frac{b}{\pi}\int_1^t \frac{(t-1)^\gamma}{t(t-\alpha)^\gamma}\,dt$$

where $b = a\alpha^\gamma$. Deduce the corresponding results for the case when $x_1 \to \infty$.

Fig. 7.18(i) depicts the flow in the z-plane. From considerations of continuity, since U is the entry velocity at the section $B_\infty C_\infty$, the exit velocity at the section $A_\infty F_\infty$ is Ua/b. At the entry section the fluid may be considered to be created by a line source, the rate at which fluid crosses the section being Ua. Similarly at the exit section the fluid may be considered to be drained away by a line sink at the same rate.

The infinite polygon $A_\infty B_\infty C_\infty D E F_\infty A_\infty$ may be regarded as a quadrilateral with $A_\infty \equiv F_\infty$ and $B_\infty \equiv C_\infty$. The internal angles at (A_∞, F_∞), (B_∞, C_∞), D, E are respectively $0, 0, (1+\gamma)\pi, (1-\gamma)\pi$. We map the four points on to the full line $\text{Im}(t) = 0$ and the interior of the polygon in the z-plane into the entire half-plane $\text{Im}(t) \geqslant 0$. Three of the points are at our disposal: let us choose the three mappings:

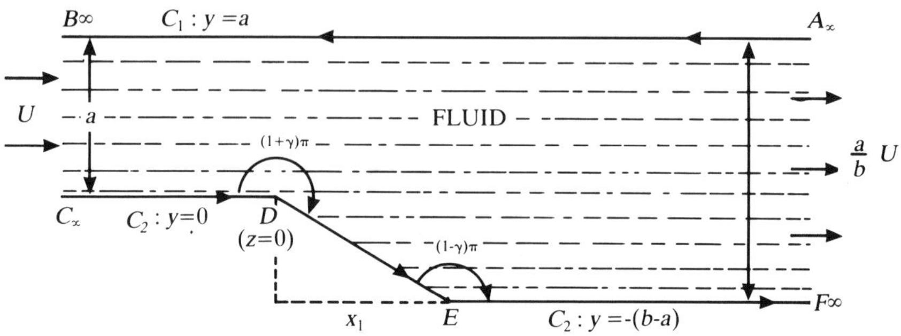

Fig. 7.18(i)

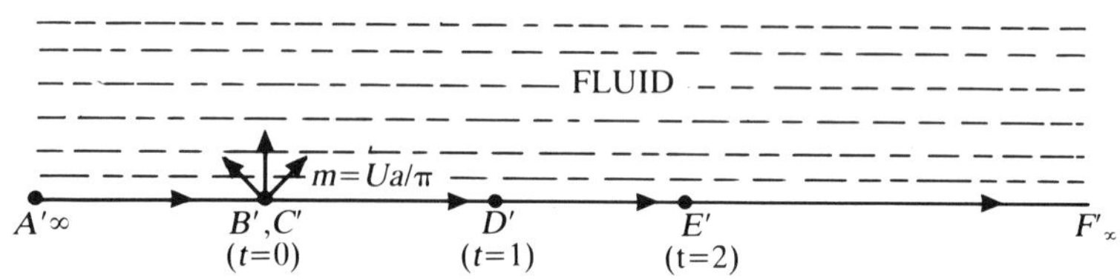

Fig. 7.18(ii)

(B_∞, C_∞) into (B', C'), where $t = 0$;

D into D', where $t = 1$;

E into E', where $t = \alpha > 1$, since $b > a$.

To produce the correct configuration, A_∞ maps into A'_∞, where $t = -\infty$ and F_∞ into F'_∞, where $t = \infty$. Fig. 7.18(ii) shows the entire mapping of Fig. 7.18(i). In both of these diagrams, if we walk along the path $ABCDEFA$, the fluid is always on our left-hand side. As A'_∞ and F'_∞ are points at infinity, the corresponding factors are suppressed and the appropriate Schwarz–Christoffel transformation assumes the form

$$\begin{aligned} dz/dt &= K(t-0)^{0-1}(t-1)^{1+\gamma-1}\,(t-\alpha)^{1-\gamma-1} \\ &= K\,t^{-1}\,(t-1)^{\gamma}\,(t-\alpha)^{-\gamma} \end{aligned}$$

or

$$z = K\int_1^t \frac{(t-1)^{\gamma}}{t(t-\alpha)^{\gamma}}\,dt + L.$$

When $t = 1, z = 0$ so that $L = 0$ and

$$z = K\int_1^t \frac{(t-1)^{\gamma}}{t(t-\alpha)^{\gamma}}\,dt.$$

The equivalent line source t $z = -\infty$ maps into a line source through (B', C') which must emit a volume of fluid Ua per unit time per unit length into the region $\text{Im}(t) \geqslant 0$. Since the region is spread over an angle π, the strength of the line source at (B', C') is $m = Ua/\pi$ per unit length. This gives rise to a complex potential in the t-plane of

$$w = (Ua/\pi) \ln t.$$

Then, regarding w as a function of z,

$$\begin{aligned} \mathrm{d}w/\mathrm{d}z &= (\mathrm{d}w/\mathrm{d}t)(\mathrm{d}t/\mathrm{d}z) \\ &= \frac{Ua}{\pi t} \cdot \frac{t(t-\alpha)^\gamma}{K(t-1)^\gamma}. \end{aligned}$$

When $z = -\infty$, $t = 0$ and $\mathrm{d}w/\mathrm{d}z = U$. Hence

$$U = (Ua/\pi K)\alpha^\gamma,$$

and so $K = a\,\alpha^\gamma/\pi$. When $z = +\infty$, $t = +\infty$, $\mathrm{d}w/\mathrm{d}z = Ua/b$, so that

$$Ua/b = Ua/K\pi$$

or $K = b/\pi$. From the two forms for K, it follows that $b = a\,\alpha^\gamma$ and

$$z = \frac{b}{\pi}\int_1^t \frac{(t-1)^\gamma}{t(t-\alpha)^\gamma}\,\mathrm{d}t.$$

When $x_1 \to \infty$, Fig. 7.18(i) shows $b \to \infty$. Thus $a\,\alpha^\gamma \to \infty$ and so $\alpha \to \infty$. From the last form

$$\begin{aligned} z &= \frac{b}{\pi\alpha^\gamma}\int_1^t \frac{(t-1)^\gamma}{t[(t/\alpha)-1]^\gamma}\,\mathrm{d}t \\ &= \int_1^t \frac{(t-1)^\gamma}{t[(t/\alpha)-1)]^\gamma}\,\mathrm{d}t \\ &\to \frac{a}{\pi}\int_1^t \frac{(1-t)^\gamma}{t} \end{aligned}$$

$\mathrm{d}t$, as $\alpha \to \infty$.

Example [7.7]

Water flows along a canal at uniform speed U. The sides of the canal are parallel and distant h apart until a junction is reached where the canal divides into two branches of widths h_1 and h_2 and at angle α to each other (Fig. 7.19(i)). The problem is to find the downstream velocities U_1, U_2 in the two canal branches.

We regard the configuration $A_\infty B_\infty C D_\infty E A_\infty$ (Fig. 7.19(ii) as a pentagon having internal angles

$$0 \text{ at } A_\infty, \quad 0 \text{ at } B_\infty, \quad 2\pi - \alpha \text{ at } C, \quad 0 \text{ at } D_\infty, \quad \pi + \alpha \text{ at } E.$$

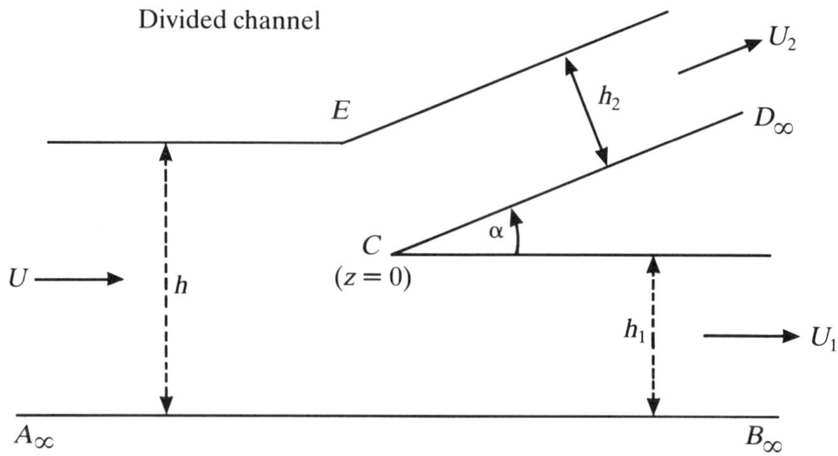

Fig. 7.19(i)

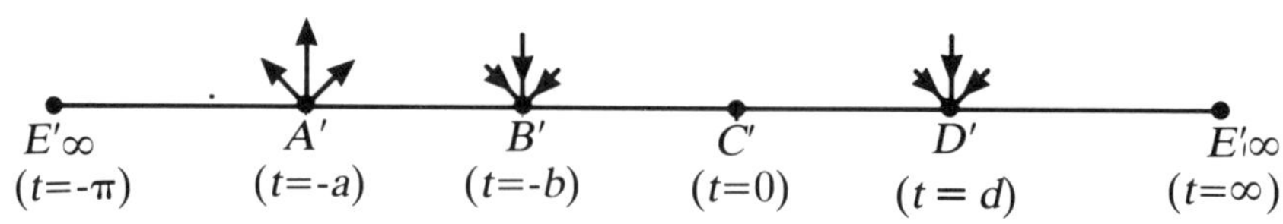

Fig. 7.19(ii)

Then the Schwarz–Christoffel transformation required to map the boundary of this polygon on to the real axis $\mathrm{Im}(t) = 0$ of the t-plane is given by

$$\mathrm{d}z/\mathrm{d}t = K(t+a)^{-1}\,(t+b)^{-1}\,(t-0)^{1-(\alpha/\pi)}\,(t-d)^{-1},$$

where $t = -a$, $t = -b$, $t = 0$, $t = d$, $t = \infty$ are the real mappings of A_∞, B_∞, C, D_∞, E on $\mathrm{Im}(t) = 0$. There is no contribution from the vertex E since $t = \infty$ at E'_∞, the mapping of E on the line. We are free to choose one of the three quantities a, b, d and the other two can, as we shall see, be found in terms of it.

The volumetric flux per unit time over a section of the canal of unit depth upstream of the junction is Uh. Thus at $t = -a$ we require a uniform line source Uh/π per unit length, since the volumetric flux Uh is spread over an angle π. Similarly line sinks of strengths U_1h_1/π and U_2h_2/π per unit length are introduced at $t = -b$ and $t = d$. The complex potential, then, for Fig. 7.19(ii) is

$$w = \{Uh\ln(t+a) - U_1h_1\ln(t+b) - U_2h_2\ln(t-d)\}/\pi,$$

from which

$$\mathrm{d}w/\mathrm{d}t = \{Uh(t+a)^{-1} - U_1h_1(t+b)^{-1} - U_2h_2(t-d)^{-1}\}/\pi.$$

Hence from $\mathrm{d}w/\mathrm{d}z = (\mathrm{d}w/\mathrm{d}t)/(\mathrm{d}z/\mathrm{d}t)$, we have

$$\frac{dw}{dz} = \frac{1}{\pi K t^{\beta}}\{Uh(t+b)(t-d) - U_1 h_1 (t+a)(t-d) - U_2 h_2 (t+a)(t+b)\},$$

where $\beta = 1 - (\alpha/\pi)$. At $t = -a = a\,e^{i\pi}$ (corresponding to A_∞), $u = U$, $v = 0$ and so $dw/dz = u - iv = U$. Thus at $t = -a$,

$$\pi K a^{\beta}\, e^{i\beta\pi} = h(a-b)(a+d).$$

At $t = -b$ (corresponding to B_∞), $u = U_1$, $v = 0$ and

$$\pi K b^{\beta}\, e^{i\beta\pi} = h_1(a-b)(b+d).$$

At $t = d$ (corresponding to D_∞), $u = U_2 \cos\alpha$, $v = U_2 \sin\alpha$ so that

$$dw/dz = u - iv = U_2\, e^{-i\alpha},$$

giving

$$\pi K d^{\beta}\, e^{-i\alpha} = -h_2(a+d)(b+d).$$

Now $-i\alpha = i\pi\beta - \pi i$ and so the last relation gives

$$\pi K d^{\beta}\, e^{i\beta\pi} = h_2(a+d)(b+d).$$

We have thus obtained the relations

$$\pi K\, e^{i\pi\beta} = \frac{h(a-b)(a+d)}{a^{\beta}} = \frac{h_1(a-b)(b+d)}{b^{\beta}} = \frac{h_2(a+d)(b+d)}{d^{\beta}}.$$

They imply that

(i) $\beta =$ real constant,

(ii) $$\frac{h}{a^{\beta}(b+d)} = \frac{h_1}{b^{\beta}(a+d)} = \frac{h_2}{d^{\beta}(a-b)}.$$

The two relations (ii) show that of the three points $t = a$, $t = b$, $t = c$, only one is at our disposal, the others being determined from (ii). For a pentagon three points are fixed: in this case we fix C' and E'_∞ and any one of the three points A', B', D'. The other two are then determined from (ii).

At the origin C in the z-plane there is a stagnation point at which $dw/dz = 0$. Since C' is $t = 0$, the form obtained for dw/dz gives

$$\frac{Uh}{a} = \frac{U_1 h_1}{b} - \frac{U_2 h_2}{d}.$$

The equation of continuity is simply

$$Uh = U_1 h_1 + U_2 h_2.$$

Thus the last two equations give

$$\frac{U_1}{U} = \frac{h}{h_1}\,\frac{b(a+d)}{a(b+d)}, \qquad \frac{U_2}{U} = \frac{h}{h_2}\,\frac{d(a-b)}{a(b+d)}.$$

From (ii), we can express the ratios h/h_1 and h/h_2 in terms of a, b, d. We then find

$$\frac{U_1}{U} = \left(\frac{b}{a}\right)^{1-\beta} = \left(\frac{b}{a}\right)^{\alpha/\pi}; \qquad \frac{U_2}{U} = \left(\frac{d}{a}\right)^{1-\beta} = \left(\frac{d}{a}\right)^{\alpha/\pi}.$$

We now introduce the dimensionless quantities

$$\lambda = h_1/h, \qquad \mu = h_2/h, \qquad \nu = (U_1 h_1)/(Uh).$$

The equation of continuity gives at once

$$1 - \nu = (U_2 h_2)/(Uh).$$

The condition at the stagnation point C is equivalent to

$$1 = \frac{U_1 h_1}{Uh}\left(\frac{a}{b}\right) - \frac{U_2 h_2}{Uh}\left(\frac{a}{d}\right)$$

or

$$1 = \lambda(\nu/\lambda)^{1-(\pi/\alpha)} - \mu[(1-\nu)/\mu]^{1-(\pi/\alpha)}.$$

This is a transcendental equation for ν and hence for U_1 and U_2.

When Fig. 7.19(i) is reflected in the streamline $A_\infty B_\infty$ the figure obtained is that of a main canal feeding two symmetrically disposed branches. The solution obtained above also solves this problem.

Example [7.8]
Uniform flow past a plate at right angles to the stream.

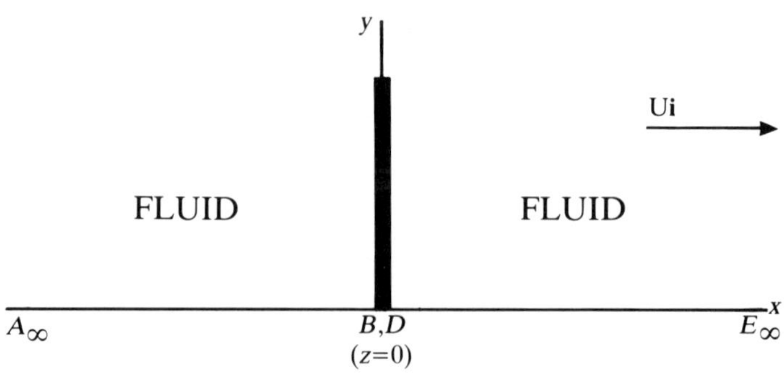

Fig. 7.20(i)

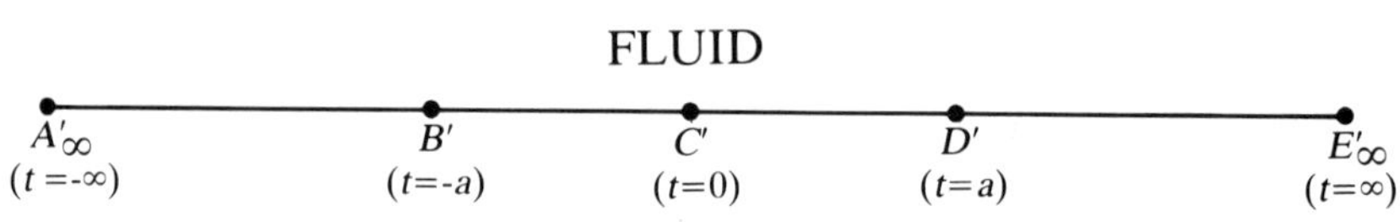

Fig. 7.20(ii)

An infinite plate of width $2a$ is placed at right angles to a uniform stream of velocity $U\mathbf{i}$. We find the complex potential.

From symmetry, $y = 0$ is a streamline and this can be replaced by a rigid boundary. Thus we need only consider the configuration of Fig. 7.20(i) with rigid boundaries along $y = 0, -\infty \leqslant x \leqslant \infty$ and $x = 0, 0 \leqslant y \leqslant a$. Fluid occupies the whole of the region $y \geqslant 0$.

We treat $A_\infty BCDE_\infty$ as a pentagon with angles 0 at A_∞, $\pi/2$ at B, 2π at C, $\pi/2$ at D, 0 at E_∞. Let us map this figure on to the t-plane, in which points denoted by primed letters in Fig. 7.20(ii) correspond to the unprimed letters in the z-plane. We take $t = -\infty$ for A', $t = -a$ for B', $t = 0$ for C', $t = a$ for D', $t = \infty$ for E'_∞. Then the appropriate Schwarz–Christoffel transformation is

$$\mathrm{d}z/\mathrm{d}t = K(t+a)^{-1/2}\,(t-0)^1\,(t-a)^{-1/2} = Kt(t^2-a^2)^{-1/2}.$$

This integrates to give

$$z = K(t^2-a^2)^{1/2} + L.$$

When $t = \pm a$, $z = 0$ and so $L = 0$. When $t = 0$, $z = ia$ and so $K = 1$. Then the transformation is

$$z = \sqrt{(t^2-a^2)}$$

or, equivalently,

$$t = \sqrt{(z^2+a^2)}.$$

This transform automatically maps A_∞ into A'_∞ and E_∞ into E'_∞.

The flow in the z-plane is equivalent to that due to a line source at A_∞ and a line sink at E_∞. Essentially, these transform into a line source at A'_∞ and an equal line sink at E'_∞. This gives rise to a uniform flow along $\overrightarrow{A'_\infty E'_\infty}$ in the t-plane. If U_0 is the uniform speed of flow in this plane, then the complex potential is

$$w = U_0 t = U_0(z^2+a^2)^{1/2},$$

from which it follows that

$$\begin{aligned}\mathrm{d}w/\mathrm{d}z &= U_0\, z(z^2+a^2)^{-1/2}\\ &= U_0\,[1+(a/z)^2]^{-1/2}\\ &\to U_0 \text{ as } |z| \to \infty.\end{aligned}$$

But $\mathrm{d}w/\mathrm{d}z = U$ at A_∞, E_∞, and so $U_0 = U$. Thus

$$w(z) = U\sqrt{(z^2+a^2)}$$

is the velocity potential for the flow in Fig. 7.20(i).

From the form of $w(z)$,

$$w'(z) = u - iv = Uz(z^2+a^2)^{-1/2}.$$

This shows that the only stagnation points of the entire field of flow are where $z = 0$, i.e. at the points B, D. Also we note that the fluid speed $|w'(z)| \to \infty$ as $|z| \to ia$ and so the pressure at $C \to 0$ resulting in the possible cavitation of the fluid.

PROBLEMS 7

1. Given that $w = z^2$, find the loci in the z-plane corresponding to $4u = 3v$, $v = 2$ and verify that these loci intersect at the same angle as the lines $4u = 3v$, $v = 2$, where $w = u + iv$.

2. $w = f(z)$ is a conformal transformation. By considering the successive transformations, $w = f(\zeta)$, $\zeta = \bar{z}$, show that $w = f(\bar{z})$ gives a transformation which preserves the magnitude but not the sense of angles. A transformation of this type is said to be *isogonal.*

3. $w = K(z + z^{-1})$ (K real). Show that $v = 0$ corresponds to the real axis and the circle $|z| = 1$ in the z-plane. Sketch a few of the curves corresponding to $u =$ const., $v =$ const.

4. $z = a \cosh w$, a being a real constant. Show that the curves $u =$ const., $v =$ const. correspond to confocal ellipses and hyperbolae in the z-plane.

5. Show that the transformation

$$w = -Vz + Az^{1/2},$$

where A is a constant to be determined, gives the complex potential for the case of a homogeneous liquid streaming past the infinite parabolic cylinder

$$r^{1/2} \cos(\tfrac{1}{2}\theta) = a^{1/2},$$

the velocity at infinity being V in the positive direction of the x-axis.

6. Show that the complex potential

$$w = V\left(\frac{\zeta - ib + c^2}{\zeta - ib}\right), \quad \zeta = \xi + i\eta,$$

gives irrotational streaming motion with speed V parallel to the ξ-axis past a fixed circular cylinder S of radius c with centre $0 + ib$.

If a is real and $z = \zeta + a^2/\zeta$, $z = x + iy$, show that

$$\frac{z - 2a}{z + 2a} = \frac{(\zeta - a)^2}{(\zeta + a)^2}$$

and, if $c^2 = a^2 + b^2$, determine the contour C in the z-plane corresponding to S.

If C represents the boundary fixed in a uniform stream with velocity V parallel to Ox, show that a circulation $-4\pi bV$ ensures that the velocity is finite at all points of C.

7. If $\zeta = z + a^2/z, (a > 0)$, sketch the curve in the ζ-plane which is the transform of the circle

$$|z - b + a| = b \quad (b > 0)$$

in the z-plane, and prove that there is a cusp at the point $\zeta = -2a$.

If the curve in the ζ-plane is the profile of a rigid body surrounded by an infinite fluid which at great distances has velocity components $[-U \cos \alpha, -U \sin \alpha]$ along the ξ- and η-axes, and if the circulation round the body is such that the velocity is everywhere bounded, prove that the fluid flows past the cusp at a speed $(Ua/b) \cos \alpha$.

8. Show that the transformation $t = \cosh (\pi z/b)$ can be used to map a semi-infinite strip of width b in the z-plane on to a half-plane in the t-plane.

A semi-infinite strip given by the boundaries $x = 0, y = b$ contains a source at $x = 0, y = 0$ sending liquid into the strip at the rate $2\pi m$ and a sink at $x = 0$, $y = b$ of equal strength. Show that the complex potential is $w = +4m \ln \tanh (\pi z/2b)$.

9. Inviscid liquid of constant density ρ flows without circulation past the fixed elliptic cylinder $x^2/a^2 + y^2/b^2 = 1$, the velocity of the undisturbed stream being $[-U \cos \alpha, -U \sin \alpha]$. Using the transformation

$$2z = (a + b)\zeta + (a - b)\zeta^{-1},$$

or otherwise, determine the complex potential w from the corresponding flow past a circular cylinder.

Show that, if $|z|$ is large,

$$w \simeq U[z\, e^{-i\alpha} + \tfrac{1}{2}(a + b)(b \cos \alpha + ia \sin \alpha)/z],$$

and hence find the couple M exerted on unit length of the cylinder.

10. Incompressible inviscid two-dimensional motion is taking place within a rectangle open at $y = +\infty$ and bounded by $y = 0, x = a, x = -a$. This motion is due to a source m at $(a, 0)$ and an equal sink at $(-a, 0)$. Show that the complex potential w of the motion is given by $\sin (\pi z/2a) + \coth (w/4m) = 0$.

Show further that the streamlines are given by $\cos (\pi x/2a) = \lambda \sinh (\pi y/2a)$, where λ is a constant.

11. If the z-plane is mapped on the t-plane by a conformal transformation, show that a doublet at the point $z = z_0$ in the z-plane will transform into a doublet at the corresponding point of the t-plane, whose strength is multiplied

by $|dt/dz|_{z=z_1}$ and whose inclination to the real axis is increased by $\arg(dt/dz)_{z=z_0}$.

By the application of a suitable transformation, find the complex potential function for the steady motion in the semi-infinite strip ($x \geqslant 0$) of the z-plane with rigid boundaries $x = 0, y = 0, y = 2b$, due to a doublet of strength μ at the point (a, b), with its axis in the positive x direction.

12. Prove that the transformation

$$t = z + l^2/z$$

maps the circle $z = k(> l)$ into an ellipse in the t-plane. Hence or otherwise show that the complext potential of the flow, without circulation, due to a uniform stream, of speed V at infinity, past an elliptic cylinder with semi-axes $a, b\,(<a)$ is

$$w = +\tfrac{1}{2}iV[t + \{t^2 - (a^2 - b^2)\}^{1/2} - (a+b)^2\ [t + \{t^2 - (a^2 - b^2)\}^{1/2}]^{-1}]$$

when the major axis is at right angles to the undisturbed stream.

13. Liquid occupies the region $y \geqslant 0$, and an infinite number of line vortices, each of strength k, are spaced at a distance a apart at a distance $\frac{1}{2}b$ from the boundary $y = 0$.

Show that the vortices move parallel to the boundary with velocity $(k/2a) \coth (\pi b/a)$.

Show also that the pressure at a point of the boundary directly opposite one of the vortices is given by

$$\frac{p_\infty - p}{\rho} = \frac{k^2}{8a^2}\left[\left(2\coth\frac{\pi b}{2a} - \coth\frac{\pi b}{a}\right)^2 - \coth^2\frac{\pi b}{a}\right]$$

where p_∞ is the pressure far from the boundary.

14. Liquid filling the space between barriers coincident with the planes $y = \pm\frac{1}{2}\pi$ is set in motion by a rectilinear vortex of strength κ along Oz. The spaces on the other side of the barriers are occupied by stationary liquid in which the pressure is equal to the pressure at infinity on Ox. Prove that the thrust exerted on the strip of either barrier between the planes $z = 0, z = 1$ is $\rho\kappa^2$.

15. Prove that the motion due to a set of line vortices each of strength κ at the points $z = \pm n\pi$ $(n = 0, 1, 2, 3, \ldots)$ is determined by the relation $w = -i\kappa \ln \sin z$.

16. The boundaries of a liquid are the planes $x = \pm a$. A vortex filament parallel to the z-axis meets the z-plane in the point (x', y') and is of strength κ. Prove that the equation of the streamlines may be written

$$\cosh\,[\pi(y - y')/2a]\ \sec\,(\pi x/2a) - \sin\,(\pi x'/2a)\tan\,(\pi x/2a) = \text{constant}.$$

17. One boundary of a channel consists of the rectilinear segments

$$y = 0, \qquad 0 \leqslant x < \infty;$$
$$x = 0, \qquad 0 \leqslant y \leqslant a - b \quad (a > b);$$
$$y = a - b. \qquad -\infty < x \leqslant 0,$$

and the second boundary is $y = a$, $-\infty < x < \infty$. An incompressible liquid whose velocity components at $x = +\infty$ are $[-U, 0]$ flows along the channel which it fills completely. Show that the flow is given by

$$w = \phi + i\psi = \frac{aU}{\pi}\ln t, \quad z = \frac{a}{\pi}\int_{a^2/b^2}^{t} (t-1)^{1/2}\, t^{-1}\, (t - a^2/b^2)^{-1/2}\, dt,$$

and that along the curves of constant pressure

$$(1-\lambda)\exp\left(\frac{2\pi\phi}{aU}\right) - 2\left(\frac{a^2}{b^2} - \lambda\right)\exp\left(\frac{\pi\phi}{aU}\right)\cos\left(\frac{\pi\psi}{aU}\right) + \frac{a^4}{b^4} - \lambda = 0,$$

where λ is a parameter.

18. An incompressible liquid moves irrotationally in two dimensions within an L-shaped region which has two rigid boundaries C_1 and C_2 defined as follows:

$$C_1 : x = 0, 0 < y < \infty; \quad y = 0, 0 < x < \infty;$$
$$C_2 : x = b, a < y < \infty; \quad y = a, b < x < \infty;$$

At $y = \infty$ the liquid is flowing in the negative y direction with speed U and it always remains in contact with the side walls. Show that, with the usual notation, the flow may be determined from the equations

$$z = A\int_0^t \frac{(\alpha - t)^{1/2}}{t^{1/2}(1-t)}\, dt, \quad w = B\ln(t-1),$$

where α is real.

Prove that $A = b/\pi$, $B = bU/\pi$ and $\alpha = 1 + (a/b)^2$.

19. The boundary $ABCDEF$ (described in a counter-clockwise direction) of an infinite liquid consists of a horizontal plane $AB(-\infty < x < \infty, y = h)$, two horizontal planes $CD(x \leqslant 0, y = h - k)$, $EF(x > 0, y = 0)$, and a vertical plane $ED(x = 0, 0 \leqslant y \leqslant h - k)$, where $h > k > 0$. Liquid is in two-dimensional irrotational motion inside this region, the velocity for x large and positive being U parallel to AB or FE. Map the boundary on to the real axis of the plane of a

complex variable t in such a way that A, B, C, D, E, F correspond to the values $-\infty, 0, 0, 1, a, \infty$ of t respectively.

Hence obtain the complex potential in the form

$$w = \frac{Uh}{\pi} \ln\left(\frac{h^2 - k^2 t^2}{1 - t^2}\right),$$

$$z = \frac{h}{\pi}\left\{\ln\left(\frac{1+t}{1-t}\right) - \frac{k}{h}\ln\left(\frac{h+kt}{h-kt}\right)\right\}.$$

20. Show that the relation

$$t = z \sin\tfrac{1}{2}\alpha \; \frac{z + a \operatorname{cosec}\tfrac{1}{2}\alpha}{z + a\sin\tfrac{1}{2}\alpha} \qquad (0 < \alpha < \tfrac{1}{2}\pi)$$

where a is real, transforms the circumference of the circle $|z| = a$ into an arc of the same circle subtending an angle 2α at the centre.

For a two-dimensional flow of inviscid liquid past such an arc, show that the lift produced by a uniform stream U flowing parallel to the chord of the arc is $2\pi\rho h U^2$, where h is the height of the arc.

21. A circle $|z| = a$ is transformed into a thin aerofoil section of chord length approximately $4a$ by the transformations

$$t = z\left[1 + \sum_{m=1}^{n} A_m (a/z)^m\right].$$

$$\zeta = t + \frac{a^2}{t}.$$

Prove that the lift and moment coefficients are

$$L = \pi(\alpha + \beta),$$

$$M = \tfrac{1}{4}L + \tfrac{1}{4}\pi\beta + \tfrac{1}{8}\pi D - \tfrac{1}{4}\pi\alpha C,$$

where α is the small angle of incidence and β, C, D are constants for the aerofoil.

8

Waves

8.1 THE OCCURRENCE OF WAVES

The dynamics of wave motion is of great importance in a wide variety of physical phenomena since wave motion is one of the principal modes of energy transmission. For instance, the energy received from the Sun is transmitted as waves of different energy forms which are recognisable as heat, light, radio waves and various other types of radiation such as X-rays. The practical applications of electromagnetic waves are widespread and range from telecommunications, including radio and television, to microwave cookery. But perhaps the forms of wave motion which have been observed and studied for the longest time are waves in fluids. These encompass sound waves, which are pressure waves propagated through a medium such as air, and water waves which may be observed on the surface of any body of water from oceans to puddles. These waves range in size from the huge tidal waves caused by earthquakes and the tides caused by the gravitational attraction between the Earth and the Moon, down to the smallest ripples on a still pool caused by a light breeze.

Although there has been an awareness of wave motions in fluids throughout the history of mankind, a theoretical description of such wave motions is by no means an easy task, and can only be undertaken after certain drastic simplifications in mathematical modelling are made on account of the complexity of the physical nature and properties of the phenomena. In this chapter, we shall present an account of some of the simpler theoretical studies which can be successfully carried out into surface wave phenomena. In this respect, ideal fluid theory provides a particularly useful and accurate method of modelling such phenomena, since there are no solid boundaries present where viscosity would have a significant effect. Within this chapter we shall be concerned with wave motions at the surfaces of liquids, particularly water, so the effect of compressibility may be ignored. The topic of wave propagation in compressible media will be studied in the companion volume to this book.

A wave in a fluid is the continuous transfer of a state or form from one part of the fluid medium to another. This does not imply that the medium itself is moving from one place to another, but merely that, through it, a particular condition or disturbance is being propagated. The fact that the medium itself is *not* undergoing any large scale movement or mass transfer is often difficult to appreciate when we observe waves on the surface of the sea, or the breaking of waves on a beach. But the fact that small bodies floating on the surface of the sea are not borne onwards by the waves as they progress is clearly an indication that the elevated masses of water are not moving forward bodily to any great extent. In fact, it is only the *disturbed shape* of the surface which is moving from place to place. As a wave passes a floating body, the body is carried forward a small distance on the crest of a wave only to retreat again when in the trough. The result is that the position of the floating body is largely undisturbed by the passage of the wave. This property of wave motion is exploited by surfers since the object in surfing is to ride the ocean waves before they break on the beach.

8.2 THE MATHEMATICAL DESCRIPTION OF WAVE MOTION

Let us consider a mass of incompressible liquid of uniform density ρ which may or may not be of infinite depth but which has a free surface S given by $z = 0$, when in an undisturbed state, with (x, y, z) rectangular Cartesian coordinates. The liquid may or may not be bounded in the x- and y-directions and in the case when the liquid being considered is for instance the sea or a lake, the surface S would be open to the atmosphere. Suppose now that the surface S is disturbed in such a way that at $t = 0$ its form is given by $z = f(x)$ and at time t, its form is $z = f(x - ct)$, then a wave motion on the surface S is defined. The wave is a *plane wave* because the section in any plane y = constant is the same, and it moves in the positive direction of the x-axis with speed c, since if t is increased by t' and x by ct', the ordinate y is unaltered. A particular example of a plane surface wave disturbance is given by

$$z = a \sin(mx - nt + \epsilon), \tag{8.1}$$

where a, m, n and ϵ are constants. This represents a *simple harmonic progressive wave* moving in the positive direction of the x-axis, with speed $c = n/m$, called the *speed of propagation*. The distance between the highest point of any two consecutive wave crests, or indeed between any corresponding points of two consecutive waves, is $2\pi/m = \lambda$. This distance is called the *wavelength*, and the reader will notice that at the surface, equation (8.1) shows that

$$z(x + \lambda, t) = z(x, t).$$

The *period* of the wave is $2\pi/n$ since the surface elevation at time $t = t' + 2\pi/n$ has the same form relative to the origin as at time $t = t'$ for each wave crest will have moved forward a distance λ in the time interval $2\pi/n$. The maximum positive value of the surface elevation above $z = 0$ is called the *amplitude* of the wave motion and its value is a. The constant m is called the *wave number* and it represents the number of waves per unit length while the constant n is called the *frequency* and it gives the number of wave oscillations in unit time. Note that

$$\text{speed of propagation} = \frac{\text{frequency}}{\text{wave number}} . \tag{8.2}$$

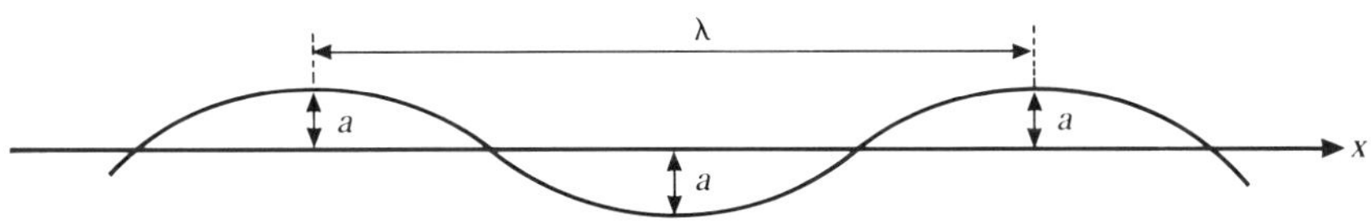

Fig. 8.1

The constant ϵ is called the *phase*. If we compare the waveforms

$$z = a \sin (mx - nt),$$

$$z = a \sin (mx - nt + \epsilon),$$

we see that each represents a wave motion and both have the same amplitude, wavelength and period but they differ in phase with one waveform displaced a distance ϵ/m relative to the other. Alternatively, in terms of time, one waveform has a start of ϵ/n relative to the other.

It should be noticed that a simple harmonic progressive wave disturbance satisfies the differential equation

$$\frac{\partial^2 z}{\partial t^2} = \frac{1}{c^2} \frac{\partial^2 z}{\partial x^2} . \tag{8.3}$$

which is the *one-dimensional wave equation*. The general solution of this equation may easily be found by transforming the variables using the substitutions

$$\theta = x - ct, \qquad \chi = x + ct,$$

whereupon

$$\frac{\partial}{\partial x} = \frac{\partial}{\partial \theta} + \frac{\partial}{\partial \chi}, \qquad \frac{\partial}{\partial t} = c \left(\frac{\partial}{\partial \chi} - \frac{\partial}{\partial \theta} \right),$$

and (8.3) transforms to

$$\frac{\partial^2 z}{\partial \theta \, \partial \chi} = 0. \tag{8.4}$$

The general solution of this equation is

$$z = F(\theta) + G(\chi)$$

for arbitrary twice differentiable functions $F(\theta)$, $G(\chi)$. Accordingly, the general solution of (8.3) is

$$z = F(x - ct) + G(x + ct). \tag{8.5}$$

The two terms on the right-hand side of (8.5) represent respectively plane waves propagating along the positive and negative directions of the x-axis with speed c.

Notice also that another example of a simple harmonic progressive waveform is given by

$$z = a \cos(mx - nt + \epsilon). \tag{8.6}$$

The reader will appreciate that (8.6) is actually of the same form as (8.1) but with a change of phase.

A study of the fluid motion associated with a simple harmonic progressive waveform is of fundamental importance in the study of surface waves because a more general waveform of the type (8.5) can be expressed as the sum of simple harmonic waves of the type (8.1) or (8.6) by the methods of Fourier analysis.

If a surface distance is the combination of two simple harmonic progressive plane waves of the same amplitude, wavelength, period and phase, but travelling in opposite directions, the resulting surface displacement is given by

$$\begin{aligned} z &= a \sin(mx - ct) + a \sin(mx + ct) \\ &= 2a \sin mx \cos nt, \end{aligned} \tag{8.7}$$

where the phase constant has been eliminated by redefining the coordinate x. Such a disturbance is called a *stationary* or *standing* wave. At any time t, equation (8.7) represents a sine curve whose amplitude $2a \cos nt$ varies continuously with time. The points of intersection of the waveform (8.7) with the x-axis are fixed points called *nodes*. In the same way, a progressive wave may be regarded as the superposition of two standing waves of the same amplitude, wavelength and period with the crests and troughs of one wave coinciding with the nodes of the other wave. The two waves are a quarter period out of phase. For example

$$\begin{aligned} z &= a \sin(mx - nt) = a \sin mx \cos nt - a \cos mx \sin nt \\ &= a \sin mx \cos nt + a \cos mx \cos n(t + \pi/2n). \end{aligned}$$

8.3 GRAVITY WAVES

With the coordinate system defined in Section 8.1, we shall now consider waves due to small oscillatory motions which take place at or near the surface of an incompressible liquid, in particular water. We shall suppose that the fluid is unbounded in the x- and y-directions and that the depth h, when undisturbed, is very much larger than either the amplitude or wavelength of the disturbance; consistent with this supposition is the assumption that squares, products and higher powers of components of fluid velocity are negligible.

We shall first of all consider a plane wave disturbance of the free surface S propagated in the positive x-direction. This means that the fluid motion is two-dimensional with the velocity components $(u, 0, w)$ independent of the y-coordinate. Since the motion in the fluid could be generated from rest, the fluid velocity $\mathbf{q}$ is irrotational as a consequence of Kelvin's theorem proved in Section 3.7. A velocity potential $\phi = \phi(x, z, t)$ thus exists which satisfies Laplace's equation

$$\frac{\partial^2 \phi}{\partial x^2} + \frac{\partial^2 \phi}{\partial z^2} = 0 \tag{8.8}$$

throughout the fluid in order that the equation of continuity div $\mathbf{q} = 0$ be satisfied with $\mathbf{q} = \nabla\phi$. At any fixed boundary to the fluid there is the kinematic condition

$$\frac{\partial \phi}{\partial n} = 0, \tag{8.9}$$

with n denoting distance along the normal to the boundary in the direction out of the fluid.

For the present, we shall ignore the effect of surface tension at the free surface S and shall assume that the effect of the atmosphere is to maintain a constant pressure Π over this surface. Within the fluid, the pressure is given by Bernoulli's equation

$$\frac{p}{\rho} + \frac{\partial \phi}{\partial t} + \frac{1}{2}\mathbf{q}^2 + gz = F(t), \tag{8.10}$$

where g is the acceleration due to gravity. However, we are assuming that $\mathbf{q}^2$ is negligible, so if Π/ρ and the arbitrary function $F(t)$ are absorbed into $\partial\phi/\partial t$, we find that at the free surface S,

$$\frac{\partial \phi}{\partial t} + gz = 0. \tag{8.11}$$

On letting the equation of the free surface be $z = \eta(x, t)$, it follows from Taylor's theorem that on S,

$$\frac{\partial \phi}{\partial t} = \left(\frac{\partial \phi}{\partial t}\right)_{z=0} + \eta\left(\frac{\partial^2 \phi}{\partial z \partial t}\right)_{z=0} + \mathrm{O}(\eta^2), \tag{8.12}$$

and since

$$\left(\frac{\partial^2 \phi}{\partial z \partial t}\right)_{z=0} = \left(\frac{\partial w}{\partial t}\right)_{z=0},$$

which is a small quantity if we assume that no rapid changes in the fluid motion take place with time, then, to the *first order* in small quantities, equation (8.11) yields

$$\frac{\partial \phi}{\partial t} + g\eta = 0, \qquad (z = 0). \tag{8.13}$$

In addition, there is the kinematic condition to be satisfied at the free surface. As was shown in Section 3.2, this condition is

$$\frac{\mathrm{D}}{\mathrm{D}t}[z - \eta(x, t)] \equiv w - \frac{\partial \eta}{\partial t} - u\frac{\partial \eta}{\partial x} = 0 \tag{8.14}$$

on the free surface $z = \eta(x, t)$. Now $\partial\eta/\partial x$ represents the slope of the free surface and if we assume that this is small and at most the same order as the surface elevation η, equation (8.14) gives, when only *first order* terms in small quantities are retained,

$$\frac{\partial \phi}{\partial z} = \frac{\partial \eta}{\partial t}, \qquad (z = 0). \tag{8.15}$$

The surface elevation $\eta(x, t)$ may be eliminated from (8.13) and (8.15) to give

$$\frac{\partial^2 \phi}{\partial t^2} + g\frac{\partial \phi}{\partial z} = 0, \qquad (z = 0). \tag{8.16}$$

The problem is therefore reduced to finding a solution of (8.8) which satisfies the two boundary conditions (8.9) and (8.16). Once the velocity potential is known, the surface elevation can be determined from (8.13). The form of boundary conditions (8.9) and (8.16) are such that the solution for ϕ will involve an *arbitrary* multiplicative constant which in turn, through (8.13) or (8.15), is proportional to the amplitude of η. There is therefore an infinity of values that may be taken by the amplitude of the surface elevation provided of course that the amplitude is sufficiently small for the first order, or linearized, equations derived above to be valid.

Usually a fixed boundary is the bottom of the lake, or whatever, when the fluid has finite depth h. Equation (8.9) is then simply

$$\frac{\partial \phi}{\partial z} = 0, \qquad (z = -h). \tag{8.17}$$

For fluid of infinite depth, this condition is replaced by

$$|\nabla \phi| \to 0, \qquad (z \to -\infty). \tag{8.18}$$

The reader should be able to verify that in the above formulation of the problem of plane waves propagated along the x-axis, the fluid could be bounded by rigid planes of the form y = constant which are perpendicular to the wave-fronts.

Let us now consider the propagation of a simple harmonic progressive wave of the type for which the elevation is

$$\eta(x, t) = a \sin (mx - nt), \tag{8.19}$$

with a, m, n constants as before. We shall suppose that this wave is propagated on the free surface S of fluid of uniform undisturbed depth h and either of unlimited extent in the horizontal x- and y-directions or confined to a channel with vertical sides at right angles to the wavefront. The form of (8.13) suggests that we try as the solution for ϕ the form

$$\phi(x, z, t) = f(z) \cos (mx - nt). \tag{8.20}$$

This satisfies (8.8) provided that

$$f''(z) - m^2 f(z) = 0.$$

Accordingly

$$f(z) = A\,\mathrm{e}^{mz} + B\,\mathrm{e}^{-mz},$$

where to satisfy (8.17), the constants A and B are such that $f'(-h) = 0$, giving

$$A\,\mathrm{e}^{-mh} = B\,\mathrm{e}^{mh} = \tfrac{1}{2}C$$

say, with C arbitrary. In which case

$$\phi = C \cosh m\,(z + h) \cos (mx - nt). \tag{8.21}$$

There remains the condition (8.16) which is satisfied provided that

$$n^2 = mg \tanh mh,$$

and since the propagation speed c is n/m and wavelength is $2\pi/m$, it follows that

$$\frac{c^2}{gh} = \frac{\lambda}{2\pi h} \tanh \frac{2\pi h}{\lambda}. \tag{8.22}$$

The constant C can be expressed in terms of the surface wave amplitude a by means of (8.15). We find that

$$-C = a(n/m) \operatorname{cosech} mh = ac \operatorname{cosech} (2\pi h/\lambda). \tag{8.23}$$

Equation (8.22) is dimensionless in form and was derived on the assumption that a is small in comparison with both λ and h i.e. $a \ll \lambda$, $a \ll h$.

A particular case of interest is that of deep liquids for which $h \gg \lambda$, in which case $\tanh (2\pi h/\lambda) \approx 1$ and (8.22) can be approximated by

$$\frac{c^2}{gh} = \frac{\lambda}{2\pi h}. \tag{8.24}$$

This equation is consequently valid for $a \ll \lambda \ll h$. The other extreme case is that of shallow liquids for which $h \ll \lambda$. Now $\tanh (2\pi h/\lambda) \approx 2\pi h/\lambda$ and (8.22) reduces to

$$\frac{c^2}{gh} = 1. \tag{8.25}$$

This limiting approximation is valid for $a \ll h \ll \lambda$. A plot of c^2/gh against $\lambda/2\pi h$ is shown in Fig. 8.2 together with the limiting cases (8.24) and (8.25). The figure shows that for liquid of constant depth, long waves travel faster than those with short wavelengths. This explains the observation that the approach of a storm across an ocean is indicated first by the arrival of very long waves of relatively small amplitude, known as *swell*, which have been generated by the storm far out at sea. These long waves may precede the arrival of the storm itself by several days. With the approach of the storm will come the waves of shorter wavelength which have been travelling much more slowly than the long waves. Fig. 8.2 also shows that for a wave of given wavelength, the speed of propagation decreases with depth, which explains why waves approaching a beach always appear to do so with their crests parallel to the beach, even though they may have originated out at sea as waves on deep water approaching the coastline at an arbitrary oblique angle. The orientation of the wave crests results from the reduction in wave speed as the depth decreases at the beach since the part of a

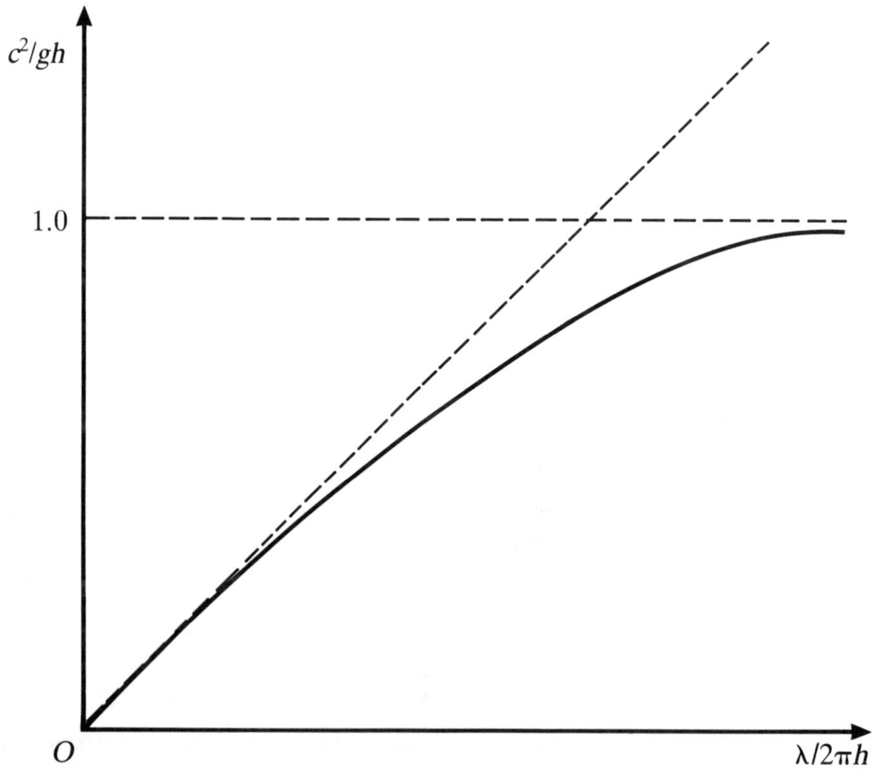

Fig. 8.2

wave crest closer to the beach moves more slowly than that further away, having the effect of twisting the wave crest so that it approaches a parallel configuration to the beach. This effect is particularly noticeable if one flies over a bay. Both of these examples illustrate the *dispersive* character of wave motion in that wave speed varies with wavelength and/or depth, as illustrated by equation (8.22) and Fig. 8.2. Indeed, equation (8.22) is often referred to as the *dispersion relation*. Strictly speaking, the waves generated in these examples are not plane waves, but the essential character of this aspect of waves in real life is captured by the simple plane wave model.

For infinitely deep liquid, the reader will have no difficulty in verifying that the solution for ϕ in this case is

$$\phi(x, z, t) = -ac\,\mathrm{e}^{mz} \cos(mx - nt), \tag{8.26}$$

and the dispersion relation corresponding to (8.22) is then

$$c^2 = \lambda g/2\pi. \tag{8.27}$$

As for liquid of finite depth, the longer the wavelength of the wave, the faster is its propagation speed. The error in calculating c from (8.27) rather than (8.22) is less than 0.5% if $h/\lambda > 0.5$, and one might conjecture that variations in bottom topography have but slight effect on a progressive wave provided that the depth is never less than about half a wavelength. Observation bears this out, so one might say that the wave is not 'aware' of the bottom unless the depth is less than half a wavelength. The above analysis may be applied to plane waveforms of arbitrary shape since such a waveform may be decomposed into a number of purely sinusoidal waves using the methods of Fourier analysis. Each Fourier component will be a plane progressive wave so the foregoing results show that the arbitrarily shaped waveform does not propagate so that its shape remains constant unless the wave is propagated on very shallow liquid. This is because the propagation speed c of each sinusoidal wave component of the arbitrarily shaped wave depends on the wavelength λ of that component, unless the shallow liquid limit formula for c, given by (8.25), applies. Thus, in general, the different Fourier components of the arbitrarily shaped wave travel at different speeds, so that the waveform continuously changes shape. This phenomenon again illustrates the dispersive character of wave motion.

8.4 THE PARTICLE PATHS

As a wave travels across the surface of an otherwise quiescent liquid, the individual particles of the liquid perform small cyclical motions. The trajectory described by a specific liquid particle may be determined quite simply since if $\mathbf{r}$ is the position vector of the particle at time t, then

$$\mathbf{r} = \mathbf{r}_0 + \mathbf{R}(t) \tag{8.28}$$

where $\mathbf{r}_0$ is the position vector of some mean position of the particle. Thus $\dot{\mathbf{R}}(t) = \mathbf{q}$ and the particle trajectory is determined on solving this latter differential equation.

Let us consider the problem of a plane progressive wave travelling on a liquid of depth h. Let $\mathbf{R}(t) = X(t)\mathbf{i} + Z(t)\mathbf{k}$ and $\mathbf{r}_0 = \mathbf{x}_0\mathbf{i} + z_0\mathbf{k}$. In which case equation (8.21) gives

$$\left.\begin{aligned} \dot{X}(t) &= \partial\phi/\partial x = -mC\cosh m(z+h)\sin(mx-nt) \\ \dot{Z}(t) &= \partial\phi/\partial z = \quad mC\sinh m(z+h)\cos(mx-nt) \end{aligned}\right\}. \tag{8.29}$$

From (8.23), the constant C is of the same order of magnitude as the small amplitude a of the surface wave elevation. Thus, to the first order in small quantities, (8.29) gives

$$\dot{X}(t) = -mC\cosh m(z_0+h)\sin(mx_0-nt),$$

$$\dot{Z}(t) = \quad mC\sinh m(z_0+h)\cos(mx_0-nt),$$

which on integration give

$$X(t) = -(mC/n)\cosh m(z_0+h)\cos(mx_0-nt),$$

$$Z(t) = -(mC/n)\sinh m(z_0+h)\sin(mx_0-nt),$$

and on using (8.23), the equation of the particle path is accordingly

$$\frac{(x-x_0)^2}{\cosh^2 m(z_0+h)} + \frac{(z-z_0)^2}{\sinh^2 m(z_0+h)} = \frac{a^2}{\sinh^2 mh}. \tag{8.30}$$

The particle path is therefore an ellipse with major and minor semi-axes of lengths $a\cosh m(z_0+h)/\sinh mh$ and $a\sinh m(z_0+h)/\sinh mh$ respectively while the distance between the foci is $2a\operatorname{cosech} mh$, which is the same for all particle trajectories. The quantity (mx_0-nt) is in effect the eccentric angle of the ellipse and it *decreases* at a constant rate with t as in an orbit described under a central force varying directly with distance. The fluid particle therefore describes its elliptic orbit *clockwise* when the wave is travelling along the direction of the positive x-axis from left to right. The particle paths described by equation (8.30) are sketched in Fig. 8.3. For particles which lie on the free surface $z = 0$, the major and minor semi-axes of the ellipse are $a\coth mh$ and a respectively while at the bottom $z = -h$, the ellipse degenerates into the straight-line segment $(x-x_0)\,a\operatorname{cosech} mh, = z = -h$.

When the fluid has infinite depth, a similar calculation may be carried out to find the particle paths using ϕ given by (8.26), but the equation of the trajectories can be deduced from (8.30) by letting $h \to \infty$. The ellipses become circles of radii $a\exp(mz_0)$, which rapidly decays to zero as z_0 decreases through negative values.

A fluid partile coinciding with the crest of a wave moves in the direction of the wave but when it is in the trough of the wave, it moves in the opposite direction to that of the wave. It will be noticed that the displacement of the

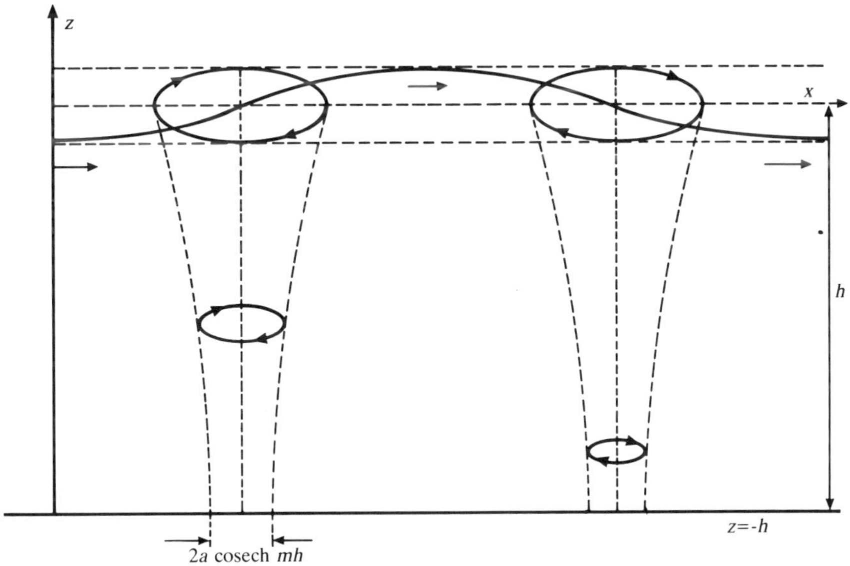

Fig. 8.3

particle from its mean position is generally very small, with maximum displacement occurring for particles in the free surface. This bears out our remarks in Section 8.1 that the main body of fluid is not transported by the wave motion. Furthermore, the rapid decay in the displacement with particle depth z_0 is consistent with the observation that submarines when submerged experience little effect from storms on the surface of the ocean.

8.5 WAVE ENERGY

Consider again a simple harmonic plane progressive wave on the surface of liquid of constant depth h. The surface elevation and velocity potential are given by

$$\eta = a \sin (mx - nt), \tag{8.31}$$

$$\phi = -\frac{ag}{n} \frac{\cosh m(z+h)}{\cosh mh} \cos (mx - nt). \tag{8.32}$$

We shall now calculate the potential and kinetic energies of the liquid in the region between two vertical planes parallel to the direction of the wave propagation and at unit distance apart. For a single wavelength, the potential energy is

$$V = \frac{1}{2}\rho g \int_0^\lambda \eta^2 \, dx = \frac{1}{4}\rho g a^2 \lambda, \tag{8.33}$$

since $\lambda = 2\pi/m$. The kinetic energy is given by

$$T = \frac{1}{2}\rho \int_0^\lambda \int_{-h}^\eta \left\{\left(\frac{\partial\phi}{\partial x}\right)^2 + \left(\frac{\partial\phi}{\partial z}\right)^2\right\} \mathrm{d}z\,\mathrm{d}x \tag{8.34}$$

$$= \frac{\rho a^2 g^2 m^2}{4n^2 \cosh^2 mh} \int_0^\lambda \int_{-h}^\eta [\cosh 2m(z+h) - \cos 2(mx - nt)]\,\mathrm{d}z\,\mathrm{d}x$$

$$= \frac{\lambda a^2 g^2 m \sinh 2m(\eta + h)}{8 n^2 \cosh^2 mh}. \tag{8.35}$$

On retaining only the leading, i.e. second order, term in small quantities, equation (8.35) gives, using Taylor's theorem and the dispersion relation (8.22):

$$T = \frac{1}{4}\rho g a^2 \lambda \frac{mg \tanh mh}{n^2} = \frac{1}{4}\rho g a^2 \lambda. \tag{8.36}$$

Hence the kinetic and potential energies of a simple harmonic progressive wave of small amplitude are equal to the order of a^2, and the total energy per wavelength is $\frac{1}{2}\rho g a^2 \lambda$ to this order of magnitude. Since this result together with (8.33) and (8.36) are independent of the depth h of the liquid, it is clear that the same formulae apply to liquid of infinite depth, which of course may be verified directly. The quantity $\frac{1}{2}\rho g a^2$ is known as the *energy density* of the wave.

As we pointed out in the introduction to this chapter, waves are an important mechanism for transmitting energy from one place to another. Although a progressive wave advances with the propagation speed c, it does not follow that this is the speed at which energy is transmitted by the wave. The rate of energy transmission is measured by determining the rate at which the fluid pressure on one side of a vertical section of the liquid, at right angles to the direction of wave propagation, is doing work on the liquid on the other side of the section.

Consider the rate at which energy is transmitted across unit width of this section. This will accordingly be the rate at which work is done by the pressure over the same unit width of section and is therefore

$$W = \int_{-h}^\eta pu\,\mathrm{d}z = \int_{-h}^\eta p \frac{\partial\phi}{\partial x}\,\mathrm{d}z. \tag{8.37}$$

From Bernoulli's equation, the leading order term in the pressure is

$$-\rho \frac{\partial\phi}{\partial t} - \rho g z.$$

On making this substitution and using (8.32), equation (8.37) gives

$$W = \frac{\rho g^2 a^2 m \sin^2 (mx - nt)}{n \cosh^2 mh} \int_{-h}^{\eta} \cosh^2 m(z+h)\, dz$$

$$+ \frac{\rho g^2 am \sin (mx - nt)}{n \cosh mh} \int_{-h}^{\eta} z \cosh m(z+h)\, dz. \tag{8.38}$$

The mean value of W over a complete period is

$$\overline{W} = \frac{n}{2\pi} \int_0^{2\pi/n} W\, dt$$

$$= \frac{\rho^2 g^2 a^2 m}{2n \cosh^2 mh} \int_{-h}^{0} \cosh^2 m(z+h)\, dz, \tag{8.39}$$

where the contribution to the integral from $z = 0$ to $z = \eta$ is ignored since it adds a term of order a^3 to $\overline{W}$. Thus

$$\overline{W} = \frac{1}{2} \rho g^2 a^2 \frac{m}{n \cosh^2 mh} \left[\frac{\sinh 2mh}{4m} + \frac{h}{2} \right],$$

and since $c = n/m$ and $n^2 = mg \tanh mh$, it follows that

$$\overline{W} = \tfrac{1}{4} \rho g a^2 (1 + 2mh \operatorname{cosech} 2mh) c. \tag{8.40}$$

The *velocity of energy transport* c^* is obtained by dividing $\overline{W}$ by the energy density, giving

$$c^* = \tfrac{1}{2} c (1 + 2mh \operatorname{cosech} 2mh). \tag{8.41}$$

For infinitely deep liquid the corresponding result is

$$c^* = \tfrac{1}{2} c, \tag{8.42}$$

showing in this case that energy is transmitted at half the wave propagation speed. In shallow liquid we see that $c^* \approx c$ and for general depths h,

$$\frac{1}{2} < \frac{c^*}{c} < 1.$$

To illustrate the immense amount of energy which is transmitted by waves on the ocean, consider a single wave with amplitude 1 m and wavelength 100 m. The energy flow rate per metre width of wave is given by (8.40) in the limit $h \to \infty$. This gives

$$\overline{W} \approx \frac{1}{4} \times 10^3 \times 10 \times \frac{10^2}{8} \approx 3.2 \times 10^4 \text{ Wm}^{-1}.$$

Thus the amount of energy transmitted every second by *this wave alone* to a coastline of 10^6 m, which is about the length of the British Isles, is

3.2×10^4 MW. The rewards from finding a device which can harness this energy are therefore very high.

8.6 GROUP VELOCITY

When waves are generated on the surface of a liquid, e.g. by dropping a stone in a pond or by the motion of a boat on a lake, the waves formed in general consist of a superposition of waves of various amplitudes and wavelengths. Since the propagation speed of a train of waves is an increasing function of the wavelength, the waves generated are subject to dispersion, and one might expect that with the passage of time the waves would be sorted into various *packets* or *groups* so that each group consists of waves having roughly the same wavelength. Let us examine this phenomenon with respect to two simple harmonic plane progressive waves of the same amplitude but slightly different wavelengths and frequencies. The surface elevation may be written as

$$\begin{aligned}\eta &= a\left\{\sin\,[mx - nt] + \sin\,[(m+\delta m)x - (n+\delta n)t]\right\} \\ &= 2a\cos\tfrac{1}{2}\,[\delta mx - \delta nt]\,\sin\tfrac{1}{2}\,[(2m+\delta m)x - (2n+\delta n)t] \qquad (8.43)\end{aligned}$$

Since δm and δn are assumed to be small, (8.43) represents a sinuous curve which may be obtained by multiplying the ordinates of the curve $\eta = 2a\sin\frac{1}{2}\,[(2m+\delta m)x - (2n+\delta n)t]$ by $\cos\frac{1}{2}\,[\delta mx - \delta nt]$. Hence the result is equivalent to a train of waves with variable amplitude A given by

$$A = 2a\,|\cos\tfrac{1}{2}\,[\delta mx - \delta nt]|. \qquad (8.44)$$

The amplitude A is periodic and varies from zero to $2a$. This wave train is illustrated in Fig. 8.4, where the solid curve is the actual waveform and the broken curves are $\eta = \pm\, 2a\cos\frac{1}{2}\,[\delta mx - \delta nt]$.

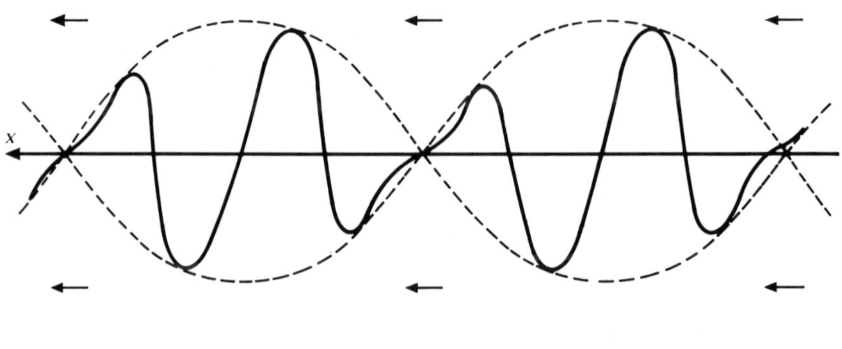

Fig. 8.4

The profile of the wave train is therefore a group of sinuous curves of amplitude gradually increasing from zero to a value not exceeding $2a$ and then decreasing to zero followed by a succession of similar groups of sinuous curves.

The distance 'between' two successive groups may be defined as the distance from a point of one group to the corresponding point of the next group. It follows that the distance between successive groups is $2\pi/\delta m$ and the time taken for the group to advance this distance is $2\pi/\delta n$. It is therefore natural to talk of a *group velocity* U which is the speed at which a group of waves advances. Thus

$$U = \frac{\delta n}{\delta m} \tag{8.45}$$

In reality, n is generally a continuous function of m so that U is then defined as the limiting value of the ratio $\delta n/\delta m$ as $\delta m \to 0$. Accordingly,

$$U = \frac{\mathrm{d}n}{\mathrm{d}m}, \tag{8.46}$$

and since $c = n/m$, equation (8.46) gives

$$U = c + m\,\frac{\mathrm{d}c}{\mathrm{d}m}, \tag{8.47}$$

or equivalently, since $\lambda = 2\pi/m$,

$$U = c - \lambda\,\frac{\mathrm{d}c}{\mathrm{d}\lambda}. \tag{8.48}$$

The group velocity is therefore in general not the same as the speed of propagation of the component waves in the group, and this agrees exactly with observation for in viewing a group of waves approaching a shore, single waves are seen to advance through the group and their amplitudes increase and then decrease as they are replaced by other waves in the group.

For the case of plane progressive waves on the surface of liquid of depth h, the dispersion relation is

$$c^2 = (g/m)\tanh mh \tag{8.49}$$

and therefore

$$U = \tfrac{1}{2}c\,(1 + 2mh \operatorname{cosech} 2mh).$$

Thus the group velocity has the same value as the velocity of energy transport. The result that energy is propagated at the group velocity may seem surprising since one velocity is determined from dynamical considerations while the other is of purely kinematic origin. However, the two velocities cease to have the same value when energy is dissipated, as happens when the viscosity of the liquid is taken into account.

In shallow liquids, the group velocity is close to the propagation speed while for deep liquids, the group velocity is then close to one-half of this speed.

8.7 THE EFFECT OF SURFACE TENSION

In the foregoing discussion of gravity waves the effect of surface tension was ignored. In a real fluid this additional stress at a free surface is always present and affects the dynamic but not the kinematic condition at the free surface when it is deformed from a flat shape, as pointed out in Chapter 3. If Π denotes atmospheric pressure outside the liquid, which as before we shall assume is constant, p is the fluid pressure, and γ is the surface tension or surface energy per unit area due to capillary forces, the dynamic boundary condition at any point P of the free surface S is

$$p \pm \gamma \left(\frac{1}{R_1} + \frac{1}{R_2} \right) = \Pi, \tag{8.50}$$

with the + or − sign applying accordingly as S is respectively concave or convex *downwards* at P and R_1, R_2 are the two principal radii of curvature of S at P.

In the case of two-dimensional waves, $R_2 = \infty$ and with $\eta(x, t)$ denoting as usual the small amplitude surface elevation of S,

$$\frac{1}{R_1} = \mp \frac{\partial^2 \eta}{\partial x^2}, \tag{8.51}$$

when squares of small quantities are neglected and the upper and lower signs apply when S is respectively concave or convex downwards at P. Equation (8.50) then gives

$$p - \Pi + \gamma \frac{\partial^2 \eta}{\partial x^2} = 0 \tag{8.52}$$

at any point of S.

Considering now the propagation of plane progressive waves of small amplitude on the free surface of liquid of constant depth h, the equations governing the fluid motion are again (8.8) to (8.15) except that (8.13) must be modified to conform with (8.52). The new dynamical free surface condition becomes

$$\frac{\gamma}{\rho} \frac{\partial^2 \eta}{\partial x^2} - \frac{\partial \phi}{\partial t} - g\eta = 0, \qquad (z = 0). \tag{8.53}$$

The velocity potential is again given by

$$\phi = -ac \frac{\cosh m(z + h)}{\sinh mh} \cos (mx - nt)$$

when the surface elevation is $\eta = a \sin (mx - nt)$ and satisfaction of (8.53) leads to the new dispersion relation. It may be easily verified that this is

$$c^2 = \left[\frac{\gamma m}{\rho} + \frac{g}{m} \right] \tanh mh, \tag{8.54}$$

or equivalently,

$$\frac{c^2}{gh} = \frac{\lambda}{2\pi h}\left[1 + \frac{\gamma}{\rho g}\left(\frac{2\pi}{\lambda}\right)^2\right] \tanh \frac{2\pi}{\lambda}. \tag{8.55}$$

As $\gamma \to 0$, the dispersion relation (8.22) is recovered.

For deep liquids, $2\pi h/\lambda \gg 1$ and (8.55) becomes

$$\frac{c^2}{gh} = \frac{\lambda}{2\pi h}\left[1 + \frac{\gamma}{\rho g}\left(\frac{2\pi}{\lambda}\right)^2\right],$$

and if in addition the effect of surface tension dominates over gravity so that

$$\frac{\gamma}{\rho g}\left(\frac{2\pi}{\lambda}\right)^2 \gg 1, \tag{8.56}$$

the propagation speed satisfies the equation

$$\frac{c^2}{gh} = \frac{2\pi\gamma}{\rho g \lambda h}. \tag{8.57}$$

Waves which satisfy these conditions and travel at the speed determined from (8.57) are known as *capillary waves*. Since the propagation speed given by (8.57) depends on the wavelength λ, capillary waves are dispersive. Fig. 8.5 shows the variation in c^2/gh with $\lambda/2\pi h$ given by (8.55). Comparison with Fig. 8.2 reveals that the effect of surface tension is significant only in the deep liquid range of $\lambda/2\pi h$ for a wave of given wavelength or for waves of short wavelength on liquid of given depth h. This is because condition (8.56) is realised only when λ is very small. Such waves are called *ripples*.

The wave propagation speed c has a minimum when $m = \sqrt{(\rho g/\gamma)}$ and its value c_0 is given by

$$c_0^2 = 2\sqrt{(\gamma g/\rho)}.$$

This minimum speed occurs for a wavelength

$$\lambda = \lambda_0 = 2\pi\sqrt{(\gamma/\rho g)}.$$

Thus

$$\frac{c^2}{c_0^2} = \frac{1}{2}\left(\frac{\lambda}{\lambda_0} + \frac{\lambda_0}{\lambda}\right),$$

and for each $c > c_0$, there are two possible values of the wavelength λ and these are reciprocals.

The group velocity is given by (8.48) as

$$U = c - \lambda\frac{dc}{d\lambda},$$

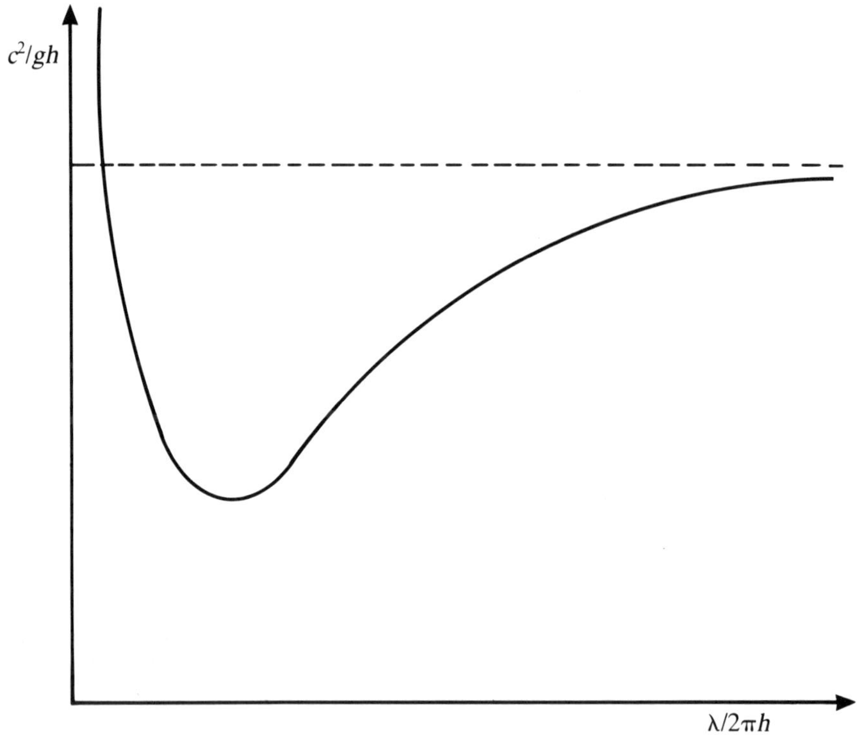

Fig. 8.5

Thus for wave motions when surface tension effects are taken into account,

$$U = c\left[1 - \frac{\lambda^2 - \lambda_0^2}{2(\lambda^2 + \lambda_0^2)}\right].$$

For ripples, when surface tension effects are dominant, $U \approx 3c/2$ and thus the group velocity exceeds the propagation speed c, while for wavelengths at which the effect of surface tension is insignificant, the deep liquid gravity wave limit of $U = \frac{1}{2}c$ is approached.

8.8 STANDING WAVES

As stated in Section 8.1, a standing wave is created by the superposition of two trains of progressive waves of the same amplitude, wavelength, period and phase but travelling in opposite directions. A typical plane simple harmonic standing wave profile is

$$\eta = a \sin mx \cos nt. \tag{8.58}$$

At a given value of t, the form of the surface is a sine curve of amplitude $a \cos nt$. The points at which $mx = k\pi$, $(k = 0, \pm 1, \ldots)$, are always fixed and are called *nodes*. When $\cos nt = \pm 1$, the surface has the form of the sine curve $\eta = \pm a \sin mx$ which represents the maximum departure of the free surface from its mean level. When $\cos nt = 0$, the free surface coincides with its mean level.

For such a disturbance of small amplitude at the free surface of liquid of constant depth h, the velocity potential ϕ has the form

$$\phi = -\frac{ac \cosh m(z+h)}{\sinh mh} \sin mx \sin nt. \tag{8.59}$$

We could alternatively derive (8.59) without reference to the progressive wave solution by starting with the assumption that ϕ has the form

$$\phi = f(z) \sin mx \sin nt. \tag{8.60}$$

Notice that dependence of ϕ on time is suggested by the kinematic free surface condition (8.15). In satisfying this condition together with (8.8) and (8.17) we arrive at the result (8.59).

For small amplitude standing waves of the type (8.59) on liquid of infinite depth, the corresponding expression for ϕ is

$$\phi = -ac\, \mathrm{e}^{mz} \sin mx \sin nt. \tag{8.61}$$

The particle paths are found in the same way as for progressive waves. Thus the coordinates $(x, 0, z)$ of a particle at a mean position $(x_0, 0, z_0)$ must satisfy

$$\frac{\mathrm{d}x}{\mathrm{d}t} = \frac{\partial\phi}{\partial x}, \qquad \frac{\mathrm{d}z}{\mathrm{d}t} = \frac{\partial\phi}{\partial z}. \tag{8.62}$$

With ϕ given by (8.59), we obtain

$$x - x_0 = a\frac{\cosh m(z_0+h)}{\sinh mh} \cos mx_0 \cos nt, \tag{8.63}$$

$$z - z_0 = a\frac{\sinh m(z_0+h)}{\sinh mh} \sin mx_0 \cos nt. \tag{8.64}$$

Thus, eliminating t from (8.63) and (8.64),

$$(z - z_0)/(x - x_0) = \tanh m(z_0 + h) \tan mx_0. \tag{8.65}$$

In contrast to the case of progressive waves, the particle trajectories for standing waves are straight lines whose slope depends on the location of the mean position of the fluid particle. In particular, for particles on a vertical through a peak or trough where $mx_0 = (k + \frac{1}{2})\pi$, with k any integer or zero, the trajectories are vertical while for particles on a vertical through a node where $mx_0 = k\pi$, the trajectories are horizontal, as illustrated in Fig. 8.6. The amplitude of particle displacement is maximum for particles in the free surface and minimum for particles on the bottom where $z_0 = -h$. For particles beneath a crest or a trough of the wave, the vertical displacement decreases in amplitude from a at the free surface to zero at the bottom, while for a particle beneath a

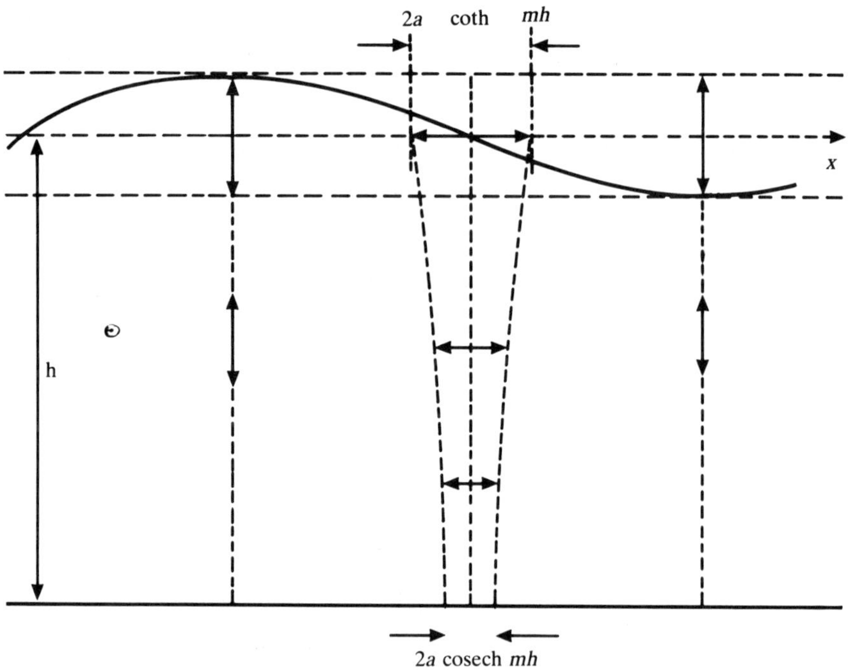

Fig. 8.6

node, the vertical displacement has amplitude a coth mh at the free surface decreasing to a cosech mh at the bottom. It is interesting to note that although the *surface* is at rest at a node, the fluid particle which has its mean position at a node is in motion.

Since we have assumed that the liquid is unbounded in the direction of the x-axis, this means that there is no restriction on the values that may be taken by the constant m. However, when the liquid is bounded in the x-direction such as in a canal with closed vertical sides at $x = \pm \frac{1}{2}l$, the values of m are then restricted to those consistent with the satisfaction of the additional kinematic conditions that $\partial\phi/\partial x = 0$ when $x = \pm \frac{1}{2}l$. With the form of ϕ given by (8.59), this requires that $ml = (2s + 1)\pi$ for any integer s and the possible wavelengths of the standing waves are then restricted to those for which $\lambda = 2l/(2s + 1)$. Physically, standing waves represent the principal or normal modes of free oscillation of the surface when the liquid is contained in a restricted region such as a vessel. We shall now consider such problems in more detail.

8.9 WAVES IN A CANAL

Consider a canal with vertical walls given by $x = 0, b$ which is filled with liquid to a depth h so that the bottom is given by $z = -h$, where b and h are constants. We shall assume that the canal is infinitely long in the direction of the y-axis. We have seen in earlier sections that both progressive and standing plane waves may be propagated along the length of such a canal since the side walls do not impose any restriction on wave propagation in that direction. We now consider the possibility of propagating two-dimensional waves *across* the canal. We shall ignore surface tension at the free surface since we have seen that its effect is significant only when the surface wave has very small wavelength.

The boundary value problem for the velocity potential $\phi = \phi(x, y, z, t)$ may therefore be stated as follows:

$$\frac{\partial^2 \phi}{\partial x^2} + \frac{\partial^2 \phi}{\partial z^2} = 0 \tag{8.66}$$

for $0 \leqslant x \leqslant b, -\infty < y < \infty, 0 \geqslant z \geqslant -h$, together with

$$\frac{\partial \phi}{\partial z} = 0, \qquad (z = -h), \tag{8.67}$$

$$\frac{\partial \phi}{\partial x} = 0, \qquad (x = 0, b), \tag{8.68}$$

$$\frac{\partial^2 \phi}{\partial t^2} + g\frac{\partial \phi}{\partial z} = 0, \qquad (z = 0), \tag{8.69}$$

when a surface wave of small amplitude a is propagated on the free surface. The presence of the walls $x = 0, b$ eliminates the possibility of a progressive wave solution. Thus a standing wave type of solution will be sought and there is no loss of generality in assuming that ϕ is of the form

$$\phi = F(x, z) \cos nt, \tag{8.70}$$

since any phase constant may be eliminated by shifting the time origin. On seeking a solution for $F(x, z) = X(x). Z(z)$ in separated variables this leads to the equations

$$\frac{X''(x)}{X(x)} = -\frac{Z''(z)}{Z(z)} = \text{constant}.$$

Obviously the solution $X(x)$ will involve circular or hyperbolic functions of x according as the constant is negative or positive and vice versa for $Z(z)$. The boundary conditions (8.68) on the canal walls lead us to set the constant $= -m^2$, with m real and non-zero for a periodic solution in x. Thus the solution for ϕ is of the form

$$\phi = (A_1 \sin mx + A_2 \cos mx)(B_1 \sinh mz + B_2 \cosh mz) \cos nt,$$

where A_1, A_2, B_1, B_2 are constants. Equation (8.67) implies that

$$B_1 \cosh mh = B_2 \sinh mh,$$

and therefore ϕ may be rewritten as

$$\phi = (A_1 \sin mx + A_2 \cos mx) \cosh m(z+h) \cos nt,$$

with the constants A_1 and A_2 redefined. The conditions (8.68) on the walls of the canal imply that for a non-trivial solution,

$$A_1 = 0, \quad \sin mb = 0.$$

Thus $mb = k\pi$ with $k = \pm 1, \pm 2, \ldots$. There remains the dynamic condition at the free surface (8.69) to be satisfied. This leads to

$$n^2 \cosh mh = mg \sinh mh,$$

or equivalently, since $c = n/m$ and $\lambda = 2\pi/m$, the dispersion relation

$$\frac{c^2}{gh} = \frac{\lambda}{2\pi h} \tanh \frac{2\pi h}{\lambda}. \tag{8.71}$$

This has the same form as the dispersion relation (8.22) for plane progressive waves, but in this problem the permissible wavelengths λ are restricted to those given by $\lambda = \lambda_k$, where $\lambda_k = 2\pi/k$.

The most general form of solution for ϕ satisfying all the equations and conditions of the problem is

$$\phi = \sum_{k=1}^{\infty} E_k \cos \frac{k\pi x}{b} \cosh \frac{k\pi\,(z+h)}{b} \cos \frac{c_k \pi t}{b}, \tag{8.72}$$

where

$$c_k^2 = \frac{gb}{\pi} \tanh\left(\frac{k\pi h}{b}\right).$$

The coefficients E_k are undetermined but may be found if the initial shape of the free surface is prescribed using the methods of Fourier analysis. Each term of the series in (8.72) is the velocity potential corresponding to a standing wave disturbance of the free surface which in turn represents one of the normal modes of oscillation of the liquid in the canal. The fundamental mode is that for which $k = 1$. The wavelength of the standing wave motion is then $2b$ and is the longest wave that can be propagated across the width of the canal. Notice that all normal modes result in a sloshing motion of the liquid so that it moves up and down the sides of the canal.

8.10 WAVES IN A RECTANGULAR TANK

If instead of a canal, the liquid is contained in a rectangular tank so that $0 \leqslant x \leqslant b$ and $0 \leqslant y \leqslant d$, the fluid motion will not now in general be of the plane wave type as the velocity potential must satisfy the three-dimensional Laplace equation

$$\frac{\partial^2 \phi}{\partial x^2} + \frac{\partial^2 \phi}{\partial y^2} + \frac{\partial^2 \phi}{\partial z^2} = 0. \tag{8.73}$$

We now look for a solution in separated variables of the form

$$\phi(x, y, z, t) = X(x) \,.\, Y(y) \,.\, Z(z) \cos nt \tag{8.74}$$

and find that to satisfy (8.67), (8.68) and the added kinetic conditions on the walls $y = 0, d$, namely

$$\frac{\partial \phi}{\partial y} = 0, \qquad (y = 0, d), \tag{8.75}$$

It is necessary that

$$X(x) \propto \cos \frac{k\pi x}{b},$$

$$Y(y) \propto \cos \frac{l\pi y}{d},$$

$$Z(z) \propto \cosh \mu_{k,l} (z + h),$$

where

$$\mu_{k,l}^2 = (k\pi/b)^2 + (l\pi/d)^2,$$

and $k, l = 0, \pm 1, \pm 2, \ldots$. Satisfaction of the dynamic condition at the free surface leads to the dispersion relation $n = n_{k,l}$, where

$$n_{k,l}^2 = g\, \mu_{k,l} \tanh \mu_{k,l} h. \tag{8.76}$$

Thus the most general solution for ϕ to this problem takes the form

$$\phi = \sum_{k=1}^{\infty} \sum_{l=1}^{\infty} E_{k,l} \cos \frac{k\pi x}{b} \cos \frac{l\pi y}{d} \cosh \mu_{k,l} (z + h) \cos n_{k,l} t. \tag{8.77}$$

As for waves on a canal, the coefficients $E_{k,l}$ may be determined, using the methods of Fourier analysis, if the initial shape of the free surface is prescribed. The general term of the double series (8.77) represents the velocity potential when the free surface profile is a two-dimensional standing wave of the form

$$\eta = a \cos \frac{k\pi x}{b} \cos \frac{l\pi y}{d} \sin n_{k,l}\, t. \tag{8.78}$$

This wave satisfies the two-dimensional wave equation

$$\frac{\partial^2 \eta}{\partial x^2} + \frac{\partial^2 \eta}{\partial y^2} = \frac{1}{c^2} \frac{\partial^2 \eta}{\partial t^2},$$

where

$$c = \frac{n_{k,l}}{\pi}\left(\frac{k^2}{h^2} + \frac{l^2}{d^2}\right)^{-1/2}.$$

When either k or l, but not both, are zero, the wave then becomes a plane wave of the type studied in earlier sections.

8.11 WAVES IN A CYLINDRICAL TANK

An analysis similar to that presented in the previous section may be applied to the propagation of waves at the free surface of liquids contained in tanks of cross-section other than rectangular. We shall consider a tank which is a circular cylinder of radius b and liquid filling the tank to depth h. Using cylindrical polar coordinates (r, θ, z) with axis $r = 0$ along the axis of the tank and $z = 0$ in the undisturbed free surface of the liquid, the equations which must be satisfied by the velocity potential $\phi = \phi(r, \theta, z, t)$ are

$$\frac{1}{r}\frac{\partial}{\partial r}\left(r\frac{\partial \phi}{\partial r}\right) + \frac{1}{r^2}\frac{\partial^2 \phi}{\partial \theta^2} + \frac{\partial^2 \phi}{\partial z^2} = 0 \tag{8.79}$$

throughout the liquid $0 \leqslant r \leqslant b$, $0 \leqslant \theta \leqslant 2\pi$, $0 \geqslant z \geqslant -h$ together with the boundary conditions

$$\frac{\partial \phi}{\partial z} = 0, \qquad (z = -h), \tag{8.80}$$

$$\frac{\partial \phi}{\partial r} = 0, \qquad (r = b) \tag{8.81}$$

$$\frac{\partial^2 \phi}{\partial t^2} + g\frac{\partial \phi}{\partial z} = 0, \qquad (z = 0), \tag{8.82}$$

where in the dynamic free surface condition we have ignored the effect of surface tension. The solutions of the boundary value problem posed by equations (8.79) to (8.82) inclusive provide the possible waveforms which may exist on the free surface of the liquid contained in the cylinder, subject to the simplifications of the linearized small amplitude wave theory.

The solution of Laplace's equation in cylindrical polar coordinates may be sought by the method of separation of variables. A standing wave is the only physically possible simple harmonic wave within the cylinder. Thus we look for a solution for ϕ of the form

$$\phi = R(r)\,.\,T(\theta)\,.\,Z(z)\sin nt. \tag{8.83}$$

Substitution of ϕ of this form into (8.79) leads to

$$\frac{r}{R}\frac{\mathrm{d}}{\mathrm{d}r}\left(r\frac{\mathrm{d}R}{\mathrm{d}r}\right) + \frac{1}{T}\frac{\mathrm{d}^2 T}{\mathrm{d}\theta^2} + \frac{r^2}{Z}\frac{\mathrm{d}^2 Z}{\mathrm{d}z^2} = 0. \tag{8.84}$$

The second term contains the only θ-dependent term and it must be a constant. We choose this constant to be $-m^2$, in which case

$$T(\theta) = B_1 \sin m\theta + B_2 \cos m\theta$$

and for single valued solutions, $T(\theta + 2\pi) = T(\theta)$. Thus m must be an integer or zero. Equation (8.84) is thus equivalent to

$$\frac{1}{rR}\frac{\mathrm{d}}{\mathrm{d}r}\left(r\frac{\mathrm{d}R}{\mathrm{d}r}\right) - \frac{m^2}{r^2} + \frac{1}{Z}\frac{\mathrm{d}^2 Z}{\mathrm{d}z^2} = 0, \tag{8.85}$$

and we now see that only the third term depends on z. Thus

$$\frac{1}{Z}\frac{\mathrm{d}^2 Z}{\mathrm{d}z^2} = \text{constant} = k^2,$$

giving

$$Z(z) = C_1 \sinh kz + C_2 \cosh kz.$$

Equation (8.85) can now be written as

$$r^2\frac{\mathrm{d}^2 R}{\mathrm{d}r^2} + r\frac{\mathrm{d}R}{\mathrm{d}r} + (k^2 r^2 - m^2)R = 0, \tag{8.86}$$

which is Bessel's equation of order m when the independent variable is taken to be kr. Thus

$$R(r) = A_1 J_m(kr) + A_2 Y_m(kr),$$

with J_m and Y_m respectively the Bessel functions of the first and second kind of order m. The first two of each of these functions for $m = 0, 1$ are sketched in Fig. 8.7.

Reference to any standard text on Bessel functions reveals that the functions $Y_m(x)$ diverge to $-\infty$ at $x = 0$ for all values of m. Thus the radial dependence of the velocity potential can only be proportional to $J_m(kr)$. The boundary condition (8.80) shows that

$$C_1 \cosh kh = C_2 \sinh kh,$$

so that ϕ has the form

$$\phi = (B_1 \sin m\theta + B_2 \cos m\theta)\cosh k(z + h) J_m(kr) \sin nt. \tag{8.87}$$

The free surface condition (8.82) gives the dispersion relation

$$n^2 = gk \tanh kh. \tag{8.88}$$

There remains the boundary condition (8.81) at the wall of the cylinder. For a non-trivial solution, this requires that

$$J'_m(kb) = 0, \tag{8.89}$$

where the prime denotes differentiation with respect to kb. This equation has an

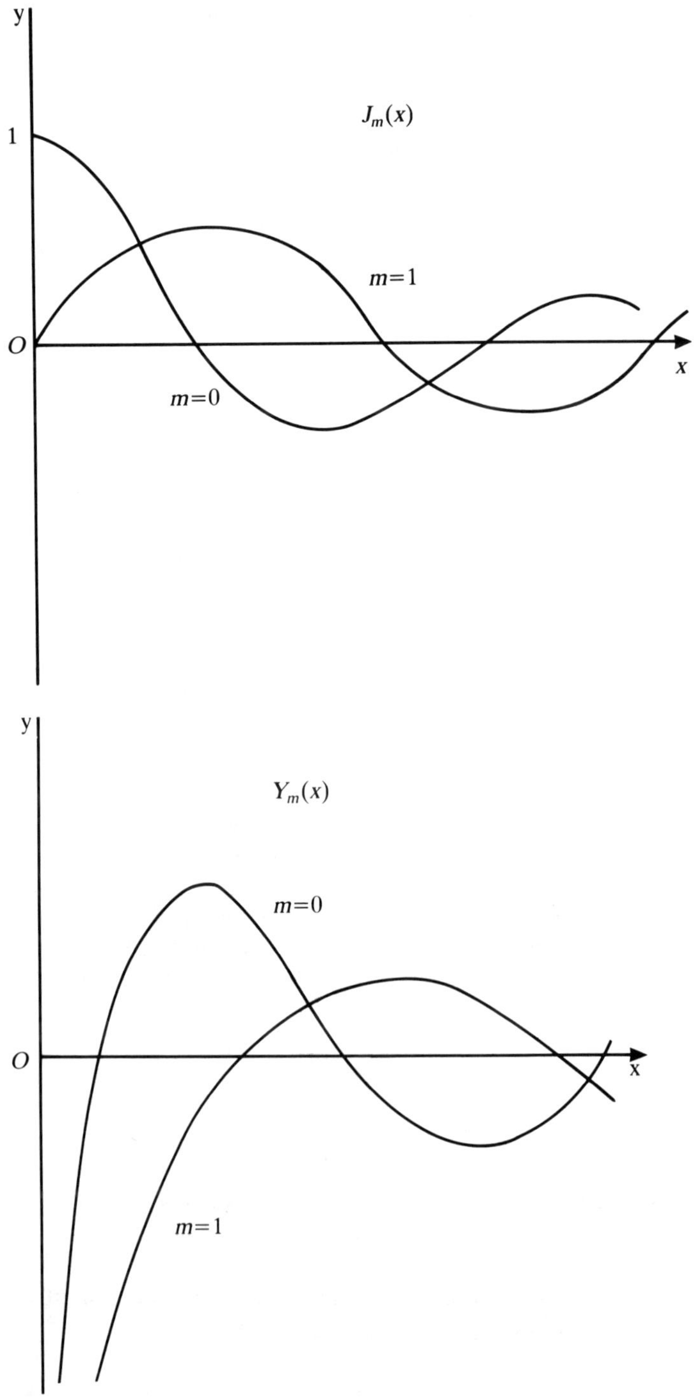

Fig. 8.7

infinity of solutions whose properties are discussed in texts on Bessel functions. The solutions may be denoted by $k_{m,l}\,b$, the first few of which have the values shown below.

Values of $k_{m,l}b$

	l		
m	0	1	2
0	3.832	7.016	17.74
1	1.841	5.331	8.536
2	3.054	6.706	9.969

Thus the most general solution for ϕ of the form (8.83) is given by

$$\phi = \sum_{m=0}^{\infty} \sum_{l=0}^{\infty} [A_{m,l} \sin m\theta + B_{m,l} \cos m\theta] \cosh k_{m,l}(z+h)\, J_m(k_{m,l}r) \sin n_{m,l}t, \tag{8.90}$$

with $n_{m,l}$ given by (8.88) when $k = k_{m,l}$. Each term of (8.90) represents a solution to the boundary value problem corresponding to a standing wave disturbance of the free surface of the form

$$\eta = [a_{m,l} \sin m\theta + b_{m,l} \cos m\theta]\, J_m(k_{m,l}\, r) \cos n_{m,l}\, t.$$

The constants $A_{m,l}$, $B_{m,l}$, and therefore $a_{m,l}$, $b_{m,l}$, are determined from knowledge of the original shape of the free surface, using the condition

$$\frac{\partial \phi}{\partial t} + g\,\eta = 0, \qquad (z=0). \tag{8.91}$$

A simple illustration of this theory is obtained by observing the surface of a cup of tea or some other liquid if the cup is jarred by striking its base symmetrically on to a table. The disturbance of the surface is axisymmetric and therefore purely radial. What is essentially observed is the fundamental radial mode given by $m = l = 0$, so the velocity potential and surface elevation are proportional to $J_0(k_{0,0}r)$. The shape of the surface will be similar to that of the J_0 Bessel function sketched in Fig. 8.7, with the peak on the axis of symmetry quite pronounced. If the cup is struck on its side, the disturbance is not then axisymmetric and the non-axisymmetric modes associated with $m \geqslant 1$ are brought into evidence.

8.12 WAVES ON THE SURFACE OF A UNIFORM STREAM

In Section 8.3 we showed that when a simple harmonic plane progressive wave with surface elevation

$$\eta = a \sin (mx - nt)$$

is propagated on the free surface of a liquid of constant depth h which is unbounded in the x-direction, the velocity potential for the fluid motion has the form

$$\phi = -\frac{ac \cosh m(z + h) \cos (mx - nt)}{\sinh mh} . \tag{8.92}$$

For this representation, the axes of reference are fixed in space and the wave travels in the positive direction of the x-axis with speed c. If, however, we choose to refer to axes which are fixed *relative to the wave*, this is equivalent to imposing on the whole system of wave, liquid and horizontal bottom a uniform velocity $-c\mathbf{i}$. In such a frame of reference, the velocity of the fluid and consequently the velocity potential ϕ are independent of time and functions only of the coordinates x and z in the moving frame. The form of ϕ is accordingly

$$\phi = -cx + [A' \cosh mz + B' \sinh mz] \cos mx, \tag{8.93}$$

where A' and B' are constants. The boundary condition at the bottom $z = -h$ of zero normal component of velocity requires that

$$\frac{\partial \phi}{\partial z} = 0, \qquad (z = -h). \tag{8.94}$$

Thus, with ϕ given by (8.93),

$$A' \cosh mh - B' \sinh mh = 0,$$

implying that

$$A' = A \sinh mh, \quad B' = A \cosh mh \tag{8.95}$$

for some constant A. Equation (8.93) then gives

$$\phi = -cx + A \cosh m(z + h) \cos mx \tag{8.96}$$

as the expression for the form taken by the velocity potential in terms of coordinates x, z in a frame which moves with the velocity of the wave. A further advantage of working in this frame of reference is that the free surface elevation is now given by

$$\eta = a \sin mx, \tag{8.97}$$

and is therefore fixed relative to the moving frame. The free surface is consequently a streamline in this frame. Since the flow is two-dimensional, the stream function ψ and the velocity potential ϕ are related through the Cauchy–Riemann equations (see Section 6.2):

$$\frac{\partial\phi}{\partial x} = \frac{\partial\psi}{\partial y}, \qquad \frac{\partial\phi}{\partial y} = -\frac{\partial\psi}{\partial x}, \tag{8.98}$$

and with ϕ given by (8.96), equations (8.98) given on integration

$$\psi = -cz - A \sinh m(z+h) \sin mx \tag{8.99}$$

as the corresponding expression for the stream function. The bottom is the streamline $\psi = ch$ and the free surface $z = a \sin mx$ is the streamline $\psi = 0$ provided that

$$ca + A \sinh mh = 0. \tag{8.100}$$

It therefore follows that

$$\phi = -cx - \frac{ac \cosh m(z+h) \cos mx}{\sinh mh}, \tag{8.101}$$

$$\psi = -cz + \frac{ac \sinh m(z+h) \sin mx}{\sinh mh}. \tag{8.102}$$

If the liquid is also moving with a uniform velocity $V\mathbf{i}$ when the surface is undisturbed, the expressions for ϕ and ψ are then given by (8.101) and (8.102) with c replaced by $c - V$. We therefore have accordingly

$$\phi = (V-c)\left[x + \frac{a \cosh m(z+h) \cos mx}{\sinh mh}\right], \tag{8.103}$$

$$\psi = (V-c)\left[z - \frac{a \sinh m(z+h) \sin mx}{\sinh mh}\right]. \tag{8.104}$$

The formula for the pressure may be found from Bernoulli's equation:

$$\frac{p}{\rho} + gz + \frac{1}{2}\mathbf{q}^2 = \text{constant},$$

where

$$\mathbf{q}^2 = \left(\frac{\partial\phi}{\partial x}\right)^2 + \left(\frac{\partial\phi}{\partial z}\right)^2 = \left(\frac{\partial\psi}{\partial z}\right)^2 + \left(\frac{\partial\psi}{\partial x}\right)^2. \tag{8.105}$$

On using (8.103) or (8.104) we find that at the free surface

$$\frac{p}{\rho} + ga \sin mx + \frac{1}{2}(V-c)^2\,[1 - 2ma \coth mh \sin mx] = \text{constant}, \tag{8.106}$$

when powers in a higher that the first are neglected.

8.13 STEADY FLOW OVER A SINUOUS BOTTOM

A problem analogous to that discussed in the previous section is posed when a uniform stream flows along a canal or river with an uneven bed. For simplicity we shall assume that the bed of the stream is corrugated and given by

$$z = -h + b \sin mx \tag{8.107}$$

where $b \ll h$, and let the undisturbed uniform stream velocity be $V\mathbf{i}$. The velocity potential ϕ satisfies the two-dimensional Laplace equation

$$\frac{\partial^2 \phi}{\partial x^2} + \frac{\partial^2 \phi}{\partial z^2} = 0, \tag{8.108}$$

and of course in this problem, it is independent of the time t. We therefore seek a solution of (8.108) of the form

$$\phi = Vx + (A \cosh mz + B \sinh mz) \cos mx. \tag{8.109}$$

The corresponding form for the stream function ψ from (8.99) is therefore

$$\psi = Vz - (A \sinh mz + B \cosh mz) \sin mx. \tag{8.110}$$

The boundary conditions of the problem require that the bed is a streamline and that the free surface is a surface of constant pressure as well as being a streamline Neglecting second order terms in b/h, the first of these conditions requires that

$$V(-h + b \sin mx) - (-A \sinh mh + B \cosh mh) \sin mx$$

be a constant for all values of x. Consequently

$$Vb = B \cosh mh - A \sinh mh. \tag{8.111}$$

The free surface is the streamline $\psi = 0$ if the elevation is $\eta = a \sin mx$, say, provided that $Va = B$ and second order terms in a/h are negligible.

For steady motion, the pressure is given by Bernoulli's equation:

$$\frac{p}{\rho} + \frac{1}{2}\mathbf{q}^2 + gz = \text{constant}, \tag{8.112}$$

and since p must be constant at the free surface, the linearized form of (8.104) implies that

$$ag \sin mx - VAm \sin mx = \text{constant}$$

for all values of x, giving $A = ag/mV$. The free surface elevation is therefore given by

$$\eta = \frac{b \sin mx}{\cosh mh - (g/mV^2) \sinh mh}. \tag{8.113}$$

Taking b to be positive, (8.113) shows that the coefficient of $\sin mx$ is positive or negative according as V^2 is greater or less than $(g/m) \tanh mh$, which is the speed of propagation of waves on the surface of liquid at rest of constant depth h with the same wavelength as the corrugations. When $V^2 > (g/m) \tanh mh$, the crests and troughs of the free surface are vertically above the ridges and hollows of the bed of the stream and when $V^2 < (g/m) \tanh mh$, the crests of the surface wave are above the hollows of the bed. It should, however, be noticed that when V^2 is close to $(g/m) \tanh mh$, the amplitude of the surface elevation becomes large and the assumption on which the derivation of (8.113) relies is no longer valid.

8.14 WAVES AT AN INTERFACE BETWEEN FLUIDS

Consider a fluid of density ρ' and depth h' flowing with constant velocity $V'\mathbf{i}$ over a layer of fluid of density ρ and depth h flowing with constant velocity $V\mathbf{i}$, the fluids being bounded above and below by fixed horizontal planes. Taking the plane $z = 0$ as the undisturbed interface between the two fluids, which are assumed to be immiscible, we shall investigate the propagation of waves of small elevation along the interface of the form

$$\eta = a \sin (mx - nt),$$

where $a \ll h, a \ll h'$. The wave propagation speed $c = n/m$.

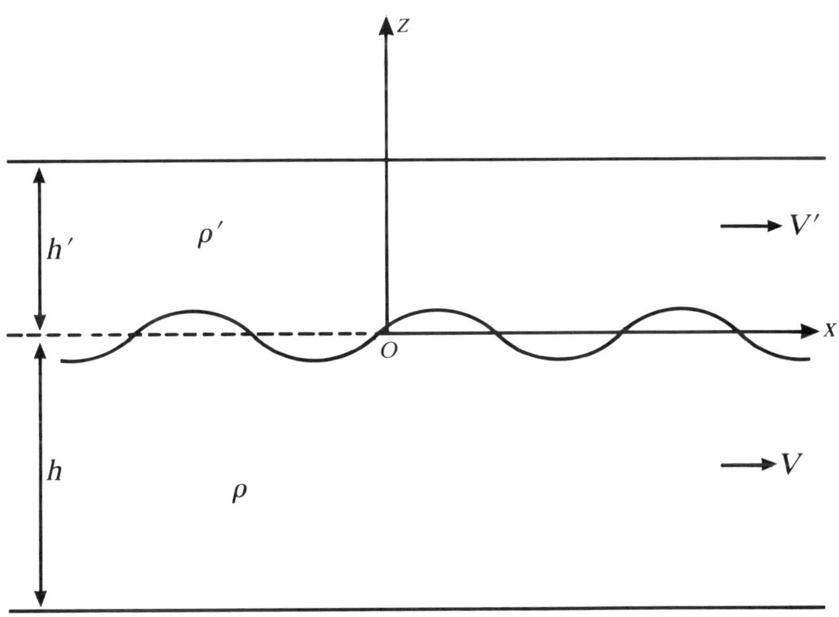

Fig. 8.8

If we impose on both fluids a velocity $-c\mathbf{i}$, this has the effect of reducing the wave profile to rest and changing the velocities of the streams to $(V-c)\mathbf{i}$ and $(V'-c)\mathbf{i}$ respectively. From (8.104) the velocity potential ϕ for the lower fluid is evidently

$$\phi = (V-c)\left[x + \frac{a \cosh m(z+h) \cos mx}{\sinh mh}\right], \qquad (\eta \geqslant z \geqslant -h), \tag{8.114}$$

and the velocity potential ϕ' for the upper fluid is obtained from (8.114) by replacing h by $-h'$ and V by V'. Thus

$$\phi' = (V'-c)\left[x - \frac{a \cosh m(z-h') \cos mx}{\sinh mh'}\right], \quad (\eta \leqslant z \leqslant h'), \tag{8.115}$$

Neglecting terms involving squares of a, the speed in the lower fluid is given by

$$\mathbf{q}^2 = (V-c)^2 - \frac{2ma\,(V-c)^2}{\sinh mh} \cosh m(z+h) \sin mx, \tag{8.116}$$

and at the interface, (8.116) gives

$$\mathbf{q}^2 = q_0^2 = (V-c)^2\,[1 - 2ma \coth mh \sin mx]. \tag{8.117}$$

Similarly, in the upper fluid, the speed at the interface is given by

$$(\mathbf{q}')^2 = (q_0')^2 = (V'-c)^2\,[1 + 2ma \coth mh' \sin mx]. \tag{8.118}$$

The dynamic condition at the interface requires that the pressures p and p' in the lower the upper fields respectively are the same. But from Bernoulli's equation applied to each fluid,

$$p + \tfrac{1}{2}\rho q_0^2 + \rho g\eta = \text{constant},$$

and

$$p' + \tfrac{1}{2}\rho'(q_0')^2 + \rho' g\eta = \text{constant}.$$

Thus for $p = p'$ at $z = \eta = a \sin mx$,

$$\rho'(q_0')^2 - \rho q_0^2 + g\eta\,(\rho' - \rho) = \text{constant}. \tag{8.119}$$

It therefore follows that after substituting from (8.117) and (8.118), the coefficient of $\sin mx$ must vanish. This leads to the equation

$$m\rho\,(V-c)^2 \coth mh + m\rho'(V'-c)^2 \coth mh' = g(\rho - \rho'). \tag{8.120}$$

Equation (8.120) is the dispersion relation for the propagation of plane waves at the interface. For the propagation of waves to be possible, the solution of

(8.120) for c must be real. A complex value of c would imply an exponentially growing dependence on t in the surface elevation. Physically this means that if the interface were slightly disturbed, then the disturbance would grow showing that the configuration of the two streams is *unstable*. This problem therefore provides us with a prototype of simple form which is an introduction to the branch of fluid mechanics known as *hydrodynamic stability*.

If the fluids have infinite depth, equation (8.120) simplifies considerably to give

$$m\rho\,(V-c)^2 + m\rho'\,(V'-c)^2 = g(\rho-\rho'), \tag{8.121}$$

which may be solved for c giving

$$c = \frac{\rho' V' + \rho V}{\rho + \rho'} \pm \left[\frac{(\rho-\rho')}{(\rho+\rho')}\frac{g\lambda}{2\pi} - \frac{\rho\rho'}{(\rho+\rho')^2}\,(V'-V)^2\right]^{1/2}. \tag{8.122}$$

Equation (8.122) shows that c may be real or complex depending on the sizes of the various parameters in the problem.

Consider first the special case when $V = V' = 0$ and $\rho' = 0$. This corresponds to two fluids at rest, apart from the wave motion, and such that the density of the upper fluid is very small in comparison with that of the lower fluid. An example would be still air over still water. Equation (8.122) then reduces to

$$c = \pm\sqrt{\left(\frac{g\lambda}{2\pi}\right)}$$

which agrees with (8.27) giving the speeds of propagation of waves on the surface of a deep liquid. The significance of the $-$ sign in one of the solutions just indicates that a wave could be propagated in the negative x-direction. Since the two solutions are real, this means that propagation of small amplitude waves is possible with either of the speeds, showing that a small disturbance of the interface will not grow exponentially with time and therefore the configuration of a very light fluid at rest on top of a much denser fluid also at rest is a *stable* configuration, which is intuitively obvious.

Consider now the case when $\rho = \rho'$ and $V \neq V'$, so that there is then a discontinuity in velocity. Equation (8.122) now gives

$$c = \tfrac{1}{2}\,[(V+V') \pm i(V-V')],$$

showing that unless $V = V'$, the speeds c have imaginary parts which will result in interfacial waves which grow exponentially in time. The configuration of two fluids of the same density having different velocities is therefore *unstable*. This form of instability is known as *Helmholtz* or *Rayleigh instability*. To see what happens when $V' = V$, we first write $V' = V(1+\epsilon)$ and ultimately let $\epsilon \to 0$. Now

$$n = mc = \tfrac{1}{2} mV(2 + \epsilon \pm i\epsilon).$$

Hence

$$\eta = a \sin m[x - Vt - \tfrac{1}{2}\epsilon(1 \mp i) Vt],$$

which for small ϵ is approximately given by

$$\eta = a \sin m(x - Vt) - \tfrac{1}{2} am\epsilon(1 \pm i) Vt \cos m(x - Vt),$$

showing that $|\eta|$ increases indefinitely with t. This special case explains the flapping of sails and flags in a wind. The air may be regarded as divided by a thin membrane in the form of the flag on both sides of which the air moves with the same velocity. The motion is unstable and a slight disturbance of the flag grows with time producing the large scale flapping movement of the flag.

If both fluids are at rest so that $V = V' = 0$ but their densities differ, then (8.122) gives

$$c = \pm \sqrt{\left[\frac{g\lambda}{2\pi}\left(\frac{\rho - \rho'}{\rho + \rho'}\right)\right]},$$

showing that when the heavier fluid is at the bottom, c is real so the configuration is stable, but when the heavier fluid is on top, the configuration is then unstable. This is an example of *Taylor instability*.

An analysis of the values of c given by the general equation (8.120) may also be carried out, but since coth mh and coth mh' are often very close to unity the above infinite depth analysis deals qualitatively with situations which may occur when the fluids have finite depths. A case we might consider in more detail is the propagation of waves on water exposed to the air. We may safely set $h' = \infty$ and, if σ is the specific gravity of air, we find from (8.120), with $V = V' = 0$, that

$$c^2 = \frac{g\lambda}{2\pi}\frac{(1 - \sigma)}{(\coth mh + \sigma)} = \frac{g\lambda}{2\pi}\tanh mh\,[1 - \sigma(1 + \tanh mh)]$$

approximately, since $\sigma \approx 1.3 \times 10^{-3}$ is small. Comparing this result with (8.22) shows that the atmosphere tends to diminish the wave velocity.

Another interesting variant of this problem occurs if we let the upper surface of the upper fluid be a free surface, e.g. a layer of oil on the surface of water. If the form of the free surface is a wave travelling with the same speed as the interfacial wave, the problem can be reduced to one of steady motion by the addition of a velocity $-c\mathbf{i}$ to the system of fluids as was done above. Taking $V = V' = 0$, the problem to solve is a combination of a gravity wave problem for fluid of depth h together with a stream flowing over a sinuous bottom discussed in Section 8.13, taking the free surface of the upper fluid to be

$$z = h' + b \sin mx.$$

where $b \ll h'$. We shall leave as an exercise to the reader the formulation of this problem. The dispersion relation is found to be

$$c^4 m^2\,(\rho \coth mh \coth mh' + \rho')$$
$$- c^2 mg\rho\,(\coth mh + \coth mh') + g^2\,(\rho - \rho') = 0, \qquad (8.123)$$

and the ratio of the amplitudes of the waves is

$$\frac{b}{a} = \frac{mc^2}{mc^2 \cosh mh' - g \sinh mh'}\,. \qquad (8.124)$$

From (8.123) we see that there are two possible velocities of propagation for a given wavelength provided $\rho > \rho'$. In the particular case when the lower fluid has infinite depth we may set $\coth mh = 1$ and the roots of (8.123) are then

$$c^2 = \frac{g}{m}, \qquad c^2 = \frac{g}{m}\,\frac{\rho - \rho'}{\rho \coth mh' + \rho'}\,. \qquad (8.125)$$

The first of these roots gives the propagation speed for gravity waves on the surface of a single deep liquid and is thus independent of the relative densities of the liquids. This wave propagates as if the upper fluid were absent. On writing $\theta = mh'$, $\sigma = \rho'/\rho$ and $\tau = g(1-\sigma)h'/c^2$, the second of equations (8.125) gives

$$f(\theta) = \sigma + \coth\theta - \tau/\theta = \sigma + (\coth\theta - 1/\theta) - (\tau - 1)/\theta = 0.$$

But $\coth\theta - 1/\theta$ is positive if $\theta > 0$ and $(\tau - 1)/\theta < 0$ if $\tau < 1$. Hence $f(\theta) > 0$ if $\tau < 1$ and there are then no positive roots of the equation $f(\theta) = 0$. But if $\tau > 1$, then $f(0) = -\infty$ and $f(\infty) = (1 + \tau > 0$. Furthermore, $f'(\theta) > 0$ when $\tau > 1$. Thus there is one and only one positive root of the equation $f(\theta) = 0$ if $\tau > 1$. It therefore follows that only one type of wave can be propagated if

$$c^2 > g(1-\sigma)h',$$

and its wavelength $\lambda = 2\pi c^2/g$, but for

$$c^2 < g(1-\sigma)h',$$

a second type of wave exists for which the ratio of the elevation at the interface to that at the free surface is given by (8.124) with c^2 taking the second of the values given by (8.125). Thus

$$\frac{a}{b} = \frac{\sigma}{1-\sigma}\,e^{mh'},$$

so that if σ is close to unity, the theory suggests that the elevation of the waves at the interface can be very large compared with the surface elevation. This result is particularly important in fiords where there is a layer of fresh water on the denser salt water. It is observed that ships sometimes experience unusually large drags in these waters and the enhanced value of the resistance may be attributed to the generation of waves of large amplitude at the interface between the fresh and salt water if relatively poor mixing of the waters has taken place.

8.15 EFFECT OF SURFACE TENSION ON WAVES AT AN INTERFACE

We shall return to the configuration of the last section in which two immiscible fluids of densities ρ and ρ' have a common interface $z = 0$ and are bounded by the fixed planes $z = -h$ and $z = h'$. The upper and lower fluids have velocities $V'\mathbf{i}$ and $V\mathbf{i}$ respectively. The analysis of the preceding section leading to equations (8.117) and (8.118) clearly applies whether or not surface tension acts at the interface. The effect of the surface tension enters the dynamic condition at the interface and this is now

$$\gamma \frac{\partial^2 \eta}{\partial x^2} + p - p' = 0, \qquad (z = 0) \tag{8.126}$$

with $\eta = a \sin mx$. The pressures are given to the first order in a/h and a/h' by

$$\frac{p}{\rho} + ag \sin mx + \frac{1}{2}(V - c)^2 \; [1 - 2ma \coth mh \sin mx] = \text{constant},$$

and

$$\frac{p'}{\rho'} + ag \sin mx + \frac{1}{2}(V' - c)^2 \; [1 + 2ma \coth mh' \sin mx] = \text{constant},$$

and, after substituting for these terms into (8.126), we obtain the dispersion relation

$$\gamma m^2 + g(\rho - \rho') = (V - c)^2 \; m\rho \coth mh + (V' - c)^2 \; m\rho' \coth mh'. \tag{8.127}$$

Consider now the case when both fluids are at rest, apart from the wave motion, and are so deep compared with the wave-length that we may set $\coth mh = \coth mh' = 1$. The speed of wave propagation is then given by $c = c_0$, where

$$c_0^2 = \frac{g}{m} \frac{\rho - \rho'}{\rho + \rho'} + \frac{\gamma m}{\rho + \rho'} . \tag{8.128}$$

We can simulate the effect of wind on deep water by setting $V = 0$, $\coth mh = \coth mh' = 1$ in (8.127). This gives

$$\frac{\gamma m}{\rho} + \frac{(1 - \sigma)g}{m} = c^2 + \sigma (V' - c)^2,$$

where $\sigma = \rho'/\rho$ is the specific gravity of air. This equation is equivalent to

$$c^2 - \frac{2\sigma c}{1 + \sigma} V' + \frac{\sigma}{1 + \sigma} (V')^2 - c_0^2 = 0, \tag{8.129}$$

with c_0, given by (8.128), the propagation speed when there is no wind. Solving (8.129) for c yields

$$c = \frac{\sigma V'}{1+\sigma} \pm \left[c_0^2 - \frac{\sigma(V')^2}{(1+\sigma)^2} \right]^{1/2}. \tag{8.130}$$

We see that for a wave of given wavelength $\lambda = 2\pi/m$, the propagation speed is greatest when it equals the wind speed, which in turn is given from (8.130) by

$$V' = (1+\sigma)^{1/2}\, c_0 - c_M. \tag{8.131}$$

It therefore follows that with wind of any speed other than that of the waves, the wave speed is less than that given by (8.131). For instance, the speed with no wind is c_0 and from (8.131) with $\sigma = 1.3 \times 10^{-3}$, we have $c_0/c_M = 0.999$.

When $V'/c_0 = (1 + 1/\sigma)^{1/2} \approx 27.753$, one of the values of c given by (8.130) is zero. Thus a static corrugation of the free surface with wavelength $2\pi/m$ can be maintained by a wind velocity of $c_0(1 + 1/\sigma)^{1/2}$, but such an equilibrium of the surface is unstable. If $V'/c_0 < (1 + 1/\sigma)^{1/2}$, the two values of c given by (8.130) have opposite signs and hence waves may travel with the wind or against it, but the wave speed is faster with the wind. If $V'/c_0 > (1 + 1/\sigma)^{1/2}$, the waves cannot travel against the wind.

When $V'/c_0 = (1 + \sigma)\sigma^{-1/2} \approx 27.771$, the two values of c given by (8.130) are then equal. For $V'/c_0 > (1 + \sigma)\sigma^{-1/2}$, both values of c are complex showing that the waves grow with time.

From (8.130), the minimum value of c_0 is given by

$$\left\{ \frac{2\sqrt{[\gamma g(1-\sigma)]}}{1+\sigma} \right\}^{1/2}.$$

Hence water with a plane level surface is unstable when the wind velocity exceeds

$$\left\{ \frac{2\sqrt{[\gamma g(1-\sigma^2)]}}{\sigma} \right\}^{1/2}.$$

Taking $\gamma = 0.073$ N/m, $g = 9.81$ m/s^2, $\sigma = 1.3 \times 10^{-3}$, the theory suggests that a flat water surface becomes unstable for wind velocities greater than about 6.42 m/s, which is close to 12.5 knots, and this is supported by observation.

PROBLEMS 8

1. Prove that

$$w = A \cos \frac{2\pi}{\lambda}(z + ih - Vt)$$

is the complex potential for the propagation of simple harmonic surface waves of small height on water of depth h, the origin being in the undisturbed free surface. Express A in terms of the amplitude a of the surface oscillations.

Prove that

$$V^2 = \frac{g\lambda}{2\pi} \tanh \frac{2\pi h}{\lambda},$$

and deduce that every value of V less than $\sqrt{(gh)}$ is the velocity of some wave.

Prove that each particle describes an ellipse about its equilibrium position. Obtain the corresponding result when the water is infinitely deep.

2. Calculate the kinetic and potential energies associated with a single train of progressive waves on deep water, and from the condition that these energies are equal obtain the formula

$$V^2 = \frac{g\lambda}{2\pi}.$$

Show how this result is modified when the wavelength is so small that the potential energy due to surface tension is not negligible.

3. Show that the wavelength λ of stationary waves on a river of depth h, flowing with velocity v, is given by

$$v^2 = \frac{g\lambda}{2\pi} \tanh \frac{2\pi h}{\lambda}.$$

Deduce that, if the velocity of the stream exceeds $\sqrt{(gh)}$, such stationary waves cannot exist.

4. In a train of waves on deep water given by

$$\phi = \tfrac{1}{2} Vh \exp(-2\pi z/l) \quad \cos \frac{2\pi}{l}(x - Vt),$$

show that, if $(h/l)^2$ is negligible, the fluid particles describe circles with uniform speed.

Prove that, to a second approximation, the surface particles have a slight mean drift in the direction of propagation.

5. When simple harmonic waves of length λ are propagated over the surface of deep water, prove that, at a point whose depth below the undisturbed surface is h, the pressure at the instants when the disturbed depth of the point is $h + \eta$ bears to the undisturbed pressure at the same point the ratio

$$1 + \frac{\eta}{h} e^{-2\pi h/\lambda} : 1,$$

atmospheric pressure and surface tension being neglected.

6. Two fluids of densities ρ_1, ρ_2 have a horizontal surface of separation but are otherwise unbounded. Show that when waves of small amplitude are propagated at their common surface, the particles of the two fluids describe circles about their mean positions; and that at any point of the surface of separation where the elevation is η, the particles on either side have a relative velocity $4\pi c\eta/\lambda$.

7. Investigate the wave motion occurring at a horizontal interface between two fluids, of which the upper one of density ρ_2 has a general stream velocity U, and the lower one of density ρ_1 is at rest except for the small motion, the fluids being otherwise unlimited.

Show that the wave velocity c of waves of length λ is given by the equation

$$g(\rho_1 - \rho_2) = \frac{2\pi}{\lambda}\,(\rho_1 c^2 + \rho_2 (c - U)^2),$$

and prove that, for a given value of U, waves below a certain wavelength cannot be propagated.

8. Two incompressible fluids of densities ρ_1, $\rho_2 (\rho_1 > \rho_2)$ are superposed. The upper fluid is moving as a whole with velocity U_2, and the lower with velocity U_1, in the direction of the axis of x, which is horizontal, that of y being vertically upwards. Show that the height η of a wave disturbance, whose velocity potentials in the two fluids are ϕ_1, ϕ_2 respectively, satisfies the following equations at the boundary:

$$\rho_1\left(\frac{\partial \phi_1}{\partial t} + U_1\,\frac{\partial \phi_1}{\partial x}\right) - \rho_2\left(\frac{\partial \phi_2}{\partial t} + U_2\,\frac{\partial \phi_2}{\partial x}\right) = (\rho_1 - \rho_2) g\eta\,,$$

$$-\frac{\partial \phi_1}{\partial y} = \frac{\partial \eta}{\partial t} + U_1\,\frac{\partial \eta}{\partial x}\,, \qquad -\frac{\partial \phi_2}{\partial y} = \frac{\partial \eta}{\partial t} + U_2\,\frac{\partial \eta}{\partial x}\,.$$

9. An infinite liquid of density σ lies above an infinite liquid of density ρ, the two liquids being separated by a horizontal plane interface. Show that the velocity c of propagation of waves of length λ along the interface is given by

$$c^2 = \frac{g\lambda}{2\pi}\,\frac{\rho - \sigma}{\rho + \sigma}\,.$$

Prove that, for any group of such waves, the group velocity is equal to one-half of the wave velocity.

10. Two-dimensional waves of length $2\pi/m$ are produced at the surface of separation of two liquids which are of densities ρ, $\rho'(\rho > \rho')$ and depths h, h' confined between two fixed horizontal planes. Prove that, if the potential

energy is reckoned zero in the position of equilibrium, the total energy of the lower liquid is to that of the upper in the ratio

$$\rho\{(2\rho-\rho')\coth mh+\rho'\coth mh'\}:\rho'\{(\rho-2\rho')\coth mh'-\rho\coth mh\}.$$

11. A layer of fluid of density ρ_2 and thickness h separates two fluids of densities ρ_1 and ρ_3, extending to infinity in opposite directions. If waves of length λ, large compared with h, be set up in the fluid, show that their velocity of propagation is either

$$\left\{\frac{g\lambda}{2\pi}\,\frac{\rho_3-\rho_1}{\rho_3+\rho_1}\right\}^{1/2}\quad\text{or}\quad\left\{gh\,\frac{(\rho_3-\rho_1)(\rho_3-\rho_2)}{\rho_2(\rho_3-\rho_1)}\right\}^{1/2}.$$

12. Obtain the conditions to be satisfied for small oscillations at the horizontal interface of two semi-infinite liquids of densities ρ, ρ' $(\rho>\rho')$ moving with general stream velocities U, U' in the same horizontal direction, the surface tension T being taken into account.

Show that there are two possible wave velocities for a wave of length λ, namely

$$c=\frac{\rho U+\rho' U'}{\rho+\rho'}\pm\sqrt{\left\{\frac{\lambda g}{2\pi}\,\frac{\rho-\rho'}{\rho+\rho'}+\frac{2\pi T}{\lambda(\rho+\rho')}-\frac{\rho\rho'(U-U')^2}{(\rho+\rho')^2}\right\}}.$$

13. The fluid in the region $0<z<h$, of density ρ_2, separates two fluids of densities ρ_1 and ρ_3, occupying the regions $h<z<\infty$ and $-\infty<z<0$, respectively, when at rest under gravity; and $\rho_1<\rho_2<\rho_3$. If waves of length λ, large compared with h, are set up in the middle layer, find the two possible velocities V_1, V_2 of propagation, showing that one value V_1 is independent of ρ_2 and such that a group of such waves of sensibly the same length advances with a velocity $\frac{1}{2}V_1$, whilst the other value V_2 is independent of λ. [The axis of z is taken vertically upwards.]

14. Two liquids, which do not mix, occupy the region between two fixed horizontal planes. The upper, of density ρ' and mean depth h', is flowing with the general velocity U over the lower, which is of density ρ and mean depth h, and is at rest except for wave motions. Prove, neglecting viscosity, that the velocity V of waves of length $2\pi/k$, travelling over the common surface in the direction of U, is given by

$$\rho V^2\coth kh+\rho'(U-V)^2\coth kh'=T_1 k+g(\rho-\rho')/k,$$

where T_1 is the surface tension.

Apply the result to discuss the stability of the surface of deep water over which a wind is blowing with a given velocity. [For numerical purposes g may be taken as 9.8, and T_1 as 74×10^{-3} units, and ρ'/ρ may be taken to be 0.0013.]

15. Two portions of a large uniform stream of liquid of density ρ, flowing with velocity U, are separated by a plane boundary of perfectly flexible fabric, of mass m per unit area, and subject to a tension T, the boundary being parallel to the stream. Show that waves of length λ can be propagated along the fabric, in the direction of the stream, with a velocity V given by

$$mV^2 - T + \frac{\lambda\rho}{\pi}(U - V)^2 = 0,$$

provided that

$$T\left(1 + \frac{m\pi}{\lambda\rho}\right) > mU^2.$$

16. Find the wave velocity of a train of simple harmonic waves, of wavelength λ, moving under the influence of gravity and capillarity on the common surface of two fluids of densities ρ and ρ', when T is the surface tension. Show that there is a minimum wave velocity; find its value and that of the corresponding wavelength. Prove that the group velocity of a group of waves of nearly the same amplitude, wavelength, and phase is greater or less than the wave velocity according as the wavelength is less or greater than that corresponding to the minimum wave velocity. Mention any phenomenon which is explained by this result.

17. A layer of liquid of density ρ_1 and height h rests upon the horizontal surface of unlimited liquid of density $\rho_2 (\rho_2 > \rho_1)$. If T_1, T_2 are the surface tensions at the upper and lower boundaries of the layer, prove that the velocity V with which waves are propagated along the layer satisfies the equation

$$\begin{aligned} &V^4 k^2 \rho_1(\rho_2 + \rho_1 \tanh kh) \\ &\quad - V^2 k[k^2\{\rho_1(T_1 + T_2) + \rho_2 T_1 \tanh kh\} + \rho_1\rho_2 g(1 + \tanh kh)] \\ &\quad + \{k^2 T_1 + \rho_1 g\}\{k^2 T_2 + (\rho_2 - \rho_1)g\} \tanh kh = 0, \end{aligned}$$

where $2\pi/k$ = wavelength.

18. Show that, if the velocity of the wind is just great enough to prevent the propagation of waves of length λ against it, the velocity of propagation of waves with the wind is $2c\{\sigma/(1 + \sigma)\}^{1/2}$, where σ is the specific gravity of air and c is the wave velocity when no air is present.

19. Find the velocity of straight ripples of length λ, on water of density ρ, surmounted by air of density ρ', as maintained by gravity and the surface tension τ, and if $\tau = 80 \times 10^{-3}$ for water, find for what wavelength the velocity of propagation is least, and also the value of this minimum velocity.

20. A stream is running with mean velocity U in the plane xy between a horizontal bottom $y = 0$ and a fixed upper boundary $y = h + a \cos mx$, where a is small. Find the character of the motion by determining its velocity potential or stream function.

Prove that, if U^2 exceeds a critical value $\frac{g}{m} \tanh mh$, the pressure on the upper boundary is in excess of the mean in its higher parts and in defect in its lower parts: and vice versa.

21. A canal, of infinite length and rectangular section, is of uniform depth h and breadth b in one part but changes gradually to uniform depth h' and breadth b' in another part. An infinite train of simple harmonic waves travelling in one direction only is propagated along the canal. Prove that, if a, a' are the heights and $2\pi/m, 2\pi/m'$ the lengths of the waves in the two uniform portions,

$$m \tanh mh = m' \tanh m'h',$$

and

$$a^2 b \operatorname{sech}^2 mh(\sinh 2mh + 2mh) = a'^2 b' \operatorname{sech}^2 m'h' (\sinh 2m'h' + 2m'h').$$

22. Find the possible periods of standing oscillations in a trough of depth h and length l, and show that, if initially the water be at rest with its free surface plane and inclined at a small angle α to the horizontal, the velocity potential and the stream function at any time are given by

$$\phi + i\psi = -\alpha \sum_{s=0}^{\infty} \frac{4l^2}{\pi^3} \frac{p_s \sin p_s t}{(2s+1)^3} \frac{\cos\{(2s+1)\pi(x+iy)/l\}}{\sinh\{(2s+1)\pi h/l\}},$$

where $p_s/2\pi$ is the frequency for the vibration of type s.

23. Prove that, if a canal of rectangular section is terminated by two rigid vertical walls whose distance apart is $2a$, and if the water is initially at rest and has its surface plane and inclined at a small angle β to the length of the canal, the altitude η of the wave at any time t is given by

$$\eta = \frac{8\alpha\beta}{\pi^2} \sum_0^{\infty} \frac{(-1)^n}{(2n+1)^2} \cdot \sin(2n+1)\frac{\pi x}{2a} \cdot \cos(2n+1)\frac{\pi ct}{2a},$$

where c is the velocity of a wave of length $4a/(2n + 1)$ in an infinitely long canal.

24. A rectangular trough, of length $2a$, is filled with liquid to a depth h, and made to oscillate in the direction of its length with velocity $u_0 \cos pt$. Show that the velocity potential of the forced oscillations is given by

$$\phi = \left\{-xu_0 + \sum_{n=0}^{\infty} A_n \sin\frac{(2n+1)\pi x}{2a}\cosh\frac{(2n+1)\pi(y+h)}{2a}\right\}\cos pt,$$

where

$$A_n = 8au_0\,(-)^n \operatorname{sech}\frac{(2n+1)\pi h}{2a}\Big/(2n+1)^2\,\pi^2(1-p_n^2/p^2),$$

p_n denoting the period of free waves of length $4a/(2n+1)$ in liquid of depth h.

25. A long rectangular tank of length $2a$, filled with water up to a small height h, is initially at rest, and is then given a small longitudinal velocity $V\sin nt$. Show that the height η of the free surface above the equilibrium level at time t and at a distance x from that end of the tank which is initially rearmost, is given by

$$c\eta/Vh = -\sin n\frac{(x-a)}{c}\cdot\cos nt$$

$$+\frac{2n}{ca}\sum_{s=0}^{\infty}\frac{\cos\{(s+\frac{1}{2})\pi x/a\}\cos\{(s+\frac{1}{2})\pi ct/a\}}{n^2/c^2+(s+\frac{1}{2})^2\pi^2/a^2},$$

where $c^2 = gh$ and s is an integer.

Appendix: Units of measurement

Since the primary purpose of this book has been to develop the theory of the dynamics of an ideal fluid, we have deliberately presented results in a form which is independent of any specific system of units. However, in carrying out calculations of physical quantities in a problem which are then to be used, perhaps to correlate with measurements determined in experiments, it is then necessary to specify the particular system of units used to measure the physical quantities.

A system of metric units, which has gained increasing use amongst the scientific and engineering communities throughout the world, is that known as the Système International d'Unités, abbreviated to SI units. In this system, the basic units of length, mass and time are respectively the metre, the kilogram and the second. These in turn are denoted by the symbols m, kg and s. From these three fundamental units, all other mechanical properties of a fluid may be expressed with suitably defined units of measurement. We list some of these in Table 1.

Table 1: Units of measurement.

Quantity	Unit	Symbol	Expression in basic units
force	newton	N	kg m/s^2
pressure, stress	pascal	Pa	N/m^2
work, energy	joule	J	N m
power	watt	W	J/s
frequency	hertz	Hz	s^{-1}

Other quantities which commonly occur in fluid mechanics are listed in Table 2.

Table 2: Quantities occurring in fluid mechanics

Quantity	Expression in other units
velocity	m/s
acceleration	m/s^2
velocity potential	m^2/s
density	kg/m^3
viscosity (dynamic)	Pa s
surface tension	N/m

The SI system of units is widely, but not universally, used at the present time. For instance, in the United Kingdom and the United States, a 'mixed' system is still in use, where some quantities are measured in English units. The basic units of the English system are the pound (lb), foot (ft) and second (s). In practice it is only necessary to know the conversion factors between the pound and kilogram and foot and metre to be able to convert from one system of units to another. Since 1 foot = 12 inches and 1 metre = 10^2 cm, and knowing that (approximately)

$$1 \text{ inch} = 2.54 \text{ cm},$$

we can derive the equivalence relation

$$1 \text{ foot} = 0.3048 \text{ m}. \tag{A.1}$$

In addition, there is the (approximate) equivalence relation

$$1 \text{ pound} = 0.4536 \text{ kg}. \tag{A.2}$$

The equivalence relations (A.1) and (A.2) permit the conversion of all physical quantities associated with a fluid from English to SI units and vice versa. Relation (A.1) may be written as

$$1 = \frac{2.54 \text{ cm}}{1 \text{ inch}} = \frac{25.4 \text{ mm}}{1 \text{ inch}} = \frac{0.3048 \text{ m}}{1 \text{ ft}}. \tag{A.3}$$

Likewise, from (A.2),

$$1 = \frac{0.4536 \text{ kg}}{1 \text{ lb}}. \tag{A.4}$$

The reader should note that one newton is the force required to give a body of mass 1 kg an acceleration of 1 m/s^2. The force required to give a body of mass 1 kg an acceleration of (approximately) 9.81 m/s^2 is one kilogram-force, or 1 kgf in abbreviated form. Such an acceleration would be given to a body of mass 1 kg at sea level by the gravitational attraction of the Earth. Similarly in the English system of units, 1 pound-force, or 1 lbf in abbreviated form, is the force required to give a body of mass 1 lb an acceleration of (approximately) 32.174 ft/s^2. This is the acceleration due to gravity in English units. The *poundal* (pdl) is the force required to give a body of mass 1 lb an acceleration of 1 ft/s^2. Thus

$$1 = \frac{4.448 \text{ N}}{1 \text{ lbf}} = \frac{32.174 \text{ pdl}}{1 \text{ lbf}}$$

Another system of units is the cgs system in which the centimetre (cm), gram (g) and second (s) are the basic units of length, mass and time respectively. Since this system is metric, the conversion factors between cgs and SI units follow immediately since

$$1 \text{ cm} = 10^{-2} \text{ m}, \quad 1 \text{ g} = 10^{-3} \text{ kg}.$$

The unit of force in the cgs system is the dyne and this is defined to be the force required to accelerate a body of mass 1 g with an acceleration of 1 cm/s^2. It follows that

$$1 \text{ dyne} = 1 \text{ g cm/s}^2 = 1 \times 10^{-3} \text{ kg} \times 10^{-2} \text{ m/s}^2 = 10^{-5} \text{ N}.$$

The reader should have no difficulty in using the various systems of units and mastering the conversion factors once he is aware of the combination of physical dimensions (mass, length and time) which make up a specific physical quantity. For instance,

$$\text{surface tension} = \text{force per unit length}$$
$$= MLT^{-2}/L = MT^{-2}$$

in dimensional notation form. It therefore follows that surface tension may be measured in pdl/ft or dyne/cm or N/m among others. The reader who requires further explanation of dimensional analysis should consult a text such as reference 15 in the bibliography.

In conclusion, we have listed (approximate) values of some commonly used physical quantities associated with water and air in SI units in Table 3.

Table 3: Approximate values of common properties of water and air

Water	
coefficient of compressibility (isothermal)	4.9×10^{-10} m^2/N
density	10^3 kg/m^3
viscosity (dynamic) at 20°C	10^{-3} Pa s
surface tension at 20°C with air	72.3×10^{-3} N/m
mercury	375×10^{-3} N/m
Air	
one atmosphere (atm)	1.013×10^5 Pa
standard temperature and pressure (s.t.p.)	0°C, 1 atm
density at s.t.p.	1 kg/m^3
viscosity (dynamic) at s.t.p.	1.7×10^{-5} Pa s
surface tension at 20°C with mercury	487×10^{-3} N/m
glycerine	63×10^{-3} N/m

Bibliography

Some of the following books and articles have been referred to by number within the text. Others are given to provide the reader with a useful, but by no means exhaustive, list of additional reading material where, in some instances, topics dealt with in this book are developed at greater length. In the case where an article appearing in a journal, e.g. the *Mathematical Gazette*, is referenced, the name of the journal is followed by the year of publication in brackets, volume number and first page in that order.

1. Batchelor, G.K. *An Introduction to Fluid Dynamics*, C.U.P. (1967).
2. Chirgwin, B.H. & Plumpton, C. *Elementary Classical Hydrodynamics*, Pergamon (1967).
3. Chorlton, F. *Vector and Tensor Methods*, Ellis Horwood (1976).
4. Chorlton, F. The torus and an associated coordinate system, *Mathematical Gazette* (1981), **65**, 434, 289.
5. Chorlton, F. Some potential problems involving spheres, *Mathematical Gazette* (1967), **LI**, 376, 120.
6. Chorlton, F. Scalar and vector fields considered fundamentally, *International Journal of Mathematical Education in Science and Technology* (1981), **12**, 1, 75.
7. Chorlton, F. Some features of regular functions of a complex variable, *International Journal of Mathematical Education in Science and Technology*, (1980), **11**, 4, 587.

8. Copson, E.T. *Theory of Functions of a Complex Variable*, O.U.P. (1950).
9. Coulson, C.A. *Waves*, Oliver & Boyd (1955).
10. Curle, N. & Davies, H.J. *Modern Fluid Dynamics*, Volume 1: *Incompressible Flow*, Van Nostrand (1968).
11. Currie, I.G. *Fundamental Mechanics of Fluids*, McGraw-Hill (1974).
12. Jeffreys, H. *Cartesian Tensors*, C.U.P. (1963).
13. Lamb, H. *Hydrodynamics*. C.U.P. (1945).
14. Lighthill, M.J. *Waves in Fluids*, C.U.P. (1978).
15. Massey, B.S. *Mechanics of Fluids*, Van Nostrand Reinhold (1983).
16. Milne-Thomson, L.M. *Theoretical Hydrodynamics*, Macmillan (1955).
17. Ramsey, A.S. *A Treatise on Hydromechanics*, Part II: *Hydrodynamics*, G. Bell & Sons (1957).
18. Rutherford, D.E. *Vector Methods*, Oliver & Boyd (1946).
19. Scorer, R.S. *Cloud Formations*, Ellis Horwood (1985).
20. Sneddon, I.N. *Fourier Transforms*, McGraw-Hill (1951).
21. Sneddon, I.N. *Special Functions of Mathematical Physics and Chemistry*, Wiley–Interscience (1956).
22. Spain, B. *Tensor Calculus*, Wiley–Interscience (1960).
23. Spiegel, M.R. *Vector Analysis*, Schaum Outline Series, McGraw-Hill (1959).
24. Stoker, J.J. *Water Waves*, Interscience (1950).
25. Titchmarsh, E.C. *The Theory of Functions*, O.U.P. (1939).
26. Weatherburn, C.E. *Elementary Vector Analysis*, G. Bell & Sons (1956).
27. Weatherburn, C.E. *Advanced Vector Analysis*, G. Bell & Sons (1957).
28. Whittaker, E.T. & Watson, G.N. *Modern Analysis*, C.U.P. (1950).

Solutions to problems

CHAPTER 1

1. (i) 7; (ii) $[\frac{2}{7}, -\frac{3}{7}, \frac{6}{7}]$; (iii) $(-\frac{5}{8}, -3, 4)$; $(1, -7, -12)$; (iv) $(0, -\frac{11}{2}, 9)$.

5. $-\frac{1}{2}$.

7. $\left[\dfrac{1, 1, -2}{\sqrt{6}}\right]$; $x + y - 2z + 5 = 0$.

25. $\nabla\phi = 5\mathbf{i} + 4\mathbf{j} + 3\mathbf{k}$; dir. deriv. = 7.

26. $\nabla\phi = \mathbf{r}$; dir. deriv. $= -\frac{1}{2}$.

27. $\nabla\phi = \phi'(R)\hat{\mathbf{R}}$ where $\hat{\mathbf{R}}$ = unit vector in perp. from line to point.

28. $\phi = \phi_0 + r_0^{-1} - r^{-1}$.

29. $[r^2\mathbf{a} - (\mathbf{a} \cdot \mathbf{r})\mathbf{r}]/r^3$.

30. Surfaces intersect orthogonally.
31. Level surfaces: $r_1 + r_2$ = const., a family of confocal ellipsoids of revolution about AB, with A, B as foci.

34. (i) $4\pi a^2 h$, div $\mathbf{F} = 2$; (ii) $6\pi a^2 h$, div $\mathbf{F} = 3$.

37. Area $= \pi a^2$; S.A. $= \pi$.

38. Element $dr \times r d\theta$ in S subtends S.A. at P of $r dr d\theta \cos\phi/(r^2+a^2)$, where $\cos\phi = a(r^2+a^2)^{-1/2}$. Total S.A. at $P = (\pi - 3)/6$.

40. $48\pi a^5/r$.

44. $\phi = Ar^{-3}$ $(\mathbf{r} \, . \, \nabla\phi = r d\phi/dr)$.

45. $\phi - Ar^{-1} + Br^{-2}$.

46. (a) 1/2; (b) $\sqrt{2}/2$.

47. 1.

48. $3\pi a^3 b/4$.

49. $\gamma = \frac{3}{4}\pi c^2 (4b^2 + c^2)$; $\gamma' = 4c^2(3b^2 + c^2)$; $\gamma/S, \gamma'/S' \to 3b^2$.

50. $\overrightarrow{OA} \times \overrightarrow{OP} \equiv a^2(l^2+m^2)\sin\theta\,(l\mathbf{i} + m\mathbf{j} + n\mathbf{k})$; $OP = a\sqrt{(l^2+m^2)}$; $\gamma = l(l^2+m^2)\pi a^2$; $\gamma/S = l$; max $\gamma/S = 1$.

59. (i) $\mathbf{0}$; (ii) 4π when P_0 within S, 0 when P_0 outside S.

66. (i)
$$\operatorname{curl}\mathbf{F} = \frac{1}{r^2\sin\theta}\left[r\left(w\cos\theta + \sin\theta\frac{\partial w}{\partial\theta} - \frac{\partial v}{\partial\phi}\right)\hat{\mathbf{r}} + r\left(\frac{\partial u}{\partial\phi} - w\sin\theta - r\sin\theta\frac{\partial w}{\partial r}\right)\hat{\boldsymbol{\theta}} + r\sin\theta\left(v + r\frac{\partial v}{\partial r} - \frac{\partial u}{\partial\theta}\right)\hat{\boldsymbol{\phi}}.\right.$$

(ii) $f(r) = Ar + Br^{-2}$.

70. $\phi = \frac{1}{2}(\omega \, . \, \mathbf{r})^2$.

71. curl $\mathbf{F} = \mathbf{0}$; $\phi = x^3yz + 2xy^2z - 3xz^2 + 2x + 5z + \text{const.}$

72. $\phi = \theta + \text{const.}$; $\theta = \tan^{-1}(y/x)$.

73. Value of $[\frac{1}{3}(x^3+y^3+z^3) - xyz]$ at P_2 $-$ that at P_1.

78. (ii) $\phi = \frac{1}{2}(\mathbf{a}\,.\,\mathbf{b})\mathbf{r}^2 + \text{const.}$; (iii) $\phi = (\mathbf{a}\,.\,\mathbf{r})(\mathbf{b}\,.\,\mathbf{r}) + \text{const.}$

81. $A_i' = l_{ir}A_r$ or $A_1' = l_{11}A_1 + l_{12}A_2 + l_{13}A_3$, etc., where $A_1 = x_1^2 = \frac{1}{9}(x_1' - 2x_2' + 2x_3')^2$; $A_2 = x_2^2 = \frac{1}{225}(14x_1' + 5x_2' - 2x_3')^2$; $A_3 = x_3^2 = \frac{1}{225}(2x_1' - 10x_2' - 11x_3')^2$.

83. $A_1' = (160x_1' + 119x_2')x_3'/169$; $A_2' = (119x_1' - 120x_2')x_3'/169$; $A_3' = (5x_1' + 12x_2')(12x_1' - 5x_2')/169$.

84. $A_{22} = A_{23} = A_{32} = A_{33} = 1$; all other comps. zero; $A_{11}' = 64/49$.

86. $\gamma_{io} = \alpha_{io} + \beta_{io} + \alpha_{ko}\beta_{ik}$; $\gamma_{ij} = \alpha_{ij} + \beta_{ij} + \alpha_{kj}\beta_{ki}$.

CHAPTER 2

1. curl $\mathbf{q} = \mathbf{0}$, hence vorticity is zero.

6. $$\frac{\partial \rho}{\partial t} + \frac{\partial}{\partial R}(Rq_R) + \frac{\partial}{\partial z}(Rq_z) = 0.$$

CHAPTER 3

2. One half.
3. 786 mm of mercury.
11. 1.83 revs per second.
21. Initial velocity proportional to OP^{-3} in direction perpendicular to PP'.

CHAPTER 4

18. $\phi = Vz + Va^3z/2r^3$, where $r^2 = x^2 + y^2 + z^2 \geqslant a^2, y \geqslant 0$.

$$\left(1 - \frac{a^3}{r^3} + \frac{a^6}{4r^6}\right) - \frac{z^2}{r^2}\left(\frac{a^3}{r^3} - \frac{3a^6}{4r^6}\right) = \text{constant}.$$

CHAPTER 5

6. $p = p_\infty + \frac{1}{2}\rho U^2 \cos^2(\frac{1}{2}\theta)\,[1 - 3\sin^2(\frac{1}{2}\theta)]$.
15. $T = \frac{1}{3}\pi\rho a^3 V^2$.
26. $\theta = \pm \cos^{-1}[(a^2 + f^2)/\sqrt{2}\,(a^4 + f^4)^{1/2}]$.

CHAPTER 6

1. (i) $2\pi i$; (ii) $3\pi i/8$.
2. (i) $\pi i/18$; (ii) $(4\pi i/\sqrt{3}) \exp(-\frac{1}{2}) \sin(\sqrt{3}/2)$.
8. Pressure at $(0, y) = -2\rho m^2 y^2/(y^2 + c^2)$. Along positive axis.
13. Uniform flow, vel. U, past a line source, strength m.
14. $a^2\kappa/(b^2 - a^2)$.
16. $w = \ln\{z(z^2 + 1)/(z - 1)^4\}$; vel. $= (-24\mathbf{i} + 7\mathbf{j})/60$.
17. $$w = m\{\ln(z - 2a) + i\ln(z + 2a) + \ln(a - 2z) - i\ln(a + 2z) - \ln z + i\ln z\} + \text{const.}$$
22. Force per unit length $= \rho\pi\kappa^2/3a$ along x-axis.

CHAPTER 7

1. $(2x + y)(x - 2y) = 0;\ xy = 1.$
6. C is circle, centre $z = i(b - a^2/b)$, radius c/b.
9. $$w = V\left\{e^{-i\alpha}\left[\frac{z + (z^2 - a^2 + b^2)^{1/2}}{a + b}\right] + e^{i\alpha}\left[\frac{z - (z^2 - a^2 + b^2)^{1/2}}{a - b}\right]\right\};$$
   $M = \pi\rho(b^2 - a^2)\sin\alpha\cos\alpha.$

CHAPTER 8

1. $A = aV/\sinh(2\pi h/\lambda).$
13. $V_1 = [g\lambda(\rho_3 - \rho_1)/2\pi(\rho_3 + \rho_1)]^{1/2},$
    $V_2 = [gh(\rho_2 - \rho_1)(\rho_3 - \rho_2)/\rho_2(\rho_3 - \rho_1)]^{1/2}.$
19. $\lambda = 4.55 \times 10^{-4}$ m, $c_0 = 0.24$ m/s.

Index

Mathematics and its Applications

Series Editor: G. M. BELL, Professor of Mathematics, King's College (KQC), University of London

Artmann, B.	**The Concept of Number***
Balcerzyk, S. & Joszefiak, T.	**Commutative Rings***
Balcerzyk, S. & Joszefiak, T.	**Noetherian and Krull Rings***
Baldock, G.R. & Bridgeman, T.	**Mathematical Theory of Wave Motion**
Ball, M.A.	**Mathematics in the Social and Life Sciences: Theories, Models and Methods**
de Barra, G.	**Measure Theory and Integration**
Bell, G.M. and Lairs, D.A.	**Co-operative Phenomena in Lattice Models Vols. I & II**
Berkshire, F.H.	**Mountain and Lee Waves**
Berry, J.S., Burghes, D.N., Huntely, I.D., James, D.J.G. & Moscardini, A.O.	**Teaching and Applying Mathematical Modelling**
Burghes, D.N. & Borrie, M.	**Modelling with Differential Equations**
Burghes, D.N. & Downs, A.M.	**Modern Introduction to Classical Mechanics and Control**
Burghes, D.N. & Graham, A.	**Introduction to Control Theory, including Optimal Control**
Burghes, D.N., Huntley, I. & McDonald, J.	**Applying Mathematics**
Burghes, D.N. & Wood, A.D.	**Mathematical Models in the Social, Management and Life Sciences**
Butkovskiy, A.G.	**Green's Functions and Transfer Functions Handbook**
Butkovskiy, A.G.	**Structural Theory of Distributed Systems**
Cao, Z-Q., Kim, K.H. & Roush, F.W.	**Incline Algebra and Applications**
Chorlton, F.	**Textbook of Dynamics, 2nd Edition**
Chorlton, F.	**Vector and Tensor Methods**
Crapper, G.D.	**Introduction to Water Waves**
Cross, M. & Moscardini, A.O.	**Learning the Art of Mathematical Modelling**
Cullen, M.R.	**Linear Models in Biology**
Dunning-Davies, J.	**Mathematical Methods for Mathematicians, Physical Scientists and Engineers**
Eason, G., Coles, C.W. & Gettinby, G.	**Mathematics and Statistics for the Bio-sciences**
Exton, H.	**Handbook of Hypergeometric Integrals**
Exton, H.	**Multiple Hypergeometric Functions and Applications**
Exton, H.	***q*-Hypergeometric Functions and Applications**
Faux, I.D. & Pratt, M.J.	**Computational Geometry for Design and Manufacture**
Firby, P.A. & Gardiner, C.F.	**Surface Topology**
Gardiner, C.F.	**Modern Algebra**
Gardiner, C.F.	**Algebraic Structures: with Applications**
Gasson, P.C.	**Geometry of Spatial Forms**
Goodbody, A.M.	**Cartesian Tensors**
Goult, R.J.	**Applied Linear Algebra**
Graham, A.	**Kronecker Products and Matrix Calculus: with Applications**
Graham, A.	**Matrix Theory and Applications for Engineers and Mathematicians**
Griffel, D.H.	**Applied Functional Analysis**
Griffel, D.H.	**Linear Algebra***
Hanyga, A.	**Mathematical Theory of Non-linear Elasticity**
Harris, D.J.	**Mathematics for Business, Management and Economics**
Hoksins, R.F.	**Generalised Functions**
Hoskins, R.F.	**Standard and Non-standard Analysis***
Hunter, S.C.	**Mechanics of Continuous Media, 2nd (Revised) Edition**
Huntley, I. & Johnson, R.M.	**Linear and Nonlinear Differential Equations**
Jaswon, M.A. & Rose, M.A.	**Crystal Symmetry: The Theory of Colour Crystallography**
Johnson, R.M.	**Theory and Applications of Linear Differential and Difference Equations**
Kim, K.H. & Roush, F.W.	**Applied Abstract Algebra**
Kosinski, W.	**Field Singularities and Wave Analysis in Continuum Mechanics**
Lindfield, G. & Penny, J.E.T.	**Microcomputers in Numerical Analysis**
Lord, E.A. & Wilson, C.B.	**The Mathematical Description of Shape and Form**
Marichev, O.I.	**Integral Transforms of Higher Transcendental Functions**
Massey, B.S.	**Measures in Science and Engineering**
Meek, B.L. & Fairthorne, S.	**Using Computers**

Mikolas, M. **Real Function and Orchogonal Series**
Moore, R. **Computational Functional Analysis**
Müller-Pfeiffer, E. **Spectral Theory of Ordinary Differential Operators**
Murphy, J.A. & McShane, B. **Compution in Numerical Analysis***
Nonweiller, T.R.F. **Computational Mathematics: An Introduction to Numerical Approximation**
Ogden, R.W. **Non-linear Elastic Deformations**
Oldknow, A. & Smith, D. **Learning Mathematics with Micros**
O'Neill, M.E. & Chorlton, F. **Ideal and Incompressible Fluid Dynamics**
O'Neill, M.E. & Chorlton, F. **Viscous and Compressible Fluid Dynamics***
Rankin, R.A. **Modular Forms**
Ratschek, H. & Rokne, J. **Computer Methods for the Range of Functions**
Scorer, R.S. **Environmental Aerodynamics**
Smith, D.K. **Network Optimisation Practice: A Computational Guide**
Srivastava, H.M. & Karlsson, P.W. **Multiple Gaussian Hypergeometric Series**
Srivastava, H.M. & Manocha, H.L. **A Treatise on Generating Functions**
Shivamoggi, B.K. **Stability of Parallel Gas Flows***
Stirling, D.S.G. **Mathematical Analysis***
Sweet, M.V. **Algebra, Geometry and Trigonometry in Science, Engineering and Mathematics**
Temperley, H.N.V. & Trevena, D.H. **Liquids and Their Properties**
Temperley, H.N.V. **Graph Theory and Applications**
Thom, R. **Mathematical Models of Morphogenesis**
Toth, G. **Harmonic and Minimal Maps**
Townend, M. S. **Mathematics in Sport**
Twizell, E.H. **Computational Methods for Partial Differential Equations**
Wheeler, R.F. **Rethinking Mathematical Concepts**
Willmore, T.J. **Total Curvature in Riemannian Geometry**
Willmore, T.J. & Hitchin, N. **Global Riemannian Geometry**
Wojtynski, W. **Lie Groups and Lie Algebras***

Statistics and Operational Research

Editor: B. W. CONOLLY, Professor of Operational Research, Queen Mary College, University of London

Beaumont, G.P. **Introductory Applied Probability**
Beaumont, G.P. **Probability and Random Variables***
Conolly, B.W. **Techniques in Operational Research: Vol. 1, Queueing Systems***
Conolly, B.W. **Techniques in Operational Research: Vol. 2, Models, Search, Randomization**
Conolly, B.W. **Lecture Notes in Queueing Systems**
French, S. **Sequencing and Scheduling: Mathematics of the Job Shop**
French, S. **Decision Theory**
Griffiths, P. & Hill, I.D. **Applied Statistics Algorithms**
Hartley, R. **Linear and Non-linear Programming**
Jolliffe, F.R. **Survey Design and Analysis**
Jones, A.J. **Game Theory**
Kemp, K.W. **Dice, Data and Decisions: Introductory Statistics**
Oliveira-Pinto, F. **Simulation Concepts in Mathematical Modelling***
Oliveira-Pinto, F. & Conolly, B.W. **Applicable Mathematics of Non-physical Phenomena**
Schendel, U. **Introduction to Numerical Methods for Parallel Computers**
Stoodley, K.D.C. **Applied and Computational Statistics: A First Course**
Stoodley, K.D.C., Lewis, T. & Stainton, C.L.S. **Applied Statistical Techniques**
Thomas, L.C. **Games, Theory and Applications**
Whitehead, J.R. **The Design and Analysis of Sequential Clinical Trials**

**In preparation*